T0342424

SATELLITE COMMUNICATIONS PAYLOAD AND SYSTEM

SATELLITE COMMUNICATIONS PAYLOAD AND SYSTEM

Second Edition

TERESA M. BRAUN
WALTER R. BRAUN

IEEE PRESS
WILEY

Registered Office
John Wiley & Sons, Inc., 111 River Street, Hoboken, NJ 07030, USA

Editorial Office
111 River Street, Hoboken, NJ 07030, USA
101 Station Landing, Medford, MA 02155, USA
9600 Garsington Road, Oxford, OX4 2DQ, UK
The Atrium, Southern Gate, Chichester, West Sussex, PO19 8SQ, UK
Boschstr. 12, 69469 Weinheim, Germany
1 Fusionopolis Walk, #07-01 Solaris South Tower, Singapore 138628

For details of our global editorial offices, customer services, and more information about Wiley products visit us at www.wiley.com.

Wiley also publishes its books in a variety of electronic formats and by print-on-demand. Some content that appears in standard print versions of this book may not be available in other formats.

Library of Congress Cataloging-in-Publication Data

Names: Braun, T. M., 1949–, author. | Braun, Walter R., author.
Title: Satellite communications payload and system / Teresa M. Braun,
    Walter R. Braun.
Description: Second edition. | Hoboken, NJ : Wiley-IEEE Press, 2022. |
    Includes bibliographical references and index.
Identifiers: LCCN 2021004388 (print) | LCCN 2021004389 (ebook) | ISBN
    9781119384311 (cloth) | ISBN 9781119384304 (adobe pdf) | ISBN
    9781119384328 (epub)
Subjects: LCSH: Artificial satellites in telecommunication. | Artificial
    satellites–Electronic equipment. | Artificial satellites–Radio
    antennas. | High altitude platform systems (Telecommunication) | Digital
    communications.
Classification: LCC TK5104 .B765 2022 (print) | LCC TK5104 (ebook) | DDC
    621.3841/56–dc23
LC record available at https://lccn.loc.gov/2021004388
LC ebook record available at https://lccn.loc.gov/2021004389

Cover Design: Wiley
Cover Image: © KA-SAT drawing courtesy of Eutelsat, color by T. M. Braun

Set in 10/12pt Times by Straive, Pondicherry, India

10 9 8 7 6 5 4 3 2 1

Dedication

To the memory of my father, Joseph C. Thesken, who always loved and encouraged me.

CONTENTS

PREFACE

This book's focus is the payload of communications satellites. Several books have been written about the satellite bus, general satellite communications, or applications of satellite communications, but the payload is covered only briefly in these books. In-depth books on how to design the various payload units, for example, an antenna or a filter, exist, but they are far more detailed than is necessary for someone who wants to understand the payload as a whole. This book presents concepts, theory, and principles of how the payload works, as well as current technology. It also presents the end-to-end satellite communications systems in more technical depth than in other books and with emphasis on the payload.

The satellite industry continues to be a global growth industry. Since at least 2007, the revenue has grown every year to 2020, doubling just from 2007 to 2014 (Bryce, 2017; SIA, 2018, 2019, 2020).

In this second edition, almost every chapter has been expanded and updated. The previous antenna chapter has been split into one chapter about antenna basics and single-beam antennas and another about multi-beam antennas and phased arrays. The previous chapter on filters and payload-integration elements has been split in two. A new chapter on satellite communications standards is added. An entire new part of the book has been added, about particular end-to-end satellite systems in all three satellite services. Its focus is on the payload and the interaction of the ground equipment with the payload. It includes a chapter on high-throughput satellite (HTS) systems. Nongeostationary spacecraft have been considered throughout the book.

As satellite system operators desire greater capability in their payloads, the payload systems engineer is called upon to deal with things he never had to before or deal with the old things in a more exact way, to squeeze out higher performance.

The engineer about to model the end-to-end communications system needs to fully understand the payload subtleties, new or old. A satellite customer new to the business may be mystified by discussions with the payload manufacturer and need more knowledge to be able to get what he wants. The writer of the payload part of the proposal needs to realize that the formulations or values of some requirements may have to be rethought. Today it takes on the order of ten years to "know" the payload. This book can accelerate the learning curve.

The intended audience of this book is the following people who work with communications satellites:

- Payload systems engineers, at all stages in their careers
- Engineers performing analysis and simulation of payload performance or end-to-end communications-system performance
- Satellite customers
- Satellite proposal-writers
- Payload unit engineers who are curious about the rest of the payload
- Satellite bus engineers who are curious about the payload.

Prerequisite for full understanding of many chapters is knowledge of the Fourier transform and the duality of the time and frequency domains. However, without that most of the book can still be understood. Knowledge of electricity and magnetism and of circuit theory is not required.

REFERENCES

Bryce Space and Technology (2017). *2017* State of the satellite industry report. *Satellite Industry Association*; Jun. On www.nasa.gov/sites/default/files/atoms/files/sia_ssir_2017. pdf. Accessed Jul. 21, 2020.

Satellite Industry Association (SIA) (2018). Summaries of annual State of the satellite industry reports. On sia.org/category/press-releases/. Accessed July 21, 2020.

Satellite Industry Association (SIA) (2019). Summaries of annual State of the satellite industry reports. On sia.org/category/press-releases/. Accessed July 21, 2020.

Satellite Industry Association (SIA) (2020). Summaries of annual State of the satellite industry reports. On sia.org/category/press-releases/. Accessed July 21, 2020.

ACKNOWLEDGMENTS

I, Teresa, want to express my deepest thanks to my husband, Walter Braun, who taught me communications theory on the job at LinCom Corp. in Los Angeles in the 1970s and 1980s and who has lovingly supported me and encouraged me in the writing of this book. I would also like to especially thank my Ph.D. advisor, Dr. Ezio Biglieri, for being so helpful and kind in the late 1980s when I was his graduate student and for his wonderful suggestion in about 2008 to write a book when I could not get a job. I will always be grateful to all the wonderful engineers I have worked with over the years, especially Richard Hoffmeister and Dr. Charles Hendrix, who were instrumental in my career development. Almost all of the engineers I have worked with have been passionate about their work and willing to help others learn, and they have made mine a fascinating career. Of all the companies I have worked at, two stand out for having provided me limitless opportunities to do good work: Space Systems/Loral (now part of Maxar Technologies) and William Lindsey's LinCom Corp. of 40 years ago. My career has spanned the time since the American equal-opportunity laws were being implemented at federal contractors, and I have gone from being an oddity in the engineering workplace to feeling at home among many women colleagues, in California, anyway.

I, Walter, want to thank my wife for talking me into contributing two new chapters, on HTS and non-GEO systems, an effort made difficult by the scarce information. Our collaboration provided many interesting conversations over years.

We wish to thank the colleagues who reviewed the book and provided corrections, suggestions, and explanations: Richard Hoffmeister, Charles Hendrix, and Riccardo De Gaudenzi for reviewing the entire first edition. Our chapter reviewers were Eddy Yee for Chapter 2, Gary Schennum for Chapter 3, Stephen Holme for

Chapter 5, James Sowers and Ben Hitch for Chapter 6, Messiah Khilla and Reinwald Gerhard for Chapter 7, Chak Chie for Chapters 9 and 10, Arun Bhattacharyya for Chapter 11, Ezio Biglieri for Chapter 12, Luis Emiliani for Chapters 14 and 17, Bingen Cortazar for Chapter 18, and Marcus Vilaça for part of Chapter 20. Special thanks go to Luis Emiliani for enthusiastically sharing his wide knowledge with us and to Eric Amyotte for generously answering questions.

ABOUT THE AUTHORS

Dr. Teresa M. Braun received the B.A. in mathematics from the University of California, San Diego, in 1970; and from the University of California, Los Angeles, the M.A. in mathematics in 1973, the M.S. in systems science in 1977, and the Ph.D. in electrical engineering in 1989 with dissertation on modulation and coding. She also took short courses on computer networks. She was employed for 23 years in satellite communications and 3 years in satellite navigation. In California, from 1973 to 1976, she worked on GPS development at The Aerospace Corp.; from 1977 to 1986 on analysis and simulation of end-to-end satellite communications at LinCom Corp. (now LinQuest Corp.); from 1989 to 1997 in development of new payload technology in communications and navigation at Hughes Space & Communications (now Boeing Satellite Development Center); from 1997 to 1999 in development of payload and ground-receiver technology and on a satellite constellation at Lockheed Martin's Western Development Laboratory (now part of LM's Management & Data Systems); and from 1999 to 2003 as a payload manager and department manager of payload systems analysis at Space Systems/Loral (now part of Maxar Technologies). After moving to Switzerland in 2003, she worked for a few years in project management, supplier management, modem algorithm development, and satellite system analysis. She was *née* Thesken and also formerly named McKenzie. She worked on NASA, defense, commercial, and ESA programs.

Dr. Walter R. Braun received the M.S. degree in electrical engineering from the Swiss Federal Institute of Technology in Zurich, Switzerland, in 1972 and a Ph.D. degree in electrical engineering from the University of Southern California in 1976, specializing in communications. He worked from 1976 to 1982 at LinCom Corp. on the modeling and simulation of satellite communication systems. From 1983 to 1992, he headed the communications group at the ABB research center in Switzerland, working on mobile radio and powerline communications systems. From 1992 until his retirement in 2012, he was product development manager and member of the management board of several companies in the field of communications in Switzerland.

ABBREVIATIONS

∘	Convolution
∗	Complex conjugation
≐	Is approximately equal to
≜	Is defined as
≪	Is much less than
≫	Is much greater than
∝	Is proportional to
α	Roll-off factor for RRC pulse-shaping
δ-function	Delta function
λ	Wavelength
$\varphi(f)$	Phase of transfer function in radians, a function of frequency f; also called "phase response"
ϕ	Azimuth angle in spherical coordinates
σ	Standard deviation
θ	Polar angle in spherical coordinates
$\theta(t)$	Signal phase in radians as a function of time t
8PSK	8-ary PSK
ACI	Adjacent-channel interference
ACM	Adaptive coding and modulation
ADC	Analog-to-digital converter
$A(f)$	Signal-amplitude multiplication function of a filter as a function of frequency f; also called "gain response"
ALC	Automatic level control
AM	Amplitude modulation

APSK	Amplitude-and-phase-shift keying
AWGN	Additive white Gaussian noise
B	Bandwidth in Hz
BCH	Bose–Chaudhuri–Hocquenghem
BER	Bit error rate
BFN	Beam-forming network
BOL	Beginning of life
BPF	Bandpass filter
bps	Bits per second
BPSK	Binary PSK
BSS	Broadcast Satellite Service
BW	Bandwidth
C	Carrier or signal power in a given bandwidth
C/3IM	Ratio of carrier power to 3rd-order IMP power when nonlinearity's input is two equal-power carriers
CAMP	Channel amplifier
CATR	Compact antenna testing range
C-band	Frequencies from 4 to 8 GHz
CC	Conduction-cooled
CDM	Code-division multiplexing
CDMA	Code-division multiple access
C/I	Carrier-to-interference ratio, a power ratio in some bandwidth
C/N	Carrier-to-noise ratio, a power ratio in some bandwidth
CONUS	Contiguous US
CP	Circular polarization
CPM	Continuous-phase modulation
CTE	Coefficient of thermal expansion
CW	Continuous wave, i.e. a sinewave
D	Antenna aperture diameter
$D(\theta,\phi)$	Antenna directivity, a function of θ and ϕ
DAC	Digital-to-analog converter
DAMA	Demand-assignment multiple access
dB	Decibels
dBm	10 times log of milliwatts
DC	Direct current
D/C	Downconverter
DEMUX	Demultiplexer
DRA	Direct-radiating array
DRC	Direct-radiation-cooled
DRO	Dielectric-resonator oscillator
DSL	Digital subscriber line
DSP	Digital signal processor
DTH	Direct-to-home

DVB	Digital Video Broadcasting
DVB-RCS	DVB—Return Channel Satellite
DVB-S2	DVB—Satellite Second Generation
DVB-S2X	DVB—Satellite Second Generation Extensions
E	Electrical field vector
E_b	Energy per bit
EIRP	Equivalent isotropically radiated power
EOC	Edge of coverage
EOL	End of life
EPC	Electronic power conditioner
E_s	Energy per modulation symbol
ESA	European Space Agency
ETSI	European Telecommunications Standards Institute
E–W	East–west; east and west
F	Noise figure
f	Frequency in Hz; focal length
FCC	Federal Communications Commission (US)
f/D	Ratio of focal length to aperture diameter, for a paraboloidal reflector antenna
FDM	Frequency-division multiplexing
FDMA	Frequency-division multiple access
FEC	Forward error-correcting
FFT	Fast Fourier transform
FGM	Fixed-gain mode
Fn	Flight number n of a series of spacecraft
FOV	Field of view
FSS	Fixed Satellite Service
G	Gain
GBBF	Ground-based beam-forming
GEO	Geostationary orbit or satellite
$G(f)$	Filter gain, a function of frequency f; also called "gain response"
GSM	Global System for Mobile Communications
GSO	Geosynchronous orbit or satellite
G/T_s	Antenna gain divided by system noise temperature
H	Horizontal linear polarization
H	Hybrid coupler
H	Magnetic field vector
HEMT	High-electron-mobility transistor
$H(f)$	Filter transfer function, a function of f
HPA	High-power amplifier
$h(t)$	Filter impulse response, a function of t
HTS	High-throughput satellite
Hz	Hertz, cycles per second

I	Interference power in a given bandwidth
I	In-phase component of a bandpass signal in I/Q representation
IBO	Input backoff
IF	Intermediate frequency
Im	Function which takes the imaginary part of a complex number
IMP	Intermodulation product
IMUX	Input multiplexer
IP	Internet Protocol
ISI	Inter-symbol interference
ITU	International Telecommunication Union
ITU-R	ITU Radio Sector
j	$\sqrt{-1}$
K	Kelvin
K-band	Frequencies from 18 to 27 GHz
Ka-band	Frequencies from 27 to 40 GHz
Ku-band	Frequencies from 11 or 12 GHz to 18 GHz
L-band	Frequencies from 1 to 2 GHz
LCAMP	Linearizer and CAMP unit
(L)CAMP	CAMP that may or may not contain the linearizer function
LDPC	Low-density parity-check
LEO	Low earth orbit
LHCP	Left-hand circular polarization
ln	Natural logarithm
LNA	Low-noise amplifier
LO	Local oscillator
log	Logarithm base 10
LP	Linear polarization
LPF	Low-pass filter
LTWTA	Linearized TWTA
(L)TWTA	TWTA that may or may not be linearized
MAC	Medium-access control
MBA	Multi-beam antenna
MEO	Medium earth orbit
MFPB	Multiple-feed-per-beam
MF-TDMA	Multi-frequency TDMA
MMA	Multi-matrix amplifier
$m{:}n$	Redundancy with m available units out of which n are active at any one time
MODCOD	Combination of modulation and coding format
MPA	Multi-port amplifier
MPEG	Moving Picture Experts Group
MPM	Microwave power module
MRO	Master reference oscillator

MSS	Mobile Satellite Service
MUX	Multiplexer
N	Noise power in a given bandwidth
N_0	One-sided RF or IF power-spectral-density of noise
N/A	Not applicable
NASA	National Aeronautics and Space Administration
NF	Noise figure
NFR	Near-field range
NGSO	Non-geostationary satellite orbit
NPR	Noise-[to-]power ratio
N–S	North–south; north and south
OBBF	Onboard beam-forming
OBO	Output backoff
OBP	Onboard processor
OMT	Orthomode transducer
OMUX	Output multiplexer
OQPSK	Offset QPSK
OSI	Open Systems Interconnect
P	Long-term average of signal power
P2dB	2-dB compression point
PA	Phased array
pdf	Probability density function
pfd	Power flux density
P_{in}	Power input to amplifier
$P_{in\,sat}$	Power input to amplifier that saturates amplifier
pk–pk	Peak-to-peak
PLL	Phase-locked loop
PLMN	Public land mobile network
PM	Phase modulation
P_{op}	Operating point
P_{out}	Power output by amplifier
$P_{out\,sat}$	Power output by amplifier when saturated
Pr	Function which takes the probability of an event
PS	Power supply
psd	Power spectral density
PSK	Phase-shift-keying
PSTN	Public switched telephone network
$P(t)$	Instantaneous power of signal as a function of time
$p(x)$	Probability density function of a random variable, a function of the value x
Q	Quality factor of filter
Q	Quadrature-phase component of a bandpass signal in I/Q representation

QAM	Quadrature amplitude modulation
QPSK	Quaternary PSK
R_b	Data bit rate of signal, in bps
RC	Raised-cosine shape of signal spectrum
Re	Function which takes the real part of a complex number
RF	Radio frequency
RHCP	Right-hand circular polarization
rms	Root mean square, namely the square root of the average squared value
RRC	Root raised-cosine shape of pulse-shaping filter's Fourier transform
RS	Reed–Solomon
R_s	Modulation-symbol rate, in sps
RSM	Regenerative Satellite Mesh
rss	Root sum-square, namely the square root of a sum of squares
RX	Receive
SAW	Surface acoustic wave
S-band	Frequencies from 2 to 4 GHz
SER	Modulation-symbol error rate
$S(f)$	Fourier transform of signal, especially when signal is a modulation pulse. A function of frequency f
$\mathcal{S}(f)$	Power spectral density of signal, a function of frequency f
SFPB	Single-feed-per-beam
SNR	Signal-to-noise ratio
sps	Symbols per second
s/s	Samples per second
SSPA	Solid-state power amplifier
$s(t)$	Signal as a function of time t
std dev	Standard deviation
T	Repetition interval of modulation symbol
T	Temperature
t	Time in seconds
TE	Transverse electric
TEM	Transverse electromagnetic
TID	Total ionizing dose
TDM	Time-division multiplexing
TDMA	Time-division multiple access
TM	Transverse magnetic
T_s	System noise temperature
TT&C	Telemetry, tracking, and command
TWT	Traveling-wave tube
TWTA	Traveling-wave tube amplifier

TX	Transmit
U/C	Upconverter
V	Vertical linear polarization
VCO	Voltage-controlled oscillator
VSAT	Very small-aperture terminal
VSWR	Voltage standing-wave ratio
WG	Waveguide
X-band	Frequencies from 8 to 12 GHz

CHAPTER 1

INTRODUCTION

1.1 END-TO-END SATELLITE COMMUNICATIONS SYSTEM

A typical **end-to-end satellite communications system** consists of a large ground station, a satellite on orbit, and user terminals on or near the ground, as shown in Figure 1.1 with only one user terminal. The **forward link** consists of signals transmitted by the **ground station** to the satellite, which sends them on to the user terminals. The **return link** consists of signals going from a user terminal, to the satellite, then to the ground station. Some systems, such as those that broadcast television, have only a forward link. A **hop** is a one-way link through the satellite. The forward link and the return link each consist of a concatenation of an **uplink** and a **downlink**. The satellite links with the ground station are called **feeder links**. Some systems allow communications directly from user terminal to user terminal, through the satellite. Other systems have direct satellite-to-satellite transmission.

There may be more than one spacecraft and more than one ground station in a satellite system where all work together. There is always more than one user terminal. The collection of spacecraft is the **space segment**, the collection of ground stations the **ground segment**, and the collection of user terminals the **user segment**.

The satellite **payload** is what the satellite is *for*, namely, the payload is what performs the functions desired of the satellite. In a nutshell, the payload is the communications antennas, receivers, and transmitters. The rest of the satellite, namely

Satellite Communications Payload and System, Second Edition. Teresa M. Braun and Walter R. Braun.
© 2021 John Wiley & Sons, Inc. Published 2021 by John Wiley & Sons, Inc.

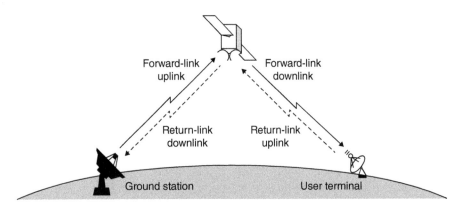

Forward-link
uplink

Forward-link
downlink

Return-link
downlink

Return-link
uplink

Ground station

User terminal

FIGURE 1.1 Typical end-to-end satellite communications system.

the **bus or platform**, supports the payload by providing a structure, power, commanding and telemetry, an appropriate thermal environment, radiation shielding, and attitude control.

There is a difference between a ground station and a gateway. A **gateway** connects the ground station's end of the satellite communications to the public telephone networks and/or the Internet. A **public switched telephone network (PSTN)** is a traditional circuit-switched telephone network (Johnson, 2020). A **public land mobile network (PLMN)** is a cellular telephone network, and one of its traditional services is Internet data connectivity (Wikipedia, 2020). The gateway may also contain an Internet router. A gateway is on-site with a ground station, and the combination is a **teleport**. Not every ground station connects with a gateway, for example, a ground station for a broadcast satellite. For a ground station that has a gateway, sometimes we call the ground station a "gateway" in this book to point up the fact of the gateway existence.

1.2 WHAT THE BOOK IS ABOUT

This book describes commercial communications satellites that are on orbit today and providing a service. These satellites made up 35% of operational satellites in 2016 (Bryce Space and Technology for the Satellite Industry Association, 2017). Military satellites are excluded because of the lack of publicly available information on them, but most of the technology and theory would be the same. Geostationary, medium-earth-orbit, and low-earth-orbit spacecraft are treated. Experimental and proof-of-concept satellites are not included. Communications for command and telemetry of the spacecraft are not included. The carrier frequencies addressed are those from 1.5 GHz (L-band) to 30 GHz (Ka-band).

The book has three parts. Part I is about the payload: its architecture, its units, and its analysis. For the payload units, their theory, architecture, technology, and specification are presented. Part II is about the end-to-end system and includes discussions on communications, the space-ground link, and simulation. Part III is about particular systems of various types, with focus on the payload and its interaction with the user and ground segments.

This book is primarily about the communications system's layer 1, the physical layer, of the Open Systems Interconnect (OSI) model (Chapter 13). Layers 2 and 3 are addressed in Part II Chapter 13 and in Part III.

1.3   CHANNEL AND CHANNEL SHARING

A payload processes signals in **channels**. (This definition of "channel" is not the same as in communications theory, where it means everything that happens to a bit stream before it is recovered in the receiver. This meaning is addressed in Chapter 12.) In a limited sense, a channel is a frequency band used to carry one or more signals that the payload processes together.

The limited resource of the satellite channel can be shared in various ways. Two ways are **frequency-division multiplexing (FDM)** and **time-division multiplexing (TDM)**, which can also be combined into TDM–FDM. FDM and TDM are illustrated in Figure 1.2. In FDM the users each have their own carrier, while in TDM the various users of the channel share the carrier in the time domain. A third, less common, way of channel sharing is by **code division multiplexing (CDM)**, in which each bit stream is multiplied by a different pseudo-random stream of *chips* of a much higher rate than the bit stream. The pseudo-random stream is usually called a *pseudo-random noise (PRN or PN) code*. The PN code is not actually random but is a sequence with very nearly the properties of a random sequence. The multiplication spreads the signal spectrum to the width of the PN code's spectrum. This is why the PN code is called a *spreading code*. The multiplication by the PN code is *direct-sequence spreading*. The resultant stream is then turned into a symbol stream and the various symbol streams are added. Each receiver multiplies the received signal by its own PN code, thereby making its bit stream emerge from the back-

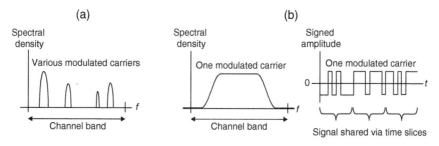

FIGURE 1.2   Two common ways of channel-sharing used in satellite communications: (a) FDM, shown in frequency domain, and (b) TDM, shown in both frequency domain and time domain.

ground. To the other receivers, this bit stream looks like noise. Globalstar is currently the only commercial satellite communications system that uses CDM (Section 19.3).

In this book, usually we will say that a channel carries a signal, not signals, with the meaning that the signal is the totality of what the channel carries.

In the fullest sense, a channel is defined not just by a frequency band but also by the uplink antenna-beam and the downlink beam that carry it. A beam is a solid angle with origin at the payload (usually identified with its footprint or coverage on the earth) and a polarization.

1.4 PAYLOAD

1.4.1 Bent-Pipe Payload

A bent-pipe payload receives, filters, frequency-converts, switches, amplifies, and transmits signals via analog electronics. Such a payload must be described on at least two levels: (1) the basic structure of the **transponder**, which is the chain of hardware **units** that supports a channel's path through the payload, where a unit is something that performs one major function of the transponder; and (2) the architecture of the overall payload.

1.4.1.1 Transponder A transponder provides a processing path through the payload for a channel. Most satellites have many transponders. A particular transponder exists only as long as the payload switches are turned so as to allow the channel to be processed wholly; when the switches are turned other ways, the transponder as an entity is gone and some of its units may be incorporated into different transponders; the action may be later reversed or followed by a different such action. Typically most of the units in a transponder are shared by other transponders; other channels merge into this channel for a while, are separated out, and so on.

The basic structure of a bent-pipe transponder is always pretty much the same. A simplified structure is shown in Figure 1.3. The transponder elements shown are fewer than in any actual transponder but they are enough to represent the major ones in any transponder. Note that the transponder is taken to include the receive and transmit antennas (which is not always the case in the literature). The signal is shown as traveling from left to right. The radio frequency (RF) lines coming out of the receive antenna and going into the transmit antenna are shown as waveguide and the rest as coaxial cable, but other mixes are possible.

The bent-pipe transponder units are the following, in order of signal flow:

- Receive antenna, which receives the signal from the uplink coverage area to the exclusion of other areas
- Preselect filter, which suppresses uplink interference

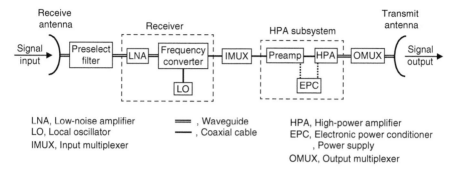

FIGURE 1.3 Simplified block diagram of a transponder of a bent-pipe payload.

- Low-noise amplifier (LNA), which boosts the received signal to a level where the noise added by the rest of the payload units will not cause serious degradation, and itself adds little noise
- Frequency converter, usually a downconverter, to convert from the receive uplink frequency to the transmit downlink frequency. The LNA and the frequency converter are together sometimes known as the receiver
- Input multiplexer (IMUX), a bank of filters for separating out and/or combining channels
- Preamplifier (preamp), which boosts the signal level to what the high-power amplifier (HPA) needs to output the desired RF power
- HPA, which boosts the signal level to what the transmit antenna needs to close the link
- Electronic power conditioner (EPC), that provides direct current (DC) power to rest of the HPA subsystem, namely the preamp and HPA
- Output multiplexer (OMUX), a bank of filters for combining and/or separating out channels
- Transmit antenna, which transmits the signal to the downlink coverage area to the exclusion of other areas.

Each unit has a large number of possible variations. For example, the HPA subsystem comes in two main varieties, one with a traveling-wave tube amplifier (TWTA) and the other with a solid-state amplifier (SSPA), as shown in Figure 1.4. The TWTA-type subsystem has three or four units as shown, including the channel amplifier (CAMP), which is the preamp. A CAMP with TWTA linearizer is a linearizer-channel amplifier (LCAMP). The SSPA usually incorporates all functions into just one unit.

Other meanings of "transponder," occurring in a statement of the number of transponders in payload, but not used in this book, are number of TWTAs and, for TV broadcast, equivalent number of 36 MHz-wide channels.

Active units are those that amplify the signal. These are the LNA, the frequency converter, the preamp, and the HPA. The rest of the units are **passive**.

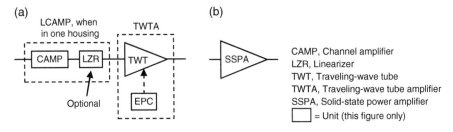

FIGURE 1.4   Two kinds of HPA subsystem: (a) TWTA subsystem, (b) SSPA unit.

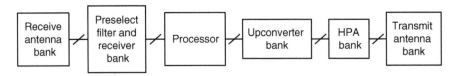

FIGURE 1.5   Simplified block diagram of a processing payload with no phased array.

1.4.1.2   Architecture   The payload consists of the antennas and the **repeater**, the payload architecture is the way in which the transponders are linked together for the purposes of reliability, selection and reuse of signal paths, best usage of the spacecraft bus capabilities, mass and cost savings, and so on. The payload architecture also includes the selection and arrangement of redundant active units. The payload units are linked together by means of integration elements, which include waveguide, coaxial cable, and switches.

1.4.2   Processing Payload

The processing payload has many elements in common with the bent-pipe payload but employs an onboard processor (OBP) to do more complex things than the bent-pipe can do. There are two classes of processing payload: (1) the non-regenerative or transparent processing class which has reconfigurable filters and switching; and (2) the regenerative class, which performs demodulation and remodulation at the very least and in practical terms also decoding and encoding.

In a digital processing payload (see below) and in some other processing payloads, the transponder has lost its meaning, as chains of hardware units have given way to signal processing in a computer or digital signal processor (DSP).

Figure 1.5 is a simplified block diagram of a processing payload, where the processor performs, at the very least, channel demultiplexing, switching, and channel re-multiplexing. A different architecture from what is shown in the figure would apply to a processing payload with at least one phased array.

The non-regenerative processing payload is far more common than the regenerative kind. Its transponders are conceptually like those of the bent-pipe payload,

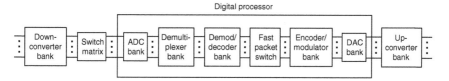

FIGURE 1.6 Example of a digital processing payload, shown without antennas and amplifiers.

but it has an OBP to help perform some of the functions. The OBP may perform some signal processing digitally or it may do it with analog hardware. If digital, the signals it works on have been converted to a low intermediate frequency (IF), much lower than either the input or output RF to the payload. This is necessary because of processing-hardware speed limitations. If the OBP does not perform digital signal processing, it may for example command analog beam-forming of a phased array.

The regenerative payload is very different. At its core is a digital signal processor which processes the signals at baseband. It decodes and re-encodes the signals, and filters and routes them. The regenerative payload demodulates the signals before they go to the baseband processor and re-modulates them at output. It performs RF functions in the same way that a bent-pipe or non-regenerative processing payload does. An example is shown in Figure 1.6, where "ADC" is an analog-to-digital converter and "DAC" is a digital-to-analog converter.

1.5 GROUND TRANSMITTER AND GROUND RECEIVER

Block diagrams of typical models of the ground transmitter, a bent-pipe payload, and ground receiver are given in Figure 1.7. There is both uplink and downlink noise, coming from the payload and ground receiver, respectively. There is interference coming in on the links. The main nonlinear element in the system is the payload HPA in this typical case. There is nonideal behavior in most elements. The antennas and atmospherics are dealt with separately.

1.6 SYSTEM EXAMPLE

An example of the complexity of the particular satellite systems covered in the book is the Intelsat 29e system, which provided a high-throughput Ku-band service. The satellite was launched in 2016. The centers of the spot beams are marked in Figure 1.8, with the large dots indicating both user beams and ground-station beams. The spot beams arcing across the north Atlantic provided service to aeronautical users. The ground stations included gateways that interfaced with

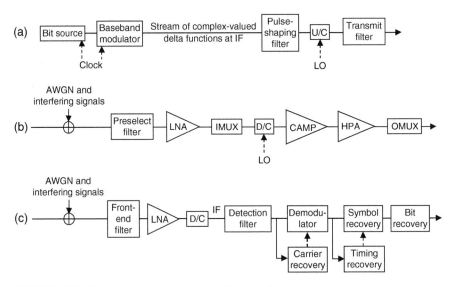

FIGURE 1.7 Typical model of a system with bent-pipe payload: (a) ground transmitter, (b) payload, (c) ground receiver.

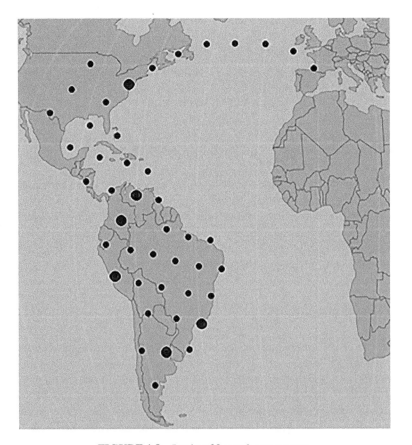

FIGURE 1.8 Intelsat 29 spot beam centers.

the Internet. The payload also had a C-band beam that covered South America. A programmable digital channelizer connected C- or Ku-band uplink channels to C- or Ku-band downlink channels in the same or other beams.

1.7 CONVENTIONS

The term "system" always refers to an end-to-end system.

The term **payload engineer** is sometimes used in the book, but the meaning is always the payload *systems* engineer, that is, the engineer responsible for making sure that the customer's payload requirements are met. This complex and rewarding job requires understanding and always keeping in mind the customer's written and unwritten requirements, negotiating with the payload-unit engineers on the allocation or flowdown of the customer requirements to unit level, working cooperatively with the unit engineers and bus engineers, being a main interface between the customer and the satellite program under the guidance of the program manager, keeping analysis at all times that shows all customer requirements will be met, formally accepting the units, and in some companies planning and supervising the payload test.

A **block diagram** consists of blocks connected by lines and represents signal flow through processing functions. Often the blocks are payload units. The flow is always left to right unless marked otherwise. In Part I of the book, single lines represent coaxial cable and double lines waveguide, but in Parts II and III all lines are single no matter if they are coax or waveguide. A line with a short slanted line through it represents multiple lines.

The frequency bands have various naming schemes worldwide, but in this book we will use the definitions adopted as a standard by the IEEE (2003) and shown in Table 1.1. Thirty and 20 GHz are now a common up- and down-link spacecraft frequency pair, and the pair is said to be Ka-band. As 20 GHz is not really Ka-band but K-band, in this book the pair is called **Ka/K-band**.

TABLE 1.1   IEEE Frequency Band Designations, L-Band to Ka-Band

Frequency (GHz)	Band
1–2	L
2–4	S
4–8	C
8–12	X
12–18	Ku
18–27	K
27–40	Ka

A term is written in bold font where it is defined or at least described for the first time, if the term is used elsewhere in the book or is just important to know. A new term is written in italic font if the term is of some but lesser importance.

In this book, "t" means time and "f" means frequency. Other lower-case letters with "(t)" after them are time-domain functions, upper-case letters with "(f)" after them are frequency-domain functions, upper-case italic are Fourier transforms, and upper-case italic script are power spectral densities. The lower-case italic letters i, k, l, m, and n are integers. The lower-case italic letter j is the square root of -1. Temperature is in Centigrade degrees unless stated to be in Kelvin.

Theoretical definitions and properties are given with an emphasis on practicality and comprehendability. Rigor in stating all conditions of applicability is de-emphasized, because in real payload life the conditions are virtually always filled. References are given that contain the rigor, though. The reader who suspects he has a pathological case on his hands must look to those references.

1.8   BOOK SOURCES

Almost all of the book has come from publicly available sources, all in the English language. English is the language of space systems, at least in the US and Europe. The book's sources in order of most-to-less frequently used are Google Scholar, the IEEE online technical library Xplore, general Google, the AIAA online technical library, books, and regulatory filings. Google enabled the finding of company catalogs and white papers, presentations, microwave encyclopedias, and patents. Wikipedia was a good source for background information whose reliability is not critical to a payload engineer. Some technology companies publish regularly and some scarcely publish, so the results of only the former are well represented in the book.

1.9   SUMMARY OF REST OF BOOK

The rest of the book is divided into three parts. Part I encompasses Chapters 2–11 and contains things that every payload engineer needs to know. Satellite customers, payload unit engineers curious about other parts of the payload, satellite bus engineers curious about the payload, and ground-station engineers would also be mainly interested in this material. Part II encompasses Chapters 12–16 and provides additional material for people who analyze or model the end-to-end system. Part III encompasses Chapters 17–20 and shows the diversity of satellite systems, optimized to achieve different goals. For most examples, the exposition includes the services provided, the ground and user segments, the frequencies and beams, the air interface, and the payload.

Chapter 2 is about the on-orbit environment that the payload sees and what generally happens to the payload because of it. The spacecraft orbit is a big driver of the spacecraft and payload layout. On-orbit effects are divided into thermal, aging, radiation, and attitude disturbances. What these do to the payload, after the spacecraft bus ameliorates them, is briefly summarized (detail is provided in the payload-unit chapters).

The rest of the chapters in Part II are on the payload units.

Chapter 3 is on antenna basics and the single-beam antenna including testing. It also presents ancillary antenna components and autotrack. The material is a prerequisite for Chapter 11.

Chapter 4 is on payload-integration elements. It presents coax and waveguide; isolators, hybrids, and other integration elements; and last switches and redundancy architectures in general. Last is a section on impedance mismatch and S-parameters, important topics in payload integration.

Chapter 5 is on the microwave filter as a unit. It is not about filters interior to units, nor about filters used in conjunction with a digital processor (the latter are addressed in Chapter 10). The chapter starts with the basics of analog filters, then specializes to the basics of microwave filters. It presents the technology of bandpass filters, then describes the payload's various filter units.

Chapter 6 is on the receiver units, namely the LNA and the frequency converter. It first addresses payload-level issues, namely architecture, redundancy, and combining. It explains the nonlinearity of both units, including intermodulation products, and phase noise. It describes both units.

Chapter 7 is on the HPA subsystem, which consists of the HPA itself, its preamplifier, any linearization circuit for the amplifier, and the power supply. The "subsystem" terminology is used in order to encompass both the TWTA subsystem and the SSPA. The chapter explains the nonlinearity of the HPA and discusses the tradeoff between TWTAs and SSPAs.

Chapter 8 explains the key analog communications parameters that are specified in the highest-level payload requirements document. It explains what these parameters are, which units and payload-integration elements contribute to them, how to approximate overall payload performance from units' performance (for predicting payload performance), and briefly how to test these parameters. It does not explain what the parameters mean to the communications signal, since this requires some knowledge of communications theory, which is provided in a later chapter.

Chapter 9 presents additional payload-level analyses, some of which are crucial to know and some of which may come in handy some time. The crucial ones include dealing with noise figure and keeping and maintaining performance-prediction budgets. Another analysis explores the different ways to think about what the HPA does to the signal.

Chapter 10 is on the processing payload and its processing operations. The closely related flexible payload is also discussed. Examples of non-regenerative and regenerative payloads are presented. The payload's digital communications parameters are discussed.

Chapter 11 is about the multi-beam antenna and the phased array. This chapter is the last in Part I because it relies on the earlier chapters. The chapter describes the various kinds of antennas of these types, their technology, beam-forming for a phased array, amplification of a phased array, and testing.

Starting Part II, Chapter 12 presents the principles of digital communications theory, including modulation and coding. The chapter is aimed at engineers who have neither the time nor the inclination to study a whole book on communications theory. On some topics, it provides descriptions rather than exact definitions. No theorems are proved, and many drawings are provided. All digital communications theory terms used in the book are addressed in this chapter.

Chapter 13 briefly describes the bottom three layers of the OSI model of telecommunications as a basis for the following material on satellite protocols and the chapters in Part III. The satellite protocols addressed in this chapter include second-generation of Digital Video Broadcasting-Satellite (DVB-S) and Digital Video Broadcasting-Return Channel Satellite (DVB-RCS).

Chapter 14 is on the communications link, that is, the signal path between the payload and the ground. Link availability is defined. The three main topics are (1) what causes the signal power on the link to vary, especially atmospherics, (2) what establishes the noise power on the link and what makes it vary, and (3) interference from self and from other systems.

Chapter 15 is on how to model and optimize the performance of a multi-beam downlink payload with atmospherics, using probability theory. The variations and uncertainties over the payload life of signal, interference, and noise are included. An example is worked through.

Chapter 16 is about modeling the end-to-end communications system with focus on payload performance. The two types of modeling are software simulation (e.g., with Matlab) and hardware emulation (with a testbed). Methods for simulating various aspects of a realistic system are discussed. The limitations of simulation are also discussed. It is shown that the goodness of numerical results depends on how closely the model reflects the actual system.

Starting Part III, Chapter 17 is about the most common types of satellite systems, those in the fixed satellite service (FSS) and the broadcast satellite service (BSS). Examples are given. The FSS examples are more or less traditional spacecraft.

Chapter 18 is about a category of FSS systems known as high-throughput systems (HTS). Their characteristic attribute is a very large number of narrow beams. HTS in Ka/K-band and those in Ku-band serve different types of users and therefore have different designs.

Chapter 19 is about systems in low or medium earth orbits. The ones providing commercial service today are based on concepts developed around the year 2000. New systems are being currently deployed, but they are not providing commercial service today (2020). We deviate from the general concept of the book and cover some systems with a good chance of going live in the near future.

Chapter 20 is about a few systems in the mobile satellite system (MSS). Most use L-, S-, and/or C-band, but the last system presented uses Ka/K-band.

The book appendices are on the decibel, the Fourier transform, elements of probability theory, and Gauss–Hermite integration, which is a way to vastly speed up some simulations.

REFERENCES

Bryce Space and Technology for the Satellite Industry Association (2017). 2017 State of the satellite industry report. Presentation package. June. On www.nasa.gov/sites/default/files/atoms/files/sia_ssir_2017.pdf. Accessed July 21, 2020.

IEEE Std 521-2002 (2003). *Standard Letter Designations for Radar-Frequency Bands*, Piscataway, NJ: IEEE.

Johnson C (2020). What is PSTN and how does it work? Nextiva blog. On www.nextiva.com/blog/what-is-pstn.html. Accessed Oct. 12, 2020.

Wikipedia (2020). Public land mobile network. July 12. On en.wikipedia.org/wiki/Public_land_mobile_network. Accessed Oct. 12, 2020.

PART I

PAYLOAD

CHAPTER 2

PAYLOAD'S ON-ORBIT ENVIRONMENT

This chapter describes the payload's on-orbit environment and the general effects of the environment on the payload. How each of the payload units is affected is addressed in detail in Chapters 3 through 7. The first section of this chapter talks about what determines the on-orbit environment. The second talks about what the on-orbit environment is for geostationary and nongeostationary spacecraft and how the spacecraft bus mitigates the environment for the payload. The third section discusses the general effects of the mitigated environment on the payload and the measures that the payload takes for further mitigation.

2.1 WHAT DETERMINES ENVIRONMENT

The payload's on-orbit environment is determined, at the highest level, by the spacecraft's orbit, layout, and orientation in space.

2.1.1 Orbit

Communications satellites can be found in various **orbits**. The satellite revolves about the center of the earth in either a nominal circle or a nominal ellipse. The nominal shape is characterized by the orbit's **eccentricity**, defined in Figure 2.1, which is zero for a circle and positive for an ellipse. The circle or ellipse is only nominally the motion because it is perturbed by the nonspherical shape of the earth, local variations in the earth's density, the gravitational pull of other heavenly bodies,

Satellite Communications Payload and System, Second Edition. Teresa M. Braun and Walter R. Braun.
© 2021 John Wiley & Sons, Inc. Published 2021 by John Wiley & Sons, Inc.

$$\text{Ellipse eccentricity} = \sqrt{1 - \frac{b^2}{a}}$$

FIGURE 2.1   Eccentricity defined.

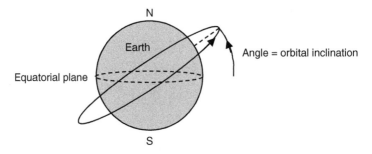

FIGURE 2.2   Orbital inclination defined.

and other effects. The satellite's path lies in a plane that contains the center of the earth. This plane makes an angle with the earth's equatorial plane, called the **inclination**, which stays nominally constant. The highest latitude that the satellite reaches in its orbit is equal to the inclination angle. Orbital inclination is illustrated in Figure 2.2. Communications satellites have an eastward component of their motion. Since the earth rotates eastward, this puts each satellite in contact with the earth longer and minimizes handovers. (Other kinds of satellites, for example earth observation, may move generally westward when their mission does not require good contact with the earth below.) Another aspect of the orbit is its period, which along with eccentricity determines its altitude profile. Yet another aspect, although dependent on the others, is the speed variation over the orbit, and a satellite in elliptical orbit will move faster while it is near the earth and more slowly while it is farther away. The various aspects of the satellite's orbit and, if there is more than one satellite in the constellation, of the other satellites' orbits are chosen by the system designer to serve a particular communications purpose. If an equatorial orbit with speed matching that of the earth's rotation will not serve the purpose well, then there is probably more than one good solution and a complex tradeoff must be made. Further description of orbits, orbit tradeoffs, and orbit perturbations is given in, for example, (Evans, 1999).

The three types of orbits in use today (2020) for communications satellites are the following:

- **Geostationary orbit (GEO)**. The altitude above the earth is about 36,000 km, and the distance from the center of the earth is 42,164 km. The satellite revolves in a circle in the equatorial plane at the same angular speed at which the earth rotates, so the satellite is very nearly stationary with respect to the earth underneath. Over the day the sub-satellite point on the earth traces a small figure-eight, where the crossing-point of the figure-eight is on the equator. The largest deviation north or south of this figure-eight is less than 1° for most communications satellites (Morgan and Gordon, 1989) and equal to the orbit plane's inclination. An exception is some of the spacecraft in the mobile satellite service (MSS), which have a varying inclination within ±3° over the spacecraft life (Chapter 20). The communications' coverage area of a GEO satellite is in the low and mid-latitudes. A constellation of just three satellites can provide coverage to the vast majority of the earth's population (Rankin, 2008). An uncommon variation on the GEO is the **geosynchronous orbit**, which always crosses the equator at the same longitude but has a few degrees of inclination. It may be an intended orbit, but rarely. Alternatively, it may be a GEO gone bad because the spacecraft has run out of station-keeping fuel. The payload can no longer fully perform its original mission. An example in 2020 is Telstar 12, which launched in 1999 and ran out of station-keeping fuel in about 2014. Telesat still operates the satellite for Ku-band mobile services, using it to augment the Telstar 12 Vantage satellite's coverage (Telesat, 2020).

- **Medium earth orbit (MEO).** The altitude of the circular orbit is between about 8,000 and 20,000 km (ITU, 2019).
- **Low earth orbit (LEO).** The altitude of this circular orbit is between 400 and 2000 km (ITU, 2019), and its orbital period is a little over an hour and a half.

A **highly elliptical orbit (HEO)** is an orbit with high eccentricity. It typically has an orbital period of 12 or 24 hours. No commercial communications satellites use this orbit in 2020.

A MEO or LEO constellation is chosen for services that are especially sensitive to propagation delay. A constellation with inclined orbits can provide coverage to the entire earth.

The orbits of some current communications satellite systems that serve different purposes are the following:

- GEO: Most capable communications satellites, providing almost all kinds of communications
- MEO: O3b ("other three billion") at altitude of 8063 km, equatorial orbit. System provides Internet backbone and voice and data communications to mobile providers
- LEO: Globalstar spacecraft at altitude of 1400 km, six satellites in each of eight orbit planes, 52° inclination. System application is global mobile telephone and low-rate data relay (Martin et al., 2007)
- LEO: Iridium Next at altitude of 780 km, 11 satellites in each of six orbit planes, 86° inclination. System application is global mobile telephone and low-rate data relay (Iridium, 2016).

In April 2020, these were the numbers of commercial communications satellites on orbit (Emiliani, 2020):

- GEO–340
- MEO–20
- LEO–645.

2.1.2   GEO Spacecraft's Layout and Attitude

By far most communications satellites today are three-axis stabilized spacecraft. A GEO satellite is usually a rectangular box with external attachments. The earth-facing side, the **earth deck**, is often square, otherwise rectangular. It has antennas mounted on it, usually more than one. There are four rectangular sides attached to the earth deck, namely the **north, south, east, and west panels**. Most spacecraft have antennas attached to the east and west panels. The solar panels or arrays are attached to the north and south panels. The face opposite the earth deck, the **anti-earth deck**, has thrusters on it. This typical overall layout of a GEO spacecraft is shown in Figure 2.3, along with its orientation on orbit which shows why the panels have the names they do.

Figure 2.4 shows two actual GEO spacecraft. Figure 2.4a shows the C- and Ku-band Intelsat 34 spacecraft, launched in 2015. It is rather typical of GEOs. Its body is a rectangular box, and it has reflectors deployed off the east and west panels and reflectors on the earth-facing deck. Figure 2.4b shows the S- and Ku-band EchoStar XXI. Its body is also a rectangular box, and it has the Ku-band reflector deployed off the east or west panel. However, what stands out is the huge S-band reflector deployed off the west or east panel. The spacecraft is in the MSS so it creates a large number of small spot beams, thereby needing such a large antenna at this low-frequency band.

A spacecraft's orientation on orbit, called its **attitude**, is determined for a GEO by two constraints. One is that the earth deck must point down to the earth. For most

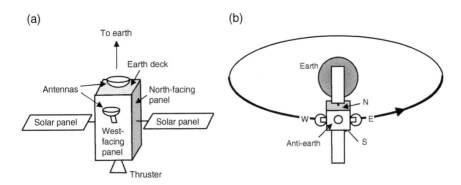

FIGURE 2.3   GEO spacecraft: (a) typical overall layout, (b) orientation on orbit.

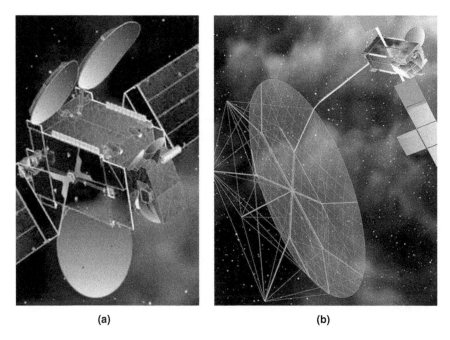

(a)   (b)

FIGURE 2.4   Geostationary spacecraft: (a) Intelsat 34 (©2013 Maxar Technologies. Reprinted with permission from de Selding (2013)) and (b) EchoStar XXI (©2006 Maxar Technologies. Reprinted with permission from Spaceflight101 (2006)).

spacecraft, it points exactly down, although there are a few for which the earth deck points just nearly down, toward the coverage area. The other is that the solar arrays need to have sun all the time except when there is no sun, during eclipse (Section 2.2.1.5). All GEO communications satellites have their solar arrays attached to the north and south panels, and the solar arrays rotate about a north–south axis. In this way, the arrays are always nearly perpendicular to the sun's rays.

2.1.3   GEO Spacecraft's Payload Configuration

Now we look inside a GEO spacecraft with bent-pipe payload (Section 1.4.1). Let us imagine it sliced and opened out as you might do with a cardboard box, so we can see the inner surfaces. The anti-earth deck has been thrown away because typically it contains no payload equipment. An example of the payload hardware layout on the spacecraft is shown in Figure 2.5. Ideally, all the payload equipment would be right at the top (in this orientation) of the panels to minimize losses from waveguide and coaxial cable runs to the antennas, but there is too much payload hardware for this, so a compromise must be made. The low-noise amplifiers (LNAs) must be close to the receive antennas, because long transmission-line runs to the LNAs would degrade the signal-to-noise ratio (SNR) unacceptably. The frequency

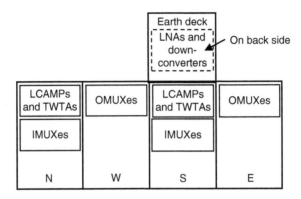

FIGURE 2.5   GEO spacecraft opened out to reveal the configuration of a bent-pipe payload with TWTAs, inside the spacecraft.

downconverters should be close to the LNAs so that the long cable runs to the input multiplexers (IMUXes) are at the lower frequency, where the cable loss is less. Now, traveling-wave tube amplifiers (TWTAs) consume the most direct current (DC) power of any payload equipment, and the radio frequency (RF) power out of the TWTAs cannot be wasted, so the TWTAs and their channel amplifiers (CAMPs) must be placed close to the output multiplexers (OMUXes) and the OMUXes must be placed close to the transmit antennas. In this example, the transmit antennas (not shown) could be on the earth deck or the east and west panels. Finally, the IMUXes go on the north and south panels, where they will be cooler than on the east and west, so they go below the TWTA subsystems. There are many transmission lines and minor passive elements which are not shown here. Another driver of the configuration is the need to minimize the total amount of coaxial cable and waveguide, for weight and space. Units or banks of units that have many RF lines in between would be preferably placed close together. The payload equipment cannot take the entire panels because bus equipment has to go on them, too.

2.1.4   Non-GEO Spacecraft Layout and Attitude

To describe how some non-GEO spacecraft are oriented in space, we must first define the **body axes** of a satellite. The axes are defined as if the spacecraft were an aircraft flying along (Figure 2.6). The aircraft's pitch axis is the axis that goes through the wings. To make this apply to a satellite, we associate the aircraft's forward direction with that of the spacecraft in orbit. (If the satellite were a GEO, the pitch axis would be north–south, and the solar panels would be aligned with and rotate about this axis.) Most spacecraft are always pitching forward in inertial space as they traverse their orbit.

A Globalstar-2 LEO spacecraft is pictured in Figure 2.7. Launch of a complete constellation of 24 second-generation spacecraft was completed in 2013. The spacecraft earth deck is rectangular instead of square as on most GEO satellites.

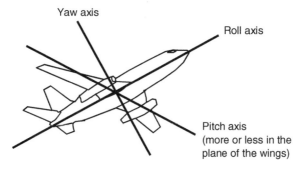

FIGURE 2.6   Body axes of aircraft.

FIGURE 2.7   LEO spacecraft Globalstar-2. (© Thales Alenia Space. From Spacewatch. Global (2016)).

The three large objects on it are a collection of phased array antennas. For the two of these objects that are pyramidal, each flat panel is a separate phased array. These panels are not parallel to the earth's surface as they would be on a GEO spacecraft. Their orientation allows the panel for each beam to face directly to its coverage area. The side panels of the spacecraft body are trapezoidal and shallower than those on a GEO. The ratio of earth-deck surface area to combined side-panels surface area (where repeater electronics are placed, inside) is larger than on a GEO.

The Globalstar-2 orbit's inclination is not near zero as for a GEO but is 52°, so if the solar panels could only be moved by the same means as on a GEO they would often be away from the sun. The solar panels do rotate about the pitch axis as on a GEO, but additionally the satellite rotates about its yaw axis (Rodden et al., 1998). This is *yaw steering*. The spacecraft cannot be said to be continually pitching forward as it moves along its orbit.

An Iridium Next LEO spacecraft looks similar to the Globalstar-2 spacecraft, except that the phased arrays lie flat. Launch of a complete constellation of 66 spacecraft was completed in 2019. Their orbital inclination is 86.4°, almost polar. If the solar panels could only rotate on one axis, during the part of the year when a spacecraft's orbital plane is nearly perpendicular to the sun's rays the spacecraft would receive little or no sunlight. These spacecraft cope with this by having two-axis motion for the solar panels. Besides the usual axis, the second axis enables the solar panels to tip toward or away from the earth (Aerospace Technology, 2018).

The earlier Iridium generation looked quite different. They were launched from 1991 to 2002. The small earth deck, with the reflector antennas on it, was triangular. Attached to the earth deck and flaring away from it were three phased-array antennas.

The O3b MEO spacecraft, shown in Figure 2.8, is also shaped much like Globalstar-2 and Iridium Next. Their launching started in 2013. Their orbit is equatorial, so one-axis rotation of the solar panels is sufficient.

The reason that the shapes of the current LEO and MEO spacecraft are similar is that this shape makes it feasible to put multiple spacecraft into one launch vehicle. O3b launched four at a time, all arranged like spokes in a wheel inside the

FIGURE 2.8 MEO spacecraft O3b in factory. (© Marie Ange Sanguy).

launch vehicle fairing, with the earth decks facing outward (Arianespace, 2013). Globalstar-2 launched six at a time, with four arranged like those of O3b and two higher up on the launch vehicle (Graham, 2011).

2.2 ON-ORBIT ENVIRONMENT AND MITIGATION BY SPACECRAFT BUS

2.2.1 Thermal

2.2.1.1 Cold of Space   The temperature of space without sunlight is 3 K or −270 °C. Part of the spacecraft experiences temperatures of over 150 °C while in sunlight and temperatures of about −150 °C when in the earth's shadow (Bloch et al., 2009).

2.2.1.2 Heat from Spacecraft Electronics   The spacecraft electronics produce heat since no device is 100% power-efficient. This is the basic source of the heat that keeps the electronics warm.

2.2.1.3 Changing Direction of Sun from Satellite   The orbit and the attitude of the satellite determine how long each of the satellite's faces will be in the sun and at what angle to the sun. When a face is in the sun it is warmer than in the shade, and the more nearly perpendicular the face or panel is to the sun's rays the warmer it is.

The earth orbits the sun in the **ecliptic plane**. The earth itself rotates in a different plane, the **equatorial plane**, which is 23.44° off the ecliptic plane. The spacecraft's orbital plane may be yet another plane. Depending on the time of year, sunlight hits the spacecraft at different angles over the course of an orbit.

For a GEO satellite, the variation of the sun's direction over the day and year are illustrated in Figures 2.9, 2.10, and 2.11. What is illustrated is how the satellite looks from the sun's viewpoint at equinox and the northern hemisphere's summer and winter solstices, respectively, at four times of the day. At equinox, no sun is on either N–S panel, so there is minimal **diurnal variation** for these panels, that is, over the course of the day. Also, for part of the day sun is on the east panel and none is on the west, and for the other half of the day sun is on the west panel and none is on the east, so these panels have a significant diurnal temperature variation. On the other hand, at summer solstice the north panel gets the same amount of sun all day long and the south panel gets none, so there is no diurnal variation. The experience of the E–W panels at summer solstice is about the same as at equinox, although the amount of sunshine falling on them is a little less because of the 23.44° tilt of the orbit plane relative to the line between sun and earth. The same is true of the earth deck. Winter solstice is similar to summer solstice except that north and south have switched places. In summary, the N–S panels never receive much sun, while the E–W panels go from full sun to full shade and back over the course of a day.

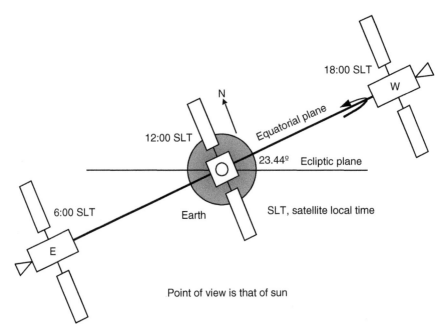

FIGURE 2.9   GEO satellite's diurnal motion at equinox.

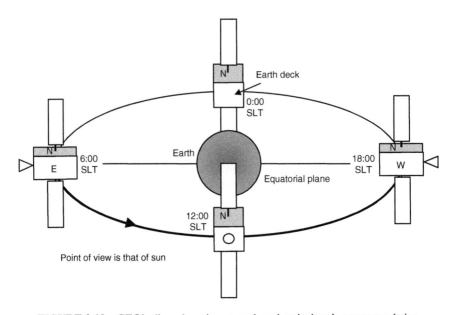

FIGURE 2.10   GEO's diurnal motion at northern hemisphere's summer solstice.

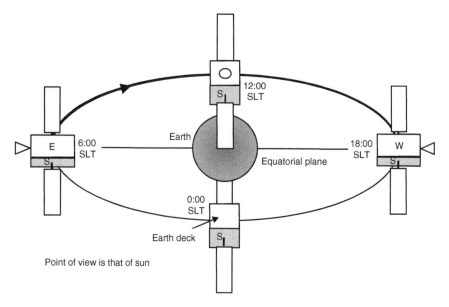

FIGURE 2.11   GEO's diurnal motion at northern hemisphere's winter solstice.

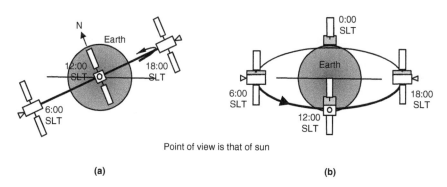

FIGURE 2.12   Equatorial MEO's orbital motion at (a) equinox and (b) northern hemisphere's summer solstice.

The N–S panels are on average cooler than the E–W panels and see a much smaller temperature swing.

For an equatorial MEO, the situation is similar to that for a GEO except that the periods of time in the sun are shorter. The situation for an O3b MEO is shown in Figure 2.12.

For the Iridium Next LEOs, in a nearly polar orbit, the situation is totally different. For example, at the time of year when a spacecraft's orbit plane is perpendicular to the sun's rays, the spacecraft has the same, long side panel in full sun all day. On the other hand, three months later, the spacecraft is in the sun only about half the day, when its anti-earth deck or one of the two smaller side panels is in full sun or both are in the sun at an angle. Only the earth deck hardly gets any sun, so it is always cool. There is no pair of side panels which is always cool. Sometimes one panel gets the sun all day long.

The situation for the Globalstar LEOs would lie between those of O3b and Iridium Next.

2.2.1.4 *Changing Distance of Sun from Earth*   The earth's orbit about the sun is slightly elliptical, with an eccentricity of 0.0167. This causes a ±3% variation in sunlight strength reaching the earth-orbiting satellite (Morgan and Gordon, 1989). The difference this makes to the payload is only in the definition of testing temperatures: **cold** is defined for no sun and **hot** for full sun at minimum distance.

2.2.1.5 *Eclipse*   Eclipse is when the satellite is out of the sunlight because the sun is blocked by the earth. During this time, the solar panels can generate no electricity, so the spacecraft batteries are drawn on.

For a GEO satellite, the time in eclipse per day varies with the time of year, as can be seen in Figures 2.9–2.11. The eclipse season is the three weeks before and the three weeks after each equinox. There are no eclipses outside of these two periods. At an equinox, eclipse lasts about 68 minutes a day, and a week and a half before or after equinox, it lasts about 60 minutes (Morgan and Gordon, 1989).

A MEO satellite is in eclipse much more of the time than a GEO is, and a LEO even more. An equatorial MEO is in eclipse during part of most days of the year. Depending on the exact altitude of the orbit, near the solstices it may have no eclipse. The eclipse lasts for the shortest time at the solstices. For example, an O3b MEO is in eclipse during at least some part of every orbit, all year long, as illustrated in Figure 2.12.

A LEO is in eclipse almost half of every day for the whole year, unless the orbit is near-polar, when during two periods of the year the LEO is in no eclipse. The spacecraft battery of a non-GEO is drawn on much more often than a GEO's is, and it charges and discharges much more often, which shortens its life.

2.2.1.6 *Mitigation by Thermal Control Subsystem*   The bus's thermal control subsystem provides the correct temperature ranges for the various elements of the payload. The range is nominally from about −20 to 65 °C (Bloch et al., 2009). The thermal control subsystem carefully balances heat removal with heat retention, so the payload equipment becomes neither too hot nor too cold. The whole temperature control of the payload and the rest of the spacecraft is a complex design problem that the thermal engineer attacks iteratively, with the payload engineer in the loop if need be. In the rare cases when the subsystem cannot provide the desired temperature ranges for the payload elements, the payload has to be partially redesigned.

Thermal control of the spacecraft is mainly achieved by isolating it from the space environment with thermal blankets and rejecting most of the internally generated heat through panels that radiate heat into space (Sharma, 2005). Also, coatings aid in heat transfer among units. Let us describe thermal control from the innermost element out to the outside of the spacecraft.

Active units and high-power RF loads create waste heat, which is drawn out of the element, fed into the spacecraft's heat transport system, and transported to where it is radiated into space. Each payload element that must dissipate heat has a **baseplate** as its bottom. The element is built so that its heat flows to the baseplate, which gets attached to a spacecraft **mounting plate**. The spacecraft's thermal control subsystem is designed to ensure that the mounting plate will be kept within a particular range of temperatures. The engineer designs the element so that when its baseplate is in contact with a mounting plate at the maximum and minimum temperatures, every component of the element will be within its required temperature range. If he cannot do that, the spacecraft thermal engineer has to adjust the thermal control system to provide a better mounting-plate temperature range.

The mounting surfaces for a GEO's payload electronics, aside from the antennas and associated waveguide, are several, shown with the very thick dashed lines in Figure 2.13: the four side panels' inside surfaces, internal equipment panels, and the outer surface of the earth deck. Figure 2.13a is what the spacecraft looks like if the earth and anti-earth decks are removed so one can look into the spacecraft. Figure 2.13b shows the spacecraft from the side without the east and west panels. The fully interior panels are optional, as are their particular width and length. (Spacecraft bus equipment is also mounted to these surfaces and to the lower part of the panels shown without thick dashed lines.)

Embedded in some of the GEO's equipment mounting panels is a matrix of **heat pipes**, which carry waste heat away from the spacecraft mounting plates. A heat

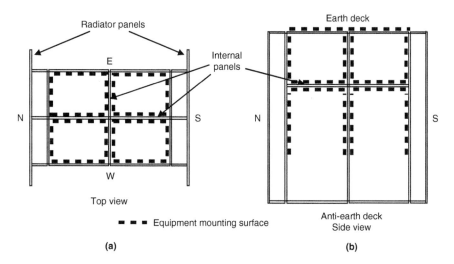

FIGURE 2.13 Typical spacecraft configuration of a GEO's payload-equipment surfaces (after (Yee, 2002)).

pipe is a passive, self-contained device that evaporates a working liquid at its hot end, carries the gas to its cooler end where the liquid dumps its heat and condenses, and carries the liquid back to the hot end by capillary or wicking action (Cullimore, 1992). Other heat pipes carry the heat from the equipment panels to the **radiator panels**. In the radiator panels are further heat pipes. As shown in Figure 2.13, radiator panels typically cover the outside of the entire north and south equipment panels (Watts, 1998). Breakthroughs in the use of heat pipes since the late 1990s have expanded the range of possibilities of the payload configured on the spacecraft. The capacity of the thermal subsystem has been increased, so the payload can use and transmit more power. Unit temperature swings are reduced, decreasing the payload-performance variation. Some important breakthroughs in heat pipes, listed in the order of date of patent application submission:

- Heat pipes to thermally couple the east and west panels (Hosick, 2000)
- Deployable radiator panels that deploy aft, off the north and south panels, as shown in Figure 2.14 (Pon, 2002)
- Crossing heat pipes to thermally couple the north and south radiator panels, as shown in Figure 2.15. They enable the equipment on north and south to have more constant temperatures (Watts, 1998) and they increase the thermal capacity of these panels by 11% (Low and Goodman, 2004)
- Loop heat pipes, which have much higher capacity and can transport heat a longer distance than regular heat pipes. Loop heat pipes are used to connect the equipment panels to the radiator panels and even to multiple radiator panels (Yee, 2002)
- Bidirectional loop heat pipes, which can transfer heat between equipment panels and also between radiators (Yee, 2003)
- Radiator panels on the east and west sides, coupled together with loop heat pipes. The east and west equipment panels contain heat pipes. This increases the heat dissipation capability of the east and west panels together by about 50%. These equipment panels can then contain equipment that tolerates high temperature, such as RF loads, feeds, switches, circulators, and OMUXes (Low and Luong, 2002)
- Thermal coupling of north and south panels to east and west, to dump heat from north and south to east and west (Jondeau et al., 2012).

Another part of the thermal control subsystem is a variety of *thermal control coatings*. One group of coatings is absorbers, namely black paint or black anodizing, which are applied to some internal units to improve heat radiation. Another group is solar reflectors, which reflect most of incident solar energy back into space and which emit infrared energy. One example is optical solar reflector, which forms the outer surface of radiator panels. It consists of glass or quartz on a layer of silver or aluminum attached to the panel with highly conductive adhesive. On a GEO the radiator panels are on the north and south spacecraft panels, near the hot TWTAs. Another example of a solar reflector is white paint, which has high thermal

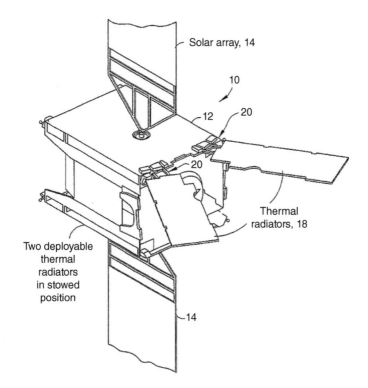

FIGURE 2.14   Deployable radiator panels on a GEO spacecraft (Pon, 2002).

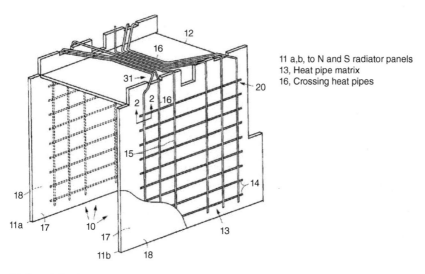

FIGURE 2.15   Crossing heat pipes in a GEO spacecraft (Low and Goodman, 2004).

emittance (Sharma, 2005). It is used on the heat radiators of direct-radiation-cooled TWTAs (Section 7.4.7.2), visible sticking out between spacecraft side panels in Figure 2.4a.

The last part of the thermal control subsystem is the **thermal blanket** which covers the whole outside of the spacecraft except for antennas, thrusters, and radiators (Wertz and Larson, 1999). It has 10–15 layers of alternating Mylar and low-conductance spacer. It is mounted on spacecraft panels with Velcro tape (Sharma, 2005).

So how does a non-GEO handle thermal control? The answer depends on the orbit. For O3b with an equatorial orbit, possibly the measures are the same as on a GEO. For Iridium Next, with a near-polar orbit, it would seem that heat pipes that connect all four side panels would be essential.

2.2.2 Aging

An active payload unit ages as it is operated (and does not age when it is not operated).

The spacecraft's thermal subsystem ages, in that the radiators' outer surfaces and other spacecraft surface-finishes age, reducing their capacity to reflect sunlight or radiate heat (Yee, 2011). The payload equipment on heat pipes becomes warmer as the years go by unless heaters are employed in the early years for compensation.

2.2.3 Radiation

Radiation is transfer of energy by means of a particle (including photons) at X-ray frequencies or higher (ECSS, 2008b). Primary radiation comes from outside the spacecraft, and secondary radiation comes from the interaction of primary radiation with the spacecraft (ECSS, 2008a). Primary radiation originates from the galaxy all around and specifically from the sun. Cosmic rays, which are actually charged particles, come from the galaxy and are 87% protons, 12% alpha particles (the nuclei of helium), and 1% heavy ions (Bourdarie and Xapsos, 2008). The solar wind ultimately provides the rest of the primary radiation. It is a continuous stream of charged particles, mostly electrons and protons, ejected from the sun. It is stronger during the more active part of the approximately 11-year solar cycle. Sometimes during especially high solar activity, bursts of solar energetic particles last from a few hours to several days. These particles are protons, electrons, and heavy ions of much higher than normal energy. The stronger the solar wind is, the more galactic cosmic-ray particles it sweeps away from the earth (ECSS, 2008a).

The earth is protected from the particles by its magnetic field, which deflects some galactic cosmic-ray particles (Wikipedia, 2010a) and most solar particles (Wikipedia, 2010b) and traps other particles away from the solid earth. To first order, the earth's magnetic field is a magnetic dipole tilted about 11° off the earth's rotation axis and with its center about 500 km to the north of the equator

(Wikipedia, 2011). The places where the dipole goes through the earth's surface are the *geomagnetic poles*. The trapped particles form the *Van Allen radiation belts*.

The Van Allen belts of high-energy electrons and protons are illustrated in Figures 2.16–2.18. For scale comparison, the GEO orbit is at 6.6 earth radii from the center of the earth, a MEO orbit at about 2.2 to 4.1, and a LEO at 1.1 to 1.3. The belts for electrons of greater than 1 MeV are shown in Figure 2.16. On the geomagnetic equatorial plane, the strongest part of the electron belts is at about 4.3 earth radii from the center of the earth. Figure 2.17 shows the belts for protons of energy greater than 10 MeV. These belts are smaller than the electron belts. The strongest part of the belt in the geomagnetic equatorial plane is at about 1.75 earth radii from the center of the earth (Daly et al., 1996). The MEO orbit lies below the worst part of the electron belt and above the worst part of the proton belt (Ginet et al., 2010). A LEO polar or near-polar orbit will be exposed to trapped particles near the magnetic poles: electrons below 1000 km, and both electrons and protons above that (Bourdarie and Xapsos, 2008). As can be seen in Figure 2.18, the proton belts come especially close to the earth's surface in the southern hemisphere, creating the *South Atlantic anomaly*. The center of the anomaly is at about 35° W longitude (Daly et al., 1996), which is about the longitude of the eastern edge of South America.

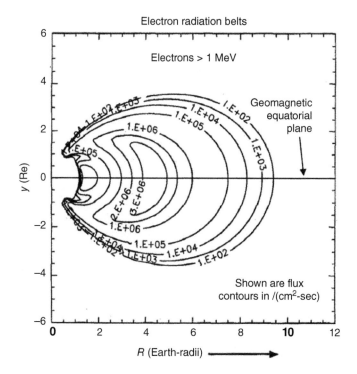

FIGURE 2.16   Van Allen electron belts. (©1996 IEEE. Reprinted, with permission, from Daly et al. (1996)).

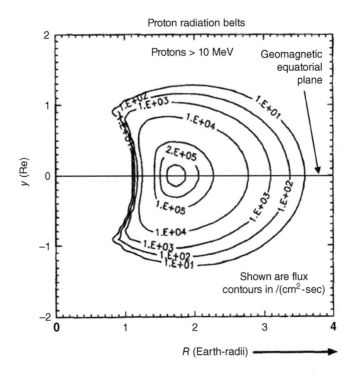

FIGURE 2.17 Van Allen proton belts. (©1996 IEEE. Reprinted, with permission, from Daly et al. (1996)).

The cause is the non-coincidence of the magnetic dipole center and the earth's center. The anomaly can impinge on a LEO orbit. This is believed to be the cause of the degradation of the S-band power amplifiers on the first-generation Globalstar satellites (de Selding, 2007).

High-energy electrons and protons can harm some of the payload electronics. Spacecraft structural enclosures are typically made of aluminum, a few millimeters thick, as shielding. In order to penetrate 2 mm, an electron must have energy greater than about 1 MeV and a proton greater than about 20 MeV (Daly et al., 1996).

High-energy electrons can be defended against by *shielding* in the payload unit or in the bus; the thicker the shielding, the larger the reduction. High-energy protons cannot be stopped by shielding because they can penetrate far, so sensitive parts must be radiation-hard (Section 2.3.2.2) (Wikipedia, 2018). For a LEO, radiation exposure depends on the orbit inclination and altitude (Ya'acob et al., 2016).

2.2.4 Orbit Disturbances

The GEO, MEO, and LEO orbits have somewhat different perturbations, and the spacecraft lifetimes differ. As a result, their **station-keeping** efforts are different.

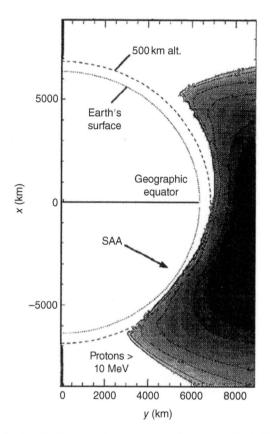

FIGURE 2.18   South Atlantic anomaly shown in earth cut at longitude 35° W.
(©1996 IEEE. Reprinted, with permission, from Daly et al. (1996)).

The GEO spacecraft is subject to perturbation forces that slowly change the longitude, the inclination, and the orbit eccentricity. The distortion of the earth's gravitational field due to the earth's non-sphericity and inhomogeneity causes the spacecraft to drift in longitude toward one of the two stable-equilibrium longitudes (Borissov et al., 2015). The gravity of the sun and moon changes the inclination at a yearly rate of between 0.75° and 0.95°. A non-zero inclination is significant in terms of the longitude variation it causes when the inclination is greater than a few degrees. The distortion of the earth's gravitational field causes either a steady increase or decrease of the orbit's semi-major axis, depending on the longitude. Solar radiation pressure also changes the eccentricity, along with a small effect from the moon. An increase in eccentricity causes a longitude variation (ITU-R S.484, 1992).

A GEO spacecraft has to always keep close to its nominal longitude. For example, the International Telecommunication Union (ITU) requires a satellite in the fixed satellite service (FSS) (Chapter 17) to keep its longitude within ±0.1°

of nominal. If the orbit is slightly inclined, it is the longitude where the orbit plane crosses the equator that must be kept to this limit (ITU-R S.484, 1992).

The GEO satellite operator may decide to let the orbit inclination vary over the years, to save north–south station-keeping fuel, as for example most of the mobile satellite-service spacecraft do that are described in Chapter 20. The original inclination is set at one end of the desired range and the inclination changes monotonically over the years to the other extreme. The ground stations and user terminals must be able to follow this (ITU-R S.743, 1994).

Onboard propulsion for GEO station-keeping used to always be chemical but is increasingly electric ion propulsion because of the large mass savings. Chemical propellant makes up about half the mass of the satellite in transfer orbit (Gopinath and Srinivasamuthy, 2003). Chemical propulsion requires burns every three to five weeks (Corey and Pidgeon, 2009). Electric propulsion requires burns more often but they are milder. One satellite manufacturer reported that on average each of the four spacecraft thrusters fires for about 40 minutes a day, the firings being in diagonal pairs (Goebel et al., 2002). Another manufacturer reported the same average burn time, for a satellite with beginning-of-life mass of 5000 kg (Gopinath and Srinivasamuthy, 2003). The more massive a spacecraft is, the longer the burns have to be (Corey and Pidgeon, 2009). Four thrusters are typical.

GEO north–south station-keeping (that is, inclination control), no matter which propulsion technology, requires an order of magnitude more fuel than east–west does (Oleson et al., 1997; Goebel et al., 2002).

For MEO satellites, the main perturbations are the same as for a GEO. Satellites in different orbit planes will experience different effects from the earth's gravitational field distortions. Station-keeping thruster burns may not be necessary. The initial orbit parameters can be set to offset values and the values allowed to drift over life to values at the other end of what is allowed (Fan et al., 2017).

For LEO satellites, the main perturbations are the distortion in the earth's gravitational field, atmospheric drag, and solar gravity. The main effects are the constant decay in the semi-major axis and a minor inclination drift (Garulli et al., 2011). Onboard propulsion for station-keeping can be either chemical or electrical.

2.2.5 Attitude Disturbances

The most important attitude disturbances external to the spacecraft are gravity gradient, solar radiation pressure, magnetic-field torques, and, for LEOs, aerodynamics. The principal disturbances internal to the spacecraft are uncertainty in the center of gravity, thruster misalignment, mismatch of thruster outputs, rotating machinery, sloshing of liquids such as fuel, and dynamics of flexible bodies such as antennas, booms, and solar panels (Wertz and Larson, 1999). Compared to larger satellites, small satellites are more susceptible to external disturbances and the actuator dynamics have more of an influence on the satellite dynamics (MacKunis et al., 2008).

The spacecraft **attitude control system** is designed to keep the attitude under control well enough so that the antenna pointing is good enough to support the payload's mission.

Onboard propulsion is used during **momentum-dumping** or *momentum-wheel uploads*. A spinning reaction wheel absorbs environmental torques such as gravity gradients, solar radiation pressure, atmospheric drag, and the ambient magnetic field. Over time it spins faster and faster but cannot spin at arbitrarily high rates. The stored angular momentum must be unloaded via the onboard thrusters (Weiss et al., 2015).

When the onboard thrusters use chemical propellant, the worst disturbances in satellite attitude are due to the brief burnings of the spacecraft thrusters for orbit correction or a momentum dump.

When the onboard thrusters use electric ion propulsion, the attitude transients are much smaller. Even if there is an anomaly during thrusting, the momentum wheels can easily absorb it and there is little or no impact on the payload (Corey and Pidgeon, 2009).

2.3 GENERAL EFFECTS OF MITIGATED ENVIRONMENT ON PAYLOAD

The on-orbit environment affects the payload in the ways summarized in Table 2.1. In the subsections below, we discuss these and what some payload units do to further mitigate the environment. How the general effects translate into performance effects for the various payload units will be discussed in Chapters 3 through 7.

2.3.1 Temperature-Variation Effects

Temperature variations of the payload units come on three time-scales for any orbit: once per orbit, yearly, and over life. For payload electronics on a panel with heat pipes, the once-per-orbit and yearly variations are not simply periodic. As the

TABLE 2.1   General Effects of On-Orbit Influences

Environment Aspect	General Payload Effects
Thermal with the exception of reflector front going in and out of sun	Temperature range of thermal interface (mounting plate) that spacecraft presents to unit
Radiation	Irreversible changes in RF elements and quartz oscillators; irreversible and reversible changes in digital circuits
Aging	Increased temperature range variation (due to heat pipes aging); irreversible changes in units
Attitude disturbances	Antenna gain variation for non-autotracked antenna

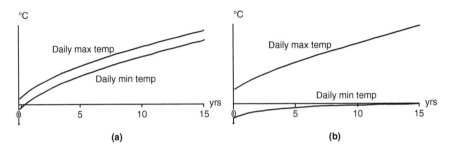

FIGURE 2.19   Examples (exaggerated) of unit temperature variation on a GEO: (a) unit on N or S panel when panels are thermally coupled; (b) unit on E or W panel when panels have independent heat pipe systems.

thermal subsystem ages, the once-per-orbit and the yearly variations may grow in size. The growth is between linear and logarithmic (Yee, 2011). In addition, the unit's average temperature drifts upward from **beginning of life (BOL)** to **end of life (EOL)**. Two exaggerated examples of lifetime variation for a unit on a GEO satellite are shown in Figure 2.19, one example for a unit on the N or S panel and the other for a unit on the E or W panel. All panels are assumed to have heat pipes, and the N–S panels are thermally coupled. The yearly temperature variation is not shown; the diurnal variation is taken as the worst over the year. The N–S panels intrinsically provide a smaller diurnal temperature variation than E–W ones do.

2.3.1.1   *On Active Units*

In general, active units have performance that varies with temperature. There are two groups of active units to consider: those that come before the preamp of the high-power amplifier (HPA) subsystem and those that come after. The preamp divides the payload into two almost independent sides. The preamp must be able to accept a range of input power, part of which range accounts for temperature variation in units before it; additionally, it must be able to output a range of power to the HPA, part of which range accounts for temperature variation in the HPA.

Some payload units perform additional thermal control beyond what the bus's thermal control subsystem does. Some units may have their own heating for BOL, when the thermal control subsystem provides its lowest temperatures. Some units such as the master oscillator have heaters to increase their thermal stability. The direct-radiation-cooled TWTA has its own radiator that extends out beyond the spacecraft edge; this provides relief to the thermal control subsystem.

Some units such as the linearizer-channel amplifier (LCAMP) correct their own performance over temperature based on uncorrected-performance curves measured in the factory.

2.3.1.2   *On Antennas*

The changing sun angle also affects the antennas. A reflector has a different amount of surface distortion depending on its temperature (how much sun is on it) or whether it is transitioning between the two extremes. This is not a large factor in antenna performance.

2.3.1.3 On RF Lines As coaxial cable's temperature goes up, its insertion loss increases linearly; the insertion phase shifts as a nonlinear function of temperature.

The behavior of waveguide over temperature depends on the waveguide material. Most materials expand linearly with increasing temperature, in which case the operating frequency range shifts down. Super Invar scarcely changes shape at all over a limited temperature range.

2.3.1.4 On Microwave Filters The three filter technologies we will study are the empty-cavity waveguide filter, the dielectric-resonator filter, and the coaxial-cavity filter. The waveguide filter behaves as its waveguide material behaves. The other two are minimally temperature-sensitive.

2.3.2 Radiation Effects

2.3.2.1 General Payload Radiation Effects Energetic charged particles can cause ionization in two ways in payload parts. One way is that as they collide with atoms on their way through material, they pull the electrons out of their orbits around the atomic nuclei and release the electrons into the surrounding matter. In solid-state matter, the result is the creation of electron–hole pairs along the paths. Another way is to produce secondary radiation. When the high-energy particles hit dense matter, they abruptly lose some energy in the braking, creating photons. This is the Bremsstrahlung effect. These photons ionize matter along their paths (Ecoffet, 2013).

The performance of some payload parts can be either disrupted or permanently damaged by the ionization caused by a single charged particle. These are *single-event effects (SEE)*. Non-destructive effects are the corruption of digital data or the placement of a device in a different operational state. Destructive effects are the inducement of high-current conditions which can destroy the device. Heavy ions cause direct ionization of sensitive regions of the semiconductor, while protons and neutrons interact with atomic nuclei within or very near the active semiconductor and result in localized charge generation (ECSS, 2008b).

The performance of other payload parts can be affected by accumulated ionization, namely the **total ionizing dose (TID)**. Such parts include quartz crystals and semiconductor devices (Ecoffet, 2013). TID degradation in microelectronics made from semiconductors results in a gradual loss of performance and eventual failure. TID effects in semiconductor devices depend on the creation of electron–hole pairs within dielectric layers by the incident radiation and subsequent generation of (1) traps at or near the interface with the semiconductor and (2) trapped charge in the dielectric (ECSS, 2010). In microelectronics, TID causes surface leakage current in semiconductors and, specifically to metal-oxide–semiconductor (MOS), gate threshold-voltage shifts. The charged particle can be either a primary-radiation proton or a secondary-radiation particle (ECSS, 2008b). TID can be reduced by shielding that absorbs most electrons and lower-energy protons (Barth et al., 2004).

With a much smaller probability than causing ionization, charged particles can also interact with the nuclei of the atoms they collide with. A fraction of the particle's energy transfers to the nucleus, which excites the nucleus and displaces it within the lattice structure, creating interstitials (an atom where there should be none) and vacancies in the lattice. The interstitials and vacancies are mobile and can cluster together or react with impurities in the lattice, creating stable defect centers. This is *displacement damage (DD)*. It affects bipolar transistors (ECSS, 2008b).

These three radiation effects for payloads are listed with their particle sources in Table 2.2.

2.3.2.2 Payload Parts Affected by Radiation

The only radiation-sensitive payload parts are quartz crystal oscillators and some integrated circuits (ICs), as shown in Table 2.3. The table also shows which of the three radiation effects applies. A crystal oscillator undergoes a frequency shift from TID (ECSS, 2010). Power MOS is used in solid-state power amplifiers (Section 7.5).

GaAs- and InP-based monolithic microwave integrated circuits (MMICs), commonly used for RF applications, are not significantly radiation-sensitive

TABLE 2.2   Summary of Radiation Effects for Payload and Causes (ECSS, 2008b)

Radiation Effect	Sources
Single-event effects (SEE)	Trapped protons, solar protons and heavier ions, cosmic-ray protons and heavier ions
Total ionizing dose (TID)	Trapped protons and electrons, solar protons; secondary particles including photons but not neutrons
Displacement damage (DD)	Trapped protons, solar protons; secondary protons and neutrons

TABLE 2.3   Summary of Radiation Effects on Payload Parts (ECSS, 2010)

Part	Technology	Radiation Effects
Crystal oscillator	Oscillator crystal	TID
Integrated circuit	Power MOS	TID and SEE
	CMOS	TID and SEE
	Bipolar	TID, SEE, and DD
	Bipolar CMOS (BiCMOS)	TID, SEE, and DD
	Silicon on insulator (CMOS) including silicon on sapphire	TID and SEE but no single-event latch-up

TABLE 2.4 Examples of CMOS IC Technology for Payload

IC Type	Examples of Usage
FPGA	On-board processing including packet processing and control, DSP, and RAM (Xilinx, 2010, 2017); channelizer, downconverter, beam-former, frequency reconfiguration (Huey, 2016)
ASIC	On-board processor (Hili et al., 2016)
Mixed-signal (i.e., analog/digital)	ADC, DAC (ECSS, 2008b)
RF/mixed-signal	Digital step attenuator, RF C-switch used in phased arrays, and phase-locked loop (Robinette, 2009)

(Kayali, 1999). One supplier advises unit designers that their GaAs MMICs can take up to 10^7–10^8 rads TID (TriQuint, 2010a). The transistors are high-electron-mobility transistor (HEMT), pseudo HEMT (pHEMT), or other field-effect transistors (FETs).

CMOS (complementary MOS) technology is known for low power and ease of integration (Robinette, 2009). However, off-the-shelf parts are susceptible to radiation effects. Nonetheless, they are used on communications satellites because they can be made **radiation-hard**. Such parts are based on their non-hardened equivalents, with some design and manufacturing variations that reduce the susceptibility to radiation damage. They are individually tested in radiation before they are used (Wikipedia, 2018). Parts that are advertised as rad-hard are usually just hard to TID unless stated otherwise (Barth et al., 2004).

Specifically, the silicon-on-sapphire (SoS) technology can be classified as silicon-on-insulator (SOI) but has linearity superior to other SOI CMOS and to GaAs (Robinette, 2009). The main advantage of SoS is that a range of devices such as RF, analog, logic, EEPROM, and high-Q passives can be integrated on one chip (Reedholm, 2017). Table 2.4 lists some examples of CMOS IC technology in use today in payloads.

2.3.3 Aging Effects

Irreversible changes in units over life cause slow changes in performance even when the unit operating conditions remain the same. Aging is from accumulated operating time (ECSS, 2006).

Most RF GaAs-device failures occur in the FET channel, where failure is defined as reaching a 1-dB degradation in RF output power. Failure is from physical and chemical processes (TriQuint, 2010b).

What ages in a crystal oscillator is the crystal itself, its mounting, its electrodes, and so on (HP, 1997).

The lifetime drift of some payload units' performance can be compensated through ground command. The supplier of any unit will have performed life test on several examples of that unit. Suppose that the supplier has observed a nonzero average drift per year. Then the future drift of a flight unit can be predicted, but with uncertainty. Over the years, maybe the operation of the unit can be partially compensated through a commanded change in some parameter. However, if the drift observed by the supplier has a zero average, nothing can be done.

2.3.4 Antenna Gain Variation

Spacecraft attitude error causes antenna pointing error for a non-autotracked (Section 3.10) antenna, which leads to gain variation (Section 3.9). Autotrack is a control system that measures antenna pointing error and uses it to continuously correct the pointing.

REFERENCES

Aerospace Technology Web site (2018). Iridium Next satellite constellation. Project information. On www.aerospace.technology.com/projects/iridium-next-satellite-constellation/. Accessed Aug. 30, 2018.

Arianespace (2013). A batch launch for the O3b constellation. Launch news item. June 25. On www.arianespace.com/news-launch-kits/2013/present-archive.asp. Accessed Sep. 25, 2013.

Barth JL, LaBel KA, and Poivey C (2004). Radiation assurance for the space environment. *IEEE International Conference on Integrated Circuit Design & Technology*; May 18–20.

Bloch M, Mancini O, and McClelland T (2009). What we don't know about quartz clocks in space. *Institute of Navigation Precise Time and Time Interval Meeting*; Nov. 16–19.

Borissov S, Wu Y, and Mortari D (2015). East-west GEO satellite station-keeping with degraded thruster response. *MDPI Aerospace*; 2 (Sep.); 581–601.

Bourdarie S and Xapsos M (2008). The near-earth space radiation environment. *IEEE Transactions on Nuclear Science*; 55 (4) (Aug.); 1810–1832.

Corey RL and Pidgeon DJ (2009). Electric propulsion at Space Systems/Loral. *International Electric Propulsion Conference*; Sep. 20–24.

Cullimore BA (1992). Heat transfer system having a flexible deployable condenser tube. US patent 5,117,901. June 2.

Daly EJ, Lemaire J, Heynderickx D, and Rodgers DJ (1996). Problems with models of the radiation belts. *IEEE Transactions on Nuclear Science*; 43; 403–415.

de Selding PB (2007). Globalstar says service sustainable until new satellites arrive in 2009. *Space News*; Feb. 12.

de Selding PB (2013). Intelsat enlists Space Systems/Loral to build IS-34 satellite. *SpaceNews*; July 23.

Ecoffet R (2013). Overview of in-orbit radiation induced spacecraft anomalies. *IEEE Transactions on Nuclear Science*; 60 (3) (June); 1791–1815.

Emiliani LD (2020). Computer program with input of Union of Concerned Scientists satellites database of 2020 Apr 1. Oct. 8.

European Cooperation for Space Standardization (ECSS) (2006). *Standard ECSS-Q-30-11A (2006): Space Product Assurance, Derating--EEE Components*. The Netherlands: ESA Publications Division.

ECSS (2008a). *Standard ECSS-E-ST-10-04C (2008a): Space Engineering, Space Environment*. The Netherlands: ESA Requirements and Standards Division.

ECSS (2008b). *Standard ECSS-E-ST-10-12C: Space Engineering, Methods for the Calculation of Radiation Received and Its Effects, and a Policy for Design Margins*. The Netherlands: ESA Requirements and Standards Division.

ECSS (2010). *Standard ECSS-E-HB-10-12A: Space Engineering, Calculation of Radiation and Its Effect and Margin Policy Handbook*. The Netherlands: ESA Requirements & Standards Division.

Evans BG, editor (1999). *Satellite Communication Systems*, 3rd ed. London: The Institution of Electrical Engineers.

Garulli A, Giannitrapani A, Leomanni M, and Scortecci F (2011). Autonomous station keeping for LEO mission with a hybrid continuous/impulsive electric propulsion system. *International Electric Propulsion Conference*; Sep. 11–15.

Ginet GP, Huston SL, Roth CJ, O'Brien TP, and Guild TB (2010). The trapped proton environment in medium earth orbit (MEO). *IEEE Transactions on Nuclear Science*; 57 (6) (Dec.); 3135–3142.

Goebel DM, Martinez-Lavin M, Bond TA, and King AM (2002). Performance of XIPS electric propulsion in on-orbit station keeping of the Boeing 702 spacecraft. *AIAA/ASME/SAE/ASEE Joint Propulsion Conference*; July 7–10.

Gopinath NS and Srinivasamuthy KN (2003). Optimal low thrust orbit transfer from GTO to geosynchronous orbit and stationkeeping using electric propulsion system. *International Astronautical Congress of the International Astronautical Federation*; Sep. 29–Oct. 3.

Graham W (2011). Soyuz 2-1A closes 2011 with successful launch of six Globalstar-2 satellites. News article. Dec. 28. On www.nasaspaceflight.com/2011/12/soyuz-2-1a-2011-launch-six-globalstar2-satellites/. Accessed July 28, 2019.

Hewlett Packard (1997). Fundamentals of quartz oscillators. Application note 200-2. On www.hpmemory.org/wb_pages/wall_b_page_01.htm. Accessed Nov. 24, 2010.

Hili L, Roche P, and Malou F (2016). ST 65nm a hardened ASIC technology for space applications. *European Space Components Conference*; Mar. 1–3.

Hosick DK, inventor; Space Systems/Loral, Inc., assignee (2000). High power spacecraft with full utilization of all spacecraft surfaces. US patent 6,073,887. June 13.

Huey K (2016). Xilinx Virtex-5QV update and space roadmap. Viewgraph presentation. On indico.esa.int/event/130/contributions/717/. Accessed Sep. 7, 2018.

Iridium Communications (2016). New platform. New possibilities. Sep. 22. On www.iridiumnext.com/2016/09/22/new-platform-new-possibilities/. Accessed Oct. 3, 2018.

ITU (International Telecommunication Union) (2019). Key outcomes of the World Radiocommunication Conference 2019. *ITU News Magazine*; no 6.

ITU-R. Recommendation S.484-3 (1992). Station-keeping in longitude of geostationary satellites in the fixed-satellite service.

ITU-R. Recommendation S.743-1 (1994). The coordination between satellite networks using slightly inclined geostationary-satellite orbits (GSOs) and between such networks and satellite networks using non-inclined GSO satellites.

Jondeau L, Flemin C, and Mena F, inventors; Astrium SAS, assignee (2012). Device for controlling the heat flows in a spacecraft and spacecraft equipped with such a device. US patent 8,240,612 B2. Aug. 14.

Kayali S (1999). Reliability of compound semiconductor devices for space applications. *Microelectronics Reliability*; 39; 1723–1736. On trs.jpl.nasa.gov/. Accessed Dec. 1, 2010.

Low L and Luong J (2002). Spacecraft radiator system and method using east west coupled radiators. US patent application 2002/0139512 A1. Oct. 3.

Low L and Goodman C, inventors; Space Systems/Loral, assignee (2004). Spacecraft radiator system using crossing heat pipes. US patent 6,776,220 B1. Aug. 17.

MacKunis W, Dupree K, Bhasin S, and Dixon WE (2008). Adaptive neural network satellite attitude control in the presence of inertia and CMG actuator uncertainties. *American Control Conference*; June 11–13.

Martin DH, Anderson PR, and Bartamian L (2007). *Communications Satellites*, 5th ed. El Segundo, CA: The Aerospace Press; and Reston, VA: American Institute of Aeronautics and Astronautics, Inc.

Morgan WL and Gordon GD (1989). *Communications Satellite Handbook*. New York: John Wiley & Sons.

Oleson SR, Myers RM, Kluever CA, Riehi JP, and Curran FM (1997). Advanced propulsion for geostationary orbit insertion and north-south station keeping. *AIAA Journal of Spacecraft and Rockets*; 34 (1); 22–28.

Pon R, inventor; Space Systems/Loral, assignee (2002). Aft deployable thermal radiators for spacecraft. US patent 6,378,809 B1. Apr. 30.

Rankin W (2008). The world's population in 2000, by latitude. On www.radicalcartography. net. Accessed July 27, 2019.

Reedholm Systems (2017). Silanna. Client profile-103a. On reedholmsystems.com/wp-content/uploads/2017/03/CP-103a.df. Accessed Sep. 10, 2018.

Robinette D (2009). UltraCMOS RFICs ease the complexity of satellite designs. *Microwave Journal*; 52 (8) (Aug.); 86–99.

Rodden JJ, Furumoto N, Fichter W, and Bruederle E, inventors; Globalstar LP and Daimler-Benz Aerospace AG, assignees (1998). Dynamic bias for orbital yaw steering. US patent 5,791,598. Aug. 11.

Sharma AK (2005). Surface engineering for thermal control of spacecraft. *Surface Engineering Journal*; 21 (3); 249–253. On www.researchgate.net/publication/233610754.

Spaceflight101 (2006). Echostar 21. Article. On spaceflight101.com/proton-echostar-21/echostar-21/. Accessed Sep. 21, 2011.

Spacewatch.Global (2016). Globalstar-second-generation-satellite-photo. Nov. 25. On spacewatch.global/2016/11/globalstar-solutions-monitor-fleets-safeguard-oil-industry-workers-tunisia/globalstar-second-generation-satellite-photo-courtesy-of-thales-alenia-space/. Accessed Aug. 30, 2018.

Telesat (2020). Telstar 12 now at 109.2° WL. Fleet satellite information. On www.telesat. com/our-fleet/telstar-12. Accessed July 19, 2020.

TriQuint Semiconductor, Inc. (2010a). Gallium arsenide products, designers' information. On www.triquint.com/prodserv/tech_info/docs/mmw_appnotes/designer_a.pdf. Accessed Dec. 7, 2010.

TriQuint Semiconductor, Inc. (2010b). Micro-/millimeter wave reliability overview. On www. triquint.com/shared/pubs/processes/Micro_Millimeter_Wave_Reliability_Overview.pdf. Accessed Dec. 7, 2010.

Watts KP, inventor; Hughes Electronics Corp, assignee (1998). Spacecraft radiator cooling system. US patent 5,806,803. Sep. 15.

Weiss A, Kalabić U, and Di Cairano S (2015). Model predictive control for simultaneous station keeping and momentum management of low-thrust satellites. *American Control Conference*; July 1–3.

Wertz JR and Larson WJ (1999). *Space Mission Analysis and Design*, 3rd ed. Hawthorne, CA: Microcosm Press; and New York: Springer.

Wikipedia (2010a). Cosmic ray. On en.wikipedia.org. Accessed Dec. 6, 2010.

Wikipedia (2010b). Solar wind. On en.wikipedia.org. Accessed Dec. 6, 2010.

Wikipedia (2011). Geomagnetic pole. Oct. 12. On en.wikipedia.org. Accessed Oct. 28, 2011.

Wikipedia (2018). Radiation hardening. Sep. 6. On en.wikipedia.org. Accessed Sep. 10, 2018.

Xilinx, Inc. (2010). Xilinx launches first high-density, rad-hard reconfigurable FPGA for space application. News release. On www.prnewswire.com/news-releases/xilinx-launches-first-high-density-rad-hard-reconfigurable-fpga-for-space-applications-98733224.html. Accessed Nov. 26, 2010.

Xilinx, Inc. (2017). Hundreds of Xilinx space grade FPGAs deployed in launch of Iridium Next satellites. News release. Jan. 16. On www.xilinx.com/news/press/2017.html. Accessed Sep. 7, 2018.

Ya'acob N, Zainudin A, Magdugal R, and Naim NF (2016). Mitigation of space radiation effects on satellites at low earth orbit (LEO). *IEEE International Conference on Control System, Computing and Engineering*; Nov. 25–27.

Yee EM, inventor; Space Systems/Loral, Inc., assignee (2002). Spacecraft multiple loop heat pipe thermal system for internal equipment panel applications. US patent 6,478,258 B1. Nov. 12.

Yee EM, inventor; Space Systems/Loral, Inc., assignee (2003). Spacecraft multi-directional loop heat pipe thermal systems. US patent 6,591,899 B1. July 15.

Yee EM, Space Systems/Loral (2011). Private communication. Nov. 7.

CHAPTER 3

ANTENNA BASICS AND SINGLE-BEAM ANTENNA

3.1 INTRODUCTION

This chapter is on the payload unit which is the most eye-catching: the antenna. Along with the solar panels, the antennas make a satellite look like a satellite. Antennas easily excite enthusiasm in some people: most antennas are big, some of them move, and antenna deployment videos are exciting. And who is not charmed when an antenna engineer demonstrates a two-axis deployment with his arm?

This book divides the topic of antennas into two chapters. Antenna basics and single-beam antennas are addressed in this chapter. Multibeam antennas and phased arrays are deferred to Chapter 11 because they draw on material in several chapters, including high-power amplification and processing. The properties of antennas that form a small number of beams can probably be inferred from this chapter and Chapter 11.

This chapter applies, as does the rest of the book, to commercial communications satellites on orbit today or about to be launched. Also included are antennas or antenna components developed by satellite manufacturers, because if a satellite manufacturer has spent the time and money to develop an engineering model, the probability is high that the manufacturer will fly it within a few short years of reporting on it.

Satellite Communications Payload and System, Second Edition. Teresa M. Braun and Walter R. Braun.
© 2021 John Wiley & Sons, Inc. Published 2021 by John Wiley & Sons, Inc.

By far most payload antennas are for communications links between satellite and ground, but a few are for satellite-to-satellite crosslinks, which we will not deal with.

Here is a brief explanation of antennas as background for the detailed description that the rest of the chapter provides. A **reflector** is commonly called a *dish*. A reflector antenna may have one or two reflectors. If it has two, the larger one is the **main reflector** and the smaller one is the **subreflector**. The (main) reflector is the one that directly receives and/or transmits power over communications distances. A reflector antenna always has a **feed**: on receive a feed collects the radio frequency (RF) power reflected into it by the reflector(s), and on transmit the feed sends out the RF power to the reflector(s). A **horn** is a piece of waveguide, flared at the open end, that is most often used as a feed but is sometimes used as an antenna in its own right. The antenna usually includes other components which can variously separate out receive and transmit signals, turn circularly polarized signals into linearly polarized (LP) signals and vice versa, and separate out or combine two LP signals.

Other single-beam antenna topics in this chapter are pointing error, autotrack, antenna inefficiencies, and testing.

3.2   EXAMPLES OF SINGLE-BEAM ANTENNA

Let us whet our appetite for single-beam antennas by viewing a variety of *antenna farms* of such antennas, where a farm is the collection of payload antennas on a spacecraft.

A typical antenna farm on a common type of satellite was depicted in Figure 2.4a. The spacecraft is a C- and Ku-band broadcast satellite, Intelsat 34. It has three fixed reflector antennas. The two Ku-band main reflectors hang on the east panel, each having its subreflector affixed to the earth deck and its feed on the panel. The C-band reflector hangs on the west panel (Nagarajah, 2015). The white stubs in a row, protruding on the edges of the east panel, are the heat-radiating elements of traveling-wave tube amplifiers (TWTAs). The payload has 24 C-band transponders and 24 Ku-band transponders and was launched in 2015 (Krebs, 2017a).

Instead of one or two dishes separately deployed off a side panel of the spacecraft, a pallet deployed off a side panel can accommodate up to four steerable reflectors (not pictured) (Scouarnec et al., 2013).

The earth-deck portion of an antenna farm with an unusually large number of antennas is depicted in Figure 3.1. The satellite is the Ka/K- and Ku-band Eutelsat 3B. It has five Ka/K-band single-reflector antennas with feeds mounted on the **tower** and one Ku-band Gregorian antenna (Section 3.4.3), all single spot-beam antennas and all independently steerable to any place on the visible earth. All antennas both receive and transmit. The spacecraft was launched in 2014 (Glâtre et al., 2015).

The antenna farm on a medium earth orbit (MEO) satellite was depicted in Figure 2.8. The satellite is in the O3b constellation. The first thing that stands out

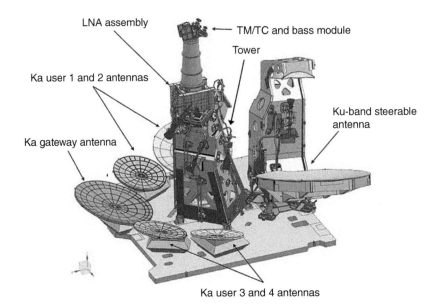

FIGURE 3.1   Large Ka/K and Ku antenna farm on earth deck of Eutelsat 3B. (©2015 IEEE. Reprinted, with permission, from Glâtre et al. (2015)).

about the spacecraft is its unusual shape compared to a geostationary orbit (GEO) spacecraft. The 12 Ka/K-band single-reflector antennas are all independently steerable over the full visible earth. Each antenna follows the earth location of its beam as the spacecraft moves. All antennas receive and transmit (Amyotte et al., 2010). The twenty first-generation O3b spacecraft were launched from 2013 to 2019 (Krebs, 2017b, 2020).

3.3   GENERAL ANTENNA CONCEPTS

3.3.1   Beams

There are, roughly speaking, five sizes of antenna-beam coverage, defined the same for transmit and receive and for all orbits:

- Global or earth-coverage of the visible earth
- Hemispherical, which covers about half the visible earth
- Zonal, which covers approximately a continent
- Regional (also, linguistic-region in Europe), which covers an area about the size of Spain
- Spot, which covers an area about the size of metropolitan New York City. The term is sometimes used to mean instead a regional or zonal beam in comparison to a larger beam on the same satellite.

The Intelsat satellite fleet alone has all these sizes of beam (Schennum et al., 1999; Intelsat, 2008).

Especially at Ku-band and higher frequencies, the antenna may be designed to compensate long-term average attenuation due to rain. The compensation is done differently for different beam sizes: for global and hemispherical, probably no compensation; for zonal and regional, compensation via the beam providing higher gain toward some areas than others; and for spot, compensation via some beams having higher gain than others.

Transmit beams have power constraints. Toward their coverage areas, they must provide enough power but not too much or they violate International Telecommunication Union (ITU) rules. Outside their coverage areas, in some cases, beams must have very low power levels so as not to interfere into other countries' telecommunications.

3.3.2 Aperture

To the family of **aperture antennas** belong reflector antennas, horn antennas, and planar arrays (Chang, 1989), namely all the kinds of payload antenna. An aperture antenna is one for which a plane can be defined on which the tangential electric or magnetic field strength distribution is known or can be well estimated; the field is significant over only a finite area in this plane, and this finite area is the **aperture** (Chang, 1989). For a reflector antenna with a main reflector which is the surface of revolution of a planar curve, the **aperture** is the flat circular area that would close off the reflector surface; the aperture is perpendicular to the curve's axis of revolution. When the main reflector's surface is just part of such a symmetric surface, the aperture is the projection of the reflector surface onto the same flat circular area. For a horn, it is the radiating opening. For a planar array, it is the part of the plane containing the array.

The aperture's tangential electric field determines the electric field everywhere in the half-space in front of the aperture, and the aperture's tangential magnetic field determines the magnetic field everywhere there (Chang, 1989). The tangential electric field is an RF vector field so has the properties of direction, phase, and amplitude. If we call the aperture plane the $z = 0$ plane, we can say that all points with $z > 0$ are in the half-space in front of the aperture. An implication is that there is no radiated electric field in the half-space behind the aperture. An actual aperture would in fact have a little bit of backward radiation, where some radiation curls back around the aperture edges, but it is negligible by design.

The amplitude of the electric field across the aperture is the **aperture illumination** [MIT-13]. The theoretical aperture with constant amplitude and phase of the tangential electric field has *uniform illumination*. A reflector's aperture illumination is normally **tapered**, meaning that it rolls off from its peak value in the aperture middle to the aperture edges. Taper is normally given as a positive dB number. This topic continues in Section 3.11.2.

3.3.3 Antenna Pattern

The antenna pattern of an aperture antenna can be derived from the two-dimensional Fourier transform of the tangential electric field in the aperture (Collin, 1985). At small off-boresight angles, the pattern very nearly equals the transform. The larger the aperture, the narrower the beam, so the higher the gain at peak of beam.

An antenna pattern has two parts, the gain pattern and the polarization pattern. The antenna pattern is defined at every direction from the antenna, at sufficient distance from the antenna (Section 3.3.3.2).

The **field of view (FOV)** of an antenna can be defined in several ways (Gagliardi, 1978), but we define it as those directions over which the antenna pattern must satisfy minimum requirements. **Edge of coverage (EOC)** is the outline of the FOV on the earth.

Passive antennas have the **reciprocity property**, whereby they have the same pattern on receive and transmit at the same frequency (Collin, 1985). A **passive antenna** is one that does not have amplification as an integral part of the antenna.

3.3.3.1 *Gain, EIRP, and G/T$_s$*   The purpose of a payload transmitting antenna is to shape the radiation of the RF power so it goes preferentially in some directions and much less in the other directions. Similarly, a payload receiving antenna shapes the receiving sensitivity of the payload to maximize it in the directions desired and suppress it in the other directions. This shaping is characterized by the antenna **directivity function** $D(\theta,\phi)$ defined on all **polar angles** θ and **azimuths** ϕ. The directivity function is proportional to the gain pattern, as we will see below in this subsection. The polar angle is more commonly called the **off-boresight angle**. The angles of a direction in the spherical coordinate system are defined in Figure 3.2, where the antenna aperture lies in the $z=0$ plane and z increases in the forward direction of the antenna pattern.

It is easiest to understand what directivity is if we first think about the ideal (theoretical, non-existent) **isotropic antenna**. It is a point source and has unity

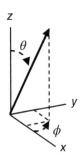

FIGURE 3.2   Spherical coordinate system.

directivity in every direction. It does not shape the radiation but sends equal amounts in all directions. Its integral over all directions is given by the following:

$$\int_0^\pi d\theta \int_0^{2\pi} D(\theta,\phi)\sin\theta d\phi = \int_0^\pi d\theta \int_0^{2\pi} \sin\theta d\phi = 4\pi \text{ for isotropic antenna}$$

For the directivity function of any other antenna, the value of the integral is the same:

$$\int_0^\pi d\theta \int_0^{2\pi} D(\theta,\phi)\sin\theta d\phi = 4\pi \quad \text{for any antenna}$$

which means that if there are directions in which an antenna's directivity is large, there must be compensating directions in which the directivity is small, less than unity. For aperture antennas, neglecting any backward radiation, D is zero for all $z<0$ or $\theta > \pi/2$.

An expression for peak directivity of a whole or partial parabolic antenna with projected-aperture radius a is as follows (Collin, 1985):

$$D_{\text{peak}} = 4\left(\frac{\pi a}{\lambda_0}\right)^2 \eta_A \eta_s$$

where λ_0 is wavelength in the same units as a, η_A is aperture efficiency, and η_s is spillover efficiency (Section 3.11.1).

Gain, however, is more useful than directivity because it characterizes the antenna as a whole, that is, as a unit that connects at its terminal to another unit. The **antenna terminal** is the interface, on the antenna side, to the rest of the payload. Many other concepts, specifications, and measurements relate to the antenna terminal, too. On transmit, directivity is referenced to the total power radiated out of the antenna, but gain is referenced to the power into the antenna terminal. Receive is similar. Gain is proportional to directivity by a factor η that accounts for antenna losses, called the **antenna efficiency**, which will be discussed in more detail in Section 3.11.1 but is simply represented here:

$$P_{\text{radiated}} = \eta P_{\text{terminal}} \quad \text{for transmitting antenna}$$
$$P_{\text{terminal}} = \eta P_{\text{received at aperture}} \quad \text{for receiving antenna}$$

Payload antennas are high-gain with the exception of earth-coverage horns on GEOs. These have a gain of about 17 dB (Schennum et al., 1999), which is between high and medium. On a low-earth-orbit (LEO) spacecraft, where the antenna gains are lower than on a GEO or MEO, the antennas are high-gain, having a gain of about 20 dB.

EIRP, equivalent isotropically radiated power, is the most important parameter of the majority of payloads, since all payloads have at least a downlink. It comes from a combination of the power into the transmitting antenna's terminal and the antenna gain. EIRP is defined on all directions. EIRP in a given direction is the total power that the antenna would radiate if it could have the gain in all directions that it has in the one direction:

$$\text{EIRP}(\theta,\phi) = G(\theta,\phi)P_{\text{terminal}}$$

G/T_s, the **figure of merit** of the payload in its function as a receive terminal, is the most important parameter for satellite uplinks. For a satellite with only a few uplinking stations, G/T_s need only be defined in their directions (allowing for antenna pointing error), but otherwise, it is defined in every direction, like EIRP. G/T_s comes from a combination of the receive antenna gain and the system noise temperature T_s (Section 8.12). The parameter sets the uplink signal-to-noise ratio (SNR).

3.3.3.2 *Far Field and Near Field*   The antenna pattern exists in the **far field** of the antenna. The far field for a high-gain aperture antenna is generally said to start at a distance of $2D^2/\lambda_0$ from the aperture, where D is the diameter of the antenna aperture and λ_0 is the wavelength of the center frequency being transmitted or received (Chang, 1989). More generally, the distance is the maximum of $2D^2/\lambda_0$, $20D$, and $20\lambda_0$ (IEEE, 2012). The wavefronts, which have constant phase by definition, seem to radiate from one spot, the antenna's **phase center**, as illustrated in Figure 3.3. In this example, the phase center lies in the aperture but this is not always the case, for example, for horns (Section 3.7).

In the far field, for a receiver at communication distances, the antenna's radiation is effectively a planar wavefront with the **transverse electromagnetic (TEM)** propagation mode. In TEM, the **electric field vector E** and its orthogonal **magnetic**

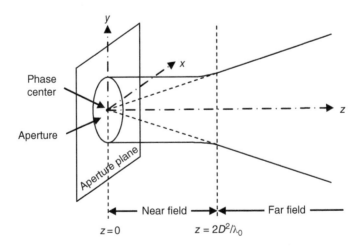

FIGURE 3.3   Near- and far-field regions for high-gain aperture antenna. (after Chang (1989)).

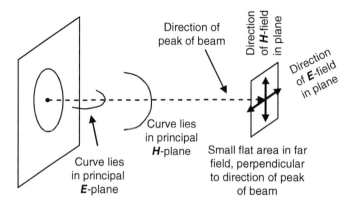

FIGURE 3.4   Principal planes for antenna measurements.

field vector *H* lie in the wavefront plane, that is, are perpendicular to the direction of propagation (Collin, 1985), and the fields' amplitudes are related by a constant (Ramo et al., 1984).

Near the antenna is the **near field** or **Fresnel region**. Here, the beam diameter increases slowly and the cross-sectional amplitude and phase distributions change little with distance z from the aperture (Chang, 1989).

Often when an antenna is tested in its far field, measurements are only taken in directions that lie in two orthogonal planes. The planes are the **principal planes**, specifically the **E-plane** and the **H-plane**. Figure 3.4 illustrates their definition, assuming that, in the far field in a plane perpendicular to the peak-of-beam direction, the *E*- and *H*-field vectors point as shown. Both planes contain the vector pointing to the peak of beam. The pattern is said to be taken in an *E*-plane or *H*-plane *cut*, that is, the set of directions from the phase center that lie in the corresponding plane. Other cuts, in other convenient planes, can be defined.

3.3.3.3   *Gain Pattern*   A gain pattern for a communications antenna has a mainlobe and sidelobes separated by nulls or near-nulls, as the example in Figure 3.5 shows. The mainlobe is much higher than the sidelobes.

There are approximate expressions for the null-to-null width of the mainlobe (Collin, 1985). For a rectangular horn whose aperture in the wide direction has dimension D, the null-to-null beamwidth in a cut in the wide direction is as follows:

$$\text{Mainlobe beamwidth} \approx \frac{\lambda_0}{D/2} \text{ rad}$$

For a circular aperture of diameter D the expression is as follows:

$$\text{Mainlobe beamwidth} \approx \frac{3.8}{\pi} \frac{\lambda_0}{D/2} \text{ rad}$$

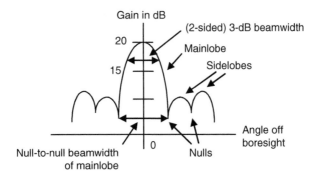

FIGURE 3.5   Definition of some antenna-gain pattern terms.

Thus, the larger the aperture is in units of wavelength, the narrower the main-lobe. The (two-sided) half-power (that is, 3 dB down from peak of beam) beam-width of a circular aperture with uniform illumination is as follows (Evans, 1999):

$$\text{Half-power beamwidth} \approx \frac{\lambda_0}{D} \ \text{rad}$$

(Compare this with the half-power width of the transform of the rectangular pulse in Section 12.8.4.2).

3.3.3.4 *Polarization*   **Polarization** of the far-field electric field is the other property of an antenna pattern, besides gain.

When an antenna radiates, at any given point in space the electric field vector moves repetitively over time with the frequency of the radiation. The general shape that the vector traces is an ellipse, and the ellipse possibilities range from a circle to a line. Looking in the direction of propagation, if the direction of rotation is clockwise the radiation is said to be *right-hand polarized* (*RHP*), and if counter-clockwise it is said to be *left-hand polarized* (*LHP*) (Collin, 1985). If the ellipse is in fact a circle, the radiation is either **right-hand circularly polarized (RHCP)** or **left-hand circularly polarized (LHCP)**. If the ellipse is merely a line that gets traced back and forth, the radiation is **LP**. The most common linear polarizations are **horizontal (H)** and **vertical (V)**, where horizontal has the *E*-field parallel to the ground (on the earth) and vertical is orthogonal to horizontal and to the direction of propagation. Sometimes, H and V just denote any two orthogonal linear polariza-tions. The **axial ratio** r of an ellipse is defined as the ratio of the minor axis of the ellipse to its major axis, which are illustrated in Figure 3.6. Thus, $r = 1$ is a circle and $r = 0$ is a line. Axial ratio in dB is $-20\log_{10} r$.

At frequencies lower than C-band, circular polarization is most often used for satellite communications, but at C-band and above both circular and linear are used (Hoffmeister, 2010). The reason is Faraday rotation (Section 14.4.3.2).

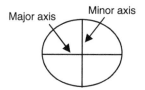

FIGURE 3.6 Ellipse axes.

For the receiving antenna to be able to receive all the radiation incident on it from a transmitting antenna, the radiating and the receiving antenna must have exactly the same polarization (Chang, 1989). If the two polarizations are orthogonal, no radiation will be received. Examples of pairs of orthogonal polarizations are (1) RHCP and LHCP and (2) vertical and horizontal. If a one-handed CP antenna receives one LP, it receives only half of the power, and vice versa.

3.4 REFLECTOR-ANTENNA BASICS

3.4.1 Paraboloidal Reflector Concepts

A reflector antenna has at least a main reflector and a feed. By far the most common form of main reflector is a paraboloid or section of a paraboloid (Chang, 1989). The general idea of a reflector antenna is that the feed either is or appears to be at the **focus** of the paraboloid, so that upon reflection the rays emerge parallel and in phase in planes perpendicular to the paraboloid's axis, as shown in Figure 3.7.

If the antenna has only a main reflector and a feed, it is a **single-reflector antenna**, for which the feed is at the focal point. If additionally it has a subreflector, it is a **dual-reflector antenna**, and the feed is no longer actually at the focus but

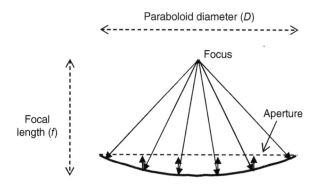

All rays are in phase at aperture.

FIGURE 3.7 Geometry of paraboloid.

virtually there. In either case, the phase center of the feed must be known so the antenna geometry can be set up correctly. A reflector antenna is inherently wideband if the phase center of the feed or virtual feed is exactly at the focus for all intended frequencies (Chang, 1989).

When an entire paraboloid forms the reflector, the reflector is said to be **center-fed** because the feed or virtual feed points down the paraboloid's central axis. If the paraboloid is partial, the antenna is **offset-fed** because the feed or virtual feed is not set to point down the paraboloid's axis, but it is still at the focus.

One of the defining characteristics of a reflector antenna is f/D, the paraboloid's **focal length f in units of the aperture diameter D**. When the paraboloid is only partial, its f and D are still defined as if the paraboloid were entire. (However, gain is a function of the actual aperture size.) There does not seem to be a standard definition of what is a short f/D and what is a long f/D, but the dividing line is about 0.7 (Legay et al., 2000).

The **primary radiation pattern** is that of the feed. The main reflector's pattern is the **secondary radiation pattern**. The tangential electric field on the feed's aperture and the antenna geometry determine the tangential electric field on the reflector's aperture and thus its antenna pattern. Details are given in Milligan (2005).

The feed and any subreflector are in the near field of the main reflector. From the point of view of the feed, the reflector is typically in its far field (Rudge et al., 1982), but in a dual-reflector antenna, the subreflector may not be in the feed's far field (Albertsen and Pontoppidan, 1984; Rahmat-Samii and Imbriale, 1998).

The sense of circular polarization is reversed every time the signal is reflected, so a RHCP single-reflector antenna requires a LHCP feed.

The only type of feed in use today for single-beam reflector antennas is the single horn, which is discussed in Section 3.7. An exception is a phased array forming the feed pattern for an antenna that creates a contoured beam (Section 11.3.1.10).

3.4.2 Single-Reflector Single-Beam Antenna

The geometry of the center-fed paraboloidal reflector is shown in Figure 3.8a. The feed, which is at the focal point of the paraboloid, must radiate energy in all directions within the angle θ_{sub} which the reflector subtends from the focus.

The center-fed reflector does not have especially low levels of **cross-polarization (cross-pol)** for LP, but it theoretically has no cross-pol with a CP feed (Chang, 1989). Cross-pol is where some amount of the undesired, orthogonal polarization is present. The center-fed reflector has the deficiency that part of the reflector is blocked by the feed, the feed's supporting struts, and the RF line to the feed (Section 3.11.3).

A solution with no blockage is the offset-fed paraboloidal reflector, with the geometry shown in Figure 3.8b. The feed is still at the focus but is now pointed in an offset direction down the middle of the reflector's angular span (Chu and Turrin, 1973). The aperture is circular although the reflector rim is elliptical (Milligan, 2005). A larger f/D is feasible than with center-fed, so the feed is farther away from the reflector, which means the feed has to have a larger aperture (to

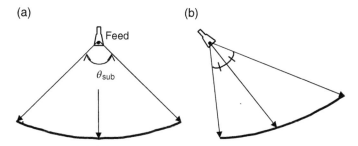

FIGURE 3.8 Reflector and feed of paraboloidal antennas: (a) center-fed, (b) offset-fed.

create a narrower beam), which can be made to have a better-shaped radiation pattern and lower cross-pol. Another advantage of the offset configuration is that the reflector does not reflect radiation back into the feed or feeds (Rudge et al., 1982). A disadvantage is that the cross-pol is increased for LP (Chang, 1989), although it decreases as f/D increases (Milligan, 2005). The feed can compensate this. The offset-fed reflector theoretically has no cross-pol with a CP feed, but the beam from the main reflector **squints**, that is, its boresight direction is off, in a direction perpendicular to the feed's pointing offset (Chang, 1989). Opposite senses of CP squint in opposite directions. Squinting can be decreased by increasing f/D or by the use of a subreflector (Milligan, 2005).

For both center-fed and offset-fed antennas, the reflector can be **shaped** to improve the beam's contour. The outline of the reflector can be changed and/or the surface can be shaped. A tradeoff study was conducted on how to obtain a 12.5-GHz CONUS beam from GEO with 1.5 dB variation of gain across CONUS to partially compensate rain attenuation, with an offset paraboloidal single-reflector antenna. It was found that surface-shaping and a single feed could accomplish this. The surface deviated from a paraboloid by 1.4 inches (about 1.5 wavelengths) at most. Without surface-shaping, a 56-element phased-array feed would have been necessary (Ramanujam et al., 1993).

3.4.3 Dual-Reflector Single-Beam Antenna

The dual-reflector antenna has a subreflector in addition to the main reflector, and the feed is not at the paraboloid's focus. The dual-reflector antenna allows more control than the single-reflector does, over the antenna's aperture field and thus the coverage contour. If shaping the main reflector alone cannot produce the desired coverage contour, the subreflector can also be shaped. The process for mutually shaping the main reflector's and subreflector's surfaces is, roughly speaking, to first shape the subreflector to control the aperture's power distribution and then to shape the main reflector to correct the aperture field's phase distribution (Collin, 1985).

3.4.3.1 *Center-Fed*   One type of center-fed dual-reflector antenna is the **Cassegrain** design, which has long been used for optical telescopes—the design was first published in 1672. The hyperboloidal subreflector has two foci, one inside the hyperboloid's curve and one outside, as shown in Figure 3.9a. A basic property of the hyperboloid is that all reflected rays from one focus seem to come from the other focus. The geometrical arrangement of the Cassegrain is that the hyperboloid's inside focus is set at the main reflector's focus and the outside focus is set at the feed's phase center, as shown in Figure 3.9b. The hyperboloid can be designed so that the feed is slightly behind the main reflector and therefore causes no blockage. The Cassegrain has other configurations besides the classical one shown in the figure (Chang, 1989).

The other type of dual-reflector antenna used is the **Gregorian** design, which has also long been used in optical telescopes, being first published in 1663. The elliptical subreflector has two foci, both inside the ellipse, as shown in Figure 3.10a.

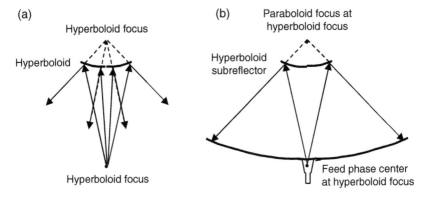

FIGURE 3.9   Hyperboloid: (a) property and (b) put to use in center-fed Cassegrain antenna.

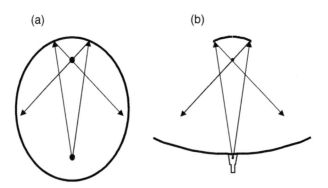

FIGURE 3.10   Ellipsoid: (a) property and (b) put to use in center-fed Gregorian antenna with subreflector of same diameter as Cassegrain in Figure 3.9b.

A basic property of the ellipse is that all reflected rays from one focus seem to come from the other focus. The geometrical arrangement of the Gregorian is that one focus of the ellipse is set at the main reflector's focus and the other focus is set at the feed's phase center, as shown in Figure 3.10b. To make visible the relative sizes of the Cassegrain and Gregorian with the same blockage by the subreflector, the subreflector in this figure was made the same size as the one in Figure 3.9b. Compared to the Cassegrain, the Gregorian's subreflector is farther away from the main reflector so requires longer supporting struts and the feed has a narrower beam so must be larger. To first order, the performance of the Gregorian is the same as that of a Cassegrain with the same f/D and D. The Gregorian has other configurations besides the classical one shown in the figure (Chang, 1989).

Every center-fed Cassegrain antenna is equivalent to a single-reflector paraboloidal antenna with the same feed but a greater f/D (Chang, 1989; Collin, 1985). The same is true of a center-fed Gregorian (Rusch et al., 1990). The properties of the equivalent paraboloid apply to the dual-reflector antenna (Chang, 1989). The dual-reflector antenna is more compact than the equivalent paraboloid, which is sometimes the main advantage.

3.4.3.2 *Offset-Fed*   Like the single-reflector antenna, the dual-reflector antenna can be made in an offset-fed configuration, as illustrated in Figure 3.11 for both Cassegrain and Gregorian.

Offset-fed reflectors are typically used because they have better performance than center-fed and package better on the spacecraft (Schennum, 2015).

The paragraph in the subsection above about offset-fed single-reflector cross-pol also applies to offset-fed dual-reflectors (Milligan, 2005). An antenna with f/D on the order of 1 and a horn with at least 10-dB aperture taper (Section 3.3.2) has little cross-pol. For a Gregorian, the cross-pol can be essentially eliminated by tilting the subreflector a little as shown in Figure 3.12 (Akagawa and DiFonzo, 1979).

Every offset, classical Cassegrain or classical Gregorian antenna, even with a tilted subreflector, is equivalent to a single-reflector paraboloidal antenna with the

(a)   (b)

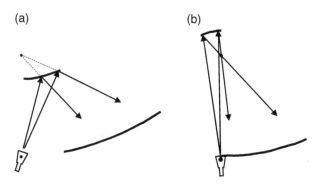

FIGURE 3.11   Offset-fed (a) Cassegrain and (b) Gregorian.

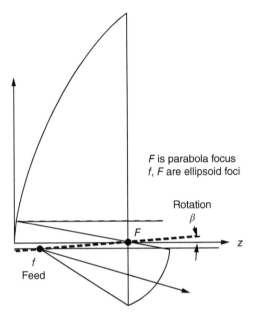

FIGURE 3.12   Gregorian subreflector tilted to correct cross-pol. (©1979 IEEE. Reprinted, with permission, from Akagawa and DiFonzo (1979)).

same feed but a greater *f/D*. Assumptions are that the subreflector is at least 10–15 wavelengths in diameter and has an illumination edge taper of at least 10 dB (Rusch et al., 1990).

Single-beam antennas on GEO satellites are typically modified Gregorian. The choice between modified Cassegrain and modified Gregorian is mainly what type fits best on the satellite (Schennum, 2014).

The subreflector of an offset-fed Gregorian is generally in the near-field region of the feed horn (Rao et al., 2013).

An example of a Gregorian antenna, most likely offset-fed, is the two Ka/K-band antennas on the earth deck of DirecTV 10, 11, and 12, launched between 2007 and 2009. Gregorian had the advantage over Cassegrain of not requiring the feed to be on a tall mast, which would have incurred line losses to the feed (Apfel, 2006).

A reconfigurable Gregorian antenna has been qualified. It has three subreflectors that can be switched among. When a different subreflector is switched in, the main reflector is repointed (Amyotte and Godin, 2017).

An example of one main reflector that serves as both a single- and dual-reflector, offset-fed antenna was on the N-Star a and b satellites, launched in 1995 and 1996. A drawing of it is given in Figure 3.13. Evident are the subreflector, the Ku-band horn, and the S-band helices. The subreflector has a *frequency-selective surface*, so that it can reflect at Ku-band and transmit at S-band (Barkeshli et al., 1995).

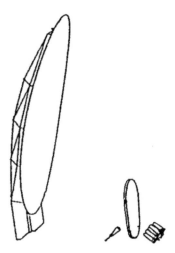

FIGURE 3.13 *N*-Star a and b Ku/S frequency-selective antenna. (©1995 IEEE. Reprinted, with permission, from Barkeshli et al. (1995)).

3.5 STEERABLE SINGLE-BEAM ANTENNAS

The three ways to steer a single-reflector antenna are to gimbal the whole antenna, just the reflector, or just the feed. The last two are mechanically simpler than the first but they incur scan loss due to the feed no longer being at the reflector focal point (Choung, 1996). One leading antenna manufacturer says that reflector-only gimbaling is an improvement over full-antenna gimbaling, especially for CP antennas, due to its simplicity (Amyotte and Godin, 2017). An example of an off-set-fed single-reflector antenna where the reflector is gimbaled and the feed is fixed is the Ka/K-band user and gateway antennas on the Eutelsat 3B earth deck (Section 3.2) (Glâtre et al., 2015). Another example, on a MEO, is the 12 Ka/K-band antennas on the O3b spacecraft (Amyotte et al., 2010). The gateway antenna of the LEO Iridium Next turns the scan loss to advantage by optimizing the feed location for the edge of the earth, where the space loss is greatest, and having the feed off-focus for nadir, where the space loss is the least (Amyotte et al., 2011; Amyotte and Godin, 2017).

The same three ways to steer also apply to dual-reflector antennas. An example of an entire offset Gregorian antenna that is gimbaled is Eutelsat 3B's Ku-band earth-deck antenna (Glâtre et al., 2015).

An example of a steerable antenna with fixed feed is OneWeb's (Section 19.5) Ka/K-band offset-fed gateway antenna. An image of a concept antenna is shown in Figure 3.14. The main reflector is parabolic. The subreflector is flat. The antenna is effectively a single-reflector offset-fed antenna. The subreflector's rotator actuator moves with the main reflector's rotator actuator in such a way as to keep the image of the feed at the main reflector's focal point. No rotary joint is needed. Both

FIGURE 3.14 Concept for fixed-feed wide-scan antenna. (Amyotte and Godin (2017) Used with permission of MDA).

reflectors are shaped. The antenna can track to more than ±60° with minimal scan loss (Amyotte and Godin, 2017; Amyotte, 2020).

3.6 REFLECTOR TECHNOLOGY FOR SINGLE-BEAM ANTENNAS

Most reflectors for C-, Ku-, and Ka/K-bands are solid, made of graphite, also known as carbon fiber. Solid reflectors cannot be collapsed for stowing for launch but are usually gimbaled toward the earth deck, if mounted there, or gimbaled up against the east or west panel if mounted there. An example of solid reflectors is those on the east and west panels of DirecTV 10, 11, and 12 (Harris Corp, 2009).

MDA introduced a new solid reflector in 2017 that is made of only three composites parts instead of more than 100 earlier. It has lower cost and mass (Amyotte and Godin, 2017).

MDA used an alternative to graphite, an aluminum alloy, on the Eutelsat 3B Ka/K-band gateway antenna. Aluminum has several advantages in this case over the usual composite graphite reflector. Reflection losses are lower. More uniform thermo-elastic distortions are possible for this large reflector. The backing structure is integrally machined with the reflector out of a block of aluminum, allowing reduced rib height. Additionally, it was less costly to produce (Glâtre et al., 2015). However, it does weigh more (Schneider et al., 2009).

MDA developed this technology further for the antennas of the large non-GEO constellations, starting with O3b user antennas and continuing on through OneWeb gateway antennas. These antennas need to be mass-produced. Process streamlining was accomplished through design for manufacturing, assembly, integration, and test (DFMAIT). The number of parts was decreased, which reduced both assembly time and assembly errors. Aluminum instead of composite offers, for reflectors of

FIGURE 3.15   Gridded offset-fed reflector geometry.

this size, advantages in procurement, mechanical performance, and RF performance (Glâtre et al., 2019).

When very low cross-pol is required for dual LP, a gridded reflector may be used. The antenna is an offset-fed single-reflector, where the reflector is separated into two surfaces. One surface is nested inside the other and turned at a slight angle so that the two feeds can be in slightly different places. The geometry is illustrated in Figure 3.15. The front surface is *gridded*, that is, it is an array of conductors parallel to one linear electric field. Radiation with that polarization is reflected but the orthogonal LP sense passes through and is reflected on the back surface. This was thought of in the 1970s (Rudge et al., 1982). The main reflector is *dual-gridded* if the back reflector is also gridded. Such a Ku-band antenna flies on Express AM4 (Camelo, 2010). The German company HPS GmbH makes such reflectors for Ku and Ka/K-bands, shaped or unshaped (HPS GmbH, 2018).

3.7   HORN FOR SINGLE-BEAM ANTENNAS

3.7.1   Horn Types

A horn can serve as an antenna in its own right when a relatively wide beam is desired or, much more commonly, as a feed for a reflector antenna or as a radiating element in an antenna array. A horn is a piece of waveguide which flares at the open end to increase the gain and improve the impedance match with free space at the aperture. The horn aperture can be rectangular or circular. Important parameters of a horn antenna are its aperture efficiency and its polarization purity.

The **rectangular-aperture or pyramidal horn** is a piece of rectangular waveguide flared in both dimensions, usually by different factors, as shown in Figure 3.16.

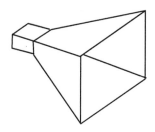

FIGURE 3.16   Rectangular horn.

The flare angle must be rather small so that the phase across the aperture is almost constant. This horn can produce dual LP (Chang, 1989), but its cross-pol performance is not as good as that of the modified conical horn (see below).

The **circular-aperture or conical horn** is more common than the pyramidal horn. It is basically a piece of circular waveguide, flared. Creating identical aperture distributions in the E- and H-planes can minimize the cross-pol (Bhattacharyya and Goyette, 2004). The simplest conical horn is smooth-walled with a linear taper, but its cross-pol is not good. There are other problems with such a horn. The sidelobes are higher in the E-plane than in the H-plane, where they are very low. When such a horn feeds a reflector, the phase centers of orthogonal polarizations are different (Milligan, 2005).

Two types of horn have been invented that solve these problems, the Potter horn and the corrugated horn. The **Potter horn** has smooth walls and a circular step between the waveguide and the flare, as shown in Figure 3.17, to excite the TM_{11} mode from the TE_{11} mode (Potter, 1963). **TM** is **transverse magnetic** and means that the magnetic field is perpendicular to the direction of propagation and the electric field is not (see Section 4.4.5 for a discussion on waveguide propagation modes). The two modes have the appropriate relative amplitude and phase to make the E-plane behave like the H-plane. Phase center coincidence follows as a result (Potter, 1963). The Potter horn is **dual-mode** because it has the two modes TE_{11} and TM_{11} at the aperture. The Potter horn has typically 70% aperture efficiency and not more since the aperture illumination is tapered to cause the magnetic field to go to zero on the horn wall, as the electric field does naturally (Bhattacharyya and Goyette, 2004). An example of a Potter horn is the medium-gain horn in the middle of the larger of Globalstar-2's phased array domes for transmit (Figure 2.7) (Croq et al., 2009).

The **corrugated circular horn** has circular corrugations on the inner wall, as depicted in Figure 3.18. The slots between corrugations are between 1/4 and 1/2 the wavelength (Lawrie and Peters, 1966). The horn supports **hybrid modes**, which are combinations of TE and TM modes. The dominant hybrid mode HE_{11} is the combination of TE_{11} and TM_{11}. (The **dominant mode** is the one with the lowest operating frequency.) The horn can be designed to have more modes than just the dominant one, that is, be **multimode**, in order to have a more uniform aperture and

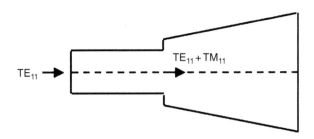

FIGURE 3.17   Potter horn. (after Collin (1985)).

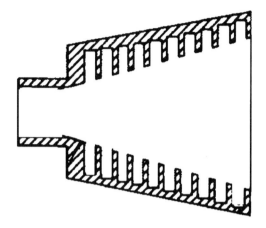

FIGURE 3.18   Corrugated horn. (after Milligan (2005)).

higher beam efficiency, but it will then also have a narrower bandwidth. The horn can provide CP or LP, single or dual. It has a circularly symmetric pattern, very low cross-polarization, high beam efficiency, and low sidelobes (Chang, 1989).

A corrugated horn with only the HE_{11} hybrid mode provides a Gaussian beam in its mainlobe. The field at the horn edge must be essentially zero. Such a horn is often used as a feed for a reflector (Rudge et al., 1982). The Gaussian beam model is assumed for other horns, for ease of calculation (for example, in Rao (2003)). See also Sections 11.3.1 and 11.4.5.

A variation on the conical horn is the smooth-walled horn with spline profile. Points are defined for specific horn radii along the z-axis of the horn, known as knots, and between the knots the shape is a spline curve, usually a cubic polynomial. Smooth-walled horns have low mass and are easy to manufacture (Simon et al., 2011).

A horn with dielectric material in it is not suitable for space antennas because the material induces electro-static discharge (Rao, 2015).

3.7.2 Horn as Antenna

Horn antennas, also called *direct-radiating horns*, can provide two kinds of beam, a global beam and, less usually, a spot beam.

A horn antenna on a GEO satellite can provide a global beam, subtending at least ±8.7°. An example of a corrugated-horn antenna no longer operational, but instructional, is the C-band global horns on the Intelsat-IX series of satellites launched from 2001 to 2003. The antennas are **dual-mode** (that is, dual-hybrid-mode, different from the dual mode of the Potter horn) corrugated horns, with the second hybrid mode HE_{12} being excited at the second step, as sketched in Figure 3.19. The presence of the second mode flattens the aperture phase distribution, leading to a relative EOC gain increase of 0.4 dB and steeper gain roll-off beyond. One antenna receives both CPs at once and the other transmits both at

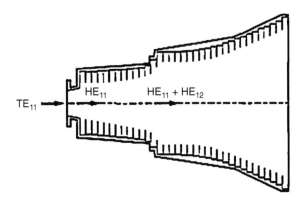

FIGURE 3.19 Dual-mode corrugated horn antenna. (©1999 IEEE. Reprinted, with permission, from Schennum et al. (1999)).

once. The port-to-port isolation of the antennas, inclusive of orthomode transducer (OMT) and polarizer, is 37 dB and their cross-polarization isolation is 41 dB (Schennum et al., 1999).

On-orbit global-horn antennas include those on the Russian satellites Express AM5 and AM6, launched in 2013 and 2014. The antennas are CP at C-, Ku-, and Ka/K-bands. Each horn produces one frequency band and one direction of transmission, that is, either receive or transmit (Grenier et al., 2012).

A second use of a horn as an antenna is to provide a high-directivity spot beam. A smooth-walled, spline-profile horn has been developed for this. The radius must monotonically increase along the z-axis, so where necessary the cubic spline is replaced by a constant-radius segment. The monotonicity ensures that if such horns are closely packed, adjacent horns will not interfere with each other. The profile of such a dual-band horn is shown in Figure 3.20, where the points are the knots at which the radius is defined and spline curves are fit in between. The horn can

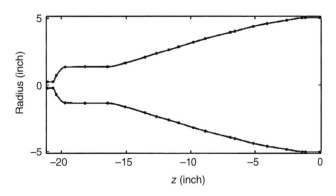

FIGURE 3.20 Spline-profile, smooth-walled, dual-band, dual-polarization horn. (Simon et al. (2015)).

support dual LP or dual CP. A similar single-band horn has a narrower throat. The diameter of the horn aperture may have to be as large as 24 wavelengths to achieve the necessary directivity (Simon et al., 2011, 2015).

3.7.3   Horn as Feed for Reflector

The electric field distribution on the horn aperture and the antenna geometry determine the reflector illumination pattern. The desired illumination pattern can be created by the proper multimode excitation in the horn. Typically the illumination is down at least 10 dB on the reflector edge compared to the middle. Some examples of illumination, not in dB, are $\cos^n$ for $n = 1$ through 4 and $1 - \varepsilon\theta^2$ for the offset angle θ. For the cosine ones, the half-power beamwidth increases and the first-sidelobe level decreases with increasing n (Pritchard and Sciulli, 1986).

For a reflector antenna creating one beam, the feed seems always to be a horn. The feed is a critical part of the antenna. The cross-pol of any reflector antenna can be no better than that of the feed (Collin, 1985). The feed can compensate some problems inherent in the reflector(s); for example, a corrugated-horn feed when excited with multiple modes can compensate the LP depolarization inherent in offset antennas, and provide dual LP (Adatia et al., 1981).

The phase center of the feed must be known so that it can be placed at the focus of the main reflector. A horn feed can be designed to have its phase center anywhere between the unflared waveguide and the aperture. However, in general, the phase center shifts with frequency (Chang, 1989).

The corrugated circular horn seems to be the favorite for a feed. It can be designed for two frequency bands simultaneously, that is, for both receive and transmit frequencies (Tao et al., 1996) and even for three bands (Uher et al., 2010). One current example is the high-power, tri-band, dual-LP horn developed for spacecraft in the fixed satellite service and the broadcast satellite service (Amyotte et al., 2013). Another is the dual-band, dual-LP Ku-band horn on the Eutelsat 3B earth deck (Glâtre et al., 2015).

3.8   OTHER ANTENNA COMPONENTS

The antenna is composed of more than just radiating elements, if one or more of the following are desired:

- Dual polarization (requires OMT)
- Circular polarization (requires polarizer)
- Both receive and transmit (requires diplexer).

The radiating element(s) together with the appropriate other components are called the **feed chain** or **feed assembly**. Sometimes, the feed chain contains an autotrack function, too (Section 3.10). Sometimes the feed chain is simply known as the feed. A feed assembly is more compact and has better RF performance than separate elements integrated (Rao, 2015).

3.8.1 Orthomode Transducer and Polarizer

This subsection describes first what the OMT and the polarizer do, then gives common implementations.

An **OMT** is a device that allows for receiving or transmitting dual LP instead of just single LP (Figure 3.21a. On receive, the OMT separates out and converts two orthogonal TE_{11} modes in one circular waveguide to two orthogonal TE_{10} modes in two rectangular waveguides (Schennum et al., 1999). **TE** is **transverse electric** and means that the electric field is perpendicular to the direction of propagation and that the magnetic field is not (for discussion on propagation modes, see Section 4.4.5). On transmit the OMT does the opposite of what it does on receive. In the figure, H and V are not necessarily meant to be horizontal and vertical, just an orthogonal pair of LPs. The OMT's important parameter is its port-to-port isolation.

If used together with a polarizer, an OMT allows for dual CP, as shown in Figure 3.21b.

A **polarizer** is a device for receiving a CP signal and converting it into a LP signal. On transmit it does the opposite. It can also act on dual polarizations. The important parameter of a polarizer is its cross-polarization isolation.

On receive the polarizer changes a single-sense CP electric field into a single sense of LP by separating out orthogonal linear components and delaying one, as illustrated in Figures 3.22 and 3.23, respectively. The component drawn horizontally is delayed by a quarter cycle between the two figures.

Let us now turn to implementations. One implementation of a combined polarizer and OMT is the common *septum polarizer*. It can handle dual CP. It is narrowband,

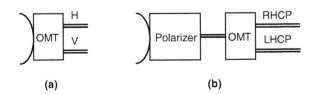

(a)   (b)

FIGURE 3.21 Orthomode transducer: (a) providing dual LP and (b) with polarizer providing dual CP.

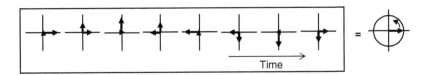

FIGURE 3.22 Resolution of one CP sense into two senses of LP.

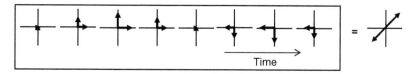

FIGURE 3.23   One LP sense delayed; combination of two senses resulting in one LP sense at 45°.

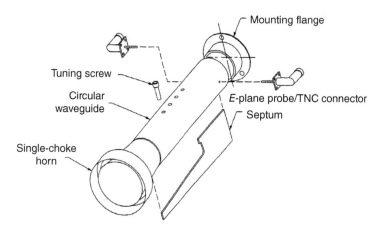

FIGURE 3.24   An implementation of a septum polarizer. (©1997 IEEE. Reprinted, with permission, from Schennum and Skiver (1997)).

that is, it can only treat one frequency band, so for a dual-band feed, there must be two (Izquierdo Martinez, 2008). A drawing of a more or less typical one is given in Figure 3.24. In others, the shape of the septum is different. The sloping septum gradually divides the circular waveguide into two separate half-circular waveguides. One half-circular waveguide contains a signal that was the RHCP signal, and the other what was the LHCP signal (Schennum and Skiver, 1997).

A type of polarizer alone is the *coaxial polarizer*, which is two coaxial pieces of circular waveguide, one smaller than the other. It can handle dual CP. It has been developed by a satellite manufacturer for L-band but can be extended to S-band (Amyotte et al., 2014). There is some kind of dielectric member over part of the length that touches both waveguides as illustrated, for example, in Figure 3.25.

A second type of polarizer is the *corrugated polarizer*, which is broadband. One with square cross-section has corrugations along one of the waveguide's transverse axes. A prototype was made for the extended Ku-band, with about 60% bandwidth. It can handle dual CP (Tribak et al., 2009).

The most common kind of OMT is based on the *turnstile junction*, which has five physical ports and is illustrated in Figure 3.26. The common port on the top would connect to the radiating element. At the common port, the two perpendicular

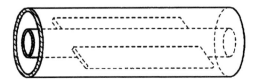

FIGURE 3.25 Example of circular polarizer. (Enokuma (2002)).

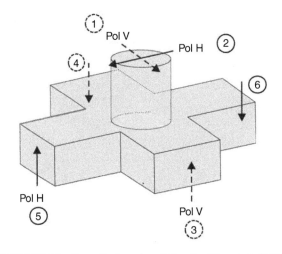

FIGURE 3.26 Turnstile junction. (after San Blas et al. (2011)).

LPs horizontal and vertical are present. The junction separates out the vertically polarized signal from the input and outputs it to ports 3 and 4. These ports each contain half of the vertical signal, with opposite phases. Similarly, the junction also separates out the horizontally polarized signal and outputs two halves to ports 5 and 6. It is evident that from one signal out of either port 3 or 4 and from one signal out of either port 5 or 6, the signal into the common port could theoretically be reconstructed. The OMT based on the turnstile junction has, besides the junction, a branch that makes the signals from ports 3 and 4 be in phase with each other and combines them, as well as another branch which does the same for the signals from ports 5 and 6 (Izquierdo Martinez, 2008). Some references equate the turnstile junction with the *orthomode junction*, but the latter has six ports.

3.8.2 Diplexer

An antenna can both receive and transmit at the same time by means of a **diplexer**. The receive and transmit signals must be in different frequency bands.

A block diagram of a diplexer is shown in Figure 3.27. In the receive signal path, a piece of waveguide sized so that the transmit frequency is below cutoff, blocks the

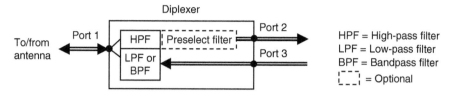

FIGURE 3.27   Diplexer for single LP.

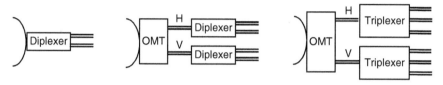

FIGURE 3.28   Various diplexer and triplexer configurations.

transmit signal, and passes the receive signal (Schennum et al., 1995). The **cutoff frequency** of a transmission line is the lowest frequency for which a mode will propagate in it. This piece of waveguide is effectively a high-pass filter (HPF). Because the transmit signal is very much stronger than the receive signal, the isolation from port 3 into port 2 must be very high. The payload's preselect filter, which is a bandpass filter (BPF), may follow the HPF in the diplexer (Kwok and Fiedziuszko, 1996) or it may be separate from the diplexer (Schennum et al., 1995). In the transmit signal path is a low-pass filter or a BPF. These filters are described in Chapter 5.

Sections 5.5.1 and 5.5.3 provide more information on the diplexer.

Various diplexer and triplexer configurations are shown in Figure 3.28. A *triplexer* outputs either two transmit frequency bands and one receive frequency band or vice versa. The radiating element must be very wideband for the triplexer case. An example is given in Uher et al. (2010).

3.9   ANTENNA POINTING ERROR

When the antenna does not autotrack (Section 3.10), the antenna pointing error comes primarily from orbital motion and spacecraft attitude error (Sections 2.2.4 and 2.2.5). Nonzero orbital inclination causes north-south displacement of a GEO spacecraft over the course of the day. Nonzero orbital eccentricity causes east-west displacement over the course of the day. Less important causes of the error are initial antenna-boresight misalignment and deformation from mechanical and thermal constraints (Broquet et al., 1985). Spacecraft attitude error is largest during stationkeeping maneuvers when the propellant is chemical.

When the antenna does autotrack, antenna pointing error does not depend on the spacecraft attitude or orbital motion, to first order.

Pointing error of a reflector antenna has two physically orthogonal dimensions, commonly called x and y. It is difficult to characterize pointing error probabilistically because, for one thing, the two errors in the two dimensions are statistically dependent. In general, they have different means and variances and there is no general expression for their probability distributions. Their means can be assumed to be zero if the antenna is autotracked. Section 15.6.2 has further discussion on pointing error.

However, the antenna pointing error must be dealt with in payload budgets. The spacecraft customer and the manufacturer must agree on exactly what measure of antenna pointing error is to be used in payload budgets, for example, 2-sigma.

There are at least two types of antenna pointing error that may be reported to the payload engineer by the manufacturing program's pointing expert. One is the "worst-case" pointing error in a dimension, ordinarily calculated as the sum of the worst-case deterministic error plus the 3σ error of the random components (Maral and Bousquet, 2002). Another is the "worst-case" pointing error combined for the two dimensions, perhaps calculated as rms of "worst-case" errors in the two dimensions.

Antenna pointing error affects the gain toward any direction of interest. Figure 3.29 shows conceptually what happens to antenna gain when the intended direction is beam center but there are azimuth or elevation pointing errors. For the same pointing error value, the degradation in gain is the same for circular beams no matter what the direction of pointing error.

Figure 3.30 shows conceptually what happens to antenna gain when the intended direction is off beam center but there are azimuth or elevation pointing errors.

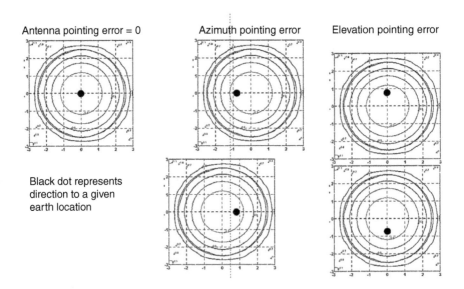

FIGURE 3.29 Antenna pointing-error effect when intended direction is beam center.

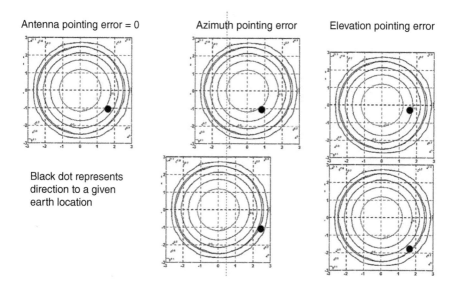

FIGURE 3.30   Antenna pointing-error effect when intended direction is off beam center.

In some cases, the pointing error causes the gain to increase, in others to decrease. The exact gain change depends on the pointing error direction.

In summary, antenna gain is a function of two-dimensional antenna pointing error. The gain is not necessarily symmetric about zero pointing error. The gain does not have the same characteristics in the two dimensions. The nominal value of gain toward a location of interest, considering antenna pointing error, is not, in general, the gain toward the intended location but must be obtained by averaging the gain over the antenna pointing error. The gain variation is not generally Gaussian-distributed.

3.10   ANTENNA AUTOTRACK

Sometimes an antenna needs to point with more accuracy than the spacecraft attitude-control subsystem alone can provide. Then the antenna will have an onboard **autotrack** system, which keeps the antenna's peak of beam pointed to a beacon or a communications signal (EMS Technologies Inc, 2002). The system is almost always a closed-loop (=feedback) control system, which measures the pointing error and feeds back the negative of the error to the antenna steering mechanism as a control signal. The pointing error has two orthogonal components, azimuth and elevation.

More precisely, in the case of a reflector antenna, a tracking feed provides two *difference signals*, one for azimuth error and one for elevation error. The difference signals are approximately proportional to the respective components of the pointing error, over the tracking range. The two angles are ideally orthogonal over the tracking range but in fact are never quite (see e.g., (Mahadevan et al., 2004)). A tracking receiver takes these signals and creates the control signals which are then input to the antenna steering mechanism.

Most often the autotrack system is of the monopulse variety, where the term *monopulse* comes from radar. A monopulse system has the following characteristics: (1) the difference signals can be continually measured; (2) the whole-signal (the *sum-signal*) strength can also be continually measured, and it is used by the tracking receiver to scale the difference signals; and (3) the antenna steering system physically turns the antenna, either the reflector, subreflector, or the feed array (Howley et al., 2008). A good description of monopulse tracking is given in (Skolnik, 1970) and its later editions. A slight variation on the monopulse system is the pseudo-monopulse one, which electronically scans the beam instead of moving the antenna (Howley et al., 2008).

There are two kinds of *tracking feed*. In one kind, easier to understand, the feed consists of multiple horns packed together. Figure 3.31 gives an example with four horns, in which each pair of opposite horns measures the difference signal along the line that joins them and all the horns together receive the sum signal. When the antenna is perfectly pointed, in each pair of horns both horns measure the same amplitude, so their difference signal is zero. When the focal point is offset, in at least one pair of horns one of the horns will receive greater amplitude than the other will, so the difference signal will be nonzero.

The other kind of tracking feed is a single multimode horn. The aperture can be square (Skolnik, 1970) or round (Yodokawa and Hamada, 1981). When the focal point is in the middle of the feed, only the conventional waveguide mode of the feed is excited, by the sum signal. When the focal point is offset, higher-order waveguide modes are additionally excited. They are coupled off and their magnitudes and signs give the orthogonal difference signals. The multimode horn is wider-band than the multi-horn feed (Skolnik, 1970). An example of predicted sum and difference contours produced by such an autotrack subsystem is shown in Figure 3.32, where the sum contours are approximately circles. It is evident that the difference contours are not exactly orthogonal.

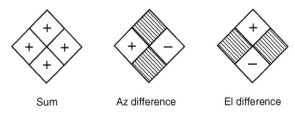

Sum   Az difference   El difference

FIGURE 3.31 Four-horn diamond feed for tracking and communications: how sum and difference signals are formed from individual signals. (after Skolnik (1970)).

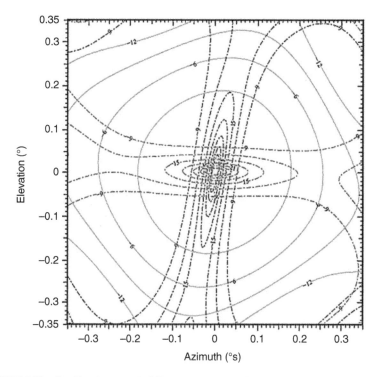

FIGURE 3.32   Predicted sum and difference contours of an autotrack subsystem for a CP offset reflector antenna. (Used with permission of SG Microwave and Maxar Technologies, from Mahadevan et al. (2004)).

Both kinds of tracking feeds can perform dual LP or dual CP with the proper hardware behind them ((Skolnik, 1970) for the four-horn diamond feed, (Prata et al., 1985) for conical multimode).

Sometimes the tracking feeds are on separate antennas from the communications antennas, as on Anik F2 (Amyotte et al., 2006), but this is uncommon, as performance is better when both tracking and communications are on the same antennas. Best performance occurs when each communications antenna has its own tracking function. Often the tracking function is incorporated in the communications feed chain via a coupler. Such a Ku-band feed is on Ciel 2, launched in 2008 (Lepeltier et al., 2007). A Ka/K-band feed flies on Eutelsat's Ka-Sat, launched in 2010 (Uher et al., 2011).

Typical total antenna pointing error is generally considered to be about ±0.13° without autotrack and about ±0.05° with autotrack (Uher et al., 2011). One antenna manufacturer's Ka-band autotrack system on a multi-spot-beam antenna contributed less than ±0.01° to a total antenna pointing error of less than ±0.04° (Amyotte et al., 2006). A patent awarded to a satellite manufacturer in 2010 would cut the total pointing error to maybe half, by adding to the autotrack control loop a feedforward component with a pointing prediction based on knowledge the system has gained on previous on-orbit days (Goodzeit and Weigl, 2010).

If the composite coverage is zonal or smaller, one beacon inside the coverage area may be enough. The closer the beacon is to the center of the coverage area, the better.

Inmarsat-4, though, has a larger, hemispherical coverage area. It uses two beacons, at the edge of the visible earth and about 90° apart. Four payload beams, ±1° from the beacon direction, are used for monopulse autotracking (Stirland and Brain, 2006) (Section 20.3.6.2).

3.11   REFLECTOR-ANTENNA INEFFICIENCIES

3.11.1   Summary

A reflector antenna has various inefficiencies or imperfections:

- *Aperture or illumination inefficiency*, resulting from deviation of aperture's tangential electric field from constant-amplitude and constant-phase across the aperture, addressed below in Section 3.11.2
- *Spillover inefficiency*, resulting from some of the power radiated from the feed missing the reflector and/or subreflector. Also called *beam efficiency*, from the viewpoint of the feed horn. There is an analog on receive, which would be some of the power collected by the reflector missing the subreflector or feed and some of the power collected by the subreflector missing the feed
- *Blockage inefficiency*, resulting from blockage of radiation by the feed, subreflector if any, and/or struts holding up the feed or subreflector, addressed in Section 3.11.3
- *Surface inefficiency*, resulting from the reflector and/or subreflector surface deviating from what was designed, due to manufacturing and alignment inaccuracies
- *Ohmic or resistive inefficiency*, resulting from losses in the conducting surfaces including the feed and reflector(s)
- *Frequency dispersion*, since the antenna is not quite as effective at frequencies other than the frequency at which it was optimized
- *Cross-polarization inefficiency*, the loss of the power that is coupled into the orthogonal polarization, addressed in Section 3.11.4
- Thermal distortion, addressed in Section 3.11.5
- *Focus-error inefficiency*, for multi-beam reflector antennas, addressed in Section 11.5.3.

Table 3.1 lists the inefficiencies and shows their effects. The effects can be summarized as follows:

- Loss in antenna gain (pattern mainlobe level) from ideal
- Altered levels of pattern sidelobes

TABLE 3.1 Antenna Inefficiencies and Their Effects

	Lowers Pattern Mainlobe	How to Characterize Loss in Pre-measurement Budget	Lowers Pattern Sidelobes	Raises Pattern Sidelobes	Adds Thermal Noise	Potentially Causes Cross-Pol Interference into Other Beams
Aperture inefficiency	x	Additive dB	x			
Spillover	x	Additive dB		x		
Blockage	x	Rss'ed dB	x	x		x
Surface inaccuracies	x	Rss'ed dB		x		
Imperfect cross-pol isolation	x	Rss'ed dB	x			x
Ohmic losses	x	Additive dB	x		x	
Frequency dispersion	x	Additive dB	x			
Thermal distortion	x	Rss'ed dB		x		

- Thermal noise added to the signal
- Creation of low-level cross-polarized signal that robs power from co-polarized signal and may cause interference to other nearby beams.

The most obvious effect of the inefficiencies is to decrease the antenna gain from the ideal gain (directivity) for the given reflector size to the actual gain. The difference between directivity and gain is the **antenna (in)efficiency**. Reduction of gain is reduction of pattern mainlobe level.

Before any measurement is made of the assembled antenna, some losses are well characterized or estimated—aperture illumination, spillover, ohmic inefficiency, and frequency dispersion. To make a budget of antenna gain at this point requires assessment of these in dB as well as of the other inefficiencies that must be estimated from previous experience. In addition, terms for computer modeling error and manufacturing tolerances must be carried in the budget. The inefficiencies that were already well characterized or estimated are entered into the budget as dB losses and summed. The others in dB are rss'ed together and are added to the first sum, to get a total antenna loss. On the other hand, once the antenna has been assembled with a good spacecraft mockup, the only uncertainty to carry in a budget is a measurement inaccuracy term.

There is also an impedance-mismatch loss (Section 4.7) between the antenna terminal and the connecting RF line that leads to the rest of the payload, but this may be considered to be a payload-integration loss and not booked to the antenna.

Besides some inefficiencies causing a reduction in gain, blockage and thermal distortion cause an increase in the pattern sidelobe levels. Depending on the amount of blockage and the aperture illumination pattern, the blockage may however decrease instead of increase the sidelobes. Raised sidelobe increases the potential for interfering into a nearby beam. One inefficiency, ohmic or resistive inefficiency, causes an increase in thermal noise in addition to its effect on gain. Lastly, cross-pol induced by blockage and imperfect cross-polarization isolation will interfere into the other polarization.

The topic of antenna temperature is not included in this chapter because it is not a unit-level topic but one of the end-to-end system (ground-satellite-ground). It is treated in Section 14.5.

3.11.2 Aperture Illumination and Spillover

The aperture efficiency and the spillover efficiency are approximately inversely related. Roughly speaking, the closer the aperture efficiency is to 100%, the larger the spillover, which means that some of the radiation from the feed and subreflector is wasted. The product of the two efficiencies, in terms of non-dB power ratios, is greatest when the aperture illumination at the edge is about 10 dB down from the illumination at the center (Milligan, 2005). This number or something near it is commonly used as the taper.

3.11.3   Blockage

Blockage and scattering issues arise for center-fed reflector antennas and for dual-gridded reflector antennas.

For center-fed antennas, the main reflector is partially blocked by the feed, any subreflector, and supporting metallic struts of the feed and any subreflector. Blockage lowers the mainlobe, raises the sidelobes, and creates cross-pol in the antenna pattern by forward-scattering the field. This effect can be reduced by coating the struts with "hard" electromagnetic surfaces composed of dielectric (Kildal et al., 1996). Significant improvement has been demonstrated (Riel et al., 2012). The uniformly illuminated aperture is affected least by blockage. The more tapered the illumination is, the more it suffers from blockage (Milligan, 2005).

For the O3b MEO satellites, with center-fed reflectors each of which can be steered over the whole earth by gimbaling the reflector, it was a complex endeavor to minimize blockage and scattering. The feed and its support struts partially block the reflector, and the struts scatter the RF field. The spacecraft manufacturer developed software to optimize the antenna over all its possible positions (Amyotte et al., 2011).

The blockage problem is different for large dual-gridded dual-LP antennas. Dielectric posts or stiffeners are needed between the front and back reflectors for structural integrity and to minimize thermal distortion of the front reflector. The posts or stiffeners scatter fields of the rear reflector and can be the main factor in the cross-pol performance of this reflector (Demers et al., 2012).

3.11.4   Cross-Polarization

In creating radiation of a desired polarization, a small amount of orthogonally polarized radiation is also inevitably created. The radiation with the desired polarization is said to be **co-polarized** or simply **co-pol**, while the undesired radiation of the orthogonal polarization is cross-pol. The only time cross-pol is a concern is when it causes interference, so cross-pol and **cross-pol interference** are practically synonymous. If the cross-pol interference and signal it is interfering with are unrelated, the interference is modeled as additive noise.

Cross-pol has several sources on the satellite. On receive, the first source is the radiating element(s). For a reflector antenna, this would be the reflector, subreflector if any, and the feed. For a horn antenna, it is the horn itself. Cross-pol could also arise from reflections off struts or anything else, but with a good bus and payload design, this is probably negligible. If the antenna is receiving dual CP, then the next cross-pol source is the polarizer. If the antenna is receiving dual pol, the next and last source is the OMT. Figure 3.33 shows the situation for reception of one polarization and Figure 3.34 for reception of dual pol, where at each element the creation of cross-pol is drawn as cross-coupling. Cross-pol arises on transmit from the same sources.

Let us quantify the cross-pol interference in the case corresponding to Figure 3.34. The two arriving signals are assumed to be perfectly orthogonal (the imperfect case is dealt with in Section 14.6.4.1). In the through paths both signals receive the same

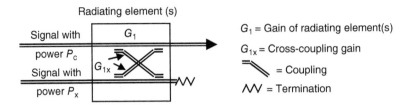

FIGURE 3.33 Cross-polarization interference shown as cross-coupling, for case where only one polarization is intended to be received.

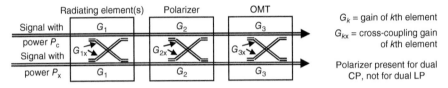

FIGURE 3.34 Cross-polarization interference on receive shown as cross-coupling.

gain. Within each of the three elements, a bit of each signal is cross-coupled into the other signal. The cross-pol performance of an element is normally quoted as relative gain or is in dBc, which are the same. The relative gains for the three elements are the gain ratios $\gamma_k = G_{kx}/G_k$ for $k = 1, 2, 3$, which are not in dB. We find that

$$\frac{P_c}{(\gamma_1 + \gamma_2 + \gamma_3)P_x} = \text{ratio of copol power to interference power}$$

from cross-pol signal

$$\frac{P_x}{(\gamma_1 + \gamma_2 + \gamma_3)P_c} = \text{ratio of copol power to interference power}$$

from cross-pol signal, for other signal

3.11.5 Thermal Distortion

The antennas are on the outside of the spacecraft body so they see the full range of temperature, namely, at different times they can be fully in the dark, fully lit by the sun at an oblique or boresight angle, and transitioning in between. These cases are called cold, hot, and **gradient**, respectively. One manufacturer always finds that the worst thermal distortion, that is, the biggest rms surface error, is in the dark (cold) (Schennum, 2010). Thermal distortion is not the largest contributor to overall antenna-gain loss (Schennum, 2011).

Reflectors are made of a material that is thermally very stable, such as graphite. A gridded reflector may be encased by a protective cover which reduces temperature extremes and gradients, as on the Eutelsat II satellites of the early 1990s (Duret et al., 1989). The reflecting surfaces of reflectors may be coated with low-insertion-loss thermal paint, as on NASA's Advanced Communications

Technology Satellite (ACTS) launched in 1993 (Regier, 1992). Thermal paint or coating makes good thermal contact with the antenna reflector, radiates heat well, and reflects light well (Wertz and Larson, 1999).

3.12   TESTING

Antenna testing has a few stages. One of the first is *photogrammetric measurement* of the reflector surface, for characterizing the reflector's thermal distortion. Figure 3.35 shows two Ka-band reflectors in a test apparatus. Measurements are taken of the points marked with light-weight targets, visible as white dots. The reflector is thermal-cycled in a high-vacuum thermal chamber, and measurements are taken at many different temperature plateaus (Wiktowy et al., 2003).

Next, either the receive or transmit pattern can be measured for a passive antenna, as both would be the same. The antenna is tested along with a good mockup of the parts of the spacecraft that have any possibility of causing blockage, reflections, or PIMs (Section 8.15). Testing is often done in a **near-field range (NFR)**, which is an anechoic chamber lined with RF absorbers. A probe performs a fine raster scan, measuring amplitude and phase of the electric field. There are two kinds of NFR for

FIGURE 3.35   Ka-band reflectors in photogrammetric test apparatus. (Used with permission of Canadian Aeronautics and Space Institution, from Wiktowy et al. (2003)).

payload antennas. The one with planar scan, illustrated in Figure 3.36, is applicable for medium- and high-gain reflector antennas and phased arrays. The linear probe is shown oriented horizontally, but the scan would also be done with the probe oriented vertically (IEEE, 2012). The NFR with spherical scan, illustrated in Figure 3.37, is appropriate for global horns and phased-array radiating elements

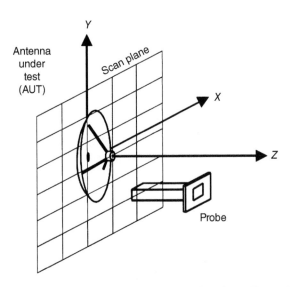

FIGURE 3.36 Near-field range with planar scan. (©2003 NSI-MI Technologies. Reprinted with permission, from Hess (2003)).

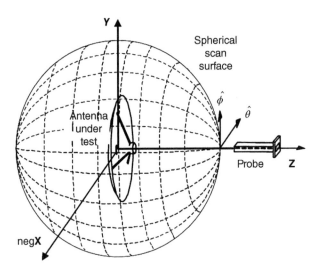

FIGURE 3.37 Near-field range with spherical scan. (©2003 NSI-MI Technologies. Reprinted with permission, from Hess (2003)).

(Hindman and Fooshe, 1998). The antenna under test (AUT) rotates on two axes while the probe stays fixed. The configuration shown is elevation over azimuth, and the AUT moves by rotation. An alternative implementation is to move the probe along the elevation and move the AUT in azimuth only (IEEE, 2012). For both NFRs, the near-field pattern is transformed via the two-dimensional Fourier transform into the far-field pattern. Planar near-field testing is as accurate as far-field. Gain, EIRP, cross-pol, and saturating flux density can be measured (Newell et al., 1988).

A flexible kind of NFR has been developed, where the scan points are on any arbitrary and irregular surface in the near field. The probe is carried on a gondola, and its position and orientation are tracked by laser (Geise et al., 2017).

One antenna manufacturer built a dedicated planar NFR for pattern-testing of large numbers of single-beam antennas for LEOs and MEOs, specifically 144 antennas for O3b and 486 for Iridium Next. The manufacturer was able to test several antennas at once and the measurements could be automated to test them in a single sequence. A big saving was in set-up time. No special alignment was required on each antenna because only master plates on the test platform needed to be accurately aligned. A time savings of 70% per antenna was estimated (Riel et al., 2015).

At spacecraft-level verification, the antenna is tested on the spacecraft in a far-field range or, more commonly, in a **compact antenna test range (CATR)**. A CATR emulates a far-field range by creating a plane wave to be received by the AUT, from the range feed's wave that has a spherical phase front, by means of a focusing element (Tuovinen et al., 1997). No mathematical transformation of measurements is required. A CATR can test a transmit antenna as well as a receive antenna. For a transmit antenna the plane wave would be transmitted and the range feed would receive (Dudok et al., 1992). The area in the range where the AUT can be placed is called the *quiet zone*. The spacecraft is mounted horizontally, that is, on its side (Scouarnec et al., 2013), on a three-axis positioner mounted on an azimuth carriage (Dudok et al., 1992).

Two kinds of CATR for payload antennas provide accurate cross-pol measurements (Fasold, 2006):

- Compensated (double-reflector) compact range (CCR), having an offset Cassegrain antenna as illumination system, as illustrated in Figure 3.38.
- Single-reflector compact ranges, having an offset parabolic reflector as illumination system.

The CCR 75/60 has long been the standard CCR. It has a quiet zone 5 m wide. More recently, because antennas are getting larger and larger, the CCR 120/100 has been developed, with an 8-m quiet zone and the CCR 150/120, with a 10-m quiet zone. The CCR 75/60 was developed by MBB GmbH of Germany, now owned by Airbus, which developed the two larger ones.

The CCR quiet zone can be extended and separated into two quiet zones of equal size, side by side in the drawing, so that the antenna can simultaneously be tested at

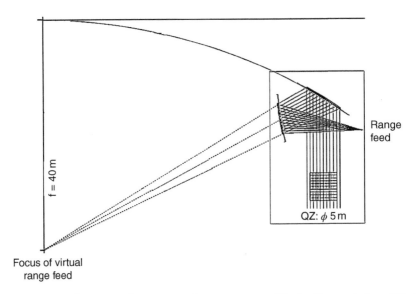

FIGURE 3.38 Geometry of compensated compact range. (Used with permission of Prof. Dr. Dietmar Fasold, author of the invited paper, and Antenna Measurement Techniques Society, from Fasold (2006)).

receive and transmit frequencies. This is accomplished by using two feeds and placing one above and one below the feed location shown in the drawing (Dudok et al., 1992; Migl et al., 2017).

The second kind of CATR has just a paraboloidal antenna for the illumination system. The configuration is similar to offset-fed but the feed is aligned with the paraboloid's axis (Rose and Cook, 2010).

Both types of CATR seem to provide good measurements. The inventor of the compensated compact range analyzed the various CATR types all with a 5-m quiet zone. He showed that co-pol performance was similar in all, but cross-pol performance was best in the compensated compact range (Fasold, 2006).

CATR measurements include the full antenna pattern, EIRP, and saturated flux density (Dudok et al., 1992).

REFERENCES

Adatia N, Watson B, and Ghosh S (1981). Dual polarized elliptical beam antenna for satellite application. *IEEE Antennas and Propagation Society International Symposium*; Vol. 19; June.

Akagawa M and DiFonzo DF (1979). Beam scanning characteristics of offset Gregorian antennas. *IEEE Antennas and Propagation Society International Symposium*; Vol. 17; June.

Albertsen NC and Pontoppidan K (1984). Analysis of subreflectors for dual reflector antennas. *IEE Proceedings, Part H: Microwaves, Optics and Antennas*; Vol. 131; June.

Amyotte E (2020). Private communication, Apr. 1.

Amyotte E and Godin M-A (2017). Antennas at MDA: innovation through cross-pollination. *European Conference on Antennas and Propagation*; Mar. 19–24.

Amyotte E, Demers Y, Martins-Camelo L, Brand Y, Liang A, Uher J, Carrier G, and Langevin J-P (2006). High performance communications and tracking multi-beam antennas. *Proceedings, European Conference on Antennas and Propagation*; Nov. 6–10.

Amyotte E, Demers Y, Hildebrand L, Forest M, Riendeau S, Sierra-Garcia S, and Uher J (2010). Recent developments in Ka-band satellite antennas for broadband communications. *Proceedings, European Conference on Antennas and Propagation*; Apr. 12–16.

Amyotte E, Demers Y, Dupessey V, Forest M, Hildebrand L, Liang A, Riel M, and Sierra-Garcia S (2011). A summary of recent developments in satellite antennas at MDA. *European Conference on Antennas and Propagation*; Apr. 11–15.

Amyotte E, Demers Y, Forest M, Hildebrand L, and Richard S (2013). A review of recent antenna developments at MDA. *European Conference on Antennas and Propagation*; Apr. 8–12.

Amyotte E, Demers Y, Hildebrand L, Richard S, and Mousseau S (2014). A review of multi-beam antenna solutions and their applications. *European Conference on Antennas and Propagation*; Apr. 6–11.

Apfel SL (2006). Optimization of the Boeing 702 for the DirecTV mission. *AIAA International Communications Satellite Systems Conference*; June 11–14.

Barkeshli S, Smith T, Luh HS, and Ersoy L (1995). On the analysis and design of the frequency selective surface for the N-Star satellite Ku/S-shaped reflector. *IEEE Antennas and Propagation Society Symposium*; June 18–23.

Bhattacharyya AK and Goyette G (2004). A novel horn radiator with high aperture efficiency and low cross-polarization and applications in arrays and multibeam reflector antennas. *IEEE Trans actions on Antennas and Propagation*; 52 (Nov.); 2850–2859.

Broquet J, Claudinon B, and Bousquet A (1985). Antenna pointing systems for large communications satellites. *AIAA Communication Satellite Systems Conference*; Mar. 7–11.

Camelo LM (2010). The Express AM4 top-floor steerable antennas. *International Symposium on Antenna Technology and Applied Electromagnetics and the American Electromagnetics Conference*; July 5–8.

Chang K, editor, (1989). *Handbook of Microwave and Optical Components, Vol. 1, Microwave Passive and Antenna Components*, New York: John Wiley & Sons, Inc.

Choung Y (1996). Dual-band offset gimballed reflector antenna. *Digest, IEEE Antennas and Propagation Society International Symposium*; Vol. 1; July 21–26.

Chu T-S and Turrin RH (1973). Depolarization properties of offset reflector antennas. *IEEE Transactions on Antennas and Propagation*; 21 (3); 339–345.

Collin RE (1985). *Antennas and Radiowave Propagation*, New York: McGraw-Hill, Inc.

Croq F, Vourch E, Reynaud M, Lejay B, Benoist C, Couarraze A, Soudet M, Carati P, Vicentini J, and Mannocchi G (2009). The Globalstar 2 antenna sub-system. *European Conference on Antennas and Propagation*; Mar. 23–27.

Demers Y, Riel M, Lopes J-L, Angevain J-C, Ihle A, Brand Y, and De Maagt P (2012). Low scattering structures for reflector antennas. *International Symposium on Antenna Technology and Applied Electromagnetics*; June 25–28.

Dudok E, Habersack J, Hartmann F, and Steiner H-J (1992). Payload test capabilities of a large compensated compact range. *AIAA International Communication Satellite Systems Conference*; Mar. 22–26.

Duret G, Guillemin T, and Carriere R (1989). The EUTELSAT II reconfigurable multi-beam antenna subsystem. *Digest, IEEE Antennas and Propagation Society International Symposium*; Vol. 1; June 26–30.

EMS Technologies Inc (2002). Autotrack combiners. Application note 44X-1. June. On www.emsdss.com/uploadedFiles/pdf/ATM.pdf. Accessed Feb. 28, 2011.

Enokuma S, inventor; Sharp Kabushiki Kaisha, assignee (2002). Circular polarizer having two waveguides formed with coaxial structure. U.S. patent. July 9.

Evans BG, editor (1999). *Satellite Communication Systems*, 3rd ed. London: The Institution of Electrical Engineers.

Fasold D (2006). Measurement performance of basic compact range concepts. *Antenna Measurement Techniques Association Europe Symposium*; May 1–4.

Gagliardi R (1978). *Introduction to Communications Engineering*. New York: John Wiley & Sons, Inc.

Geise A, Fritzel T, and Paquay M (2017). Ka-band measurement results of the irregular near-field scanning system PAMS. *Antenna Measurement Techniques Association Symposium*; Oct. 15–20.

Glâtre K, Renaud PR, Guillet R, and Gaudette Y (2015). The Eutelsat 3B top-floor steerable anntennas. *IEEE Transactions on Antennas and Propagation*; 63 (4); 1301–1305.

Glâtre K, Hildebrand L, Charbonneau E, Perrin J, and Amyotte E (2019). Paving the way for higher-volume cost-effective space antennas. *IEEE Antennas and Propagation Magazine*; 61 (5) (Oct.); 47–53.

Goodzeit NE and Weigl HJ, inventors; Lockheed Martin Corp, assignee (2010). Antenna autotrack control system for precision spot beam pointing control. U.S. patent 7,663,542 B1. Feb. 16.

Grenier C, Fontaine M, Langevin J-P, Sierra-Garcia S, Michel N, Bussières F, and Maltais S (2012). Express AM5 and AM6 satellite antennas--design and realization overview. *AIAA International Communications Satellite Systems Conference*; Sep. 24–27.

Harris Corp (2009). Ka-band antennas (DirecTV satellites). Product information. On www. govcomm.harris.com/solutions/products/000073.asp. Accessed Nov. 27, 2009.

Hess DW (2003). Readily made comparison among the three near-field measurement geometries using a composite near-field range. *Proceedings, Antenna Measurement Techniques Association Symposium*; Oct. 19–24.

Hindman G and Fooshe DS (1998). Probe correction effects on planar, cylindrical and spherical near-field measurements. *Antenna Measurement Techniques Association Conference*.

Hoffmeister R, Space Systems Loral (2010). Private communication, Mar. 5.

Howley RJ, Daffron WC, Hemlinger SJ, and Gianatasio AJ, inventors; Harris Corp, assignee (2008). Monopulse antenna tracking and direction finding of multiple sources. U.S. patent application publication 2008/0122683 A1. May 29.

HPS GmbH (2018). Reflector antennas. Product information. On www.hps-gmbh.com/portfolio/subsystems/reflector-antennas/#ank-01. Accessed Aug. 4, 2018.

IEEE (2012). IEEE recommended practice for near-field antenna measurements. Standard 1720. Dec. 15.

Intelsat (2008). Intelsat satellite guide. On www.intelsat.com/network/satellite. Accessed Jan. 2010.

Izquierdo Martinez I (2008). Design of wideband orthomode transducers based on the turnstile junction for satellite communications. Dissertation submitted to the Universidad Autónoma de Madrid. Nov. On repositorio.uam.es. Accessed Aug. 1, 2018.

Kildal P-S, Kishk AA, and Tengs A (1996). Reduction of forward scattering from cylindrical objects using hard surfaces. *IEEE Transactions on Antennas and Propagation*; 44 (11) (Nov.); 1509–1520.

Krebs GD (2017a). Intelsat 34 (Hispasat 55W-2). Gunter's Space Page; Dec. 11. Accessed July 25, 2018.

Krebs GD (2017b). O3b 1,. . ., 12. Gunter's Space Page; Dec. 11. Accessed July 26, 2018.

Krebs GD (2020). O3b 13,. . ., 20. Gunter's Space Page; Feb. 4. Accessed Mar. 28, 2020.

Kwok RS and Fiedziuszko SJ (1996). Advanced filter technology in communications satellite systems. *Proceedings, International Conference on Circuits and System Sciences*; June 20–25.

Lawrie RE and Peters, Jr L (1966). Modifications of horn antennas for low sidelobe levels. *IEEE Transactions on Antennas and Propagation*; 14 (5) (Sep.); 605–610.

Legay H, Croq F, and Rostan T (2000). Analysis, design and measurements on an active focal array fed reflector. *Proceedings, IEEE International Conference on Phased Array Systems and Technology*; May 21–25.

Lepeltier P, Maurel J, Labourdette C, Croq F, Navarre G, and David JF (2007). Thales Alenia Space France antennas: recent achievements and future trends for telecommunications. *European Conference on Antennas and Propagation*; Nov. 11–16.

Mahadevan K, Ghosh S, Nguyen B, and Schennum G (2004). TX, RX & Δx/ Δy-autotrack CP feed for multiple beam offset reflector antennas. *AIAA International Communications Satellite Systems Conference*; May 9–12.

Maral G and Bousquet M (2002). *Satellite Communications Systems*, 4th ed. Chichester, UK: John Wiley & Sons Ltd.

Migl J, Habersack J, and Steiner H-J (2017). Antenna and payload test strategy of large spacecraft's in compensated compact ranges. *European Conference on Antennas and Propagation*; Mar. 19–24.

Milligan TA (2005). *Modern Antenna Design*, 2nd ed. New Jersey: John Wiley & Sons and IEEE Press.

MIT (2013). Chapter 11, Common antennas and applications. of material for course 6.013 Electromagnetics and Applications. MIT OpenCourseWare. November 1. Accessed August 26, 2018.

Nagarajah B (2015). Intelsat 34 continues to pass the test. Intelsat launches blog. On www.intelsat.com/news/blog/intelsat-34-continues-to-pass-the-test-3. Accessed July 25, 2018.

Newell AC, Ward RD, and McFarlane EJ (1988). Gain and power parameter measurements using planar near-field techniques. *IEEE Transactions on Antennas and Propagation*; 36 (6) (June); 792–803.

Potter PD (1963). A new horn antenna with suppressed sidelobes and equal beamwidths. Technical report no 32-354. Jet Propulsion Laboratory. Feb. 25.

Prata Jr A, Filho EA, and Ghosh S (1985). A high performance – wide band – diplexing – tracking – depolarization correcting satellite communication antenna feed. *IEEE MTT-S International Microwave Symposium Digest*; June.

Pritchard WL and Sciulli JA (1986). *Satellite Communications Systems Engineering*, Englewood Cliffs, NJ: Prentice-Hall.

Rahmat-Samii Y and Imbriale WA (1998). Anomalous results from PO applied to reflector antennas: the importance of near field computations. *Digest, IEEE Antennas and Propagation Society International Symposium*; Vol. 2; June 21–26.

Ramanujam P, Lopez LF, Shin C, and Chwalek TJ (1993). A shaped reflector design for the DIRECTV™ direct broadcast satellite for the United States. *Digest, IEEE Antennas and Propagation Society International Symposium*; Vol. 2; June 28–July 2.

Ramo S, Whinnery JR, and Van Duzer T (1984). *Fields and Waves in Communication Electronics*, 2nd ed. New York: John Wiley & Sons, Inc.

Rao SK (2003). Parametric design and analysis of multiple-beam reflector antennas for satellite communications. *IEEE Antennas and Propagation Magazine*; Vol. 45; Aug.

Rao SK (2015). Advanced antenna systems for 21st century satellite communications payloads. Viewgraph presentation. On s3.amazonaws.com/sdieee/1820-DL_Rao_SanDiego_12Mar2015_D1.pdf. Accessed Dec. 22, 2017.

Rao S, Shafai L, and Sharma S, editors, (2013). *Handbook of Reflector Antennas and Feed Systems, Vol. III, Applications of Reflectors*. Boston: Artech House.

Regier FA (1992). The ACTS multi-beam antenna. *IEEE Transactions on Microwave Theory and Techniques*; 40 (6); 1159–1164.

Riel M, Brand Y, Demers Y, and De Maagt P (2012). Performance improvements of center-fed reflector antennas using low scattering struts. *IEEE Transactions on Antennas and Propagation*; 60 (3); 1269–1280.

Riel M, Arsenault P, Lemelin-Auger B, and Amyotte E (2015). Pattern testing of low-cost antennas for LEO and MEO satellites at MDA. *European Conference on Antennas and Propagation*; Apr. 13–17.

Rose CA and Cook Jr JH (2010). High accuracy cross-polarization measurements using a single reflector compact range. Technical paper. On www.mi-technologies.com. Accessed Jan. 2010.

Rudge AW, Milne K, Olver AD, and Knight P, editors, (1982). *The Handbook of Antenna Design, Vol. 1*. London: Peter Peregrinus, Ltd.

Rusch WV, Prata Jr A, Rahmat-Samii Y, and Shore RA (1990). Derivation and application of the equivalent paraboloid for classical offset Cassegrain and Gregorian antennas. *IEEE Transactions on Antennas and Propagation*; 38 (8); 1141–1149.

San Blas AA, Pérez FJ, Gil J, Mira F, Boria VE, and Gimeno B (2011). Full-wave analysis and design of broadband turnstile junctions. *Progress in Electromagnetics Research Letters*; 24; 149–158.

Schennum GH, retired from Space Systems/Loral (2010). Private communication, Sep. 23.

Schennum GH, retired from Space Systems/Loral (2011). Personal communication, Mar. 23.

Schennum GH, retired from Space Systems/Loral (2014). Private communication, July 20.

Schennum GH, retired from Space Systems/Loral (2015). Private communication, Jan. 30.

Schennum GH and Skiver TM (1997). Antenna feed element for low circular cross-polarization. *Proceedings, IEEE Aerospace Conference*; Vol. 3; Feb. 1–8.

Schennum GH, Lee E, Pelaca E, and Rosati G (1995). Ku-band spot beam antenna for the Intelsat VIIA spacecraft. *Proceedings, IEEE Aerospace Applications Conference*; Vol. 1; Feb. 4–11.

Schennum GH, Hazelwood JD, Gruner R, and Carpenter E (1999). Global horn antennas for the Intelsat-IX spacecraft. *Proceedings, IEEE Aerospace Conference*; Vol. 3; Mar. 6–13.

Schneider M, Hartwanger C, Sommer E, and Wolf H (2009). The multiple spot beam antenna project "Medusa." *European Conference on Antennas and Propagation*; Mar. 23–27.

Scouarnec D, Stirland S, and Wolf H (2013). Current antenna products and future evolution trends for telecommunication satellites application. *IEEE Topical Conference on Antennas and Propagation in Wireless Communications*; Sep. 9–13.

Simon PS, Kung P, and Hollenstein BW (2011). Electrically large spline profile smooth-wall horns for spot beam applications. *IEEE International Symposium on Antennas and Propagation*; July 3–8.

Simon PS, Kung P, and Hollenstein BW, inventors; Space Systems/Loral, assignee (2015). Electrically large step-wall and smooth-wall horns for spot beam applications. U.S. patent 9,136,606 B2. Sep. 15.

Skolnik MI, editor, (1970). *Radar Handbook*. New York: McGraw-Hill, Inc.

Stirland SJ and Brain JR (2006). Mobile antenna developments in EADS Astrium. *European Conference on Antnnnnas and Propagation*; Nov. 6–10.

Tao ZC, Mahadevan K, Ghosh S, Bergmann J, Sutherland D, and Tjonneland K (1996). Design & evaluation of a shaped reflector & 4-port CP feed for dual band contoured beam satellite antenna applications. *Digest, IEEE Antennas and Propagation Society International Symposium*; Vol. 3; July 21–26.

Tribak A, Mediavilla A, Cano JL, Boussouis M, and Cepero K (2009). Ultra-broadband low axial ratio corrugated quad-ridge polarizer. *Proceedings, European Microwave Conference*; Sep. 29–Oct. 1.

Tuovinen J, Vasara A, and Räisänen A (1997). Compact antenna test range. U.S. patent 5,670,965. Sep. 23.

Uher J, Demers Y, and Richard S (2010). Complex feed chains for satellite antenna applications at Ku- and Ka-band. *IEEE Antennas and Propagation Society International Symposium*; July 11–17.

Uher J, Richard S, Beyer R, Sieverding T, and Sarasa P (2011). Development of advanced design software for complex multimode antenna feeding systems. *IEEE Antennas and Propagation Magazine*; 53 (6) (Dec.); 70–82.

Wertz JR and Larson WJ (1999). *Space Mission Analysis and Design*, 3rd ed., 10th printing 2008. Hawthorne, CA: Microcosm Press; and New York City: Springer.

Wiktowy M, O'Grady M, Atkins G, and Singhal R (2003). Photogrammetric distortion measurements of antennas in a thermal-vacuum environment. *Canadian Aeronautics and Space Journal*; 49 (2) (June); 65–71.

Yodokawa T and Hamada SJ (1981). An X-band single horn autotrack antenna feed system. *IEEE Antennas and Propagation Society International Symposium*; June.

CHAPTER 4

PAYLOAD-INTEGRATION ELEMENTS

4.1 INTRODUCTION

This chapter presents the elements used to integrate payload units to form the payload. The most important ones are coaxial cable and waveguide. Other elements such as isolators and hybrids are also described. Since switches and hybrids are among these elements, it seems natural to discuss in general how active payload units are integrated into redundancy rings for the accommodation of failures. Finally, the integration topics of impedance mismatch and S-parameters are presented. All payload units, integration elements, and partial integrations of the payload are characterized by their measured S-parameters.

4.2 COAXIAL CABLE VERSUS WAVEGUIDE

The types of **radio frequency (RF) line** we discuss in this chapter are **coaxial cable or coax** and **waveguide**. These are the only kinds of line used to integrate the payload. There is no agreement on a term that includes both coaxial cable and waveguide. In some professional books and articles, they are both called *transmission lines* or *transmission structures*. In textbooks, coaxial cable is a transmission line when it is propagating transverse electromagnetic (TEM) mode (Section 4.3.5), and both coaxial cable and waveguide are said to be waveguides. We use the term transmission line, along with RF line.

Satellite Communications Payload and System, Second Edition. Teresa M. Braun and Walter R. Braun.
© 2021 John Wiley & Sons, Inc. Published 2021 by John Wiley & Sons, Inc.

When to use coax and when to use waveguide is usually clear. Waveguide has a higher mass per meter of run length than coax does for frequencies below about 10 GHz, while above about 10 GHz the masses are roughly the same, as will be seen in the tables below. Waveguide is larger in size than coax at all frequencies, and the lower the frequency the greater the difference. The insertion loss (Section 4.7.2) per meter of waveguide is much lower than that of coax. The power-carrying ability of waveguide is higher. The length tolerance of coax can be greater than for waveguide, since coax can be bent a little but waveguide cannot. In general, coax is used at low frequencies and low power, while waveguide is used at the higher frequencies or when the power is high.

4.3 COAXIAL CABLE

4.3.1 Coax Construction

Coax consists electrically of three elements, as shown in Figure 4.1: a center conductor, an outer conductor, and a dielectric filling the space in between. The dielectric has the main purpose of altering the electrical properties but it also serves the mechanical purpose of keeping the conductors properly separated. The inner conductor is usually silver-plated copper wire. At GHz frequencies, only the outer surface of the inner conductor and the inner surface of the outer conductor conduct. The outer conductor is made of copper or aluminum alloy. Coating or finish of the outer conductor is on the outside and does not affect the electrical performance. Semi-rigid coax is the kind used for payload integration except for applications such as a gimbaled antenna, which call for flexible coax (W L Gore and Associates, 2003). We do not discuss flexible coax further.

The dielectric is always some form of polytetrafluoroethylene (PTFE), known otherwise by the DuPont brand name Teflon (Wikipedia, 2011). There are three forms of it in use in coax:

- Solid, the oldest form — Highest in mass.
- Low-loss low-density — Intermediate in mass.
- Expanded polytetrafluoroethylene (ePTFE) or microporous — Lowest in mass, least rugged (Teledyne Storm Products, 2010).

Tables 4.1, 4.2, and 4.3 provide examples of coax structures for the three dielectrics in order, in various sizes. In all cases, the center conductor is silver-plated

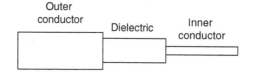

FIGURE 4.1   Coaxial cable electrical construction.

TABLE 4.1 Examples of Coax Dimension and Mass, Solid PTFE (Carlisle Interconnect Technologies, 2017)

Outer-conductor Outer Diameter (in)	Max Operating Freq (GHz)	Outer Conductor Material	Nominal Mass (kg/10 m)
0.250	18	Cu; Al	1.56; 0.928
0.141	26.5	Cu; Al	0.494; 0.290
0.085	50	Cu; Al	0.213; 0.108
0.047	90	Cu; Al	0.060; 0.032

TABLE 4.2 Examples of Coax Dimension and Mass, Low-Loss Low-Density PTFE (Carlisle Interconnect Technologies, 2017)

Outer-conductor Outer Diameter (in)	Max Operating Freq (GHz)	Outer Conductor Material	Nominal Mass (kg/10 m)
0.250	18	Cu	1.41
0.141	26.5	Cu; Al	0.477; 0.275
0.086	65	Cu; Al	0.209; 0.104
0.070	65	Cu	0.113
0.047	90	Cu; Al	0.059; 0.030

TABLE 4.3 Examples of Coax Dimension and Mass, Expanded PTFE (W L Gore and Associates, 2003)

Outer-conductor Outer Diameter (in)	Max Operating Freq (GHz)	Outer Conductor Material	Nominal Mass (kg/10 m)
0.210	18, phase-stable	Cu	0.627
0.190	18 or 30 (two options)	Cu	0.558
0.140	40	Cu	0.328
0.120	26.5	Cu	0.295
0.085	65	Cu	0.131
0.047	65	Cu	0.065

copper wire. One column in each table gives the nominal mass of a coax run in kg per 10 m, for comparison purposes with other coax and with waveguide. Coax with the aluminum-alloy outer conductor is lower in mass than with copper.

The manufacturer preconditions coax by heating it. After the coax is bent for payload integration and before installation, it may need to be conditioned again by three thermal cycles, where, in each cycle, the temperature goes from ambient to maximum to ambient to minimum to ambient. The conditioning will make the dielectric retract, so the cable should be cut ¼ inch longer than its design length. After conditioning, the cable is cut to its final length and the connectors are put on (Carlisle Interconnect Technologies, 2017).

TABLE 4.4   Velocity of Propagation in Coax with Various Dielectrics (Teledyne Storm Products, 2010)

Coax Dielectric	Dielectric Constant	Velocity of Propagation (% of c)
Solid PTFE	2.02	70
Low-loss low-density PTFE	1.6–1.8	75–79
Expanded PTFE	1.3–1.5	82–88

4.3.2   Coax Performance

According to the coax's dielectric material, the RF performance is summarized as follows:

- Solid — Highest loss.
- Low-loss low-density — Intermediate in loss. Exhibits greater variation in dielectric constant over the length and among lots than solid does.
- Expanded or microporous — Least lossy. Exhibits the greatest variation in dielectric constant over length and among lots (Teledyne Storm Products, 2010).

As a general rule, the lower the dielectric constant, the lower the loss (Carlisle Interconnect Technologies, 2017). Dielectric constants are given in Table 4.4.

Cable comes in a family of sizes, each of which can be used from direct current (DC) to some upper frequency. The higher the upper frequency is, the higher the cable's insertion loss. Smaller-size coax is not necessarily made for higher frequencies, as is true for waveguide, but for lower mass; it has higher insertion loss. For a given piece of coax, the insertion loss increases with frequency in a nonlinear fashion (Teledyne Storm Products, 2010) and the insertion phase increases linearly. Table 4.5 gives some examples of coax insertion loss at ambient. The typical impedance of coax is $50\,\Omega$ but lower- and higher-impedance coax is available.

4.3.3   Coax Environmental

The temperature sensitivity of coax depends on the dielectric material and the coax size (Teledyne Storm Products, 2010). The temperature performance of the three dielectrics is as follows:

- Solid — Most temperature-sensitive.
- Low-loss low-density — Intermediate in temperature-sensitivity.
- Expanded or microporous — Least temperature-sensitive. Most phase-stable, is used when a set of coax must track in phase and gain over temperature.

As the temperature goes up, coaxial cable's insertion loss increases. One cable manufacturer reports that the percentage change in dB loss varies nearly linearly

TABLE 4.5   Examples of Typical Coax Insertion Loss at Ambient

Outer-conductor Outer Diameter (in)	Dielectric	Outer Conductor Material	At 1 GHz, Insertion Loss Per 10 m (dB)	At 10 GHz, Insertion Loss Per 10 m (dB)	At 26.5 GHz, Insertion Loss Per 10 m (dB)
0.141[a]	Solid PTFE	Cu; Al	3.7; 3.8	14; 14	25; 26
0.141[b]	Low-loss low-density PTFE	Cu	3	11	19
0.141[b]	Expanded PTFE	Cu	3	9	15
0.086[b]	Low-loss low-density PTFE	Cu	5	20	33
0.086[b]	Expanded PTFE	Cu	6	20	34
0.070[a]	Low-loss low-density PTFE	Cu	6	21	35
0.047[a]	Low-loss low-density PTFE	Cu	10	33	54

[a] Carlisle Interconnect Technologies (2017).
[b] Teledyne Storm Products (2010).

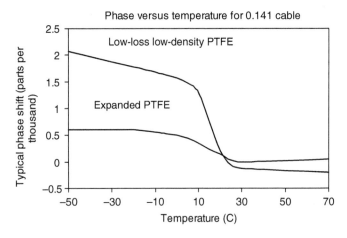

FIGURE 4.2   Example of coax electrical length variation versus temperature. (After Teledyne Storm Products (2010))

with temperature, relative to the dB loss at 25 °C. The variation is −22% at −100 °C and 18% at 150 °C, for 18 GHz. The variation is a weak function of frequency. The primary cause is the change in conductivity of silver with temperature (W L Gore and Associates, 2003).

In addition, the insertion phase changes with temperature. An example of this is shown in Figure 4.2. The insertion phase also changes with handling: bending the cable, even a little bit by removing it and reinstalling it, will cause a phase change (W L Gore and Associates, 2003).

A set of ePTFE cables can be relatively phase-matched or even absolutely phase-matched. Being able to phase-match is an important consideration when the cables must track over temperature in regard to insertion loss and/or phase. The cables have to have the same length and be at the same temperature. The tightest tracking may be achieved when the coax is all from the same material batch (W L Gore and Associates, 2003).

In space, there will be a brief period of time during which the dielectric vents. This is *out-gassing*. Spaceflight coax meets standards on the amount of outgassing (Micro-Coax, 2001; W L Gore and Associates, 2003).

Spaceflight coax is designed to be radiation-tolerant (Micro-Coax, 2001; W L Gore and Associates, 2003).

4.3.4   Connector and Adapter

There are many kinds of **connector** available to put on the ends of coax runs. The choice depends partly on the coax size and the operating frequency range. Even then, each type has several configurations, for example straight or at a right angle. Care must be taken to use connectors with the appropriate recommended operating frequency range. Some connectors of different types can mate with each other. Connectors also go on payload units. Connectors are described in Teledyne Storm Products (2010) and W L Gore and Associates (2003).

An **adapter** converts between waveguide with a given flange and coax with a given connector. It converts between the propagation modes.

4.3.5   Propagation in Coax

Only the TEM mode propagates in coax. The electric field vector E and its orthogonal magnetic field vector H lie in the wavefront plane, that is, are perpendicular to the direction of propagation, and the fields' amplitudes are related by a constant. The fields are illustrated in Figure 4.3. Recall that TEM is also the mode in which radiation propagates in space (Section 3.3.3.2). Higher-order modes than TEM could exist in coax, but the ratio of the conductor diameters is chosen so that the cutoff frequencies of the higher-order modes (Section 4.4.5) are well above the operating frequencies (Ramo et al., 1984).

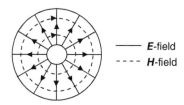

FIGURE 4.3   TEM propagation mode in coax.

Coax is a non-dispersive medium, that is, all frequencies propagate at the same speed. Light in coax has a **velocity of propagation** that is slower than in vacuum. It equals the speed of light c in vacuum divided by the square root of the dielectric constant. This equals about 70% of c for solid PTFE, 75–79% for low-loss low-density PTFE, and 82–88% for expanded PTFE, as shown in Table 4.4 (Teledyne Storm Products, 2010).

4.4 WAVEGUIDE

Rectangular, not circular, waveguide is most often used in integration. However, circular waveguide is of interest because its propagation modes are the basis of the resonance modes in some kinds of filters (Section 5.4). Besides, some antenna horns (Section 3.7.1) are flared circular waveguide.

4.4.1 Rectangular Waveguide Construction

Some defining characteristics of rectangular waveguide are shown in Figure 4.4. The waveguide's inner dimensions are width a and height b, where in normally proportioned waveguide a is about equal to twice b. The second part of the figure shows the E- and H-planes.

Table 4.6 gives the specifications of rectangular waveguide in its various standard sizes. The size designation is "WR" followed by the width a in hundredths of inches. Most of the waveguides have normal or nominal dimensions. However, those with "R/H" at the end of their designation have *reduced height*, about half the normal height. Reduced-height waveguide has a little lower mass than normal waveguide, and its small volume may be a benefit in a tight fit. There are also quarter-height waveguide and double-height waveguide (not listed). There is a normal or nominal wall-thickness. However, *thin-wall waveguide*, with half the normal thickness, is typically used on spacecraft for the reduced mass (Mele, 2011). The thin-wall thickness is shown in the table for some waveguide sizes. Aluminum is the material commonly used on spacecraft instead of copper,

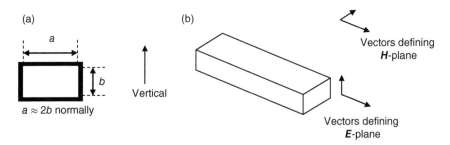

FIGURE 4.4 Rectangular waveguide: (a) dimensions, (b) E- and H-planes.

TABLE 4.6 Rectangular Waveguide Specifications, Thin-Wall (Cobham and Continental Microwave Division, 2006)

EIA WG Designation	Recommended Operating Band for TE$_{10}$ Mode (GHz)	Material Alloy	Theoretical Insertion Loss, Lowest to Highest Frequency (dB/10 m)	a (inches)	b (inches)	Thin-wall Thickness (inches)	Mass for Al Alloy, kg/10 m
WR430	1.70–2.60	Al; Cu	0.194–0.129; 0.129–0.086	4.300	2.150	—	—
WR340	2.20–3.30	Al; Cu	0.263–0.183; 0.175–0.122	3.400	1.700	—	—
WR284 R/H	2.60–3.95	Al; Cu	0.585–0.442; 0.389–0.294	2.840	0.670	—	—
WR284	2.60–3.95	Al; Cu	0.366–0.251; 0.243–0.167	2.840	1.340	—	—
WR229	3.30–4.90	Al; Cu	0.467–0.331; 0.310–0.220	2.290	1.145	—	—
WR187	3.95–5.85	Al; Cu	0.688–0.477; 0.458–0.317	1.872	0.872	—	—
WR159	4.90–7.05	Al; Cu	0.766–0.572; 0.503–0.381	1.590	0.795	—	—
WR137	5.85–8.20	Al; Cu	0.969–0.770; 0.652–0.512	1.372	0.622	—	—
WR112	7.05–10.00	Al; Cu	1.369–1.062; 0.911–0.707	1.122	0.497	0.040/0.045	2.54
WR102	7.00–11.00	Al; Cu	1.734–1.094; 1.154–0.727	1.020	0.510	—	—
WR96	7.00–17.00	Al; Cu	2.878–1.701; 1.914–1.130	0.965	0.320	—	—
WR90 R/H	8.20–12.40	Al	3.435–3.036	0.900	0.200	—	—
WR90	8.20–12.40	Al; Cu	2.135–1.477; 1.390–0.983	0.900	0.400	0.030/0.035	1.56
WR75 R/H	10.00–15.00	Al; Cu	2.561–1.952	0.750	0.200	—	—
WR75	10.00–15.00	Al	2.526–1.764; 1.680–1.174	0.750	0.375	0.020/0.025	0.93
WR67	11.00–17.00	Al; Cu	2.029–1.354; 1.349–0.900	0.668	0.340	—	—
WR62	12.40–18.00	Al; Cu	3.182–2.340; 2.116–1.556	0.622	0.311	0.020/0.025	0.78
WR51	15.00–22.00	Al; Cu	4.347–3.149; 2.891–2.094	0.510	0.255	—	—
WR42	18.00–26.50	Al; Cu	6.804–4.997; 4.528–3.323	0.420	0.170	0.020/0.025	0.51
WR34	22.00–33.00	Al; Cu	8.317–5.784; 5.531–3.848	0.340	0.170	—	—
WR28	26.50–40.00	Al; Cu	11.306–7.740; 7.552–5.174	0.280	0.140	0.020/0.025	0.40
WR22	33.00–50.00	Cu	10.643–7.234	0.224	0.112	—	—
WR19	40.00–60.00	Cu	13.061–9.383	0.188	0.094	—	—

when possible, because aluminum has about one third the mass of copper. Copper has a lower insertion loss. In this table, the smallest waveguides are not available in aluminum, perhaps because of the very high insertion loss. The recommended operating band of frequencies depends on the dimensions a and b. So does the theoretical insertion loss, which also depends on the metal and the exact operating frequency. For a given piece of waveguide, the insertion loss decreases with increasing frequency.

Waveguide can also be made of graphite fiber-epoxy composite (GFEC), consisting of layers of graphite fiber bonded with interleaved layers of cured epoxy resin. The waveguide receives a silver-plating (Kudsia and O'Donovan, 1974). Its mass is much lower than that of aluminum waveguide. It was used on the Sirius Radio satellites at S-band to combine the outputs of 32 traveling-wave tube amplifiers (TWTAs) in phase (Briskman and Prevaux, 2001). Its disadvantages are that it is expensive and fragile and it cannot be dent-tuned (Section 4.4.4).

4.4.2 Rectangular Waveguide Performance

Table 4.7 gives the insertion-loss specification of one waveguide manufacturer for the various sizes of rectangular waveguide (Cobham and Continental Microwave Division, 2006). The loss is the same for both wall thicknesses. Reduced-height waveguide has about double the dB insertion loss of normal-height waveguide. The specification is for aluminum and oxygen-free high thermal-conductivity (OFHC) copper. Silver-plating aluminum is a way to reduce the insertion loss to close to that of copper. However, silver-plating is not possible in the smallest waveguide sizes. The specified insertion losses range from between one and two times the theoretical insertion loss. Waveguide insertion loss varies more over a frequency band than coax insertion loss does. The dB insertion loss of rectangular waveguide decreases nonlinearly with frequency over its recommended operating band, and the insertion phase is nonlinear (Ramo et al., 1984).

4.4.3 Waveguide Environmental

An aluminum or copper waveguide run or *tube* expands as the waveguide material does, with temperature. The frequency response shifts to lower frequencies. In situations where this is not acceptable, for example where multiple high-power amplifier (HPA) outputs must track in phase, graphite waveguide can be used. The graphite layers have a positive *coefficient of thermal expansion (CTE)*, that is, they expand as the temperature increases, while the epoxy layers have a negative CTE. Graphite waveguide's CTE is like that of the filter material Invar (Section 5.5.5.2) (Kudsia and O'Donovan, 1974), namely almost as small as one-twentieth (NIST, 2004) of the CTE of the aluminum alloy used by one manufacturer for thin-wall aluminum waveguide (Cobham and Continental Microwave Division, 2006).

TABLE 4.7   Example of Rectangular Waveguide Insertion-Loss Specification, Worst over Recommended Operating Band (Cobham and Continental Microwave Division, 2006)

EIA WG Designation	Rigid-aluminum Insertion Loss (dB/10 m)	Rigid-OHFC Copper[a] Insertion Loss (dB/10m)
WR430	0.4	—
WR340	0.4	—
WR284 R/H	0.8	—
WR284	0.4	0.4
WR229	0.4	0.4
WR187	0.8	0.4
WR159	0.8	0.8
WR137	1.2	0.8
WR112	1.6	1.2
WR102	1.6	1.2
WR96	3	2.0
WR90 R/H	4	—
WR90	2.0	1.6
WR75 R/H	5	—
WR75	2.4	2.0
WR67	3	2.4
WR62	3	2.4
WR51	5	4
WR42	8	6
WR34	12	6
WR28	16	8
WR22	—	12
WR19	—	12

[a] Oxygen-free high thermal-conductivity copper.

4.4.4   Flange and Waveguide Assemblies

Waveguide runs receive **flanges** on their ends, which correspond to connectors for coax. Flanges come in many sizes and shapes. Some payload units also have flanges.

In order to be able to connect waveguide runs or units to waveguide runs, other waveguide elements are needed such as the following:

- Bend, in either the *E*-plane or the *H*-plane.
- Transition, which is used to connect two different sizes of waveguide.
- Twist, to change the field orientation.

Dent-tuning is sometimes performed on a waveguide assembly to improve its voltage standing-wave ratio (VSWR) (Section 4.7.1). A dent is hammered into the appropriate place by a very experienced person. Graphite waveguide cannot be dent-tuned.

4.4.5   Propagation in Rectangular and Circular Waveguide

The two forms of waves or modes that propagate in waveguide are transverse electric (TE) and transverse magnetic (TM). In TE, the electric field is transverse to the direction of propagation, while in TM, the magnetic field is transverse to the propagation direction. There is a family of TE modes and a family of TM modes, both different for rectangular and circular waveguides. The modes have the same notation for both shapes, though, namely TE_{mn} and TM_{mn}. For each index pair (m,n), that is, for each mode, there is a **mode cutoff frequency**, dependent on the waveguide dimension a and the dielectric filling the waveguide, below which there is no propagation. The waves are **evanescent** there, meaning that the fields decay exponentially along the waveguide axis. The dielectric is air on the ground and vacuum in space. The dominant mode is the one with the lowest operating frequency.

Specifically, for rectangular waveguide of normal proportions (Figure 4.4), the TE_{mn} mode has m half-wave variations of the electric field across a and n half-wave variations across b. There is a similar family TM_{mn}. The dominant mode is TE_{10}. For circular waveguide, the TE_{mn} mode has m full-wave variations of the electric field circumferentially and n half-wave radially. There is a similar family of TM modes, TM_{mn}. The dominant mode is TE_{11}. The first few modes for rectangular waveguide of normal proportions are shown in Figure 4.5 and for circular waveguide in Figure 4.6.

The fact that waveguide has a cutoff frequency is used to ensure that there is no ring-around of the payload's transmit signal into the payload's receiver. The receive antenna's terminal is made, wholly or in part, of waveguide selected in size to cut off the transmit band, which is lower than the receive band (Schennum, 2011).

A *hybrid mode* is a combination of a TE mode and a TM mode. In circular waveguide, the dominant hybrid mode HE_{11} is the combination of TE_{11} and TM_{11}. A *degenerate mode* is one of a set of propagation modes that have the same phase velocity but different field configurations in any transverse cross-section.

The **group velocity** of the propagating wave is the speed at which power and the modulation progress down the waveguide (Chang, 1989). For each waveguide size, the group velocity varies from roughly $0.6\,c$ at the low end of the recommended operating frequency band to about $0.85\,c$ at the high end, where c is the speed of light in vacuum (Cobham and Continental Microwave Division, 2006). Thus, group velocity is dependent on frequency; waveguide causes dispersion. The *phase velocity* is the speed at which a constant phase of a wave travels down the waveguide. It is greater than c, and the product of phase velocity and group velocity is a constant independent of frequency. A good article explaining these two concepts with little embedded video examples is to be found in Wikipedia (2018).

See Section 9.4.2 for a way of approximately phase-compensating a length of waveguide with a length of coax.

For further information, see for example Chang (1989) or Ramo et al. (1984).

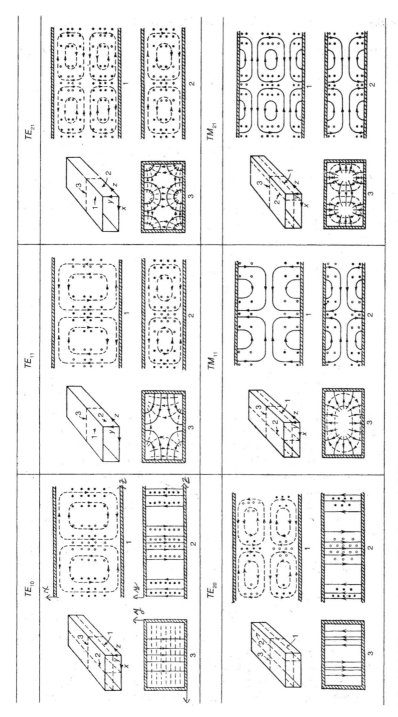

[a] Electric field lines are shown solid and magnetic field lines are dashed.

FIGURE 4.5 Propagation modes in rectangular waveguide. Copyright 1965, 1984, John Wiley & Sons. Reprinted, with permission, from Ramo et al. (1984).

Wave type	TM_{01}	TM_{02}	TM_{11}	TE_{01}	TE_{11}
Field distributions in cross-sectional plane, at plane of maximum transverse fields		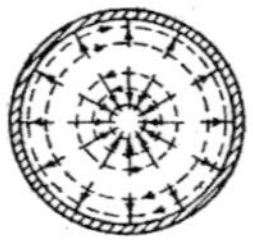	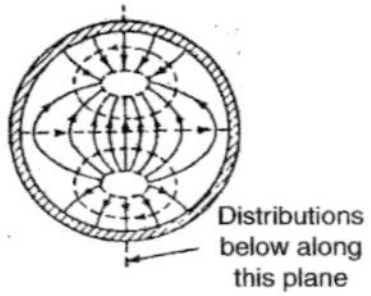	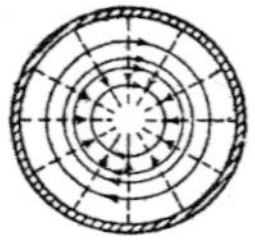	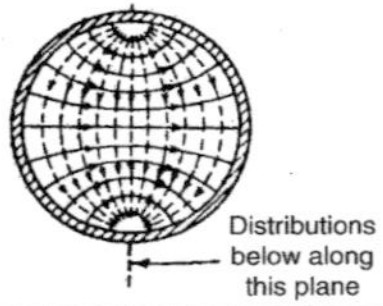
Field distributions along guide	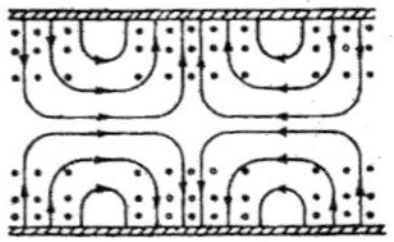	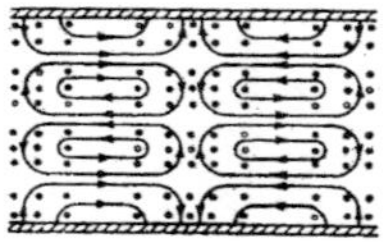	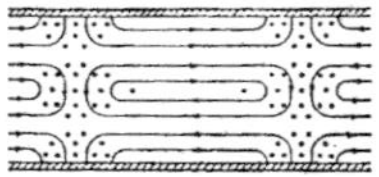	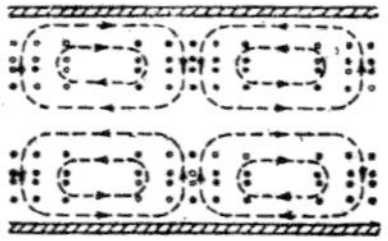	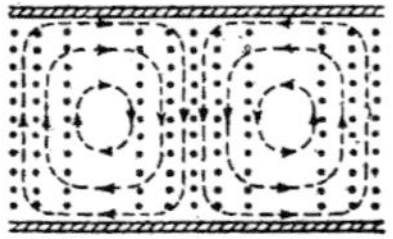
Field components present	E_z, E_r, H_ϕ	E_z, E_r, H_ϕ	$E_z, E_r, E_\phi, H_r, H_\phi$	H_z, H_r, E_ϕ	$H_z, H_r, H_\phi, E_r, E_\phi$
p_{nl} or p'_{nl}	2.405	5.52	3.83	3.83	1.84
$(k_c)_{nl}$	$\dfrac{2.405}{a}$	$\dfrac{5.52}{a}$	$\dfrac{3.83}{a}$	$\dfrac{3.83}{a}$	$\dfrac{1.84}{a}$
$(\lambda_c)_{nl}$	$2.61a$	$1.14a$	$1.64a$	$1.64a$	$3.41a$
$(f_c)_{nl}$	$\dfrac{0.383}{a\sqrt{\mu\varepsilon}}$	$\dfrac{0.877}{a\sqrt{\mu\varepsilon}}$	$\dfrac{0.609}{a\sqrt{\mu\varepsilon}}$	$\dfrac{0.609}{a\sqrt{\mu\varepsilon}}$	$\dfrac{0.293}{a\sqrt{\mu\varepsilon}}$
Attenuation due to imperfect conductors	$\dfrac{R_s}{a\eta}\dfrac{1}{\sqrt{1-(f_c/f)^2}}$	$\dfrac{R_s}{a\eta}\dfrac{1}{\sqrt{1-(f_c/f)^2}}$	$\dfrac{R_s}{a\eta}\dfrac{1}{\sqrt{1-(f_c/f)^2}}$	$\dfrac{R_s}{a\eta}\dfrac{(f_c/f)^2}{\sqrt{1-(f_c/f)^2}}$	$\dfrac{R_s}{a\eta}\dfrac{1}{\sqrt{1-(f_c/f)^2}}\left[\left(\dfrac{f_c}{f}\right)^2+0.420\right]$

[a] Electric field lines are shown solid and magnetic field lines are dashed.

FIGURE 4.6 Propagation modes in circular waveguide. Copyright 1965, 1984, John Wiley & Sons. Reprinted, with permission, from Ramo et al. (1984).

4.5 OTHER INTEGRATION ELEMENTS

The payload-integration items addressed here are used in close conjunction with filters and are sometimes part of filter assemblies.

The unexciting but crucial **termination** is a load designed to absorb the power sent to it and not reflect it back. Its impedance must be matched to whatever it is terminating. A termination must be designed to absorb the highest power that could be sent to it. There are low- and high-power terminations, of coax and waveguide. We will see that terminations are used with hybrids and switches to terminate the unused ports; the purpose is to improve the performance of the other ports.

A **circulator** is a device with three ports that sends a signal input on one port out on the next port in circular order, as shown in Figure 4.7. Circulators are used in most input multiplexers. The circulator must be able to accommodate the RF power that it could see. Circulators come in both coax and waveguide versions. The coax versions are either designed for an extremely wide range of frequencies, with VSWR that increases with frequency, or over a lesser range but with better VSWR on that range. A waveguide circulators' frequency range is restricted by its flange, which mates with a particular size of waveguide.

An **isolator** is equivalent, RF-circuit-wise, to a circulator with a termination on one port. However, an isolator and a circulator-with-termination have slightly different implementations that can matter in a high-power application: the termination that is part of an isolator is right with the rest of the isolator, while the termination on a circulator can be some distance away from the circulator, making it easier to site the termination on a heat pipe. The high-power terminations attached to the circulators at the output of TWTAs may become so hot that they have to be placed on heat pipes (Section 2.2.1.6). An isolator is used to improve the impedance match of a unit or to protect a unit from backward reflections into it. Figure 4.8a,b shows examples of an isolator used for impedance matching. Recall that a double line indicates waveguide and a single line coax. Figure 4.8c shows an isolator placed so as to protect a HPA from a backward-propagating signal that could result from a mis-switch past the HPA.

A hybrid coupler or simply **hybrid** is a device that combines two inputs, yielding two outputs, where each output is half the sum of both inputs in some particular phase relationship. A *power divider* divides one input into two equal ones. It can be

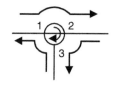

1, 2, 3 = Port numbers

➤ = Possible signal flow

FIGURE 4.7   Circulator.

(a)   (b)   (c)

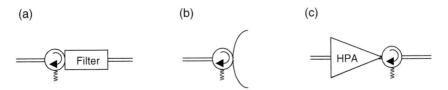

FIGURE 4.8   Isolator usage examples.

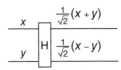

FIGURE 4.9   Lossless hybrid, showing input and output RF signals.

used in reverse so that instead two inputs are combined into one output, making it a *power combiner*. Even though their implementations may be different, a power divider is conceptually the same as a hybrid with one port terminated. Usually in this book, we use a hybrid symbol "H" to mean either a hybrid or a power divider. A hybrid is a 3-dB coupler. A hybrid's or power divider's insertion loss specification is usually given relative to the unavoidable 3-dB loss. The hybrid must be able to carry the right amount of power.

There are various kinds of hybrid. There is an overlap between the kinds implemented in waveguide and the kinds implemented in coax. The difference among hybrids is the difference between input and output phases. Figure 4.9 shows the phase relationships for one kind of hybrid. For another kind, see Section 11.12.13. By means of adding 90° phase-shifters, one can make all of them do the same thing. When only one signal, say x in the figure, is input to the hybrid, then half the power of the signal goes to each output port. When the two input signals x and y are identical and in phase, then the hybrid has only one non-zero output, equal to x with double power.

Directional couplers are integrated into the payload for testing the payload while bypassing the antennas. There is a directional coupler between each output port of a receive antenna and the low-noise amplifier (LNA) connected to it, and one between each output multiplexer (OMUX) output port and the transmit antenna input port connected to it. If there is no OMUX after the HPA, then the coupler is between the HPA and a transmit antenna input port. The *coupling factor* is high, meaning that, at the OMUX, the ratio of power into the coupler to the power out of the coupled port is large in dB. (A hybrid has a coupling factor of 3 dB.) Two representations of a directional coupler are shown in Figure 4.10. A directional coupler can be used to inject a signal into the repeater through the coupled port. There are various forms of a waveguide directional coupler, and some have a coaxial connector on the coupled port instead of a waveguide flange (Cobham and Continental Microwave Division, 2006), the better for connecting test equipment.

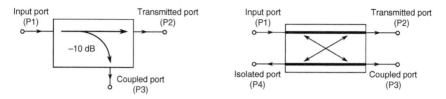

FIGURE 4.10 Two representations of directional coupler. By Courtesy Spinningspark at Wikipedia, CC BY-SA 3.0. From en.wikipedia.org/w/index.php?curid=31959064.

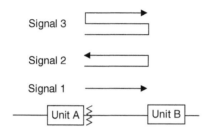

FIGURE 4.11 Fixed attenuator improving impedance matches.

Fixed-value attenuators, also called **pad attenuators**, are coaxial. Each is usable from DC to some GHz frequency, for example for one manufacturer to either 18 or 32 GHz (Aeroflex Weinschel, 2010). Available attenuation values are from 0 dB in 0.5-dB steps. These attenuators can carry 2 W. Pad attenuators are used in payload integration to align the operating signal-level ranges of successive units. They also have the purpose of improving impedance match. Such a use can be seen in Figure 4.11. There are units A and B separated by an attenuator, say a 3-dB attenuator. The intended signal, signal 1, has a level at input to unit B that is 3 dB lower than it would be without the attenuator. If there is a reflected signal from unit B back to unit A, the signal arriving at unit A's output port is 6 dB lower than it would be without the attenuator, so it is less harmful to unit A's performance. Furthermore, if there is a doubly reflected signal, signal 3, when it arrives at the input to unit B it is 9 dB lower than it would be without the attenuator, namely 6 dB lower than the intended signal. The attenuator could as well be at input to unit B instead of at output of unit A.

Phase-shifters are coaxial and are adjustable during payload integration. One example of a phase-shifter is designed for DC to 18 GHz and has a tunable range of up to 10° per GHz, namely up to 180° (Radiall, 2015). The phase-shifter varies the length of a transmission line (Radiall, 2017a). Phase-shifters are needed to put TWTA outputs that will be combined, in phase with each other at the combiner. The method for this is as follows. The TWTA manufacturer measures the phase shifts of the TWTAs. The payload integrator measures the phase shift in the waveguide from a TWTA to the post-TWTA combiner. The calculated target phase shift for the coax run from the (linearizer-)channel amplifier to TWTA input, including the phase

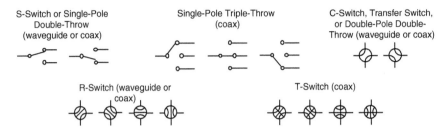

FIGURE 4.12   Switch types with all possible positions.

shifter, is then computed and the phase shifter is accordingly adjusted by means of a vector network analyzer.

Switches play a crucial role in the payload, since they allow changes in signal routing. They also allow selection of units out of a bank of units. Various kinds of switch are used in payload integration, as shown in Figure 4.12. Some kinds are only made in coax and some in both waveguide and coax. For the *C-switch*, the *R-switch*, and the *T-switch*, in some of their positions two signals can flow through the switch. If two signals flow, the switch must be able to accommodate the RF power in the two signals. The C-switch is also known as a *transfer switch* and a *baseball switch*. The R-switch has two positions in which two signals can flow through it, while all four of the T-switch positions have this feature. Two of the T-switch positions are equivalent. The T-switch paths are not all physically in one plane. Almost all switches are *latching*, meaning that the switch maintains a chosen RF contact path whether a voltage is maintained or not after switching is accomplished.

Ferrite-based waveguide switches are an alternative to the electromechanical waveguide switches in the payload receiver for Ka-band and above; at lower frequencies, their size and insertion loss are greater than those of electromechanical switches (Kroening, 2016). The ferrite switches have no moving parts; they are entirely solid-state for higher reliability. They are especially good for high-rate switching. Such a switch is directional (that is, port 1 is only for input) and its schematic is shown in Figure 4.13a. It consists of a switching circulator at port 1 and two isolators at the output ports 2 and 3 (not labeled) (Honeywell Aerospace, 2018). The Ka-band ferrite S-switch was reduced in size by 2/3. Figure 4.13b shows how the same switching circulators and isolators can be used to make an R-switch. Ten of these R-switches and eight LNAs have been integrated for LNA redundancy switching (section below) at half the size and mass of the mechanical R-switch solution (Kroening, 2016).

For coaxial switches, there are low- and high-power switches, where for one vendor low power means 5–10 W RF and high power means 33–102 W RF (Radiall, 2017b). Like coax itself, a coaxial switch works from DC to a GHz frequency. Its power-carrying ability decreases with increasing frequency. Insertion loss and VSWR increase with increasing frequency. Waveguide switches can in general accommodate higher RF power than coax switches can, because of their

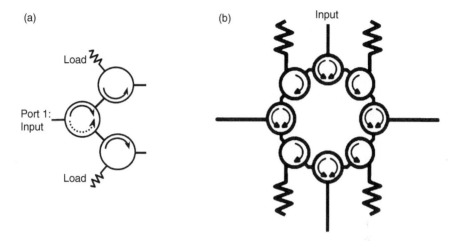

FIGURE 4.13   Ferrite switching circulators and isolators used to make (a) S-switch and (b) R-switch. ©2016 IEEE. Reprinted, with permission, from Kroening (2016).

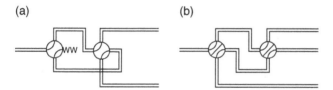

FIGURE 4.14   Redundant switches: (a) fully redundant, two-output C-switch assembly; (b) partially redundant, three-output R-switch assembly.

lower insertion loss. However, there is a coax T-switch that can handle 260 W RF at 4 GHz (Dow-Key Microwave, 2011). This vendor also makes coax and waveguide **switch modules**, integrating several switches and perhaps power dividers into a block in order to provide redundancy switching at the payload input.

For especially crucial functions, switches themselves can be put in a redundant configuration. Figure 4.14 shows two examples: in Figure 4.14a, a fully redundant one-input, two-output C-switch assembly and in Figure 4.14b a partially redundant one-input, three-output R-switch assembly (Briskman and Prevaux, 2001). See the next section for how the latter is used.

4.6 REDUNDANCY CONFIGURATIONS

A bank of active units almost always includes spare, inactive unit(s) which can replace failed active unit(s). The replacement is enabled by switches and/or hybrids. The units initially active on orbit are the **primary units** and the others are the **spare or redundant units**.

All payload units have sensors on them and send the data to the *onboard computer* (also called the *spacecraft computer*), which sends the data down to the *spacecraft operations center* (*SOC*). This data is *telemetry*. For example, LNAs, frequency converters, preamplifiers, and SSPAs (solid-state amplifiers) continuously report data on component-amplifier biases; the preamplifier's AGC (automatic gain control) reports power-level readings; the TWTA's EPC (electronic power conditioner) reports currents and voltages. The SOC monitors the telemetry data, and when an active unit is behaving badly and needs to be replaced, the SOC sends *commands* to the onboard computer. The onboard computer passes the commands on to the designated payload units and to any switches that are needed to switch out the bad unit and switch in a redundant one. Another type of command would be to turn on a cold (i.e., not powered on) redundant unit.

A bank of primary and redundant units contains either identical units, in the case of low-power units, or identical units differently tuned, in the case of TWTA subsystems, where a TWTA subsystem is the TWTA plus its preamplifier.

Redundancy is described differently in the United States and Europe. For example, suppose that there are six units of which at any time four are to be active. In the United States, this is "6-for-4 redundancy" or simply "6 for 4" and written "6:4," while in Europe, it may be written "4:2" or "4/6." The US terminology is used in this book.

Figure 4.15 shows that there is more than one way to access a spare unit when the primary unit fails. The most common way is with switches, in which case the switch collection and the unit bank together form a **redundancy ring**. However, hybrid couplers can also be used, or a mixture of hybrids and switches. The switches or hybrids need not be external to the units: in some cases of low-power units, a redundant unit and the switch or hybrid is a package with the primary unit.

We will see below that the RF line-and-switch structure in front of the unit bank is always the reverse of the RF line-and-switch structure at the output of the unit bank. Additionally, the collection of switch positions at any time in front of the unit bank and the collection at the output have this property. Typically, the switches are *ganged*, that is when a redundancy-ring switch in front of the unit bank is commanded to turn, the corresponding switch on output automatically turns in the reverse direction.

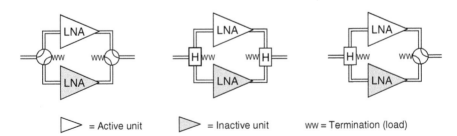

FIGURE 4.15 Redundancy via switches, hybrids, and mixture, respectively.

(a)

(b)

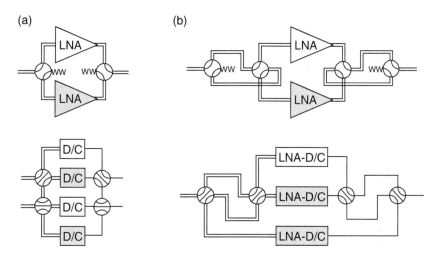

FIGURE 4.16 Low-power unit redundancy rings.

For a broadcast satellite, redundancy rings for the low-power units, that is the LNAs and the frequency converters, have a different structure from the TWTA rings (SSPA redundancy is discussed in Section 11.12). The reason is that there are only a few uplinks, so only a few LNAs. The low-power units are typically identical, or there may be a small number of banks in each of which the units are identical. The TWTAs, however, while perhaps identical in design, are typically optimized for different, narrow frequency bands. In addition, the low-power unit banks are typically small, while the TWTA banks can contain more than a hundred TWTAs.

Figure 4.16 gives four examples of redundancy rings for low-power units. The two examples in Figure 4.16a show how C- and R-switches can be used. The two in Figure 4.16b use the redundant switches introduced earlier.

For a TWTA ring, there are additional concepts and terms. Redundancy can be complete or partial. Suppose the redundancy is $m : n$. In *complete redundancy*, the failure of any $m-n$ units can be accommodated by other units in the bank. In *partial redundancy*, there is at least one failure of $m-n$ units that cannot be accommodated. The percentages of double failures, triple failures, etc., up to and including $m-n$ failures that can be accommodated, are all of interest in the calculation of the probability that the payload will meet its reliability specification.

The *first-redundant* TWTA for a primary TWTA is the one that would be chosen in case the primary fails, assuming that no other has failed. The second-redundant is the unit that would be chosen if both primary and first-redundant fail, assuming that no other has failed. The first-redundant TWTA for a band may be optimized for the same band as the primary. The second redundant is typically optimized for a larger band that includes the band of the primary, so its performance over the primary's band is not quite as good as with the primary and first-redundant TWTAs.

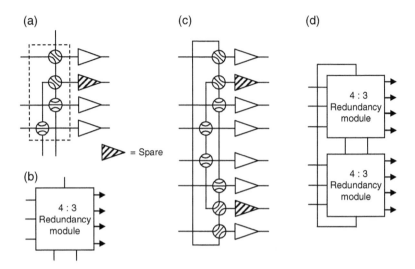

FIGURE 4.17 TWTA redundancy-ring modules, with non-interrupting first redundant. After Liang and Murdock (2008).

A redundancy ring for a TWTA bank of more than a few TWTAs has a regular or nearly regular structure, namely, it is built up of blocks which are identical or similar. In each block are the primary and first-redundant units for each input line to the block, while the second-redundant units may be in a different block. The TWTAs in a block are connected, and the blocks are interconnected. The larger the redundancy ring is, the higher the reliability (Kosinski and Dodson, 2018). How to construct rings with redundancy as close to complete as possible, with as few TWTAs as possible, with as few switches in paths (primary or redundant) as possible, with minimum interruption to existing signal paths when redundant units are switched in, and with provable redundancy has long been a subject of research.

An advance was made in 2008 (Liang and Murdock, 2008). Figure 4.17 shows modular redundancy rings with the property that the first redundant can be switched in for any one TWTA and there is no interruption to the other paths. Figure 4.17a shows a module with 4:3 redundancy. All modules in this scheme are m:m−1. This module has three connections (vertical) to connect to other modules with. Figure 4.17b is an abstract representation of the module. Figure 4.17c shows how two such modules can be connected to form a small ring, shown abstractly in Figure 4.17d. Every input path can go to either module.

Figure 4.18 shows rings made of four such modules, each module with 6:5 redundancy and each ring 24:20. Figure 4.18a,b shows two different fixed ways of interconnecting the modules. Each way allows partial 24:20 redundancy, covering different sets of failures. Figure 4.18c shows a switched interconnection method

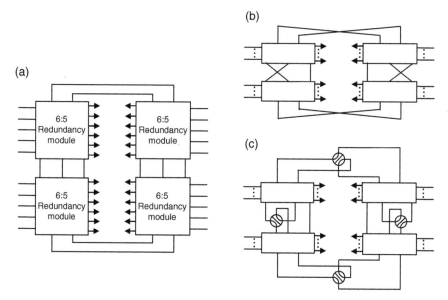

FIGURE 4.18 Integrated TWTA redundancy ring with non-interruptible modules, three different ways of connecting modules. After Liang and Murdock (2008).

that is superior. The scheme is thought to provide complete redundancy when the number of modules is four or less. For any $m:n$, there is a way to constitute a ring of $m-n$ modules that each includes one spare (Liang and Murdock, 2008).

4.7   IMPEDANCE MISMATCH AND SCATTERING PARAMETERS

4.7.1   Impedance Mismatch

Impedance mismatch or simply **mismatch** is a topic that must be mastered for proper accounting of the losses accrued during payload integration, which are not assignable to the units and integration elements themselves but to the fact that they are connected with each other. When two elements are connected, power is transferred from the first element (the **source**) to the other (the **load**) without loss if and only if the load impedance equals the complex conjugate of the source impedance. However, the match is never perfect, so some power reflects backward. RF elements are designed with the intention of making a good match with adjacent elements. Two specifications on all multi-port elements are **input return loss**, to limit reflected power, and **output return loss**, to limit reflected and doubly reflected power.

Let us see what happens at a mismatch location. The upper half of Figure 4.19 depicts the situation, at one instant in time, of a *voltage traveling wave* (left to right) hitting the mismatched connection on the right and a weaker wave reflected

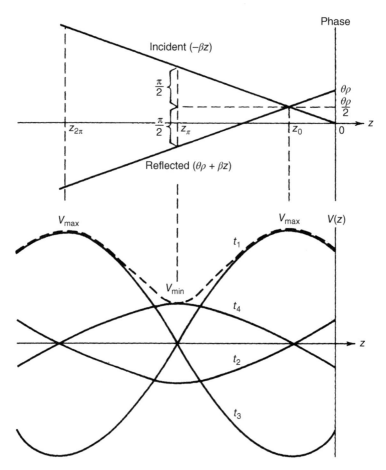

FIGURE 4.19   Impedance mismatch causes partial reflection; time fixed. Copyright 1965, 1984, John Wiley & Sons. Reprinted, with permission, from Ramo et al. (1984).

backward. Since time is fixed, the phase of the traveling wave decreases left to right in the figure. Similarly, the phase of the reflected wave decreases right to left. There is a z location where the phases of the two voltage waves are the same, marked z_0, and a location where the phases are opposite each other, marked z_π. Such locations occur every $2z_\pi$. The reflected wave is weaker than the traveling wave (the majority of the traveling wave gets past the mismatch, but this is not shown). The solid lines in the lower half of Figure 4.19 show the signed voltage, along the line, of the combined wave at various times. The dashed line is the voltage amplitude along the line (Ramo et al., 1984).

At a different time, the phases are again equal at the same locations along the line, and similarly for phases opposite each other. This is illustrated in Figure 4.20.

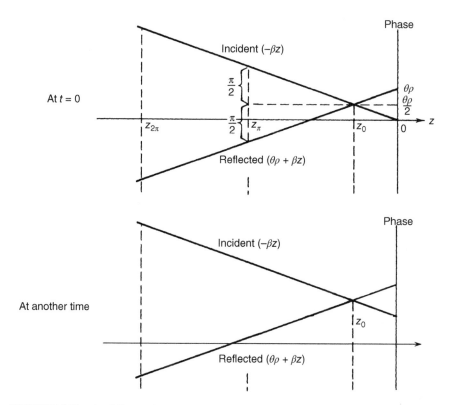

FIGURE 4.20   At different times, forward and backward waves are in or exactly out of phase at the same places. Adapted from Ramo et al. (1984).

The maximum voltage at negative z locations is

$$V_{max} = |V^+| + |V^-|$$

where V^+ is the complex voltage of the traveling wave and V^- is the voltage of the reflected wave, in complex phasor notation (Section 12.2.1). The minimum voltage at negative z locations is

$$V_{min} = |V^+| - |V^-|$$

The **VSWR**, or simply *standing-wave ratio*, is the ratio of the maximum to the minimum voltage amplitudes:

$$S = V_{max}/V_{min}$$

It can be shown that the combined wave to the left of the mismatch location is the sum of a traveling wave and a *standing wave*. A standing wave has fixed locations along the line where its signed voltage equals zero (Ramo et al., 1984).

Why do these reflected signals matter? One of two things could happen at a mismatch. A backward reflected signal at a mismatch could hit a sensitive element behind it and damage it. This is prevented by the use of terminations, high-power if necessary, on isolators, circulators, and switches. Or the backward reflected wave could meet another mismatch, where part of the signal is then reflected forward. This would represent a weaker, out-of-phase version of the intended forward signal. This occurrence is also prevented by terminations as well as mitigated by pad attenuators. Passing through the pad attenuator two times more than the intended forward signal did weakens the multipath signal. Thus, good design prevents or at least mitigates the effect of mismatches.

Only if something is broken will the effect of the multipath be seen, in the form of $|S_{21}|$ (see below) having a periodic component. If the broken thing is a piece of waveguide, the length of the piece can be estimated from the frequency period of the ripple, which usually helps locate the piece. (This is not expected to happen on orbit, just in the factory.) If the frequency ripple period is f_0, then the one-way travel time of the reflected signal from one end of the waveguide to the other is $1/f_0$. One needs only to apply the group velocity of the signal in the waveguide to obtain the waveguide length.

4.7.2 Scattering Parameters

Each two-port payload element can be characterized in terms that will allow an estimate to be made of the size of the undesired signals that could be created at mismatches at its ports. The characterization is by **scattering parameters or S-parameters**. Suppose that the voltage traveling wave with complex phasor V_1^+ is incident at port 1 of the device under test (DUT). A voltage V_2^- is transmitted from port 2, and the voltage V_1^- is reflected out of port 1. The output line is terminated in such a manner that there is no incident voltage on port 2. This situation is depicted in Figure 4.21a. Similarly, suppose that the incident traveling wave is incident on port 2 and that port 1 is terminated so that there is no incident voltage on it. This situation is depicted in Figure 4.21b. In both parts of the figure, the two-port element is the DUT.

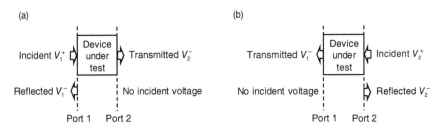

FIGURE 4.21 Voltages in definition of S-parameters: (a) for S_{21} and S_{11}, (b) for S_{12} and S_{22}.

Two of the complex-valued S-parameters can be defined now:

$$S_{21} = \frac{V_2^-}{V_1^+} = \text{Forward transmission coefficient}$$

$$S_{11} = \frac{V_1^-}{V_1^+} = \text{Input-port reflection coefficient}$$

The other two complex-valued S-parameters are similarly defined:

$$S_{12} = \frac{V_1^-}{V_2^+} = \text{Reverse transmission coefficient}$$

$$S_{22} = \frac{V_2^-}{V_2^+} = \text{Output-port reflection coefficient}$$

The short-hand definition of S_{mn} is that it is the output of port m from input on port n, relative to the input. The impedance setups for measuring the S-parameters are shown in Figure 4.22. The DUT has impedance Z_1 at input and Z_2 at output. S_{21} and S_{11} are obtained using a source at a reference impedance Z_0 and an ideal output termination, that is, with impedance Z_2^* equal to the complex conjugate of Z_2. S_{12} and S_{22} are obtained in a complementary fashion. For coaxial input, the reference impedance is typically $50\,\Omega$.

The complex value of the S-parameter S_{mn} is usually stated as a gain $|S_{mn}|$ and a phase. The gain $|S_{mn}|$, not in dB, may be any positive number. In particular, $|S_{21}|$ is the **insertion gain**. For passive devices $|S_{21}| < 1$, actually representing a loss; the term then is **insertion loss**, equal to $1/|S_{21}|$, which is greater than 1. Normally, $|S_{11}|$ and $|S_{22}|$ are less than 1. Their inverses are respectively the input and output return losses. For all these gain or loss terms, their normal representation is not as a simple ratio but in dB, for example $-20\log_{10}|S_{21}|$ for the insertion loss of a passive device.

We can now express impedance mismatch in terms of the S-parameters. The complex-valued *voltage reflection coefficient* Γ_1 at port 1 equals S_{11}. There is a standing wave on the input RF line to the DUT, consisting of the complex sum of the incident and reflected waves. The VSWR at port 1 is the ratio of the maximum

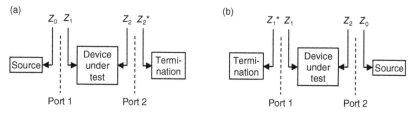

FIGURE 4.22   Conceptual impedance setups for obtaining S-parameters: (a) for S_{21} and S_{11}, (b) for S_{12} and S_{22}.

TABLE 4.8   VSWR with Corresponding Return Loss and Mismatch Loss

VSWR	Return Loss (dB)	Mismatch Loss (dB)	VSWR	Return Loss (dB)	Mismatch Loss (dB)
1.00	∞	0	1.24	19.40	0.050
1.02	40.09	0.0004	1.26	18.78	0.058
1.04	34.15	0.0017	1.28	18.22	0.066
1.06	30.71	0.0037	1.30	17.69	0.075
1.08	28.30	0.0064	1.35	16.54	0.974
1.10	26.44	0.0099	1.40	15.56	0.122
1.12	24.94	0.014	1.45	14.72	0.149
1.14	23.69	0.019	1.50	13.98	0.177
1.16	22.61	0.024	1.60	12.74	0.238
1.18	21.66	0.030	1.70	11.73	0.302
1.20	20.83	0.036	1.80	10.88	0.370
1.22	20.08	0.043	1.90	10.16	0.430

voltage amplitude $\left(\left|V_1^+\right|+\left|V_1^-\right|\right)$ on the RF line to the minimum voltage amplitude $\left(\left|V_1^+\right|-\left|V_1^-\right|\right)$. It is not given in dB; sometimes, to reinforce that point, it is stated in the format "x:1" where x is the VSWR. Thus,

$$\text{VSWR at port } 1 = \frac{1+\left|\Gamma_1\right|}{1-\left|\Gamma_1\right|}, \qquad \left|\Gamma_1\right| = \frac{\left(\text{VSWR at port } 1\right)-1}{\left(\text{VSWR at port } 1\right)+1}$$

Similarly, the voltage reflection coefficient Γ_2 at port 2 equals S_{22}. **Mismatch loss** is the ratio of power transmitted to power incident in the forward direction, namely $1/(1-\left|S_{11}\right|^2)$ or equivalently $1/(1-\left|\Gamma_1\right|^2)$. It is typically stated in dB, equaling $-10\log_{10}(1-\left|\Gamma_1\right|^2)$. Table 4.8 gives a list of VSWR values each with corresponding return loss and mismatch loss.

REFERENCES

Aeroflex Weinschel (2010). Fixed coaxial attenuators. Catalog. Jan. On www.aeroflex.com/AMS/Weinschel/PDFiles/fixedatten.pdf. Accessed Sep. 30, 2011.

Briskman RD and Prevaux RJ (2001). S-DARS broadcast from inclined, elliptical orbits. *International Astronautical Congress*; Oct 1–5; 503–518.

Carlisle Interconnect Technologies (2017). Microwave & RF cable, semi-rigid & flexible microwave cable, rev 092217. Product specifications. On www.carlisleit.com/resources/product-literature. Accessed Sep. 17, 2018.

Chang K, editor (1989). *Handbook of Microwave and Optical Components*, Vol. 1, *Microwave Passive and Antenna Components*. New York: John Wiley & Sons, Inc.

Cobham PLC and Continental Microwave Division (2006). *Waveguide Component Specifications and Design Handbook*, 7th ed. Nov. On www.cobham.com/about-cobham/aerospace-and-security/about-us/antenna-systems/exeter/products-and-services/passive-rf-components.aspx. Accessed Sep. 15, 2011.

Dow-Key Microwave (2011). Space products. Product brochure On www.dowkey.com/_upload/0/SpaceSection_spread.pdf. Accessed Sep. 28, 2011.

Honeywell Aerospace (2018). Ferrite based RF switches. Product information. On aerospace.honeywell.com/en/products/space/satelite-payload-subsystems. Accessed Sep. 19, 2018.

Kosinski B and Dodson K (2018). Key attributes to achieving > 99.99 satellite availability. *IEEE International Reliability Physics Symposium*; March 11–15.

Kroening AD (2016). Advances in ferrite redundancy switching for Ka-band receiver applications. *IEEE Transactions on Microwave Theory and Techniques*; 64 (6).

Kudsia CM and O'Donovan MV (1974). A light weight graphite fiber epoxy composite (GFEC) waveguide multiplexer for satellite application. *4th European Microwave Conference*; Sep. 10–13; 585–589.

Liang S and Murdock G (2008). Integrated redundancy ring based on modular approach. *AIAA International Communications Satellite Systems Conference;* June 10–12; 1–7.

Mele S, Cobham PLC, Exeter, New Hampshire (2011). Private communication. Sep. 19.

Micro-Coax, Inc. (2001). Space capabilities. Brochure. Feb. On www.micro-coax.com/pages/products/pdfs/Spacecapabilities.pdf. Accessed Sep. 23, 2011.

National Institute of Standards and Technology, US, Mechanical Metrology Div (2004). Engineering metrology toolbox, temperature tutorial. Nov. 9. On emtoolbox.nist.gov/Temperature/Slide14.asp. Accessed Sept 20, 2011.

Radiall SA (2015). RF coaxial phase shifter, SMA—DC to 18 GHz. detail specification. Dec. 4. On www.radiall.com/document-library/space-qualified-components/space-coaxial-phase-shifters.html. Accessed Sep. 23, 2018.

Radiall SA (2017a). Space brochure. Product summary. On www.radiall.com/document-library/space-qualified-components.html. Accessed Sep. 23, 2018.

Radiall SA (2017b). Space qualified switches. Product specifications.On www.radiall.com/document-library/space-qualified-components/catalogs.html. Accessed Sep. 19, 2018.

Ramo S, Whinnery JR, and Van Duzer T (1984). *Fields and Waves in Communication Electronics*, 2nd ed. New York: John Wiley & Sons, Inc.

Schennum GH, Space Systems/Loral (2011). Private communication. Oct. 24.

Teledyne Storm Products Co (2010). Microwave: high performance interconnect products. Apr. On www.teledynestorm.com/resource.asp. Accessed Sep. 17, 2018.

Wikipedia (2011). PTFE. Sep. 15. On en.wikipedia.org. Accessed Sep. 25, 2011.

Wikipedia (2018). Group velocity. On en.wikipedia.org. Accessed Sep. 24, 2018.

W L Gore and Associates (2003). Gore spaceflight microwave cable assemblies. Feb. On www.gore.com/MungoBlobs/883/162/gore_spaceflight_microwave_cable_assemblies_catalog.pdf. Accessed Sep. 23, 2018.

CHAPTER 5

MICROWAVE FILTER

5.1 INTRODUCTION

This chapter discusses the microwave filter units of a non-processing transponder. Besides a diplexer filter, which is part of the antenna and thus discussed in Chapter 3 instead, the first filter that the signal sees is the preselect filter. Later, filter units along the signal path are the multiplexers (MUXes), which are channelizers. We assume that the input multiplexer (IMUX) separates out frequency channels and that the output multiplexer (OMUX) combines channels (sometimes they have the opposite functions). The filters in the MUXes that correspond to individual channels are called **channel filters**. This chapter does not address filters that are components in non-filter units.

Before we can discuss the filter units, though, we must study analog filters in general and the specifics of microwave filters.

Two types of payload filters are not discussed in this chapter but in Chapter 10 on processing payload units. One type is the surface acoustic wave (SAW) filter, which is passive and operates up to about 2 GHz. It is used in conjunction with a digital processor. The other type is the digital filter.

Satellite Communications Payload and System, Second Edition. Teresa M. Braun and Walter R. Braun.
© 2021 John Wiley & Sons, Inc. Published 2021 by John Wiley & Sons, Inc.

5.2   BASICS OF ANALOG FILTERS

5.2.1   Filter Bandforms

A filter allows certain frequency bands to pass through it and blocks out others. The five filter bandforms are the following:

- **Low-pass filter (LPF)**, which passes a contiguous band of frequencies including direct current (DC), that is, zero frequency.
- **High-pass filter**, which passes all frequencies above some particular frequency. These are uncommon except that waveguide itself is a high-pass filter.
- **Bandpass filter (BPF)**, which passes a contiguous band of frequencies not including DC and rejects the others. This is the most common type of filter in the payload. Most channel filters are BPFs.
- *Band-stop filter*, which passes all frequencies but a contiguous band of frequencies that do not contain DC. An especially narrow band-stop filter is a *notch filter*.
- **All-pass filter**, which passes all frequencies without attenuation but has a phase shift that varies with frequency.

A filter lightly attenuates the part of the input signal in its passband(s), due to small resistive losses.

A filter **rejects** (sends backward) the part of the input signal that is in the filter's *stop-band(s)*. The filter's input return loss, a positive number in dB, is very low there.

5.2.2   Filter Representation

A microwave filter is an example of an analog filter. An analog filter has a continuous representation in both frequency and time.

Payload engineers usually think about a filter in terms of its **(Fourier) transfer function** or **frequency response** $H(f)$ in the **frequency domain**, where f is the frequency in Hz. The transfer function is in general complex-valued. It consists of two parts, the **gain response** $A(f)$ and the **phase response** $\varphi(f)$:

$$H(f) = A(f)e^{j\varphi(f)} \text{ where } A(f) \text{ is positive and } \varphi(f) \text{ is in rad}$$

Often the gain response is given as $G(f)$ in dB, where $G(f) = 20\log_{10}A(f)$, and $\varphi(f)$ is converted to degrees.

The transfer function is the Fourier transform of the filter's **impulse response**, which is the representation of the filter in the **time domain** (see for example McGillem and Cooper, 1991).

Unlike payload engineers, filter engineers think of filters in more than one way. One way is of course the transfer function. Another is to replace the phase response with the **group delay** $\tau_g(f)$:

$$\tau_g(f) = -\frac{1}{2\pi}\frac{d\varphi(f)}{df}\ \sec$$

Note that a linearly increasing phase corresponds to a constant group delay over frequency. Another representation, the **Laplace transfer function** $H(s)$, is used by filter engineers during the design of a filter. It is a function of the complex variable $s = \sigma + j\omega$ where σ and is real and $\omega = 2\pi f$. It is the Laplace transform of the impulse response (McGillem and Cooper, 1991). It is important to note that the (Fourier) transfer function equals the Laplace transfer function on the $j\omega$-axis.

These are all linear representations of a filter, that is, they all assume the filter response is the same no matter what the power of the input signal. Such filtering is a linear operation on the input signal, which induces **linear distortion**. The linear representation is in general valid for all passive filters and, in their linear range, for active filters. For active filters, the operational range corresponds to the linear range and perhaps a small part of the nonlinear range.

Section 12.3 provides more information on filter representation.

5.2.3   Bandpass-Filter Bandwidth

There are various definitions of the bandwidth of a BPF, based on the transfer function, applicable to different circumstances. The most common definitions for satellite communications are the **3-dB bandwidth** and the **(equivalent) noise bandwidth**, illustrated for an RF filter in Figure 5.1. We assume that the filter has transfer function $H(f)$ and is centered about the frequency f_c. (The filter may in fact

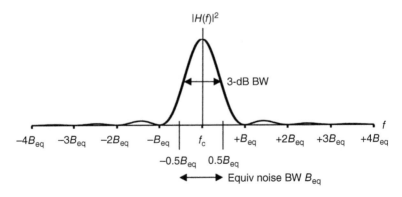

FIGURE 5.1   Definitions of two types of filter bandwidth for RF bandpass filter.

be asymmetric about f_c and thus not have a true center, in which case f_c is just a convenient frequency.) The 3-dB bandwidth is the maximum bandwidth between two frequencies where the filter's gain response falls no more than 3 dB down from its maximum value inside the band (Couch II, 1990). This definition is useful for describing roughly the bandwidth over which the signal is passed, outside of which the signal is mostly not passed.. The noise bandwidth of an RF filter is defined as follows:

$$\text{noise BW} = \frac{1}{\left|H(f_c)\right|^2} \int_0^\infty \left|H(f)\right|^2 \, df$$

This definition is useful in the formula for noise power passed by a filter when the noise power spectral density (psd) is flat and wider than the filter. Other perfectly good bandwidth definitions, for other purposes, are given in Couch II (1990).

5.2.4   Filter Response Classes

Two filter-response classes are by far the most often used in payload units: (1) the *elliptic* or Cauer and (2) the **pseudo- or quasi-elliptic** or *generalized Chebyshev* (Levy et al., 2002). The pseudoelliptic filter is something between the Chebyshev and the elliptic filter. It is instructive to look at both in relation to the classical Chebyshev filter.

The gain response of a Chebyshev LPF is shown in Figure 5.2a. It has an *equiripple* gain response in the passband, which means that the ripple has a constant size. At band edge, the gain response falls monotonically to zero.

The elliptic filter also has an equiripple gain response in-band, but it also has an equiripple gain response out of band, as seen in Figure 5.2b. The two ripples can be of different heights. At band edge, the gain response falls off more sharply than that of a Chebyshev does.

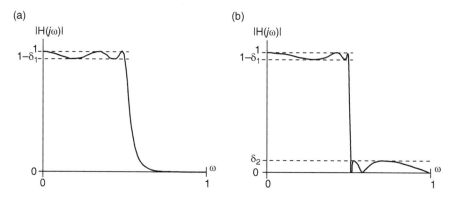

FIGURE 5.2   Gain response, not in dB, of 5th-order LPFs: (a) Chebyshev, (b) elliptic. (After Wikipedia (2011)).

The two gain responses depicted in Figure 5.2 are fifth-order, which can be told by imagining the negative-ω side of the response, also, where each ripple peak corresponds to one order. The Laplace transfer functions of Chebyshev and elliptic filters are defined in terms of two different families of polynomials (Sorrentino and Bianchi, 2010). Both can have any positive-integer order. For more information on filter responses, see for example Chang (1989).

The pseudoelliptic filter starts off as an elliptic design but the filter designer modifies it slightly to obtain the desired response. Its polynomial belongs to no particular class (Thompson and Levinson, 1988). A pseudoelliptic filter does not necessarily have the equiripple property (Kallianteris et al., 1977).

5.2.5   Network Synthesis

A filter is a network that processes signals in a frequency-dependent manner. It is most often two-port, with input and output ports. The usual way to design a filter is by *network synthesis*, that is, define a prototype LPF that meets the gain- and phase-response requirements, approximate it by a realizable network, and then transform the LPF to the desired bandform (Maxim Integrated Products, 2008). This applies to bandwidths of typically less than 2% (Cameron et al., 2018).

The method of transformation is given in Maxim Integrated Products (2008). It must be noted that in this source, brackets, and parentheses are missing in the section on notch and all-pass filters. Where these should be can be seen by reference to the preceding section in the source.

The prototype LPF has a Laplace transfer function equal to a ratio of polynomials with real coefficients. The coefficients are real because the LPF has a lumped-element (discrete resistors, inductances, and capacitors) design (Levy et al., 2002). The Laplace transfer function $H(s)$ of a realizable LPF has the following form:

$$H(s) = \frac{N(s)}{D(s)}$$

where $N(s)$ and $D(s)$ are polynomials with real coefficients.

Suppose that the degree of the numerator polynomial $N(s)$ is m and the degree of the denominator polynomial $D(s)$ is n. A polynomial has a number of roots equal to its degree, where some may be repeated. The transfer function $H(s)$ can be written as follows:

$$H(s) = \frac{N(s)}{D(s)} = K \frac{(s - z_1) \cdots (s - z_m)}{(s - p_1) \cdots (s - p_n)}$$

where K is a real constant.

In general, not all the roots of the polynomial will be real. Complex roots come in complex-conjugate pairs because the polynomial coefficients are real. So the transfer function can be written as the product of first- and second-degree polynomials with real coefficients.

The roots z_i of $N(s)$ are the **zeros** of $H(s)$, and the roots p_k of $D(s)$ are the **poles** of $H(s)$. The **order** of a filter is the larger of n and m. A LPF that rolls off with increasing frequency, which is what we want, has a higher-degree polynomial in the denominator than in the numerator. Since then $n > m$, the order of the filter is the degree n of the denominator polynomial. As $|s|$ goes to infinity, $|H(s)|$ goes to zero, proportionally to $|K/s^{n-m}|$, so infinity is said to be a zero of order $n-m$. Thus, $H(s)$ is said to have n poles and n zeros. Sometimes in a filter description, the infinite zeros are not mentioned. It is useful to know that in the filter literature, a transmission zero is the same as an attenuation pole.

The poles and finite zeros of $H(s)$ can be plotted in s-space, making the **pole-zero diagram or plot**. Figure 5.3a shows such a diagram for a fourth-order low-pass Chebyshev filter. This filter has four poles (with infinity as a quadruple zero). The two-sided baseband gain response would have four peaks in the passband ripple because of the four poles.

Figure 5.3b shows the pole-zero diagram of a fourth-order low-pass elliptic filter. It has an equal number of poles and zeros (all finite), which is a property of elliptic filters. The four finite zeros on the imaginary axis add four zeros to the out-of-band gain response of the filter, which sharpens the rolloff of the out-of-band

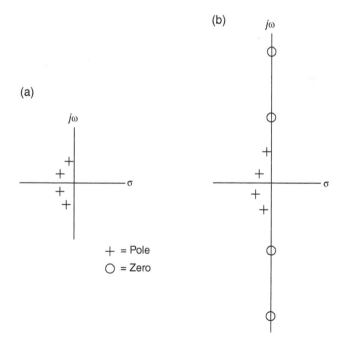

FIGURE 5.3 Pole-zero diagram for 4th-order (a) Chebyshev filter and (b) elliptic filter. (After Maxim Integrated Products (2008)).

gain response compared to that of the Chebyshev (Maxim Integrated Products, 2008). The imaginary parts of the poles lie within the prototype LPF's passband, while the zeros lie outside the passband.

The pseudoelliptic response is like the Chebyshev's but has at least one finite zero. It is usually better to minimize the number of finite poles, rather than keeping their number equal to the number of zeros as in the elliptic filter (Levy et al., 2002).

A filter is partially described by the notation n-p-r, where n is the order of the filter, p is the number of finite zeros on the imaginary axis, and r is the number of zeros on the real axis (Kudsia, 1982).

An actual filter causes a small amount of dissipation. This shifts the ideal locations of poles and zeros to the left on the s-plane by a small amount (Cameron et al., 2018).

Exceptions to the network synthesis method of ideal filter design are the practical LPF and high-pass filter (Cameron et al., 2018), the all-pass filter for phase equalization, and the filter with asymmetric frequency response (Levy et al., 2002).

5.2.6 All-Pass Filter for Phase Equalization

It is normally desired that a filter have a flat phase response in-band, to minimize signal distortion. One way to equalize the phase or group delay response of a filter is to cascade a separate all-pass filter with it. The separate filter is an *external equalizer*. This equalization method was common up until at least 1992 (Kudsia et al., 1992). The all-pass filter's Laplace transfer function is the product of terms whose poles and zeros are as shown in Figure 5.4, where the poles and zeros are placed symmetrically on the two sides of the imaginary axis. Figure 5.4a corresponds to a first-order filter, known as a C-section, and Figure 5.4b to a second-order filter, known as a D-section. The poles are in the left half-plane and zeros in the right half-plane. These sections have a gain response (on the imaginary axis) of unity. The family of all such filters has group delay responses which are nearly the opposite of the group delay responses of filters to be equalized (Cameron et al., 2018).

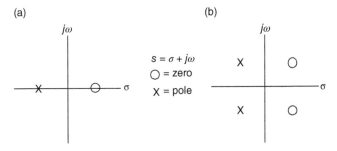

FIGURE 5.4   Pole-zero diagram of component filters of an all-pass filter for phase equalization: (a) first-order and (b) second-order.

5.3   BASICS OF SPECIFICALLY MICROWAVE FILTERS

5.3.1   Resonator

A filter is composed of coupled **resonators**. For the filters, we are interested in, a resonator has the shape of either a rectangular solid or a cylinder. The resonator boundary contains all or practically all of the electromagnetic fields. Theoretically, the resonator has an infinite number of resonant electromagnetic modes, and each mode corresponds to a particular resonant frequency. The modes are determined by the shape, dimensions, and dielectric constant of the material (including vacuum) of the resonator. If a signal is introduced into the resonator, the portion of the signal at the resonant frequencies will excite the resonator, causing a standing wave. The mode with the lowest frequency is the dominant or fundamental mode. Because a resonator always has some RF losses, each resonant frequency is actually a narrow band of frequencies (Chang, 1989).

The naming convention for resonator modes is derived from the naming convention for modes of RF lines of the same general shape, rectangular or circular (Section 4.4.5). A third index is added to the two indices of the RF-line mode. Examples are TE_{011} and TE_{11n} (Yassini and Yu, 2015).

A resonator meant to resonate in only the dominant mode is a **single-mode** resonator. A **dual-mode** resonator is meant to resonate in two physically orthogonal modes of the same natural frequency at once. In this latter case, the modes are degenerate. Triple-mode filters exist but they are usually too complicated to use (Levy et al., 2002). The resonant mode need not be fundamental – quite often it is not, when a higher Q is being pursued. A Super-Q (Section 5.3.6) dual-mode filter designed and measured by a major manufacturer in 2015 used a 13th-order mode (Yassini and Yu, 2015).

Each mode of a resonator has a *tuning device*, usually a screw, that protrudes into the plane of the resonation. Adjusting the screw varies the electromagnetically effective dimensions of the resonator (Thompson and Levinson, 1988). An alternative to a screw for some filters is dent-tuning (Section 4.4.4) (Hsing et al., 2000). Most filters with stringent requirements will employ a tuning mechanism of some kind, for compensation of the limitations of design and machining accuracy (Holme, 2019a).

Each mode of a resonator creates one pole in the Laplace transfer function of the filter it will be part of. The two modes of a dual-mode resonator can be tuned independently of each other, so they can realize separate poles (Fiedziuszko and Fiedziuszko, 2001).

Dual-mode resonators are far more common than single-mode resonators because the number of resonators in a filter is reduced by half. Such filters have been the industry standard since 1976 (Kudsia et al., 2016). The order of these filters is even; however, an extra resonator at the end of the filter enables odd-order transfer functions to be generated (Atia and Williams, 1972). Such filters have the limitation from their available couplings that at least two zeros have to be at infinity

(Kudsia, 1982) – although it is theoretically possible to realize the additional zeros by direct coupling of the input and output (Holme, 2019a). Dual-mode filters can be made in the two main filter technologies, empty-cavity waveguide and dielectric resonator, described in Section 5.4 (Fiedziuszko, 1982; Kwok and Fiedziuszko, 1996).

Single-mode resonators have been used in a dielectric-resonator (DR) filter that is part of a C-band OMUX, in order to provide better power-handling than a dual-mode (Lundquist et al., 2000).

5.3.2 Resonator Coupling and Canonical Filter

The most general two-port network is composed of single-mode resonators that are all coupled to each other (Atia and Williams, 1971). A dual-mode resonator can be thought of as two single-mode resonators, in this context. Any realization of the most general two-port network is a *canonical filter*. A canonical filter is often used to compare other filters to.

A single-mode canonical waveguide filter is shown in Figure 5.5. The input port is at the left on top, and the output port is at the left on the bottom. Adjacent cavities are coupled by a slot. Cavities one on top of the other are coupled by a circular hole.

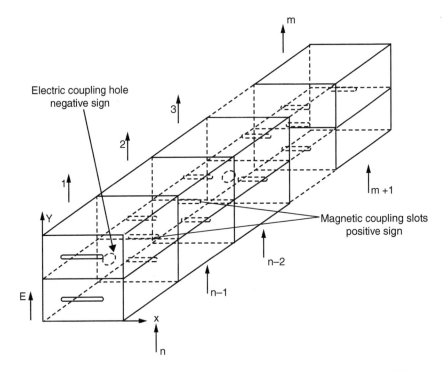

FIGURE 5.5 Canonical single-mode coupled-cavity waveguide filter. (©1974 IEEE. Reprinted, with permission, from Atia and Williams (1974)).

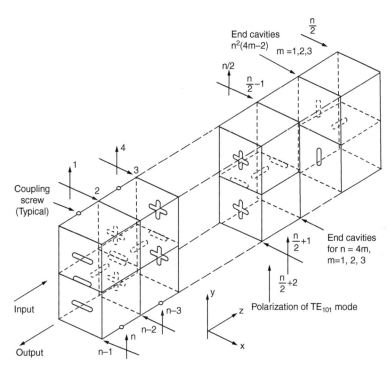

FIGURE 5.6   Canonical dual-mode coupled-cavity waveguide filter. (©1972 IEEE. Reprinted, with permission, from Atia and Williams (1974)).

Other cavities are coupled by a slot at the edge of the cavities (Atia and Williams, 1974).

Cavity *adjacency* is defined as if the cavity sequence is not folded. Perhaps, a more descriptive word for such cavities is *sequential*. When non-adjacent cavities are coupled, they are said to be *cross-coupled*.

Cross-coupling gives alternative paths for a signal to take between filter input and output. The multipath causes transmission zeros to appear in the transfer function. Oppositely phased partial signals will cause finite zeros on the imaginary axis, and equally phased partial signals will flatten the inband group delay (and gain assuming a finite resonator Q (Holme, 2019a)). Both can occur together (Levy and Cohn, 1984; Pfitzenmaier, 1982).

With dual-mode empty-cavity waveguide, the canonical filter can be realized as shown in Figure 5.6 for an even-order filter with response symmetric about band center (Atia and Williams, 1974). The filter input is on the left, as is the filter output from a different cavity. This is the *folded* filter configuration. When the number of modes is not a multiple of four, the two cavities on the rightmost end of the filter use different coupling mechanisms from the other cavities. Adjacent cavities are coupled by means of a dual-mode iris in the shape of a cross. Non-adjacent cavities are coupled by a slot, and the two modes in each cavity are coupled by means of tuning screws. Non-adjacent cavities are cross-coupled by means of an edge slot.

The canonical filter can be realized in other ways besides these two. The canonical filter can realize Chebyshev and elliptic filters, as well as filters with jointly optimized group delay and gain response (Atia and Williams, 1974).

5.3.3 Extracted Pole Technique

An alternative way to realize transmission zeros is the *extracted pole technique* (Holme, 2011). The pole here is an attenuation pole, that is, a transmission zero. This technique deals with imaginary-axis conjugate pairs of zeros. These zeros can be realized by the addition of a simple resonator, for each zero, to the main filter's input or output and by removal of cross-couplings between non-adjacent cavities in the main filter (Rhodes and Cameron, 1980). A filter with extracted poles has the further advantage that the two conjugate zeros of a pair can be independently tuned, to make a filter with an asymmetric passband. Figure 5.7 shows a filter with four cavities on the right-hand side of the drawing, sequential couplings between these cavities, and two simple resonators for extracted poles (note that these two have only one port each). The absence of cross-couplings allows a filter to have an in-line configuration.

5.3.4 Phase Self-Equalization

A **self-equalized** filter has a linear phase response or, equivalently, constant group delay across most of the passband. The filter is said to have **linear phase**. Self-equalization is almost never employed in filters which require low insertion loss — the equalization is not due to the passband extremes being improved in delay, but instead to the band center having increased delay, which means storing more energy and thus having higher loss (Holme, 2019a). The filter cannot then also have an optimal gain response. However, the phase and gain can be mutually optimized, and often that is what is meant. For a comparison of 12-pole optimal gain response, optimal phase response, and mutually optimal gain and phase response filters, see Atia and Williams (1974).

Instead of a filter being self-equalized, its phase can be equalized by a separate all-pass filter (Section 5.2.6), but this requires greater mass and volume and has worse thermal stability (Ezzeddine et al., 2013).

A canonical filter, either single- or dual-mode, can realize an optimal gain, optimal phase, or mutually optimal gain and phase filter. Flattening the phase requires positive cross-couplings (Atia and Williams, 1974). A dual-mode in-line configuration, while not being canonical, can also realize a filter with linear phase or, alternatively, nearly linear phase with better gain response (Kallianteris, 1977).

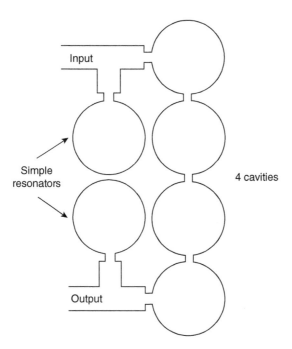

FIGURE 5.7   Example of the filter with two extracted poles. (After Cameron (1998)).

5.3.5 Q Factor

A figure of merit for resonators and filters is the (**unloaded**) **Q factor**, defined when the resonator or filter is not connected to a load:

$$Q = \frac{\omega_0 W}{P_d}$$

where ω_0 is the resonant angular frequency, W is the total average energy stored in the resonator in one resonation period $2\pi/\omega_0$, and P_d is the average power dissipated in the resonator in one period (Sorrentino and Bianchi, 2010). Thus, Q is inversely proportional to the insertion loss as a non-dB power ratio less than one. Q is proportional to the resonator's stored energy, so making the resonator larger is also a way to increase Q (Hunter et al., 2002).

It is desirable to use resonators with large Q since this reduces the filter's insertion loss and improves its selectivity performance (Cameron et al., 2018). A high Q factor of at least 8000 is usual for IMUX channel filters (Yu et al., 2004).

The higher the order of a filter, the higher the insertion loss as a positive dB number, and, thus, the lower the Q (Hunter et al., 2002).

Q depends on the filter technology (Cameron et al., 2018). Q is in general largest for DR cavity waveguide filters, followed by empty-waveguide filters, followed by coaxial-cavity filters.

5.3.6   Resonators with Improved Q Factor

There are resonators with unusually high Q factor that are sometimes used for IMUXes and OMUXes.

Low-Q resonators of an IMUX channel filter can be made to emulate resonators of very much higher Q by means of *adaptive predistortion*. The real part of each pole is altered so as to reach a desired Q. The resultant filter is much smaller and lighter but has a much higher insertion loss. This approach also reduces the variation of inband insertion loss and group delay. Such a technique may be applicable to the channel filters in an IMUX since the extra insertion loss can possibly be compensated by a higher-gain low-noise amplifier (LNA) (Yu et al., 2003). A Ku-band 10-pole self-equalized channel filter with Q of 8000 was enhanced to 16,000 with predistortion (Choi et al., 2006). The word "adaptive" is a misnomer since the predistortion does not adapt to changing circumstances. Some disadvantages of predistortion are that the filter then requires very precise tuning and that, since it inherently has poor input and output return loss, it requires very strong isolation from other cascaded components, often more than realizable circulators can provide (Holme, 2019a).

For OMUX channel filters, a recent development is the *super-Q* resonator. It uses a different dual resonant mode from usual. It offers dramatic benefits in insertion loss and power-handling at K-band (De Paolis and Ernst, 2013). One filter supplier has flight-qualified two configurations of a 17-GHz OMUX with such waveguide BPFs. One configuration allows 270 W per channel and the other 500 W per channel, and the unloaded Q factor of both is 33,700 (Fitzpatrick, 2013). Another supplier has shown the feasibility of a 20-GHz dual-mode OMUX with an unloaded Q of 31,000 (Yassini and Yu, 2015). A third vendor is offering a 20-GHz OMUX with 200 W per channel (Flight Microwave, 2019).

5.4   TECHNOLOGY FOR BANDPASS FILTERS

5.4.1   Empty-Cavity Waveguide Filter

The **empty-cavity waveguide filter**, or sometimes just **waveguide filter**, has long been used. It is still used in some situations, such as when the filter needs to carry very high power, as some OMUX channel filters do (see for example Thales Alenia Space, 2012).

Waveguide filters come in both rectangular and circular waveguide, as well as in square waveguide as we have seen in Figure 5.6. They also come in various configurations. We have seen the square-waveguide folded configuration, which is a canonical filter (Section 5.3.2).

The simplest arrangement of the waveguide filter's resonators is in a line (called the *in-line*, *longitudinal*, *coaxial*, or *axial* configuration in the literature), as illustrated in Figure 5.8 with a circular waveguide (it can also be implemented in square waveguide (Atia and Williams, 1972)). This configuration in circular waveguide is

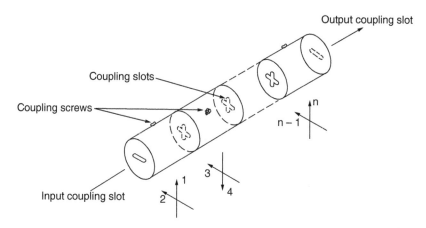

FIGURE 5.8 Circular-waveguide dual-mode filter with in-line configuration. (©1977 IEEE. Reprinted, with permission, from Williams and Atia (1977)).

common for OMUX channel filters of X-band and above (Thales Alenia Space, 2012; Tesat-Spacecom GmbH, 2016). Sequential resonators are coupled by an iris between the cavities. The example of Figure 5.8 has dual-mode resonators. The irises have a cross-shape so that two orthogonal modes in one cavity can each couple to their corresponding modes in the adjacent cavity. (An iris coupling one mode would be just a slot.) Such a filter is not canonical, as the maximum number of finite zeros it can have is half the filter order (Cameron and Rhodes, 1981). Three mode-coupling screws are shown, where each screw couples the two orthogonal modes within one cavity. The screws are at 45° from the plane of the input radiation, and the two different cases of 45° correspond to the two signs of cross-coupling. An actual filter will typically have additional 0 and 90° screws to correct for inherent differences in frequency between the two resonant modes. An alternative is to use elliptical cavities, but with a considerable increase in fabrication complexity (Holme, 2019a).

A configuration that makes cross-coupling different cavities more mechanically feasible, and is a canonical filter, is the folded configuration as shown in Section 5.3.2. In addition, there are other means for cross-coupling cavities diagonal from each other (not shown) (Atia and Williams, 1974).

There are two properties of empty-cavity waveguide filters that one should keep in mind while studying test results. For one thing, the filter's center frequency is shifted down on the ground relative to what it will be on orbit (Kudsia et al., 2016). The reason is that waveguide's cutoff frequency is inversely proportional to the square root of the dielectric constant of what is inside the cavities (Ramo et al., 1984), and air and vacuum have slightly different dielectric constants. The dielectric constant of dry air is 1.00059 at one atmosphere of pressure and 20 °C (Young and Freedman, 2012), while that of vacuum is exactly 1.0. This means a frequency shift of about 300 ppm, which for example at 12 GHz is about 4 MHz.

An additional factor is humidity on the ground, which can shift the 12-GHz filter response down by 0.8 MHz for 50% relative humidity (for this reason testing often includes a dry nitrogen backfill) (Holme, 2019a; Choi and Kim, 2013).

5.4.2 Dielectric-Resonator Filter

The **DR loaded-cavity filter**, or simply **DR filter,** has much in common with the empty-cavity waveguide filter. It is, however, much smaller than the corresponding waveguide filter would be (Fiedziuszko, 1982). DR filters are able to achieve a high Q factor between 8,000 and 15,000 (Yu and Miraftab, 2010).

The main difference between the DR filter and the waveguide filter is that here the resonators are pucks, that is, short cylinders, of a suitable dielectric material. Each dimension of the resonator is reduced compared to an empty-cavity resonator, namely divided by the square root of the dielectric constant (Zaki and Atia, 1983). The electromagnetic fields exist only in and in the near vicinity of the dielectric material (Fiedziuszko and Holme, 2001). Fields are evanescent (that is, they exponentially decay and do not propagate) outside the resonator (Holme, 2011). Each resonator is placed in a waveguide cavity, sized so that the resonator's resonance frequencies are below the waveguide cutoff (Fiedziuszko, 1982). Even though the waveguide cavity surrounding the puck does not resonate, it may affect the resonant frequency to some degree. If the evanescent fields are low, then the Q of the overall resonator is almost all due to the ceramic loss tangent — thus, higher-order modes are not usually beneficial (Holme, 2019a). The resonator is held in the middle of the cavity and does not touch the cavity walls (Fiedziuszko, 1982; Yu et al., 2004). This reduces losses in the walls (Zaki and Atia, 1983), as the boundary between the dielectric and the air/vacuum is a magnetic wall due to the difference in dielectric constant, and thus there are no resistive losses at that boundary (Holme, 2020). The waveguide is there to prevent radiation of the fields and resultant degradation of resonator Q (Fiedziuszko and Holme, 2001).

Ceramic dielectrics (such as zirconium tin titanate) are available with a wide variety of dielectric constants and effective temperature characteristics. Temperature stability is achieved by a mixture of two different ceramics, which offsets the change in the size of the material with its dielectric constant. Dielectric constants are typically 24–40, as compared to 1 of vacuum. Twenty-four or so is good due to the higher Q factor it provides. Higher dielectric constants than about 36 usually reduce Q and/or temperature stability (Holme, 2011, 2019b).

The dual-mode DR filter, pioneered by Fiedziuszko (1982), has been the standard DR filter since then (Kudsia et al., 1992). At frequencies up through Ku-band, the DR filter in single- or dual-mode form is ubiquitous (Berry et al., 2012). The DR filter is the most common technology for IMUX channel filters in X-, Ku-, and K-bands (Tesat-Spacecom GmbH, 2016).

The DR filter can create the same types of frequency responses that a similarly shaped empty-cavity waveguide filter can (Pfitzenmaier, 1982).

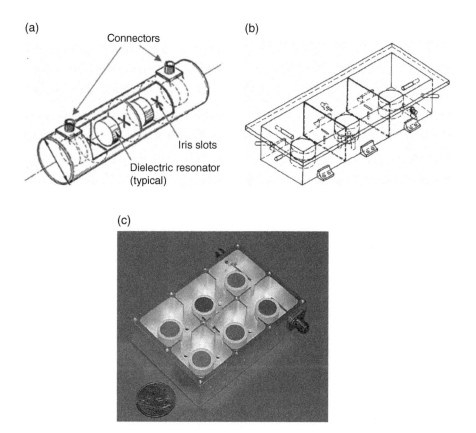

FIGURE 5.9   Dielectric-resonator filter configurations: (a) in-line, dual-mode (©1982 IEEE. Reprinted, with permission, from Fiedziuszko (1982)); (b) planar, dual-mode (©1992 IEEE. Reprinted, with permission, from Kudsia et al. (1992)); (c) folded, single-mode, without lid (©2001 IEEE. Reprinted, with permission, from Fiedziuszko and Holme (2001)).

The DR filter comes in three different configurations. One is the *in-line* or *longitudinal*, shown in Figure 5.9a. Coupling is as for the empty-cavity waveguide filter of Figure 5.8, and the same filter responses can be realized (Fiedziuszko, 1982). The screws to couple the orthogonal modes in a dual-mode cavity are not shown here but are similar to those of the waveguide filter. The filter input and output connectors are for coaxial cable. The second and third configurations are both *planar* (Tang et al., 1987). One is the *linear planar* (sometimes also called "in-line") shown in Figure 5.9b, which has T-shaped dual irises. The other is the *folded planar* shown in Figure 5.9c, which has shaped slots for coupling (Cameron et al., 1997). This configuration is often used for single-mode resonators (Holme, 2011). For them, diagonal cross-couplings can be used (Cameron et al., 1997).

5.4.3 Coaxial-Cavity Filter

The **coaxial cavity (CC) filter** is used at the lowest frequencies, from L-band to C-band. It is different from the waveguide and DR filters, whose resonating modes are related to waveguide modes — this filter's resonators have transverse electromagnetic (TEM) as the dominant and only mode. The coaxial resonator consists of a metal rod as the inner conductor, waveguide as the outer conductor, and vacuum as the dielectric (Yao et al., 1995). This filter technology is less expensive to build than the DR filter (Yu et al., 2003).

The CC filter consists of coupled TEM-mode transmission lines, typically shorted at one end and open on the other, with resonator length less than 90° (Holme, 2019a). If the resonators are all aligned the same way it is a *combline* filter, while if they are alternating it is an *interdigital* filter. Figure 5.10a shows the electric and magnetic fields of two adjacent resonators and how to couple the resonators. The total coupling has both magnetic and electric components, with the magnetic being stronger and the electric being out of phase with the magnetic. The total coupling is the magnetic coupling minus the electric coupling. The magnetic coupling is decreased by a wall between the rods, and the electric coupling is decreased by a screw. Two ways of cross-coupling resonators are shown in Figure 5.10b,c (Thomas, 2003).

The CC filter can create the same types of frequency responses that a similarly shaped waveguide or DR filter can (Pfitzenmaier, 1982).

The CC filter by itself cannot achieve a Q of 8000. At C-band, the Q can be improved from 3000 to well above 8000 (20,000 was achieved) by adaptive predistortion (Section 5.3.6). It creates a CC filter which is much smaller and lighter than a single-mode DR filter. However, the insertion loss is much higher than that of the DR filter (Yu et al., 2003).

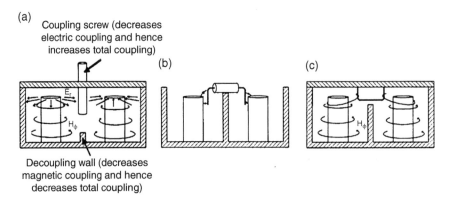

FIGURE 5.10 Coaxial cavity resonators: (a) adjacent resonators, field lines, and coupling mechanisms; (b) capacitive coupling mechanism for non-adjacent resonators; (c) inductive coupling mechanism for non-adjacent resonators. (©2003 IEEE. Reprinted, with permission, from Thomas (2003)).

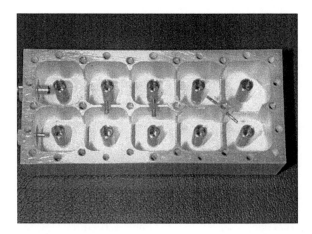

FIGURE 5.11   Folded coaxial cavity filter without lid. (©2003 IEEE. Reprinted, with permission, from Yu et al. (2003)).

A folded CC filter with two rows and non-sequential couplings is shown without a lid in Figure 5.11.

5.5 FILTER UNITS

The diplexer filter, which separates the receive and transmit signals for an antenna used for both, is regarded as part of the antenna so was first discussed in Section 3.8.2, although there is some information in Sections 5.5.1 and 5.5.3.

5.5.1 Preselect Filter

The preselect filter is the first filter unit that the signal sees, after the antenna and before the LNA (Section 6.5). If the antenna has a diplexer, the preselect filter may be part of the diplexer. The preselect filter does two things to protect the LNA and the rest of the payload. One is to reject interfering signals that might saturate the wideband LNA and drive it into compression. Another thing is to reject interfering signals that may be stronger than the signals of interest, to ensure that the interferers are not passed by the IMUX channel filters. Insertion loss is paramount, as any loss degrades the payload noise figure usually one-for-one in dB.

5.5.2 Multiplexer in General

A MUX is a unit which is an assembly of filters and payload-integration elements. Most of the filters in it are BPFs for channel filters. The MUX may contain other filters, too, for additional purposes.

Of all the filters that a signal passes through in the payload (including ones that are components in non-filter units), the only two that provide significant bandlimiting are the IMUX and OMUX. Together, the IMUX and OMUX must provide channel selectivity. A question is how to divide up the necessary amount of selectivity between the IMUX and the OMUX. The sharper a filter is, the higher its insertion loss. The IMUX and not the OMUX does most of the filtering because the OMUX carries the post-high-power amplifier (HPA) signal while the IMUX carries only a low-level signal. HPA output power must not be wasted on filter-insertion loss. Also, narrow-band filters that can accommodate high power are difficult to make, so the fewer the OMUX resonators the better. Finally, means must be provided to carry away the heat in an OMUX filter resonator, so again the fewer the OMUX resonators the better. The consequence is that the IMUX BPFs are sharper, that is, have more poles, than the OMUX BPFs do, 8–10 poles versus typically 4–6 (Berry et al. (2012) and Tesat-Spacecom GmbH (2016) for IMUX, Thales Alenia Space (2012) for OMUX).

5.5.3 Special Channel Filters

There are occasions when something besides a symmetric, one-channel BPF is preferred in a MUX.

It may be desired that a filter have an asymmetric frequency response, that is, have higher out-of-band rejection on one side of its passband than on the other. Such a filter is called an *asymmetric filter*. An example of this is the BPFs for the end channels in an IMUX or OMUX. They need greater rejection on their outer sides than on their inner sides, since there is no next-door BPF to provide additional rejection (Kudsia et al., 1992). Another example is the diplexer filter, where much stronger signals on one side must be rejected (Bila et al., 2003). A filter with an asymmetric frequency response has either more out-of-band zeros on one side of its passband than the other side or has out-of-band zeros differently spaced on the two sides.

Sometimes it makes sense in an OMUX to replace two channel filters over different bands by one filter with two passbands, called a *dual-passband filter*. For example, two noncontiguous channels may be amplified by the same TWTA but need to be separated out to go down on different beams. A solution is to design a filter, starting out with one wide passband that encompasses both bands, then adding zeros that divide the band into the desired two narrower ones. This simplifies the filter and reduces mass and volume. There can also be more than two passbands (Holme, 2002).

5.5.4 Input Multiplexer

The IMUX comes after the receiver and divides out the channels for separate processing. Two IMUX architectures are shown in Figure 5.12. The first one is the most common of all architectures, the **channel-dropping scheme** (Cameron and Yu, 2007). The signal from the receiver enters the first BPF. That filter passes one

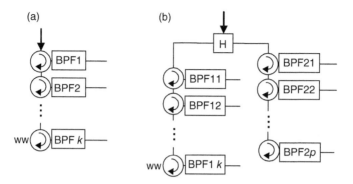

FIGURE 5.12   IMUX architectures: (a) channel-dropping and (b) channel-dropping with a power splitter.

channel band and rejects the rest of the frequencies, which go back out the first circulator and through the second circulator to the second BPF. This continues until the frequencies rejected by all the BPFs get absorbed by the load on the last circulator. The farther down the chain it is, the larger the loss the channel band sees. The second architecture, shown in part Figure 5.12b of the figure, is similar to Figure 5.12a but has a power-divider that divides the signal into two copies, each copy going into a channel-dropping scheme (Choi et al., 2006). The loss distribution in this architecture is different from that of scheme (*a*). This scheme is good if the channels are contiguous in frequency (Choi et al., 2006) but this is rarely the case (Holme, 2011). Both architectures depend on the receiver's gain being high enough to accommodate the IMUX loss.

IMUX channel filters have elliptic or pseudoelliptic responses. One IMUX manufacturer states that the most common technology for L-, S-, and C-bands is coaxial-cavity resonators, while for X-, Ku-, and Ka-bands it is dielectric resonators. Waveguide filters can be used at Ka-band for extra-challenging requirements (Tesat-Spacecom GmbH, 2016). Another IMUX manufacturer offers DR technology for C-, X-, Ku-, and Ka-bands. The channel filters are single-mode and self-equalized. These BPF's unloaded Q factor is at least 16,000 for X-band, 13,000 for Ku-band, and 9000 for Ka-band (Thales Alenia Space, 2012). A satellite manufacturer that makes IMUXes in-house reports that both single- and dual-mode are common for DR filters (Berry et al., 2012). These BPFs are self-equalized (Holme, 2011).

5.5.5   Output Multiplexer

5.5.5.1   OMUX Architecture   The OMUX is the last payload filter unit that the signal sees, usually after the HPAs. It combines channels amplified by different HPAs so that they can be transmitted on one antenna beam. The OMUX is an assembly of BPFs (channel filters), a manifold, a harmonic filter, and perhaps a notch filter. A common OMUX architecture is illustrated in Figure 5.13.

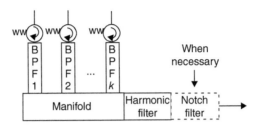

FIGURE 5.13   Common OMUX architecture.

The **manifold** is the device which combines, in a nearly lossless fashion, the outputs of the BPFs. It is most often a length of waveguide short-circuited at one end (Fiedziuszko et al., 2002). The BPFs may be attached all on one side of the manifold (*comb* or *combline* arrangement), as shown, or on both sides (*herringbone* arrangement). In either configuration, an additional BPF can be attached where the short circuit would otherwise be (Cameron et al., 2018).

There are actually two kinds of manifold. The more common one, just described, is a piece of RF line, namely, rectangular waveguide, coaxial cable, or some other low-loss structure. Waveguide is used for C-band and higher frequencies and coaxial cable for lower frequencies. A waveguide interface from the BPF to the manifold can be provided by a coupling iris and a coaxial interface by a coaxial probe (Lundquist et al., 2000). BPFs may join the manifold in either the E- or H-plane (Cameron et al., 2018). The RF-line manifold has the freedom for OMUX design optimization offered by the phase lengths between channel launch points. In manifold design, it is recommended to start off with channel spacing of an integer multiple of half a guided wavelength (Cameron et al., 2018). In addition, quite often a short, half-wavelength section is added between the BPF and the manifold main line. This gives an additional variable for optimization and allows for an inductive filter coupling (which would normally require a non-realizable negative length at the manifold) (Holme, 2019a).

The second kind of manifold, used at lower frequencies, is a star junction, essentially a manifold with zero spacing between channel filters but with short half-wavelengths between filter and junction (Holme, 2019a). At L-band a waveguide manifold would be very big, so a coaxial 6-port junction has been used (Fiedziuszko et al., 1989).

Every channel filter should have a good phase response. There are two ways to accomplish this. The much more common way is for the BPFs to be self-equalized, which yields nearly flat group delay over most of the passband. The uncommon way is to add to the OMUX an external phase (or group-delay) equalizer and a circulator. The problem with this scheme, though, is that the circulator would be so sensitive to temperature that usually phase instability arises (Holme, 2011).

The early OMUXes could only handle channels that were non-contiguous in frequency, that is, successive channels had to be separated by an empty band as wide as a channel. The first published report of a contiguous OMUX was in Chen

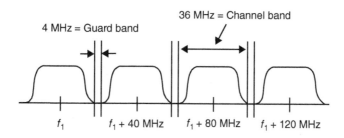

FIGURE 5.14   Example of contiguous channels, separated by 10% guard bands.

et al. (1976). Later, more practical designs were presented. First came C-band in 1978. Then the more difficult Ku-band came, a 5-channel one in 1982 and a 12-channel one, also by Chen, in 1983. (This history was provided by Holme (2011).) Contiguous channels separated by the usual 10% guard bands are illustrated in Figure 5.14. OMUXes can now contiguously cover most 500–1000 MHz bands with a single manifold, with over 18 combined channels (Holme, 2019a).

At the manifold output, there is a **harmonic filter**, which rejects the second and third harmonics of the HPAs' outputs as well as the HPAs' output noise in the payload's receive band (Saad, 1984). The harmonic filter is a LPF. The OMUX architecture in Figure 5.13 is feasible when one harmonic filter can accommodate the combined output power of the channel filters. When this power is too great, each channel filter must have a harmonic filter at its input (Cameron et al., 2018). However, with the advent of particle-tracking software, the multipactor risk of a harmonic filter with many channels can now be more accurately analyzed, so most OMUXes now use a single output harmonic filter. This has the additional benefit of strongly reducing passive intermodulation products (PIMs) (Section 8.15) from the more complex OMUX (Holme, 2019a).

At the manifold output, there may additionally be a notch filter to reject the near-out-of-band transmission in a protected frequency band, such as an astronomy band. A notch filter can be realized in the empty-cavity waveguide, coaxial cavity, or DR technologies (Cameron et al., 2005). The notch must be wide enough to accommodate the filter's shift with on-orbit temperature variation.

OMUXes are commercially available in waveguide and DR technologies: waveguide for X-band to K-band (Thales Alenia Space, 2012) and DR for L-band (Lundquist et al., 2000) to C-band (Thales Alenia Space, 2012).

5.5.5.2 *OMUX Thermal*

An OMUX, unlike an IMUX, commonly sees high power. The heat flow in the OMUX is as follows. The RF signals dissipate power in the channel filters, which means that heat arises in them. The channel filters connect to the manifold. The manifold is low-loss but sees the combined signal power, so heat also arises in it. The manifold attaches to the OMUX baseplate, to which the excess heat in the OMUX flows. The baseplate attaches to a spacecraft mounting plate, underneath of which run heat pipes. If the heat pipes of the spacecraft panel connect to heat pipes of other panels, so that heat from the hotter panel can go to the cooler panel to be more efficiently dumped, the temperature range of the mounting plate will be smaller than otherwise (Section 2.2.1).

The OMUX with filter cavities loaded with dielectric resonator is not capable of carrying such high-power signals as the OMUX with empty waveguide cavities. The thermal issues are different for the two BPF technologies.

The DR OMUX typically has aluminum cavity walls, manifold, and baseplate. Since the spacecraft panel is also typically made of aluminum, there is no thermal stress between them (Lundquist et al., 2000).

For such an OMUX, the thermal stability of the channel filter is determined by that of the dielectric ceramic (Fiedziuszko, 1982). The properties of the ceramic can be adjusted so that its coefficient of thermal expansion is zero (Fiedziuszko and Holme, 2001). The problem is to get the heat out of the dielectric resonators. One solution, for a filter with in-line arranged single- or dual-mode resonators, is to mount the resonator in its cavity on a support of thermally conductive, electrically insulating material (Holme et al., 1996). The trick here is to still accommodate inter-cavity irises. The dielectric resonators in DR channel filters may operate in single mode (Lundquist et al., 2000) or dual mode (Thales Alenia Space, 2012).

When the OMUX BPFs are of empty-cavity waveguide, the situation is more complex. The BPFs, the manifold, and the OMUX baseplate may be made of different materials with different coefficients of thermal expansion (CTE), which can cause stress at material interfaces. The OMUX baseplate and the spacecraft mounting plate are typically aluminum (Lundquist et al., 2000).

The materials currently in use for the high-power, waveguide-BPF OMUX are only two, Invar and aluminum:

- Invar, thin-wall, silver-plated. An alloy of mostly iron and 36% nickel (High Temp Metals, 2011). Heavier than aluminum. Its CTE is on a par with that of temperature-compensated aluminum. It has low thermal conductivity (Lundquist et al., 2002).
- Aluminum, mechanically temperature-compensated. (Without compensation, the CTE is about 15 times that of Invar (Wolk et al., 2002).) It is light-weight and has high thermal conductivity so conducts heat away better than Invar does (Lundquist et al., 2002).

Invar is the traditional material and still dominates the market (Tesat-Spacecom GmbH, 2016). Temperature-compensated aluminum OMUXes are capable of carrying higher power than the Invar, although this may not be true for extreme temperature environments, as the temperature-compensated designs are often only stable over a limited temperature range (Holme, 2019a). Both materials expand linearly with temperature.

There are at least four ways in which the various manufacturers conduct away the heat from their high-power waveguide OMUXes. The first way is to use all Invar with heat straps. The heat straps conduct heat from the BPFs to the baseplate and from the manifold to the baseplate. The heat straps can be brackets (Thales Alenia Space, 2012) but for extra high power, they can be made of braided aluminum cables (Fiedziuszko et al., 2011). The second way is for the BPFs to be of Invar and the manifold of aluminum. The aluminum manifold may be compensated (European Space Agency, 2014) or not (Thales Alenia Space, 2012). The third way is for the entire OMUX to be of aluminum and the manifold to be temperature-compensated (Thales Alenia Space, 2012). The fourth way is for the entire OMUX to be of aluminum and the channel filters to be temperature-compensated. Both end walls of each channel-filter cavity move inward with increasing temperature, to keep the center frequency constant (Lundquist et al., 2002; Arnold and Parlebas, 2016).

5.5.6   Technology Application to Multiplexers

Table 5.1 shows which BPF technology is used for which commercial MUX application, based on information from suppliers. For the IMUX, in order of ascending frequency, the technologies used are coaxial-cavity filter, DR filter, and empty-cavity waveguide filter. For the OMUX, in order of ascending frequency, the technologies used are DR filter and empty-cavity waveguide filter.

5.6   BANDPASS FILTER SPECIFICATION

Table 5.2 gives an example of the parameters of a BPF specification. A parameter not seen for active units is the frequency shift over temperature in units of ppm per degree of temperature rise. A **mask** is a piecewise-linear function, usually over frequency, defining the upper bound.

TABLE 5.1   Technology Application to Multiplexers

	IMUX	OMUX	
BPF Technology	Frequency Bands	Frequency Bands	RF Power Per Channel
Empty-cavity waveguide filter	Ka-band[a]	X-band to K-band[b]	350 W at Ku-band[c] 200 W at K-band[d]
Dielectric-resonator filter	C-band to Ka-band[b]	L-band[e] C-band[b]	35 W at L-band[e] 180 W at C-band[b]
Coaxial-cavity filter	L-band to C-band[a]	Not used	Not used

[a] Tesat-Spacecom GmbH (2016).
[b] Thales Alenia Space (2012).
[c] Holme (2019b).
[d] Flight Microwave (2019).
[e] Lundquist et al. (2000).

TABLE 5.2   Example of Parameters in a BPF Specification

Parameter in BPF Specification	Units
Frequency range	GHz
Insertion loss	dB
Return loss — input, output	dB
Gain variation over the band, defined by a mask	dB pk-pk
Group delay variation over the band, defined by a mask	ns pk-pk
Power handling per channel (for OMUX only)	W
Frequency shift over temperature	ppm/°C
Operational temperature range	°C

REFERENCES

Arnold C and Parlebas J, inventors; Tesat-Spacecom, assignee (2016). Generic channel filter. U.S. patent application publication US 2016/0064790 A1. Mar. 3.

Atia AE and Williams AE (1971). New types of waveguide bandpass filters for satellite transponders. *Comsat Technical Review*; 1 (1) 21–43.

Atia AE and Williams AE (1972). Narrow-bandpass waveguide filters. *IEEE Trans on Microwave Theory and Techniques*; 20; 258–265.

Atia AE and Williams AE (1974). Nonminimum-phase optimum-amplitude bandpass waveguide filters. *IEEE Trans on Microwave Theory and Techniques*; 22; 425–431.

Berry S, Fiedziuszko SJ, and Holme S (2012). A Ka-band dual mode dielectric resonator loaded cavity filter for satellite applications. *IEEE Microwave Symposium Digest;* June 17–22.

Bila S, Baillargeat D, Verdeyme S, Seyfert F, Baratchart L, Zanchi C, and Sombrin J (2003). Simplified design of microwave filters with asymmetric transfer functions. *European Microwave Conference;* Oct. 7–9.

Cameron RJ, inventor, Com Dev, assignee (1998). Dispersion compensation technique and apparatus for microwave filters. U.S. patent 5,739,733. Apr. 14.

Cameron RJ and Rhodes JD (1981). Asymmetric realizations for dual-mode bandpass filters. *IEEE Transactions on Microwave Theory and Techniques*; 29; 1.

Cameron RJ, Tang W-C, and Dokas V (1997). Folded single mode dielectric resonator filter with cross couplings between non-sequential adjacent resonators and cross diagonal couplings between non-sequential contiguous resonators. U.S. patent 5,608,363. Mar. 4.

Cameron RJ, Yu M, and Wang Y (2005). Direct-coupled microwave filters with single and dual stopbands. *IEEE Transactions on Microwave Theory and Techniques*; 53; 11.

Cameron RJ and Yu M (2007). Design of manifold-coupled multiplexers. *IEEE Microwave Magazine;* Oct; 46–59.

Cameron RJ, Kudsia CM, and Mansour RR (2018). *Microwave Filters for Communications Systems: Fundamentals, Design and Applications*, 2nd ed., Hoboken: John Wiley & Sons.

Chang K, editor, (1989). *Handbook of Microwave and Optical Components,* Vol. 1, *Microwave Passive and Antenna Components.* New York: John Wiley & Sons, Inc.

Chen MH, Assal F, and Mahle C (1976). A contiguous band multiplexer. *COMSAT Technical Review*; 6; 285–307.

Choi JM and Kim TW (2013). Humidity sensor using an air capacitor. *Korean Institute of Electrical and Electronic Material Engineers, Trans on Electrical and Electronic Materials;* 14; 4; Aug. 25. On kpubs.org/article/articleMain.kpubs?articleANo= E1TEAO_2013_v14n4_182&viewType=article. Accessed July 16, 2019.

Choi S, Smith D, Yu M, and Malarky A (2006). C and Ku band multiplexers using predistortion filters. *AIAA International Communications Satellite Systems Conference;* June 11–14.

Couch, II LW (1990). *Digital and Analog Communication Systems,* 3rd ed. New York: Macmillan Publishing Company.

De Paolis F and Ernst C (2013). Challenges in the design of next generation Ka-band OMUX for space applications. *AIAA International Communications Satellite Sytems Conference;* Oct. 14–17.

European Space Agency (2014). Tesat's new high-power Ka-band output multiplexer. On telecom.esa.int/telecom/www/object/index.cfm?fobjectid=33210. Accessed Oct. 15, 2014.

Ezzeddine H, Bila S, Verdeyme S, Pacaud D, Puech J, Estagerie L, and Seyfert F (2013). Design of compact and innovative microwave filters and multiplexers for space applications. *Mediterranean Microwave Symposium;* Sep. 2–5.

Fiedziuszko SJ (1982). Dual-mode dielectric resonator loaded cavity filters. *IEEE Transactions on Microwave Theory and Techniques;* 30; 1311–1316.

Fiedziuszko SJ, Doust D, and Holme S (1989). Satellite L-band output multiplexer utilizing single and dual mode dielectric resonators. *IEEE MTT-S International Microwave Symposium Digest;* 2; June 13–15; 683–686.

Fiedziuszko SJ and Holme S (2001). Dielectric resonators: raise your high-Q. *IEEE Microwave Magazine;* Sep.

Fiedziuszko SJ, Fiedziuszko GA, inventors; Space Systems/Loral, Inc, assignee (2001). General response dual-mode dielectric resonator loaded cavity filter. U.S. patent 6,297,715 B1. Oct. 2.

Fiedziuszko SJ, Holme SC, and O'Neal NL, inventors; Space Systems/Loral, Inc, assignee (2002). Microwave multiplexer with manifold spacing adjustment. U.S. patent 6,472,951 B1. Oct. 29.

Fiedziuszko G, Lee H, Howell A, and Holme S (2011). Recent advances in high power/high temperature satellite multiplexers. *AIAA International Communications Satellite Systems Conference;* Nov. 28–Dec. 1.

Fitzpatrick B of ComDev (2013). High power Ka-band output multiplexer. Project no 70332. ESA contract final report. Aug. 12. Accessed May 15, 2017.

Flight Microwave Corp (2019). Multiplexers. Product information. On flightmicrowave. com/products/omuxes. Accessed Jan. 21, 2019.

High Temp Metals, Inc. (2011). Product technical data. On www.hightempmetals.com/ technicaldata.php. Accessed Oct. 20, 2011.

Holme S (2002). Multiple passband filters for satellite applications. *AIAA International Communications Satellite Systems Conference;* May 12–15; 1–4.

Holme SC, Space Systems/Loral (2011). Private communication. Nov. 8.

Holme S (2019a). Private communication. June. 7.

Holme S (2019b). Private communication. July 11.

Holme S (2020). Private communication. Sep. 11.

Holme SC, Fiedziuszko SJ, and Honmyo Y, inventors; Space Systems/Loral, Inc, assignee (1996). High power dielectric resonator filter. U.S. patent 5,515,016. May 7.

Hsing CL, Jordan JE, and Tatomir PJ, inventors; Hughes Electronics Corp, assignee (2000). Methods of tuning and temperature compensating a variable topography electromagnetic wave device. U.S. patent 6,057,748. May 2.

Hunter IC, Billonet L, Jarry B, and Guillon P (2002). Microwave filters—applications and technology. *IEEE Trans on Microwave Theory and Techniques*; 50; 3.

Kallianteris S (1977). Low-loss linear phase filters. *Digest, IEEE International Microwave Symposium;* June 21–23.

Kallianteris S, Kudsia CM, and Swamy MNS (1977). A new class of dual-mode microwave filters for space application. *European Microwave Conference*; Sep. 5–8; 51–58.

Kudsia CM (1982). Manifestations and limits of dual-mode filter configurations for communications satellite multiplexers. *AIAA International Communications Satellite Systems Conference;* Mar. 8.

Kudsia C, Cameron R, and Tang W-C (1992). Innovations in microwave filters and multiplexing networks for communications satellite systems. *IEEE Trans On Microwave Theory and Techniques*; 40; 1133–1149.

Kudsia C, Stajcer T, and Yu M (2016). Evolution of microwave filter technologies for communications satellite systems. *AIAA International Communications Satellite Systems Conference;* Oct. 18–20.

Kwok RS and Fiedziuszko SJ (1996). Advanced filter technology in communication satellite systems. *Proceedings of International Conference on Circuits and System Sciences;* June 20–25; 1–4. On www.engr.sjsu.edu/rkwok/Publications/Adv_Filter_Proc_ICCASS_p155_1996.pdf. Accessed Oct. 11, 2011.

Levy R and Cohn SB (1984). A history of microwave filter research, design, and development. *IEEE Transactions on Microwave Theory and Techniques*; 32; 9.

Levy R, Snyder RV, and Matthaei G (2002). Design of microwave filters. *IEEE Transactions on Microwave Theory and Techniques*; 50; 3.

Lundquist S, Mississian M, Yu M, and Smith D (2000). Application of high power output multiplexers for communications satellites. *AIAA International Communications Satellite Systems Conference;* 1; Apr. 10–14; 1–9.

Lundquist S, Yu M, Smith D, and Fitzpatrick W (2002). Ku-band temperature compensated high power multiplexers. *AIAA International Communications Satellite Systems Conference*; May 12–15; 1–7.

Maxim Integrated Products, Inc. (2008). A filter primer. Application note 733. Oct. 6. On www.maxim-ic.com/app-notes/index.mvp/id/733. Accessed Oct. 10, 2011.

McGillem CD and Cooper GR (1991). *Continuous and Discrete Signal and System Analysis*, 3rd ed. New York: Oxford University Press.

Pfitzenmaier G (1982). Synthesis and realization of narrow-band canonical microwave bandpass filters exhibiting linear phase and transmission zeros. *IEEE Trans on Microwave Theory and Techniques*; 30; 1300–1311.

Ramo S, Whinnery JR, and Van Duzer T (1984). *Fields and Waves in Communication Electronics*, 2nd ed. New York: John Wiley & Sons, Inc.

Rhodes JD and Cameron RJ (1980). General extracted pole synthesis technique with applications to low-loss TE_{011} mode filters. *IEEE Transactions on Microwave Theory and Techniques*; 28; 1018–1028.

Saad AMK (1984). Novel lowpass harmonic filters for satellite application. *IEEE MTT-S International Microwave Symposium Digest;* May 30–June 1; 292–294.

Sorrentino R and Bianchi G (2010). *Microwave and RF Engineering*. Chichester: John Wiley & Sons, Ltd.

Tang W-C, Siu D, Beggs B.C., and Sferrazza J, inventors; Com Dev Ltd, assignee (1987). Planar dual-mode cavity filters including dielectric resonators. U.S. patent 4,652,843. Mar. 24.

Tesat-Spacecom GmbH (2016). Passive microwave products. Brochure. On www.tesat.de/en/media-center/downloads. Accessed Sep. 17, 2018.

Thales Alenia Space (2012). IMUX & OMUX. Product datasheets. On www.thalesgroup.com/sites/default/files/database/d7/asset/document/Imux_Omux.pdf. Accessed Sep. 17, 2018.

Thomas JB (2003). Cross-coupling in coaxial cavity filters—a tutorial overview. *IEEE Transactions on Microwave Theory and Techniques*; 51; 1368–1376.

Thompson JD and Levinson DS, inventors; Hughes Aircraft Co, assignee (1988). Microwave directional filter with quasi-elliptic response. U.S. patent 4,725,797. Feb. 16.

Wikipedia (2011). Chebyshev filter. Sep. 14. On en.wikipedia.org. Accessed Sep. 14, 2011.

Williams AE and Atia AE (1977). Dual-mode canonical waveguide filters. *IEEE Transactions on Microwave Theory and Techniques*; 25; 2.

Wolk D, Damaschke J, and Schmitt D, inventors; Robert Bosch GmbH, assignee (2002). Frequency-stabilized waveguide arrangement. US patent 6,433,656 B1. Aug. 13.

Yao H-W, Zaki KA, Atia AE, and Hershtig R (1995). Full wave modeling of conducting posts in rectangular waveguides and its applications to slot coupled combline filters. *IEEE Transactions on Microwave Theory and Techniques*; 43; 12.

Yassini B and Yu M (2015). Ka-band dual-mode super Q filters and multiplexers. *IEEE Transactions on Microwave Theory and Techniques*; 63; 10.

Young HD and Freedman RA (2012). *University Physics*, 13th ed. San Francisco: Addison-Wesley.

Yu M, Tang W-C, Malarky A, Dokas V, Cameron R, and Wang Y (2003). Predistortion technique for cross-coupled filters and its application to satellite communication systems. *IEEE Transactions on Microwave Theory and Techniques*; 51; 12.

Yu M, Smith D, and Ismail M (2004). Half-wave dielectric rod resonator filter. *IEEE MTT-S International Microwave Symposium Digest;* 2; June 6–11; 619–622.

Yu M and Miraftab SV, inventors; Com Dev International, assignee (2010). Cavity microwave filter assembly with lossy networks. U.S. patent 7,764,146 B2. July 27.

Zaki KA and Atia AE (1983). Modes in dielectric-loaded waveguides and resonators. *IEEE Transactions on Microwave Theory and Techniques*; 31; 1039–1045.

CHAPTER 6

LOW-NOISE AMPLIFIER AND FREQUENCY CONVERTER

6.1 INTRODUCTION

This chapter is about the **low-noise amplifier (LNA)** and the **frequency converter**, two units in the front end of the payload. We discuss their architecture, redundancy scheme, technology, and environmental considerations and give an example of their communications-related specification parameters.

The LNA sets, or chiefly sets, the noise figure of the payload. The LNA's essential features are its low noise figure and high gain. Any resistive losses before the LNA add one-to-one to the payload noise figure in dB, so the LNA must be placed as close as possible to the antenna terminal.

The frequency converter changes the carrier frequency of the signal. Uplink and downlink frequencies are always different. The essential features of the frequency converter are good spurious response, intermod performance, and linearity. Most commonly for non-processing payloads, there is only one frequency conversion, a **downconversion (D/C)**, from the radio frequency (RF) at payload input to a lower RF for payload output. However, depending on the specific frequencies, spurious requirements, and D/C performance, there may first be a D/C from RF to an **intermediate frequency (IF)** at which the signal is filtered, followed by an **upconversion (U/C)** to the output RF. For a processing payload, the D/C would be to baseband or near baseband, followed by an U/C. The D/C is almost always

Satellite Communications Payload and System, Second Edition. Teresa M. Braun and Walter R. Braun.
© 2021 John Wiley & Sons, Inc. Published 2021 by John Wiley & Sons, Inc.

located right next to the LNA to reduce the losses on the following transmission line to the input multiplexers (Section 5.5.4), since lower frequencies have less loss in waveguide and coaxial cable.

A LNA and a D/C together are sometimes called a **receiver**. This name especially makes sense when each active LNA is paired with one particular active D/C, which is a common architecture in non-processing payloads.

6.2   LNAS AND FREQUENCY CONVERTERS IN PAYLOAD

6.2.1   Architecture in Payload

Figure 6.1 illustrates some LNA and downconverter architectures at payload level. (Recall that a double line indicates waveguide and a single line coax.) Figure 6.1a shows an architecture for one uplink beam containing various signals on different carrier frequencies that are being downconverted with different local oscillators. Figure 6.1b shows an architecture for four uplink beams from two antennas where each antenna has an orthomode transducer (Section 3.8.1) to separate out two linearly polarized signals. Figure 6.1c shows an architecture for four uplink beams where each beam comes from a different antenna feed. This would be for a multibeam antenna scheme with a single feed per beam (Section 11.6).

6.2.2   Redundancy Scheme

Reliability followed by lowest mass is the main drivers of the redundancy scheme for the LNAs and frequency converters in the payload. Reliability may be defined differently for different payloads. For example, when there are only a few uplink beams the loss of one beam may be regarded as catastrophic, while when there are dozens the loss of one or two may not be disastrous.

Figure 6.2 shows various redundancy schemes for when there are only a few active receiver paths. These examples have just two active paths. All four schemes have four LNAs and four D/Cs. Schemes (a) and (b) use internally redundant units, while schemes (c) and (d) put single units into redundancy rings. Scheme (a) provides the lowest reliability, schemes (b) and (c) provide higher reliability than (a), and scheme (d) provides the highest. This does not necessarily mean that

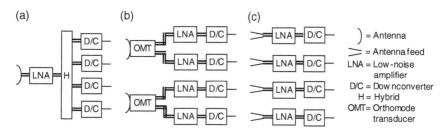

FIGURE 6.1   Examples of LNA-D/C architecture in payload (redundancy not shown).

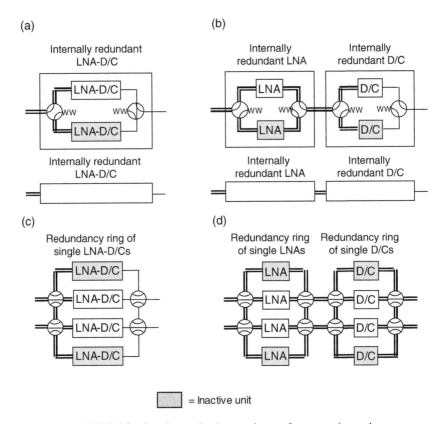

FIGURE 6.2 Receiver redundancy schemes for two active paths.

(b) is better than (a), and so forth. If the reliability provided by (a) is high enough when considered as a contributor to the overall payload reliability, then it is unnecessary and wasteful to employ the additional switches and waveguide of the other schemes. Redundancy of 2:1 or $2k:k$ for integer k (Section 4.6) is most common for the case when there is a small number of uplink beams.

An active receive phased array does not have LNA redundancy right at the antenna elements. Each antenna element has one LNA. Redundancy is provided by the extra-large number of antenna elements (Section 11.12.1).

6.2.3 Combining

LNA or receiver outputs are not normally combined but they are combined in an active, receive phased array. Let us suppose we are dealing with LNAs, but the following would equally apply to receivers. In such an array, each antenna element is connected to its own LNA. To form each beam, all the LNA outputs must be properly phased and perhaps gain-controlled before being combined, so phase and gain stability over temperature and life of the LNAs and the beam-forming network are crucial. The LNAs must individually exhibit stability, and the difference between any pair over a limited temperature range and life must exhibit stability (Yeung et al., 1993). The lower the frequency band is, the longer the guided wavelength, which makes it easier to achieve stability in the beam-forming network.

6.3 NONLINEARITY OF LNA AND FREQUENCY CONVERTER

6.3.1 Intermodulation Products

Amplifiers and frequency converters are inherently nonlinear elements, meaning that their response to a signal cannot be represented by a filter. The LNA may be slightly nonlinear since the amplification and phase shift it provides may slightly depend on the power of the input RF signal. The frequency converter is nonlinear for that reason, too, and because it changes the frequency band.

When two tones combine in a nonlinear element, the output is many tones. If the input tone frequencies are f_1 and f_2 in Hz, then the frequency and **order** of the (m,n)th **intermodulation product (IMP)** are given by:

IMP (m, n) has frequency $= |mf_1 - nf_2|$ and order $= |m| + |n|$ for integers m and n

Note that IMP(m,n) and IMP$(-m,-n)$ are identical. The outputs at f_1 and f_2 are at the **fundamental frequencies**; the outputs at multiples of f_1 and multiples of f_2 are **harmonics**; and the other outputs are **cross-products** of f_1 and f_2.

The zeroth-order IMPs are not produced by RF devices because RF devices do not couple to **direct current (DC)**, which is 0 Hz.

Figure 6.3 gives an example of the IMPs through fourth order when f_1 and f_2 are close together and $f_1 < f_2$. This example applies primarily to the LNA and to component amplifiers. The IMPs' magnitudes are not shown since they depend on the magnitudes of the fundamentals and the characteristics of the nonlinear element. All IMPs except the third-order are **out-of-band**, meaning that they lie outside the frequency band inhabited by the signal of interest. Of course, they may lie in other frequency bands and cause a problem for signals in those bands.

Other numbers of signals besides two can be input to a nonlinear element. A one-tone input yields output at the fundamental and in general all harmonics. A three-tone input at $f_1, f_2,$ and f_3 yields IMPs in general at all frequencies of the form $|mf_1 - nf_2 + kf_3|$, and similarly for any number of tones input.

A modulated-signal input can be thought of as consisting of many tones very close together; thus, the output of the nonlinear element can be roughly conceived of. Some third-order IMPs of a modulated signal will be **inband**, meaning occupying the same frequency band as the signal, and some will be in **near out-of-band**, that is, in the neighboring channels. Section 9.5.1 provides more discussion on this topic.

Figure 6.4 shows another example of IMPs, this time for two tones widely spaced in frequency. This example applies primarily to a frequency converter, in

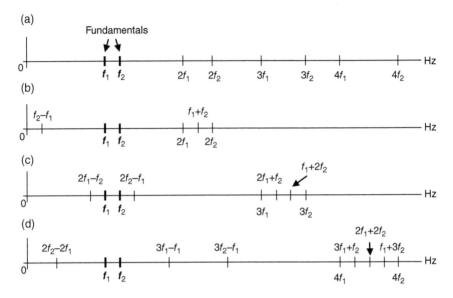

FIGURE 6.3 IMPs of two tones close together: (a) harmonics through 4th-order, (b) 2nd-order products, (c) 3rd-order products, and (d) 4th-order products.

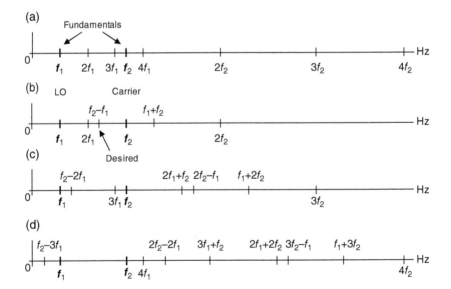

FIGURE 6.4 IMPs of two tones widely separated in frequency: (a) harmonics through 4th-order, (b) 2nd-order products, (c) 3rd-order products, and (d) 4th-order products.

which the input carrier is multiplied by a tone (label "LO" in the figure), thereby shifting the frequency of the carrier (Section 6.6.1). All the IMPs can pose a problem to the signal of interest or other signals. This is why frequency conversions must be designed and built with care. Figure 6.4b shows that the conversion creates two second-order products, not including harmonics. One is marked as desired, the other not. In fact, either one may be the desired one (but in most cases it is the one marked). This topic is addressed in Section 6.6.3.

6.3.2 Amplifier Nonlinearity Characterization

The LNA is designed to operate in a nearly linear amplification region. The LNA does not have explicit IMP specifications but instead has the extent of its nonlinear behavior characterized by one of the two following parameters.

One parameter is the **1-dB compression point**. It is the point at which the actual gain of the LNA with a CW input is 1 dB lower than the linear gain, as illustrated in Figure 6.5. It is characterized by both an input power and an output power. As the signal power into the element increases, even well below saturation, the element no longer operates in a linear fashion but its gain starts to decrease. This is **gain compression**. The LNA starts to create harmonics along with the intended output signal. Amplifiers are normally run with input power at least 10 dB below their 1-dB compression point input (Galla, 1989). This is considered to be the **linear and small-signal region of operation** (Anritsu, 2000).

The other parameter is the **3rd-order intercept point (IP3)**, whose definition is illustrated in Figure 6.6. IP3 is characterized by both an input power and an output

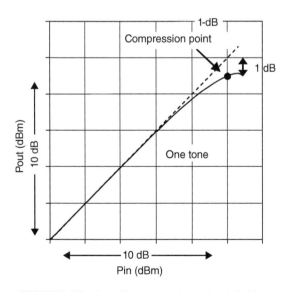

FIGURE 6.5 One-dB compression-point definition.

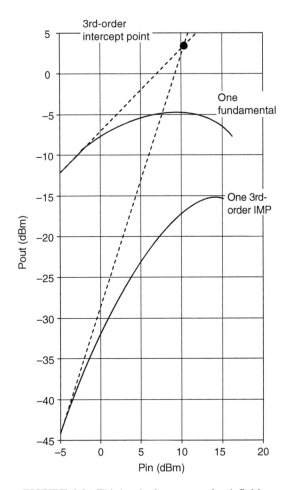

FIGURE 6.6 Third-order intercept-point definition.

power. With knowledge of this point and the power levels of two equal input tones, the levels of the 3rd-order IMPs can be calculated. To measure IP3, two equal-power tones, whose combined input power is in the linear-amplification region of the amplifier, are input to the amplifier. The tones are at frequencies not far from the middle of the amplifier's passband. The power of one output fundamental and the power of one of the nearby 3rd-order IMPs are measured. The results are sensitive to the fundamental frequencies used, so they must be stated. Measurements must be made down to at least 10 dB below the 1-dB compression point (Anritsu, 2015). The amplification curve of one fundamental is extrapolated from the measured point with the slope of 1 dB/dB. The amplification curve of one 3rd-order IMP is extrapolated from its measured point with the slope of 3 dB/dB. The independent axis is the power of one input tone (Keysight Technologies, 2018; Anritsu, 2015).

IP3 is more important than the other orders of intercept point because the 3rd-order IMPs are the largest IMPs and the closest in frequency to the fundamental, thereby a larger threat especially for multicarrier systems.

For an amplifier (but not a frequency converter (Henderson, 1981a)) the 3rd-order intercept-point input is generally 10 dB higher than the 1-dB compression-point input (Henderson, 1990).

6.3.3 Frequency Converter Nonlinearity Characterization

Ignoring the fact that the frequency converter translates the signal's frequency, a nonlinear operation, the amplitude compression of the frequency converter must be characterized. The parameters 1-dB compression point or IP3 can be used or **carrier-to-3rd-order-IM ratio (C/3IM)**. The latter is measured with two (not counting the local oscillator (LO)) equal-power inband tones at some stated power input to the converter. C/3IM is the ratio at output of the power in one fundamental to the power in one 3rd-order IMP. The independent axis is usually the input power of one carrier, but some people use the power of both tones together. The end of Section 6.6.4 contains further discussion on the frequency converter nonlinearity. C/3IM is defined somewhat differently for simply an amplifier (Section 7.2.2).

6.4 NOISE FIGURE

Every two-port electronic element can be characterized by its **gain** G and **noise figure** F, as portrayed in Figure 6.7. The two small circles for the ports are present for emphasis. One port is for input and the other is for output.

Both the input signal and the input noise experience the gain G as they go through the element.

If the input to the electronic element is noise at temperature T_0, the noise temperature at element output consists of two parts: T_0 amplified by the gain G and the thermal noise T_{int} generated internally by the element.

The noise figure F, which is not expressed in dB, is defined as the ratio of the combined output-noise temperature to the output-noise temperature due only to input noise, when $T_0 = 290$ and G is a power ratio not in dB:

$$\text{Noise figure } F = \frac{T_{0\,out} + T_{int}}{T_{0\,out}} \text{ when } T_0 = 290 \text{ K}$$

$$= \frac{G\,290 + T_{int}}{G\,290} = 1 + \frac{T_{int}}{G\,290}$$

Where o is a port

FIGURE 6.7 Characterization of general two-port electronic element.

NF is always defined on the basis of $T_0 = 290\,K$, to make the definition independent of the noise temperature actually input to the element. F is always greater than 1. The contribution to the output-noise temperature T_{int} that the element generates internally is related to F by the following:

$$T_{int} = (F-1)G\ 290$$

For the general input noise temperature T_0, the total output-noise temperature is $GT_0 + T_{int}$

There is further discussion on noise elsewhere in this book:

- System noise temperature and payload noise figure in Section 8.12.1.
- How to deal with noise figure in Section 9.2.
- Noise budget in Section 9.3.
- Noise level on communications links in Section 14.5.

6.5   LOW-NOISE AMPLIFIER

6.5.1   LNA Unit Architecture and Technology

Because the LNA's noise figure is the basis of the entire payload's noise figure, a manufacturer with the lowest noise figure has a key advantage, so work is always going on to improve low-noise transistor technology.

The LNA consists of a single chain of component amplifiers. An example is shown in Figure 6.8, with two extra-low-noise devices for the first two stages and two low-noise devices for the last two stages.

There are currently three technologies used for the low-noise transistors in payload LNAs. They are all a *high-electron-mobility transistor* (*HEMT*) or a derivative. All use GaAs, a gallium-arsenide compound semiconductor, and/or InP, an indium-phosphide compound semiconductor. InP is newer and has lower noise figure.

The HEMT is a type of *field-effect transistor* (*FET*). A FET has a *source*, which is a conductive-metal terminal. Charge carriers (which in a HEMT are electrons) enter the FET through the source. The carriers move through the FET's *channel* to

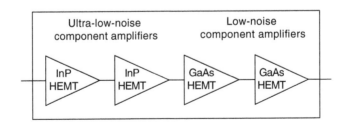

FIGURE 6.8   Example of LNA architecture.

the FET's exit terminal, the *drain*. The *gate* terminal controls the conductivity of the channel, according to the difference in voltage applied to the gate and source. The gate *length* has a major effect on the FET's maximum operational frequency: the shorter the gate the higher the maximum frequency (Microwave Encyclopedia, 2006). The *substrate* (also called *body* or *base*) of the FET is the bulk of the semiconductor in which lie the gate, source, and drain (Wikipedia, 2017a). It has two main functions: (1) to provide the starting point for the active-layer formation (growth), and (2) to have low conductivity and thereby confine the current path of the FET to the active layers nearer the surface and the gate control electrode (Sowers, 2019).

A drawing of a basic HEMT is given in Figure 6.9. The HEMT's channel is at the junction between two semiconductors of different band gaps; thus it is called a *hetero-junction*. A material's *band gap* is the amount of additional energy an electron must have to jump from a bound state to a free state, where it is able to move and carry charge (Wikipedia, 2017b). The two semiconductors that border the channel are the *barrier* and the substrate, where the substrate has the narrower band gap of the two. The gate extends part way into the barrier. In the drawing, the channel is labeled the "two-dimensional electron gas," in which electrons are highly mobile. The HEMT is also known as a *heterostructure field effect transistor (HFET)* (Wikipedia, 2017c).

An improved version of the HEMT used in some LNAs is the *pseudomorphic high-electron-mobility transistor (pHEMT)*. To improve the performance by matching the slightly different-size lattices of the barrier and substrate, the pHEMT incorporates an extra, extremely thin layer of one of the materials. This layer is sometimes called the *buffer*. The buffer lies right on top of the substrate, below the channel (Paine et al., 2000).

An even further-improved HEMT is the *metamorphic high-electron-mobility transistor (mHEMT)*. It uses high indium-content semiconductors on a GaAs substrate. This allows the more-robust substrate technology with the high performance afforded by InP materials (Sowers, 2011; Smith et al., 2003). The French company

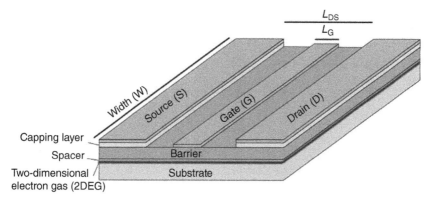

FIGURE 6.9 Drawing of basic high-electron-mobility transistor (HEMT) (Wikipedia, 2017c). (Credit: Marcin Białek).

OMMIC has space-qualified this technology and uses it in a 25–43 GHz LNA (Ommic, 2012).

In the naming convention of a HEMT, if a number starts the name, it is the gate length. For the GaAs and InP HEMTs, the GaAs or InP is respectively the substrate material. Material for other layers is optionally named. The material for the barrier and buffer and the material for the channel may be named, separated by a slash, in either order.

The older of the two technologies, GaAs, is used for GHz frequencies up through about 30 GHz. In recent years InP has at least partly taken over at 30 GHz and is the lowest-noise technology for frequencies up through 100 GHz and even higher (Chou et al., 2003).

6.5.2   LNA Environmental

Some LNAs have temperature compensation and some do not. Temperature compensation is achieved either by altering the amplifier bias or by means of a variable attenuator. An active element requires a **bias**, which is a steady voltage or current, to set its operating point (Wikipedia, 2020). A LNA without temperature compensation will exhibit higher gain and lower noise figure, the lower the temperature (Microwave Encyclopedia, 2010). In a payload without an active receive phased array, the gain variation can be accommodated by a sufficient dynamic range of the high-power amplifier's preamplifier. In a payload with such an array, temperature compensation is necessary.

An amplifier's gain decreases as the transistor ages (Paine et al., 2000; Chou et al., 2002). This can be compensated by a sufficient dynamic range of the preamp. Also, its noise figure goes up with aging.

6.5.3   LNA Specification

Table 6.1 is an example of the specification parameters for a LNA that is not part of an active receive phased array, that is, where no phase or power matching is required.

6.5.4   LNA Performance

Examples are given in Table 6.2 of noise figures of state-of-the-art LNAs.

6.6   FREQUENCY CONVERTER

6.6.1   Frequency Conversion Architecture

There are various ways to architect the collection of frequency converters in the payload. Reliability followed by lowest mass and cost are the main drivers of the choice. The choice is partly driven by the number of conversions required.

TABLE 6.1 Example of Parameters in a LNA Specification

Parameter in LNA Specification	Units
Input frequency range	GHz
1-dB compression point input	dBm
Noise figure — EOL over temperature	dB
Gain min and max — BOL over temperature; EOL over temperature	dB
Gain variation — over input frequency range; over any 100 MHz	dB p-p
Gain slope	dB/MHz
Gain stability — BOL over temperature; BOL over any 15 °C; EOL over temperature	dB p-p
Phase linearity for input carrier in some stated power range — over input frequency range; over any 100 MHz	deg p-p
Return loss — input; output	dB
Operational temperature range	°C

TABLE 6.2 Noise Figures of State-of-the-Art LNAs

Frequency Band	Bandwidth (MHz)	Max NF at Ambient and BOL (dB)	Max NF Over Temp at EOL (dB)
L-band	34	0.65	0.85
S-band	30	0.9	1.15
C-band	580	1.2	1.45
X-band	500	1.3	1.5
Ku-band (FSS) 14 GHz	750	1.2	1.5
Ku-band (BSS) 17 GHz	800	1.5	2.0
Ka-band	500	—	1.9[a]
Ka-band	3000	1.9	2.2

[a] Thales Alenia Space (2012), all others NEC Space Technologies (2019).

Typical architectures for the frequency conversion are shown in Figure 6.10, with the more typical in Figure 6.10a. Each converter has its own internal **LO**, which provides a tone. The LOs may have different frequencies. All the LOs lock to the **master reference oscillator (MRO)**. The MRO is critical to the entire payload so it almost always has 3:1 redundancy (Frequency Electronics, Inc., 2008). One MRO at a time is active. It is distributed to the individual converters by a passive network because that is more reliable than an active network. An alternative architecture for self-contained frequency converters is shown in Figure 6.10b. There is no MRO. Instead, each frequency converter contains its own **reference oscillator** from which its local oscillator derives (for example Mitsubishi Electric, 2015).

The architecture of Figure 6.10a can be altered to have the local oscillators external to the frequency converters. This is a good idea when the same LO frequency is

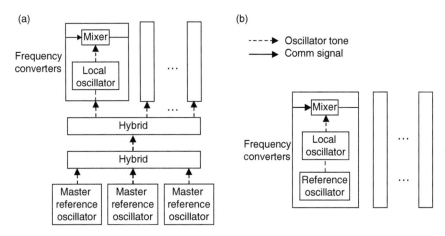

FIGURE 6.10 Typical architectures for frequency converters with internal local oscillators (redundancy not shown).

used for several converters. Then frequency-stability performance over temperature and aging is consistent across all frequency converters using the same MRO. Figure 6.11 shows two such architectures:

    a. LOs external to downconverters and fed into multiple downconverters.
    b. Dual conversions, namely a first set of frequency converters that perform D/C to IF and a second set of frequency converters that perform U/C to downlink RF, all fed by the same active MRO.

6.6.2 Frequency-Converter Unit Packaging

What we have been labeling a "frequency converter" may in fact come in two different packagings. One is a self-contained unit. The other is one element in an assembly of such elements. Both options are shown in Figure 6.12.

Payloads with dozens of conversions need the converters to be smaller and lighter than the self-contained ones. Since at least 2012, manufacturers have been producing multiple Ka-band frequency-converter **slices** that they incorporate into assemblies. One or two LOs serve all the slices in the assembly. The LOs may have dedicated reference oscillators or they may be fed by a MRO. One currently offered assembly allows between 5 and 10 converters with external LO. It can be used on either up or down conversion. Its DC/DC power converter serves multiple assemblies (Thales Alenia Space, 2012). Something like this is illustrated in Figure 6.12b. A different assembly allows between 4 and 14 downconverters and accommodates a redundant reference oscillator and LO in its housing. It can also have channel filters integrated (L3 Narda Microwave West, 2019). The payload could have many such assemblies.

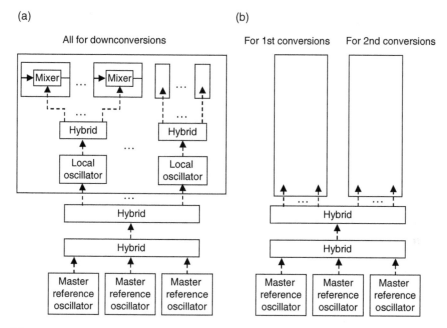

FIGURE 6.11 Architectures for frequency converters with external local oscillators (redundancy not shown).

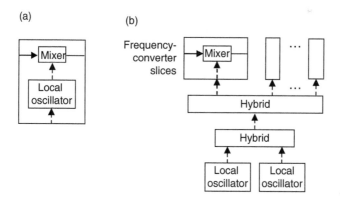

FIGURE 6.12 Frequency-converter packagings: (a) self-contained frequency converter, simplified; (b) assembly of multiple frequency-converter slices and redundant local oscillator.

6.6.3 Frequency-Converter Unit Architecture and Function

The basic architecture of a frequency converter is shown in Figure 6.13. The architecture is the same for a downconverter from RF to IF as for an upconverter from IF to RF. The local oscillator produces a **continuous wave (CW)**, that is, a tone.

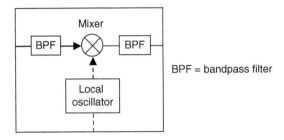

FIGURE 6.13   Frequency converter basic architecture.

The mixer multiplies the RF (IF) signal by the oscillator signal and creates two signals at different frequencies, only one of which is the desired IF (RF) output signal. Suppose $s(t) = \cos(2\pi f_c t + \theta(t))$ is the converter's input signal and $r(t) = \cos(2\pi f_{LO} t)$ is the LO signal, ignoring signal amplitudes. Then the mixer output is

$$s(t)r(t) = \cos\left(2\pi f_c t + \theta(t)\right)\cos\left(2\pi f_{LO} t\right)$$
$$= \frac{1}{2}\cos\left(2\pi(f_c - f_{LO})t + \theta(t)\right) + \frac{1}{2}\cos\left(2\pi(f_c + f_{LO})t + \theta(t)\right)$$

The frequency-difference term is described as being at the frequency $f_c - f_{LO}$ if that is a positive frequency or at $f_{LO} - f_c$ otherwise. The input and output filters suppress unwanted signals. The LO may be internal to the unit as shown or external. In the following sections, we discuss each element in some detail.

Most of the time $f_{LO} < f_c$, in which case the conversion is a *low-side conversion* as illustrated in Figure 6.14a. The mixer creates two signals, and if the frequency converter is a downconverter the higher-frequency output at $f_c + f_{LO}$ is filtered out, and if it is an upconverter the lower-frequency output at $f_c - f_{LO}$ is filtered out. There are times when a low-side conversion may not be a good idea, for example when a LO harmonic may lie very near f_c, and then a *high-side conversion* may be the answer.

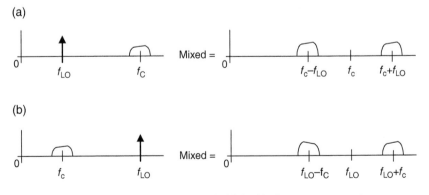

FIGURE 6.14   (a) Low-side and (b) high-side frequency conversions.

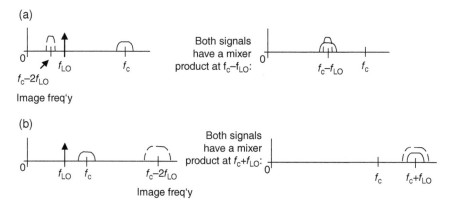

FIGURE 6.15 Examples of image frequency for low-side conversion: (a) low-side downconversion and (b) low-side upconversion.

Then $f_c < f_{LO}$. The situation is represented in Figure 6.14b. Notice that the lower-frequency mixer output has been inverted about its output carrier frequency. For some applications, such frequency inversion is undesirable.

There are two input frequencies that can mix to the desired frequency, so the undesired one at the *image frequency* must be removed before the mixing. Figure 6.15 illustrates where the image frequency is for low-side conversions. For both low-side and high-side conversions, in D/C the image frequency is at $|f_c - 2f_{LO}|$, and in U/C it is at $f_c + 2f_{LO}$. There are two ways of removing signal at the image frequency: filtering it out with the frequency converter's input bandpass filter or having the mixer cancel it. Such a mixer is an *image-reject mixer* (Sorrentino and Bianchi, 2010).

The two output frequencies of an ideal low-side mixer are the mixer products $(1,1)$ and $(1,-1)$, which have order 2. For a D/C, $(1,1)$ is the desired one and $(1,-1)$ the undesired one, and vice versa for an U/C. The leaked signals at output are $(1,0)$ and $(0,-1)$. Harmonics are $(m,0)$ and $(0,n)$ for $m > 1$ and $-n > 1$, respectively. The LO harmonics are much stronger than the input-signal harmonics (see below). The unwanted cross-products are all other (m,n).

6.6.4 Mixer

The mixer itself suppresses unwanted IMPs and leaked signals to a large extent. The possible port-to-port leakages are illustrated in Figure 6.16. The input frequency is labeled as RF and the output as IF, but the output could be the downlink RF, and other combinations are possible. There are four types of mixer circuits: single-ended, single-balanced, double-balanced, and triple-balanced. The single-balanced consists of two single-ended mixers, the double-balanced of two single-balanced, and so on. The most common type is the double-balanced. It has very good LO-to-RF isolation and LO-to-IF isolation. Port-to-port isolation is the

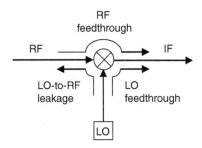

FIGURE 6.16 Mixer leakage definitions. (After Keysight Technologies (2000)).

ratio of the amount of power fed into one port relative to the amount of power that comes out another port. The double-balanced mixer theoretically creates only one fourth of all IMPs, the ones that involve an odd multiple of f_{RF} and an odd multiple of f_{IF} (Henderson, 1981b).

Several aspects of good mixer performance depend on the RF input power being well below the LO power. In most cases, it is at least 20 dB below (Henderson, 1981a).

Port-to-port isolation depends on LO level and temperature. Normally only the LO-to-RF and LO-to-IF isolations are specified because the RF signal level is so much lower than the LO level (Henderson, 1981a).

One vendor offers a mixer with a special design to minimize the 2LO harmonic for the Ka-to-K D/C, in which the 2LO frequency approximately equals the K-band frequency (Rodgers and Montauti, 2013).

The nonlinearity of the mixer is characterized only in terms of the RF signal level(s), for a given LO level and temperature. Generally, the 1-dB compression point input is 5–10 dB lower than the LO input power (Henderson, 1981a). Unlike for an amplifier, the 1-dB compression-point output is generally less than 10 dB below the 3rd-order intercept-point output (Henderson, 1990). The 3rd-order intercept output equals the intercept input plus the mixer conversion gain, which is negative in dB for a passive mixer and positive for an active one (Henderson, 1981a).

6.6.5 Reference Oscillator

There are two cases of reference oscillator. When all local oscillators derive from one reference oscillator, it is a MRO. Otherwise, it is simply a reference oscillator.

The reference oscillator is an ultra-stable quartz **crystal oscillator (XO)** (Fruehauf, 2007). The frequency of a MRO is usually 10 MHz (Frequency Electronics, Inc., 2019). When the reference's frequency gets multiplied up, the rms phase noise gets multiplied up approximately proportionately (Section 6.6.6.2). As carrier frequencies trend upward, a higher-frequency reference that would result in lower phase noise is desirable. A 128-MHz one has been space-qualified (Reddy et al., 2012).

The reference oscillator is a circuit that takes a voltage signal from a vibrating crystal, amplifies it, and feeds a portion of the amplified signal back to the crystal, which keeps the crystal expanding and contracting at one of its resonant frequencies. A voltage applied across the crystal causes it to contract, and the contraction creates a voltage across the crystal — this is the *piezoelectric effect* (Hewlett Packard, 1997).

The crystal for a reference oscillator is cut in the form of a plate (Fruehauf, 2007). The cut determines the fundamental frequency and what *overtones* (that is, harmonic or near-harmonic signals) and non-harmonic signals the oscillator circuit will generate. A non-harmonic signal close to the desired overtone poses a risk because a small environmental change can shift the non-harmonic one on top of the desired overtone, thereby weakening it and creating an *activity dip* (Hewlett Packard, 1997). An activity dip produces an undesirable abrupt change of frequency over a narrow temperature range. The thicker the plate, the higher is the order of the overtone that the crystal can vibrate in (Bloch et al., 2002).

The oscillator circuit is subject to environmental influences that change the crystal vibration frequency. The major influence would be temperature change. The second most significant influence is time, during which the oscillator ages and radiation bathes the spacecraft. Another influence is the oscillator circuit being turned off and on, which causes an offset in the frequency but does not affect the aging (Hewlett Packard, 1997).

An effective means is normally taken against temperature variation. One option is temperature-compensation. The *temperature-compensated crystal oscillator (TCXO)* uses thermistors and varactors to compensate for the crystal's frequency variation over temperature. (A thermistor is a type of resistor whose resistance strongly depends on temperature, and a varactor is a voltage-controlled capacitor (Wikipedia, 2019a; Wikipedia, 2019b).) Better than temperature compensation is an oven. The *oven-controlled crystal oscillator (OCXO)* has a heater and heater control in the oscillator circuit, as well as the temperature-sensitive elements in a thermally insulated container. An OCXO requires more DC power than a TCXO due to the power consumption of the heater (Hitch, 2019a). A superior option is a double oven (Fruehauf, 2007). This requires even more DC power than an OCXO (Hitch, 2019a). With all options, the idea is to keep the crystal at its temperature of minimum temperature-sensitivity (Hewlett Packard, 1997).

An active oscillator's frequency change with aging may be positive or negative, at a rate which is usually logarithmic with respect to time (Hitch, 2019a). The direction is ascertained from testing. The frequency change arises primarily from two things. One is mass transfer of contamination inside the resonator enclosure to the crystal. The other is stress relief in the resonator's mounting and bonding structure, the electrodes, and perhaps in the quartz. The frequency change is minimized through superior manufacturing technology (Fruehauf, 2007). A dormant oscillator may or may not have a linear frequency change over time (Bloch et al., 2009a).

The total ionizing dose of radiation on orbit will cause the oscillator frequency to change at a relatively constant negative rate. The severity of the radiation effect is minimized when the crystal material is radiation-hardened by high-temperature heating and when the crystal plate is thick enough to vibrate in its 5th overtone. The effect is partially compensated by selecting crystals that have shown positive frequency change in aging testing (Bloch et al., 2002).

Another form of radiation, high-energy particles from solar flares, can cause permanent frequency offsets (Bloch et al., 2009b).

6.6.6   Phase Noise and Local Oscillator Generation

6.6.6.1 *Phase Noise Basics*   **Phase noise** is an unwanted variation in the signal phase that comes from the communications system's oscillators. These oscillators comprise the carrier that is initially modulated on the ground, the payload frequency-conversion LO(s), the oscillator in the receiving ground station's carrier recovery, and any conversion LOs in the ground stations.

Phase noise has a spectrum (Section 12.2.3) as a function of the *offset frequency*, which is the frequency relative to the carrier frequency. A phase noise spectrum plot has *y*-axis labeled $\mathcal{L}$, the *single-sideband (SSB) phase noise*:

$$\mathcal{L}(f) = \frac{1}{2} S_\phi(f) \text{ where } S_\phi(f) \text{ is one-sided spectrum of phase noise}$$

$$\text{for offset frequency } f > 0$$

It is one side of the double-sided phase noise spectrum (Gardner, 2005). $\mathcal{L}$ is usually given in units of dBc/Hz, meaning power integrated over 1 Hz, stated as relative to the carrier power in dB. The frequency axis is shown in a logarithmic scale (so cannot go down to zero Hz). When plotted in this way, $\mathcal{L}(f)$ has decreasing slope with increasing f.

The rms phase noise with frequencies between f_1 and f_2 is then given by

$$\text{Rms phase noise with offset frequencies between } f_1 \text{ and } f_2 = 2\int_{f_1}^{f_2} \mathcal{L}(f) \mathrm{d}f$$

A formula for approximating such an integral is given in the Appendix 6.A.

The spectrum is smooth except for some small discrete components (that is, tones), called **spurious phase modulation (PM)**, at low offset frequencies. Most often these arise in synthesized LOs (Hitch, 2019b). These frequencies are sometimes so low (close to the carrier) that this spurious PM does not matter — they lie outside the frequency converter's specification band. However, since this is not always the case, the spurious PM must be measured as part of the oscillator testing (Hitch, 2019a). An example of a spectrum with spurious PM is shown in Figure 6.17. Section 12.10.2 provides a description of what part of the phase noise spectrum should have specifications on it and how phase noise affects the signal.

6.6.6.2 *Phase-Locked Oscillator*   A LO is created by *phase-locking* a *voltage-controlled oscillator* (VCO) to a reference oscillator. A voltage input drives the VCO so that it matches the frequency and phase of the reference oscillator or some

$\mathcal{L}$ (dBc/Hz)

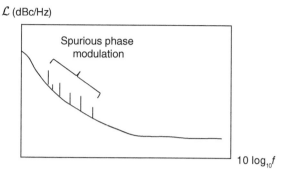

FIGURE 6.17 Example of a phase noise spectrum with spurious phase modulation.

multiple of it. There are at least two reasons to employ phase-locking: (1) to create a LO from an ultra-stable reference oscillator that is at a much lower frequency and (2) to obtain a phase noise spectrum for the LO that combines the best parts of the phase noise spectra of the reference oscillator and the VCO.

A simple *phase-locked loop* (*PLL*) to create a LO is shown in Figure 6.18. (This type is not actually used but employs fundamentally the same concept as current PLLs.) A PLL is a feedback circuit. The VCO output frequency is divided down by a factor N and that mixes with the reference oscillator's tone. The mixer serves as a phase detector (the double-frequency term is suppressed) in that its DC output voltage is proportional to the sine of the phase error between the reference tone and the divided-down LO. The loop filter is something like an integrator. The VCO integrates the error control voltage. This drives the VCO to bring the frequency and phase errors to zero. Whether the phase error is positive or negative, if it is not too big the error voltage will drive the VCO in the right direction (Gardner, 2005).

A PLL has a closed-loop transfer function, so a *loop bandwidth* B_L can be defined. It is a one-sided baseband bandwidth (Section 12.3.1). It actually has various definitions but they all provide roughly the same value on a logarithmic scale (Gardner, 2005).

The LO's phase noise is related to the phase noise of both the reference oscillator and the VCO. Roughly speaking, at offset frequencies less than the loop bandwidth the phase noise spectrum is that of the multiplied-up reference oscillator, and at higher frequencies the spectrum is that of the VCO if it were free-running (Gardner, 2005). A free-running VCO is without control-voltage input. The loop bandwidth of a PLL that creates a LO has a typical value of 100 KHz (API Technologies, 2019). This is all illustrated in Figure 6.19, where the VCO is a dielectric-resonator oscillator (DRO) (see below).

The LO of Figure 6.18 is a version of the reference oscillator multiplied by N. If the reference oscillator's phase is represented by $2\pi f_{ref}t + \varphi(t)$, where $\varphi(t)$ is the phase noise, then the phase of the tone multiplied-up by a factor of N is $2\pi N f_{ref}t + N\varphi(t)$, for offset frequencies less than roughly the loop bandwidth.

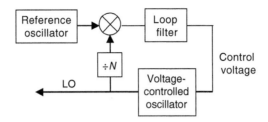

FIGURE 6.18 Simple phase-locked loop to create local oscillator.

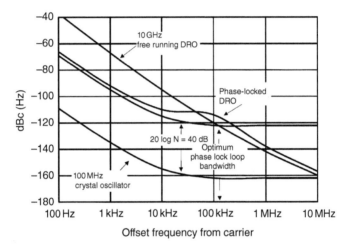

Offset frequency from carrier

FIGURE 6.19 Phase noise spectrum of reference oscillator, free-running VCO, and phase-locked VCO. (Used with permission of API Technologies. From API Technologies (2019)).

The LO's phase noise spectrum is N^2 times that of the reference oscillator's in this region. For the multiplied phase noise to be low, the reference oscillator must be ultra-stable, meaning it has ultra-low phase noise.

6.6.6.3 Phase Noise from Frequency Conversion
A non-regenerative payload does not phase-lock to the uplink signal to perform carrier recovery (Section 12.10.2), so such a payload's frequency converter simply adds its phase noise to that of the signal transmitted from the ground. Thus, payload LOs must have low phase noise.

The total phase noise added by a dual conversion (down followed by up) is minimized when the same reference oscillator is used for both LOs. Suppose that the D/C LO is a version of the reference oscillator multiplied by κ and the U/C LO is a version of the reference oscillator multiplied by λ, where $\kappa > \lambda$ and both conversions are low-side. The phase of the two LOs combined is $2\pi(\kappa - \lambda)f_{ref}t + (\kappa - \lambda)\varphi(t)$, for offset frequencies less than roughly the LOs' loop bandwidth. In this region, the rms phase noise added to the signal by the dual conversion is only $\kappa - \lambda$ times that of the reference oscillator. The subtraction indicates a cancellation of most of the low-frequency phase noise. In practice, the subtraction only occurs for offset frequencies up to 100 Hz or so from the carrier, but it is still significant (Hitch, 2019a). On the other hand, if unrelated LOs are used for the two conversions and their respective phase noises are $\kappa \varphi_1(t)$ and $\lambda \varphi_2(t)$, then the rms phase noise of the combined phase is the average value of $\sqrt{\kappa^2 \varphi_1^2 + \lambda^2 \varphi_2^2}$, which is much larger than in the case of a common reference oscillator.

6.6.7 Local Oscillator Technology

There are three technologies for the LO. When the LO frequency is fixed, the LO is said to be *fixed-frequency*, and when the LO frequency can take on more than one value, the LO is said to be *agile*. The terminology carries up to the frequency converter itself.

6.6.7.1 Dielectric-Resonator Oscillator and Coaxial-Resonator Oscillator
The most common way of deriving the LO is to phase-lock a *DRO* to the reference oscillator (Fiedziuszko, 2002). This method has replaced the old method of a simple frequency multiplication of the reference oscillator because the new method makes for a smaller, less expensive, and more reliable LO that has lower spurious outputs (Hitch and Holden, 1997). If the LO frequency is below about 2 GHz, instead of a DRO a *coaxial-resonator oscillator* (*CRO*) is used (Frequency Electronics, Inc., 2008; AtlanTecRF, 2019; Hitch and Holden, 1997).

The DRO resonates in only a narrow range of frequencies. The dielectric resonator is a short ceramic cylinder, a *puck*, supported inside a larger conductive enclosure (Skyworks, 2017). The puck is coupled to a microstrip line which is connected to a transistor (Hitch and Holden, 1997). The high-performance ceramics yield DROs with excellent frequency stability over temperature and life and little sensitivity to radiation.

The CRO has a coaxial resonator as shown in Figure 6.20. It is a TEM-mode coaxial line with a ceramic dielectric. The same or similar ceramics are used as in DROs. The line is shorted at one end and a parallel-plate microwave capacitor is used to resonate it. The circuit is connected to a transistor (Hitch and Holden, 1997).

The PLL for the LO, shown in Figure 6.21, is somewhat different from the simple PLL of Figure 6.18 because now a sampling phase detector replaces the divide-by-N and the mixer phase detector (Hitch and Holden, 1997). A DRO is shown in the figure here but it could just as well be a CRO. The sampling phase detector creates a large number of harmonics of the TCXO frequency (Andrews et al., 1990) (up to 170 (Hitch and Holden, 1997)), mixes the harmonics with the DRO's output tone, and puts out a DC voltage when one of the harmonics matches the DRO's frequency. The amplitude of this DC voltage is proportional to the sine of the phase

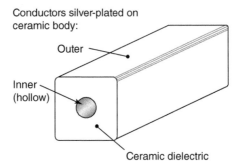

FIGURE 6.20 Coaxial resonator. (©2016 IEEE. Reprinted, with permission, from Reddy (2016)).

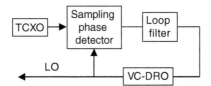

FIGURE 6.21 Dielectric-resonator oscillator phase-locked to multiple of reference oscillator. (After Hitch and Holden (1997)).

error. The DC voltage drives the DRO to phase-lock. Therefore, the DRO's free-running frequency must be close to the TCXO's desired harmonic so that it can be corrected to that harmonic.

6.6.7.2 Frequency Synthesizer A synthesizer also employs phase-locking but is able to create a large number of different frequencies (Gardner, 2005). Besides the flexibility this offers, another advantage is that each functional frequency synthesizer in a bank can serve as a backup for the others.

An example of synthesizer usage is the 19 identical synthesizer modules on the JCSAT-5A satellite launched in 2006. Each can synthesize a frequency over a range 100 MHz wide in one-MHz steps and is on-orbit programmable by ground command. A synthesizer module is shown in Figure 6.22, along with the single active OCXO, auxiliary reference, and comb module that feed it. The payload has 9:6 redundancy of the synthesizer modules and 3:1 redundancy of the other elements shown. The synthesizer module creates a UHF frequency by command setting of the dividers R and N. The auxiliary reference multiplies the OCXO's 10 MHz by ten, and the comb module multiplies that by a factor of m, creating the intermediate LO. The intermediate LO plus the UHF frequency equals the desired LO frequency for signal downconversion (Dayaratna et al., 2005).

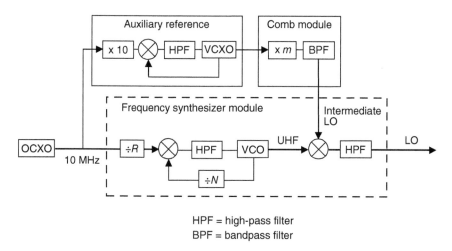

HPF = high-pass filter
BPF = bandpass filter

FIGURE 6.22 JCSAT-5A frequency synthesizer. (After Dayaratna et al. (2005)).

6.6.8 Frequency Converter Environmental

Besides the sensitivity of the reference oscillator to the environment, described above in Section 6.6.5, the frequency converter itself is sensitive to the environment. Its amplifiers show gain variation. The main sensitivity is to temperature. If necessary, the frequency converter can incorporate temperature-compensated gain control (for example Tramm, 2002). The amplifiers' gains also degrade over life. The high-power amplifier's preamplifier needs enough range to accommodate the frequency converter's gain variation.

6.6.9 Frequency Converter Specification

Table 6.3 gives an example of the communications-related parameters in a specification for a space-qualified Ka-to-K-band downconverter. Items separated by semicolons represent separate parameters. The 15 °C temperature range in some of the parameters usually represents a monthly temperature variation (Hitch, 2019a). The gain stability over temperature and life would be the sum of the separate temperature and life terms. A mask is a piecewise linear upper bound, usually as a function of frequency. Another frequency-converter specification could have somewhat different parameters that represent basically the same information.

6.7 RECEIVER

A self-contained receiver can consist of a single chain of LNA, reference oscillator, LO, and frequency converter (Thales Alenia Space, 2012; Mitsubishi Electric, 2015; L3 Narda Microwave West, 2017) or be without reference oscillator and LO (Thales

TABLE 6.3   Example of Parameters in a Frequency Converter Specification (L3 Narda Microwave West 2019)

Parameter in Frequency-Converter Specification	Units
Input frequency range	GHz
Translation frequency range	GHz
3rd-order intercept point output	dBm
Max operational power	dBm
Noise figure — EOL over temperature	dB
Gain min and max — BOL over temperature; EOL over temperature	dB
Gain flatness — over various bandwidths	dB p-p
Gain slope	dB/MHz
Gain stability — over 15 °C; over temperature; over life	dB p-p
Group delay variation over various bandwidths	ns p-p
Translation frequency stability — over various-size temperature ranges; over life at constant temperature	±ppm
Phase noise on output carrier — upper-bound mask for the phase noise spectrum	dBc/Hz
In-band mixer intermodulation products	dBc
In-band other spurs	dBc
Out-of-band spurious outputs[a]	dBc
LO harmonics[a]	dBm
Return loss — input; output	dB
Acceptance temperature range	°C

[a] In reference bandwidth of 4 KHz for frequencies below 15 GHz, of 1 MHz for frequencies above 15 GHz (Recommendation, ITU-R, SM.1541-6 2015).

Alenia Space, 2012). The best use of self-contained receivers with their own reference oscillator is when the number of receivers is not large. Payloads with dozens of receivers need the receivers to be smaller and lighter than self-contained receivers. Since at least 2012, manufacturers have been producing multiple Ka-band receiver slices in assemblies. One or two reference oscillators and LOs and one redundant DC/DC converter serve all the receiver slices in the assembly. One currently offered assembly has six receiver slices with external LO (Thales Alenia Space, 2012). Another assembly allows between 4 and 14 receiver slices and accommodates a redundant reference oscillator and LO in its housing. It can also have channel filters integrated (L3 Narda Microwave West, 2019).

6.A   APPENDIX. FORMULA FOR INTEGRATING PHASE NOISE SPECTRUM

Sometimes the payload engineer may want to integrate the payload's measured SSB phase noise spectrum $\mathcal{L}$ defined in Section 6.6.6.1 over a frequency range, by hand. If, for example, the phase noise exceeds its specification at one frequency, he

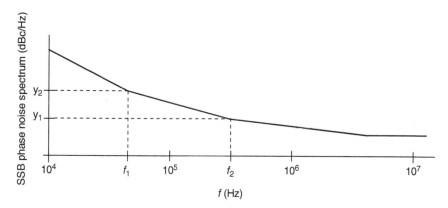

FIGURE 6.23 A piecewise-linear approximation to a single-sideband phase noise spectrum.

may want to check if nonetheless the spectrum fulfills the unstated specification underlying the specification he was given. The underlying specification, the one that matters, is whether the integral of the phase noise over a certain band is small enough (Section 12.10.2).

Figure 6.23 shows an example of a piecewise-linear approximation to a SSB phase noise spectrum. The engineer can upper-bound any measured spectrum in this way, to make it easy to integrate by hand. The phase noise specification itself may also be given in terms of such a mask.

The rms phase noise contribution that occurs in the interval between frequencies f_1 and f_2 in the figure, on which the spectrum is a straight line, is given by the following, where in this instance $\mathcal{L}(f)$ is not in dBc/Hz but y_i is, $i = 1,2$:

$$\sigma_\phi^2 \text{ contribution between } f_1 \text{ and } f_2 = 2\int_{f_1}^{f_2} \mathcal{L}(f)\,df = \begin{cases} \dfrac{2A}{m+1}\left(f_2^{m+1} - f_1^{m+1}\right) \text{ if } m \neq -1 \\ 2A \ln\left(f_2/f_1\right) \text{ if } m = -1 \end{cases}$$

where

$$A = \frac{10^{y_1/10}}{f_1^m}, \quad m = \frac{y_2 - y_1}{10\log_{10}\left(f_2/f_1\right)}, \quad y_i = 10\log_{10}\mathcal{L}(f_i), \quad i = 1,2$$

REFERENCES

Andrews J, Podell A, Mogri J, Karmel C, and Lee K (1990). GaAs MMIC phase locked source. *IEEE Microwave Theory and Techniques Symposium Digest*; 2; 815–818.

Anritsu Co (2000). Intermodulation distortion (IMD) measurements using the 37300 series vector network analyzer. Application note. On www.electrotest.co.nz/site/etl/Papers/Intermod_Measurements.pdf. Accessed Feb. 25, 2019.

Anritsu Co (2015). IMD measurements with IMDView™, MS4640B series vector network analyzer. Application note. On dl.cdn-anritsu.com/en-us/test-measurement/files/Application-Notes/Application-Note/11410-00859B.pdf. Accessed Feb. 22, 2019.

API Technologies (2019). Phase-locked DRO characteristics. DRO application note D-104. On micro.apitech.com/pdf/databook_dro.pdf. Accessed Feb. 22, 2019.

AtlanTecRF (2019). Phase locked oscillators, external reference APL – 03 series. Product data sheet. On www.atlantecrf.com/products/oscillators/phase_locked_oscillators_apl-03. htm. Accessed Feb. 15, 2019.

Bloch M, Mancini O, and McClelland T (2002). Performance of rubidium and quartz clocks in space. *Proceedings of the 2002 IEEE International Frequency Control Symposium*; May 29–31; 505–509.

Bloch M, Ho J, Mancini O, Terracciano L, and Mallette LA (2009a). Long-term frequency aging for unpowered space-class oscillators. *IEEE Transactions on Ultrasonics, Ferroelectrics, and Frequency Control*; 56; 10.

Bloch M, Mancini O, and McClelland T (2009b). What we don't know about quartz clocks in space. *Institute of Navigation Precise Time and Time Interval Meeting*; Nov. 16–19. On apps.dtic.mil/dtic/tr/fulltext/u2/a518055.pdf. Accessed Jan. 17, 2014.

Chou YC, Leung D, Lai R, Grundbacher R, Eng D, Scarpulla J, Barsky M, Liu PH, Biedenbender M, Oki A, and Streit D (2002). Evolution of DC and RF degradation induced by high-temperature accelerated lifetest of pseudomorphic GaAs and InGaAs/InAlAs/InP HEMT MMICs. *IEEE International Reliability Physics Symposium;* Apr. 7–11; 241–247.

Chou YC, Barsky M, Grundbacher R, Lai R, Leung D, Bonnin R, Akbany S, Tsui S, Kan Q, Eng D, and Oki A (2003). On the development of automatic assembly line for InP HEMT MMICs. *International Conference on Indium Phosphide and Related Materials;* May 12–16; 476-479.

Dayaratna L, Ramos LG, Hirokawa M, and Valenti S (2005). On orbit programmable frequency generation system for JCSAT 9 spacecraft. *IEEE MTT-S International Microwave Symposium Digest;* June 12–17; 1191–1194.

Fiedziuszko SJ (2002). Satellites and microwaves. *International Conference on Microwaves, Radar and Wireless Communications;* 3 (May 20–22); 937–953.

Frequency Electronics, Inc. (2008). Over 45 years of high-rel space experience. On freqelec. com/pdf/FEI%20Space%20Products%20Brochure.pdf. Accessed Oct. 10, 2014.

Frequency Electronics, Inc. (2019). Space qualified master clocks. Product data sheet. On freqelec.com/space_master_clocks.html. Accessed Feb. 6, 2019.

Fruehauf H (2007). Presentation on Frequency Electronics, Inc, technical literature: precision oscillator overview. Apr. On www.freqelec.com/tech_lit.html. Accessed Sep. 28, 2010.

Galla TJ (1989). TriQuint semiconductor technical library: cascaded amplifiers. WJ Tech-note. On www.triquint.com/prodserv/tech_info/WJ_tech_publications.cfm. Accessed Sep. 28, 2010.

Gardner FM (2005). *Phaselock Techniques*, 3rd ed. New Jersey: John Wiley & Sons, Inc.

Henderson BC (1981a). TriQuint semiconductor technical library, WJ technical publications: mixers: part 1, characteristics and performance. WJ Tech-note. Revised 2001. On www.rfcafe.com/references/articles/wj-tech-notes/watkins_johnson_tech-notes.htm. Accessed Feb. 22, 2019.

Henderson BC (1981b). TriQuint semiconductor technical library: mixers: part 2, theory and technology. WJ Tech-note. Revised 2001. On. www.rfcafe.com/references/articles/wj-tech-notes/watkins_johnson_tech-notes.htm. Accessed Feb. 22, 2019.

Henderson BC (1990). TriQuint semiconductor technical library: mixers in microwave systems (part 2). WJ Tech-note. Revised 2001. On radiosystemdesign.co.uk/doc/Mixers_in_systems_part2.pdf. Accessed Feb. 22, 2019.

Hewlett Packard (1997). Fundamentals of quartz oscillators. Application note 200-2. On www.hpmemoryproject.org/ressources/resrc_an_03.htm. Accessed Feb. 22, 2019.

Hitch B (2019a). Private communication. June 7.

Hitch B (2019b). Private communication. July 25.

Hitch B and Holden T (1997). Phase locked DRO/CRO for space use. *Proceedings of IEEE International Frequency Control Symposium;* May 28–30; 1015–1023.

Keysight Technologies (2000). formerly Agilent Technologies, formerly Hewlett Packard. Agilent PN 8753-2, RF component measurements—mixer measurements using the 8753B network analyzer. Product note. Nov. 1. On www.keysight.com/main/techSupport.jspx?pid=1000002288:epsg:pro. Accessed Feb. 22, 2019.

Keysight Technologies (2018). Performance spectrum analyzer series, optimizing dynamic range for distortion measurements. Application note. Mar. 12. On literature.cdn.keysight.com/litweb/pdf/5980-3079.pdf. Accessed Feb. 22, 2019.

L3 Narda Microwave West (2017). Downconverters/receivers. Product data sheets. On www.nardamicrowavewest.com/products/converters.htm. Accessed Mar. 15, 2017.

L3 Narda Microwave West (2019). Downconverter/receiver multipack assembly unit. Product data sheets. On www.nardamicrowavewest.com/products/converters.htm. Accessed Feb. 10, 2019.

Microwave Encyclopedia (2006). Microwave FET tutorial. Jan. 22. On www.microwaves101.com. Accessed Dec. 7, 2010.

Microwave Encyclopedia (2010). Power amplifiers. Oct. 15. On www.microwaves101.com. Accessed July 1, 2011.

Mitsubishi Electric (2015). Product data sheets. On www.mitsubishielectric.com/bu/space/satellite_components/rf_equipment/index.html. Accessed June 25, 2017.

NEC Space Technologies (2019). Product data sheets. On www.necspace.co.jp/en/products/index.html#payload. Accessed Jan. 31, 2019.

OMMIC (2012). Preliminary datasheet CGY2122XUH/C2, 25–43 GHz ultra low noise amplifier. Sep. 12. On www.ommic.com/produits/wlna. Accessed Feb. 25, 2019.

Paine B, Wong R, Schmitz A, Walden R, Nguyen L, Delaney M, and Hum K (2000). Ka-band InP HEMT MMIC reliability. *Proceedings of GaAs Reliability Workshop;* Nov. 5; 21-44.

Recommendation, ITU-R, SM.1541-6 (2015). *Unwanted emissions in the out-of-band domain.* Geneva: International Telecommunications Union, Radio Communication Sector.

Reddy M (2016). Design and simulation of L-band coaxial ceramic resonator oscillator. *IEEE Annual India Conference;* Dec. 16–18.

Reddy MB, Swarna S, Priskala, Chandrashekar M, Vinod C, Dhruva PM, and Singh DK (2012). High frequency OCXO for space applications. *IEEE International Frequency Control Symposium;* May 21–24.

Rodgers E and Montauti F (2013). Integrated multi-channel Ka-band receiver subsystem for communication payload. *Ka and Broadband Communications, Navigation and Earth Observation Conference;* Oct.14–17.

Skyworks (2017). Introduction and applications for temperature-stable dielectric resonators. Application note. On www.skyworksinc.com/TechnicalDocuments.aspx?DocTypeID=11. Accessed Feb. 15, 2019.

Smith PM, Dugas D, Chu K, Nichols K, Duh KG, Fisher J, MtPleasant L, Xu D, Gunter L, Vera A, Lender R, and Meharry D (2003). Progress in GaAs metamorphic HEMT technology for microwave applications. *Technical Digest, IEEE Gallium Arsenide Integrated Circuit Symposium*; Nov. 9–12.

Sorrentino R and Bianchi G (2010). *Microwave and RF Engineering*. Chichester: John Wiley & Sons, Ltd.

Sowers J of SSL (2011). Private communication. Nov. 16.

Sowers J of SSL (2019). Private communication. July 11.

Thales Alenia Space (2012). Receiver – LNA – DOCON. Product data sheets. On www.thalesgroup.com/sites/default/files/asset/document/Receiver-LNA-Docon102012.pdf. Accessed Sep. 17, 2018.

Tramm FC (2002). Compact frequency converters for a Ka-band telecommunications satellite payload. *AIAA International Communication Satellite Systems Conference;* May 12–15. On www.as.northropgrumman.com/products/tech_publications/pdfs. Accessed Oct. 4, 2010.

Wikipedia (2017a). Field-effect transistor. Accessed July 20, 2017.

Wikipedia (2017b). Band gap. Accessed July 20, 2017.

Wikipedia (2017c). High-electron-mobility transistor. Accessed July 20, 2017.

Wikipedia (2019a). Thermistor. Accessed July 7, 2019.

Wikipedia (2019b). Varicap. Accessed July 7, 2019.

Wikipedia (2020). Biasing. Accessed Sep. 19, 2020.

Yeung TK, Gregg H, and Morgan I (1993). Lightweight low noise space qualified L-band LNA/filter assemblies. *European Conference on Satellite Communications;* Nov. 2–4; 122–127.

CHAPTER 7

PREAMPLIFIER AND HIGH-POWER AMPLIFIER

7.1 INTRODUCTION

The high-power amplifier (HPA) amplifies the radio-frequency (RF) signal to a high level for the payload downlink. To do so, it needs a direct-current (DC) power supply, which is incorporated into the HPA. Before the signal even goes to the HPA, a preamplifier has to boost the signal to a level proper for input to the HPA.

We will call the preamplifier and the HPA together the **HPA subsystem**. There are two types of HPA subsystem: the **traveling-wave tube amplifier (TWTA)** subsystem and the **solid-state power amplifier (SSPA)**. A subsystem named for the TWTA is not normally identified, but doing so makes it easier to compare with the SSPA. TWTA subsystems are more common than SSPAs. One of the leading global satellite manufacturers reported that it had installed more than twice as many TWTAs as SSPAs on its satellites (Nicol et al., 2013a).

The HPA subsystem has the following functions:

- Channel preamplification for the HPA, with the flexibility to make the downlink power independent of the uplink power over a wide range of uplink power

Satellite Communications Payload and System, Second Edition. Teresa M. Braun and Walter R. Braun.
© 2021 John Wiley & Sons, Inc. Published 2021 by John Wiley & Sons, Inc.

- Pre-distortion (optional) to counteract the HPA's nonlinear amplification characteristics
- High-power amplification
- DC power provision.

7.2 HPA CONCEPTS AND TERMS

7.2.1 HPA Nonlinearity Description

Both types of HPA are sometimes operated in a region of nonlinear amplification. The SSPA and the typical-bandwidth TWTA are said to provide a *bandpass nonlinearity*, which we define. Suppose that the RF signal (including noise, interference, and signal distortions) into the amplifier is represented as follows:

Amplifier input is $\sqrt{2P_{in}(t)} \cos(2\pi f_c t + \theta(t))$ where f_c is the carrier frequency

$P_{in}(t)$ is the **instantaneous power** of the input signal at time t, equal to the signal's square magnitude. We now drop the t reference in order to make the amplifier functioning more transparent. As a function of P_{in}, the amplifier produces a power P_{out} and shifts the signal's phase by ϕ:

$$\text{Amplifier output is } \sqrt{2P_{out}} \cos(2\pi f_c t + \theta + \phi)$$
$$\text{where } P_{out} \text{ and } \phi \text{ are functions of only } P_{in}$$

The P_{out} **versus** P_{in} **curve** and the **phase shift versus** P_{in} **curve** completely define the bandpass nonlinearity. These curves are by definition frequency-independent. The case where the frequency dependence is significant, for TWTAs that are at least wideband (3% bandwidth or more), and is characterized in an augmented way (Section 16.3.7).

The HPA cannot put out an indefinitely large amount of power, so at some point, its gain starts to decrease as P_{in} increases. This is gain compression and was illustrated in Section 6.3.2. The range of P_{in} values that are so small there is no (significant) gain compression, that is, where the amplification is very nearly linear, is the **small-signal** region of amplification. The HPA **saturates** when it puts out its maximum RF output power. The input power at this point is $P_{in\,sat}$ and the output power is $P_{out\,sat}$. When P_{in} is below $P_{in\,sat}$, the HPA is said to be **backed off**. **Input backoff (IBO)** is the ratio $P_{in\,sat}/P_{in}$ in dB, which means it is usually positive. Sometimes, though, IBO is defined as the inverse of this fraction; it is almost always clear what is meant. When P_{in} exceeds $P_{in\,sat}$, the HPA is said to be **in overdrive** or **overdriven**. **Output backoff (OBO)** is the ratio $P_{out\,sat}/P_{out}$ in dB and is positive.

$P_{in\,sat}$ and $P_{out\,sat}$ depend on the particular type of signal input to the HPA. $P_{out\,sat}$ is the largest for a noiseless continuous wave (CW); such $P_{out\,sat}$ is the most frequently

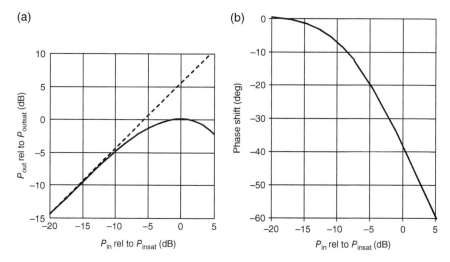

FIGURE 7.1   Example of TWTA's (a) P_{out} versus P_{in} curve and (b) phase shift versus P_{in} curve.

quoted number for HPA performance. If the input signal is not stated, CW is almost always meant. For any other kind of signal input, not only is $P_{out\ sat}$ smaller but $P_{in\ sat}$ is also different. This is illustrated in the following subsection.

Figure 7.1 shows an example of the P_{out} versus P_{in} and phase shift versus P_{in} curves for a typical TWTA with CW input. (Curves for SSPAs are given in Sections 7.5.3 and 7.5.4.) As is usual, both axes of the P_{out} versus P_{in} curve are logarithmic. For the phase shift versus P_{in} curve, the y-axis is in degrees while the x-axis is logarithmic, as is usual. The phase shift is normalized, as always, to be $0°$ at small signal. As the input signal level increases, the phase shift becomes increasingly negative, indicating an increasing delay in the output signal. In this example, powers are shown relative to saturation power. Note that the gain starts to compress at P_{in} well below $P_{in\ sat}$, a characteristic of a TWTA. At saturation, the gain compression is about 6 dB and the phase shift is about $-40°$.

The HPA's **operating point** P_{op} or **NOP** corresponds to the P_{in} that is equal to the long-term average power of the operational input RF signal, where the average is taken over a time much, much longer than the inverse of the signal's noise bandwidth. If the input signal is not simply a CW, the instantaneous power varies in time about P_{in} of the operating point. Noise will cause amplitude variation, and almost any modulated signal will have amplitude variation. So the instantaneous P_{in} will run up and down relative to the operating-point P_{in}. Even if the operating point is a few dB below saturation, there will be instants in which the HPA is in overdrive. Conversely, even if the operating point is at saturation, there will be instants in which the HPA is in its small-signal region. This is illustrated in Figure 7.2.

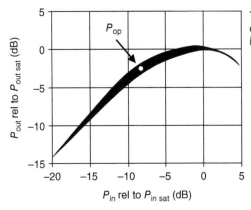

The thicker the line at a point on the curve, the more often the TWTA is instantaneously amplifying there

FIGURE 7.2 Example of how instantaneous P_{in} varies around operating-point, for realistic signal input to TWTA.

The intermodulation products (IMPs) that the HPA creates are mostly inband or near-out-of-band, except for the ones at double and triple the carrier frequency. Since the inband ones cannot be removed, they must be low enough not to be a problem. IMPs are discussed extensively in Sections 6.3 and 9.5.

The HPA's *instantaneous bandwidth* is the bandwidth that the HPA is currently meant to be operated over, without retuning or re-voltaging (Menninger, 2017).

7.2.2 HPA Nonlinearity Specification Parameters

There are four common ways in which the HPA nonlinearity is specified (but see caveats after the list):

- P_{out} versus P_{in} curve and phase shift ϕ versus P_{in} curve for CW input. Usually, each curve is specified by a mask, that is, a curve made of straight segments forming the upper bound.
- Derivative of P_{out} versus P_{in} curve and derivative of phase shift versus P_{in} curve, the **AM/AM (amplitude modulation) conversion** and **AM/PM (phase modulation) conversion** curves, respectively. (See, for example, Agilent (2000). These terms are sometimes used for other things.) The units of the former are dB/dB and of the latter deg/dB. AM/AM conversion equals 1 at small signal and 0 at saturation, and AM/PM conversion equals 0 at small signal. Usually, just maximum AM/AM conversion and maximum absolute value of AM/PM conversion are specified, not the entire curves.
- C/3IM—the curve of combined power in the two output fundamentals relative to combined power in the two nearest 3rd-order IMPs, versus combined power of the two equal-power input tones. (Sometimes a slightly different definition is used.) Input power is in dBm and the output power ratio is in dB. The HPA's $P_{out\ sat}$ for two-tone input is lower than it is for one-tone input, and

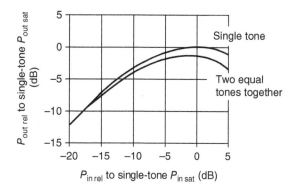

FIGURE 7.3 Different saturation points for one-tone and two-tone TWTA input.

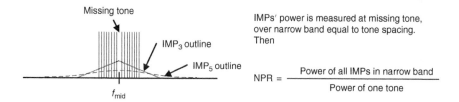

FIGURE 7.4 Noise-power ratio definition.

$P_{\text{in sat}}$ is lower. A typical example for a TWTA is illustrated in Figure 7.3. The C/3IM curve may depend on the particular tone frequencies used, for a wide band. (The C/3IM curve is defined somewhat differently for a frequency converter—Section 6.3.3.)

- **NPR (noise-[to-]power ratio)**—the power ratio of all the IMPs in one narrow band inside the signal band, to the signal power in the same narrow band, at HPA output. It is normally measured by applying a large number of equal-power, randomly phased, evenly spaced tones to the HPA input, but leaving one out. Alternatively, thermal noise of flat spectrum with a notch filtered out can be applied. The worst-case ratio over the signal band is reported, which is usually to be found midband. The definition is illustrated in Figure 7.4.

The NPR test is similar to **noise-loading** the HPA. The HPA is noise-loaded when it has no signal to amplify, only noise. The noise can be simply the noise that arises from the payload when no signal is input. HPA noise-loading can occur by mistake during testing if no signal is input to the payload and the channel amplifier's (CAMP's) automatic level control (ALC) (Section 7.4.1) highly amplifies the front-end noise.

Knowing both the AM/AM conversion curve and AM/PM conversion curve is conceptually equivalent to knowing both the P_{out} versus P_{in} and ϕ versus P_{in} curves, by integration of the former set. The integrated AM/AM conversion curve could be shifted to its proper level from knowledge of the small-signal gain, which can be found by other means. The integrated AM/PM conversion curve could be shifted properly because the phase shift at small signal is taken to be 0°. However, the accumulated errors in the integrations are large, so in fact the conversion curves do not provide the same information. The conversion curves are also inadequate in the case when TWTAs must be matched in their absolute phase delay.

The C/3IM and NPR specifications are each useful when the operational signals into the HPA will be much like the test signals, but they are otherwise not very useful (see Section 9.4.1 for detailed discussion on this).

7.2.3 Other HPA Specification Parameters

Other parameters are important in specifying the HPA. They can be grouped into out-of-band parameters and parameters that relate to performance over inband frequency. Spurious outputs, which are near the carrier, are treated in Section 8.9.

The out-of-band parameters are as follows:

1. Levels of second and third harmonics at CW saturation. These harmonics could possibly interfere into the payload input signal. Even though their levels are low compared to the HPA inband output power, they could be high compared to the very weak payload input signal. These harmonics could, as well, possibly interfere into other signals desired to be received on the earth.
2. Rejection, over uplink band, relative to inband gain. This parameter allows completion of the calculation of how much the second and third harmonics could interfere into the payload input.

The parameters that are defined over inband frequency are as follows:

1. Specification bandwidth. This is the size of the frequency band over which the HPA amplifies at a nearly constant level, to within 3 dB, for example.
2. Saturated gain flatness over the specification band. This parameter is measured by sweeping over frequency a CW strong enough to saturate the HPA across the band.
3. Small-signal gain flatness and phase variation over the specification band. This parameter is measured by sweeping over frequency a CW whose IBO is at least 15 dB everywhere on the band. The small-signal gain flatness always shows more variation across the band than the saturated gain flatness does.
4. Accuracy of commanded gain, for fixed gain mode (FGM) (Section 7.4.1). It is measured on all gain steps, at just one frequency.
5. Accuracy of commanded output-signal level, for ALC mode (Section 7.4.1). It is measured on all output signal levels, at just one frequency.

7.2.4 Power Efficiency

The HPAs are the biggest consumer of satellite DC power for a non-processing payload. Recall that the HPA includes a DC power supply. It is wise to get as much RF output power as possible from the DC power required, since the latter is limited. The ratio $P_{\text{RFout}}/P_{\text{DC}}$ of a HPA's RF output power P_{RFout} to its DC input power P_{DC} is the **power efficiency** of the HPA. An operating point at saturation is the most power-efficient but is unsuitable for a signal with much amplitude variation. However, the farther below saturation, the lower the power efficiency.

A performance parameter of the HPA which may be important is its **power-added efficiency (PAE)**, equal to $(P_{\text{RF out}} - P_{\text{RF in}})/P_{\text{DC}}$. The TWTA's power supply is the **electronic power conditioner (EPC)**. The TWTA's PAE is the product of the traveling-wave tube's (TWT's) PAE with the EPC's power efficiency and similarly for a SSPA's HPA section. For a TWTA, however, the input RF power is so small compared to the RF output power that the PAE basically equals the power efficiency.

In product literature for a TWTA, the PAE is typically stated only at CW saturation. For a SSPA, the power efficiency is stated at the CW saturation point and/or at the backoff at which NPR is about 15 dB. For a true efficiency comparison, the TWTA PAE and the SSPA efficiency would ideally be given by the manufacturers at the intended operating point.

For the entire HPA subsystem, which includes the preamplifier, the relevant parameter is simply **power efficiency** because the RF power input to the subsystem is negligible. The HPA's power supply usually also supplies DC power to the preamplifier (only a small number of Watts—Section 7.4.6). A formula for the efficiency of a TWTA subsystem for commonly given parameters is as follows (used to calculate numbers in Tables 7.4 and 7.5):

TWTA subsytem power efficiency

$$= \frac{P_{\text{out}} * \text{TWT efficiency} * \text{EPC efficiency}}{(\text{L})\text{CAMP DC power draw} * \text{TWT efficiency} + P_{\text{out}}}$$

7.3 TRAVELING-WAVE TUBE AMPLIFIER VERSUS SOLID-STATE POWER AMPLIFIER

7.3.1 General Tradeoff

The tradeoff between the TWTA and the SSPA changed somewhat in the mid-2010s with the advent of the gallium nitride (GaN)-based SSPA. Traditional SSPA technology is gallium arsenide (GaAs)-based. The GaN SSPAs perform better than GaAs SSPAs in several ways, so the SSPA now has wider applicability than it used to.

At L- and S-bands, SSPAs are more often used than TWTAs because of their size (dimensions) advantage. The advantage is great for active phased-array applications, where the SSPA is located in the limited space right behind its corresponding radiating element, thus minimizing the post-SSPA loss. At Ku-band and above, almost exclusively TWTAs are used for their high PAE (Kaliski, 2009)).

TABLE 7.1   Boeing Comparison of GaN SSPA and TWTA

Trade Element	GaN SSPA	TWTA	Comment
Size/weight	+		No difference for high-power applications
RF power		+	At low frequency +3 dB for TWT, more at higher frequencies
FIT[a] rate	−	−	LSSPA and LTWTA about equal
Power efficiency/ thermal/DC		+	SSPA competitive at L/S/C bands over normal NPR ranges, else LTWTA
Bandwidth	+		
Linear power		+	SSPA at C-band almost = LTWTA
System complexity	−	−	
Performance over temperature	−	−	SSPA range equal to TWTA range
Cost/schedule	+		SSPA has large advantage in competitive power/frequency range

[a] FIT, failures in time, number of failures in 10^9 hours of operation.
Source: ©2013 IEEE. Reprinted, with permission, from Nicol et al. (2013a).

The most informative study comparing TWTAs and SSPAs is the long-running study by Boeing on the HPAs onboard the satellites it has built. Every few years Boeing produced a new comparison report. The 2013 report said that usually the choice between TWTA or SSPA is clear when one considers the issue at spacecraft level, taking into account mission objective, RF power, amplification nonlinearity, DC power, thermal dissipation, frequency, bandwidth, cost, schedule, and more (Nicol et al., 2013a). At L-band, Boeing HPAs are all SSPAs. At S-band and C-band there are considerably more SSPA on-orbit hours than TWTA hours, but at Ku-band and higher frequencies, virtually all hours are TWTA hours. Table 7.1 reproduces the comparison chart for the GaN SSPA and the TWTA.

The comparative reliability of TWTAs and SSPAs is no longer a concern. For a long time, it was thought that SSPAs were more reliable than TWTAs, but the 2004 Boeing study showed this not to be the case (Weekley and Mangus, 2004). The running trend of on-orbit per-unit failure rate continually improved from the year 2000 to the year 2012 for both TWTAs and SSPAs, with the per-unit failure rate of the SSPAs decreasing to almost that of the TWTAs (Nicol et al., 2013a).

It was also long thought that the SSPA had the advantage of often degrading gracefully, while the TWTA failed catastrophically. This was also shown not to be true in any significant manner: they both fail catastrophically (Weekley and Mangus, 2005).

A more detailed analysis at C-band than in the conference paper can be found in the corresponding viewgraph presentation, from which the following paragraphs are taken (Nicol et al., 2013b).

The parts count for the SSPA is about the same as for a linearized (Section 7.4.1) TWTA. The TWTA's power supply and preamplifier-linearizer have similar technology and complexity as the power supply and linearizer of the SSPA.

The SSPA has the advantage in operational bandwidth. For the TWTA, it is about 15%, with the power efficiency optimal over 7–8%. For the SSPA, the operational bandwidth is about 14–36%, with the power efficiency optimal over 10%.

Most applications require NPR of 15–20 dB, and in this case, the nonlinearized SSPA is basically as good a choice as a TWTA.

7.3.2 Gallium Arsenide SSPA versus TWTA

In 2003, it was reported that with GaAs technology, output power of 20–40 W at C-band was about the only case where TWTAs and SSPAs could compete head-on (Bosch et al., 2003). The spacecraft-level tradeoff study showed even at 60 W no overall mass advantage of the GaAs SSPA compared to the linearized traveling-wave tube amplifier (LTWTA). When the supporting hardware (heat sink, heat pipe, battery, solar array, etc.) mass was included with the HPA mass, each SSPA was found to need 2 kg more mass than each LTWTA (Bosch et al., 2003). The GaAs SSPA has worse power efficiency than the TWTA. In addition, the GaAs SSPA is more sensitive to temperature, so it needs to be operated at lower mounting temperatures (Nicol et al., 2013b).

7.3.3 Gallium Nitride SSPA versus TWTA

The GaN technology development has shifted some comparison measures in favor of the SSPA where formerly the TWTA was equal or had the advantage. The biggest advantages of the GaN SSPA over the TWTA, at C-band and lower frequencies, is the much shorter delivery time and the lower cost, size, and mass. The GaN SSPA can be operated as close to saturation as the TWTA can. It can be operated at the same mounting temperatures as the TWTA. For some applications, the GaN SSPA may not need a linearizer when a TWTA would (Nicol et al., 2013b).

7.4 TRAVELING-WAVE TUBE SUBSYSTEM

7.4.1 Introduction

Figure 7.5 shows a functional breakdown of a TWTA subsystem. We must know the breakdown before we can look into the subsystem's payload-level issues in Section 7.4.2. In Section 7.4.3, we look at the variations in subsystem architecture. The functions and their terminology are as follows:

- **CAMP**, the preamplifier. When it incorporates the linearizer function, it is the **linearizer-channel amplifier (LCAMP)**. When referring to the CAMP and it does not matter whether or not it contains a linearizer function, we write

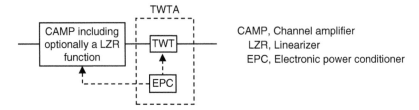

FIGURE 7.5   Functional breakdown of TWTA subsystem.

(L)CAMP. The CAMP has a wide range of gain, often 30 dB, which enables it to take almost any level of signal input to any desired TWTA-input level. (In the signal path before the CAMP there are only fixed levels of amplification.) The CAMP has two operation modes, **FGM** and **ALC** mode. In FGM, it provides the commanded amount of gain, no matter what the input signal level is as long as it is not too high. In ALC mode, it provides the commanded level of output signal, no matter what the input signal level is as long as it is within the capability range. In all cases, the CAMP puts out very much less than 1 W.

- **Linearizer** (optional). It predistorts the signal inversely to the TWTA's non-linearity, in gain and phase shift. A LTWTA can be operated closer to saturation and thus with higher PAE.
- **TWT**. The TWT and EPC (see immediately below) together form the TWTA, the HPA in this subsystem. A **LTWTA** is a TWTA together with a linearizer function (in the CAMP). A **(L)TWTA** is a TWTA irrespective of whether or not there is a linearizer in the subsystem. For TWTAs on commercial pay-loads today, the TWTA's gain ranges from 35 to 60 dB, and its saturated CW P_{out} ranges from 12 to 500 W (Thales, 2013a, 2013b, 2013c, 2013d, 2013e; Will, 2016; Dürr et al., 2014).
- **EPC**. It provides DC power at the required voltages to the TWT and the (L) CAMP.

7.4.2   TWTA Subsystems in Payload

7.4.2.1   *Architecture in Payload, Traditional, and Flexible*   In a traditional payload architecture, typically every active TWTA amplifies a set of signal(s) separate from the sets of signals that the other TWTAs amplify. This is the most common architecture. Newer architectures, those using multiport amplifiers (MPAs) or multimatrix amplifiers (MMAs), are described in Section 11.12. These are flexible in that they are part of a flexible payload architecture which allows signals entering a set of TWTAs to have varying proportions to each other.

7.4.2.2   *Combining*   Sometimes **TWTA-combining** is implemented to create a greater P_{out} than one TWTA alone can produce. The RF signal goes through one (L)CAMP and is then divided to go to the TWTAs, which are matched in their gain

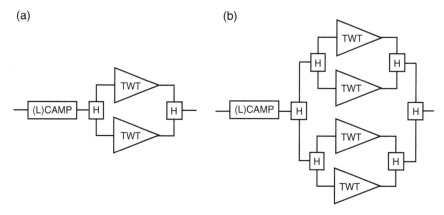

FIGURE 7.6   Examples of TWTA combining: (a) two TWTAs combined, (b) four TWTAs combined (EPCs not shown).

and phase-shift performance over the range of RF input levels that the TWTAs will see in operation. Then, the amplified signals are combined. Combining of two and four TWTs is shown in the two parts of Figure 7.6. An example of large-scale combining is the Sirius XM Radio FM-6 satellite, launched in 2013, which combines the outputs of 32 S-band, single TWTAs (Briskman, 2010).

A different type of combining occurs in a passive or semiactive phased array (Section 11.12.1), where signals from various TWTAs are combined and input to each radiating element. The same considerations apply (Kubasek et al., 2003).

How well can TWTs match in gain and phase, over RF input level? A study was made on 35 S-band TWTs at saturation. The standard deviation of gain compression at saturation was found to be 0.22 dB, and the standard deviation at saturation of phase shift to be 2.6°. The saturation point was chosen because at saturation the gain and phase shift vary the most among TWTs. The study also showed that Ku-band TWTs behave in a similar way (Kubasek et al., 2003).

When TWTAs are combined, not just the TWTAs must track over temperature but also the output waveguide or coax. The easiest way for the output waveguide or coax to track is for the pieces to be of the same length and in the same spacecraft environment. The lower the frequency, the easier it is to combine TWTAs because of the longer guided wavelength. Sections 4.3.3 and 4.4.3 provide behavior over temperature, for coax and waveguide, respectively. Section 9.4.2 gives a tip on how to ease combining.

When the outputs of multiple TWTs are combined, the power combiner (shown as a hybrid in the figure) must be able to handle the dissipation from its nonideal combining (Section 4.5).

7.4.2.3   *Redundancy Scheme*

In a traditional payload architecture, typically all the TWTAs of one frequency band, for example Ku-band, belong to one redundancy ring (Section 4.6), in which they back each other up in case of failures.

The primary, that is, first-choice, TWTA for any particular channel in the band has been optimized for that channel. The first-redundant TWTA for the channel, that is, the first choice to replace the primary TWTA if it fails, may also have been optimized for the same channel. But for a payload that carries a lot of channels, this is not often the case for second- and higher order-redundant TWTAs, which have probably been optimized for a larger bandwidth that covers all the channels it is likely to back up. Then if the second-redundant TWTA has to be employed at some point in payload life, its performance will not be quite as good as the primary TWTA's was. Sometimes the payload specifications for a channel are required to hold when the primary TWTA is used and when the first-redundant TWTA is used but not when the second-redundant is used.

Redundancy switching has operational issues. TWTA performance parameters are monitored, so that if a parameter starts to degrade, the TWTA may be switched out for a redundant one before a failure occurs. The switch-over can be scheduled for a low-traffic time. Another consideration is how the satellite operators know which switches to set in what way, especially when there has already been at least one failure in a ring. There are several things to consider. The operators have payload reconfiguration tools (Kosinski and Dodson, 2018).

The question remains, what the redundancy scheme for the (L)CAMPs is. In the case where one (L)CAMP serves one TWTA or a dual TWTA, the (L)CAMP is in the redundancy ring along with its corresponding TWTA(s). In the case where the (L)CAMP feeds many TWTAs, it is in its own separate redundancy ring, as in the Sirius XM Radio satellites, where the LCAMP had 3:1 redundancy (Briskman, 2010).

7.4.3   TWTA Subsystem Architecture

The TWTA subsystem has a rather wide range of embodiments and terminology to describe them. The particular embodiments selected for a payload depend on spacecraft-level considerations.

The (L)CAMP possibilities are illustrated in Figure 7.7. The most common, (c), is the single unit of combined linearizer and CAMP, the LCAMP.

There are two TWTA possibilities, as illustrated in Figure 7.8. They are the single and dual TWTAs. The **single TWTA** is one TWT and one EPC. The **dual TWTA** includes a **dual EPC** to power the two TWTs. The two leading global suppliers of space TWTAs, Tesat-Spacecom and L3Harris (formerly L-3

FIGURE 7.7   (L)CAMP realizations as units.

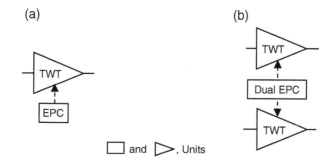

FIGURE 7.8 TWTA integrated forms: (a) single, (b) dual.

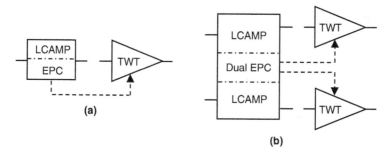

FIGURE 7.9 TWTA subsystem integrated MPM forms: (a) single, (b) dual.

Communications), both sell these integrated forms. The dual EPC is not two EPCs; the L3Harris dual EPC has only 1.1 times the parts count of a single EPC. There is a rare failure in the dual EPC that brings down both TWTs, but most of the failures bring down only one TWT (Phelps, 2008). The two TWTs that go into the dual TWTA are chosen as a matching set in their nonlinear performance. TWTA combining usually makes use of dual TWTAs.

The entire TWTA subsystem, including the LCAMP, can be integrated. There are the single and dual **microwave power modules (MPMs)** (Tesat-Spacecom term) or **linearized channel-amplifier traveling-wave tube amplifier (LCTWTA)** (L3Harris term). The two configurations are shown in Figure 7.9. A photo of a Tesat-Spacecom dual MPM is shown in Figure 7.10.

Tesat-Spacecom performed a study of the reliability of its integrated forms of TWTA. The failure rate of one channel in the dual TWTA is almost twice that of the single TWTA; similarly, the failure rate of one channel in the dual MPM is almost twice that of the single MPM. The LCAMP failures contribute little to the failure rate of either form of MPM. Despite the higher failure rates of the dual products, if the failure rates are acceptable in the overall payload, it is often a good idea to use

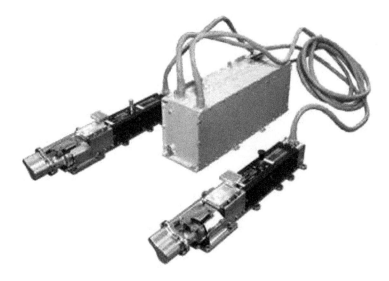

FIGURE 7.10   Photo of dual MPM. (Reprinted with permission of Tesat-Spacecom, from Tesat-Spacecom (2016)).

the dual products because of the lower cost, size, and mass compared to two single products. The study also showed the general trend of decreasing on-orbit failure rates as the years go by, due to design and workmanship having reached a very high level of maturity (Jaumann, 2015).

7.4.4   Channel Amplifier

7.4.4.1   (L)CAMP Unit Architecture and Technology   The architecture of a typical LCAMP unit is shown in Figure 7.11. A CAMP, missing the linearizer, would also be missing the linearizer-associated circuitry. This particular LCAMP is not part of a flexible TWTA subsystem (Section 7.4.10). The unit has two modules, the RF module and the control-and-DC module. The RF module has three variable-gain amplifier (VGA) sections. Immediately after the first VGA section is the first power detector, whose reading the control module uses in ALC mode [this feature does not come from Khilla et al. (2002) but from Thales (2012)]. The control module commands both the first and second VGA sections differently, depending on whether the LCAMP is in ALC mode or FGM. After the first detector but still before the second VGA section is the linearizer. Between the second and third VGA sections is a second power detector, whose reading the control module uses to prevent overdriving the TWTA; if the reading is too high, the control turns down the gain in the first VGA section. The last element in the RF module is the third VGA section, whose purposes are (1) to accommodate gain

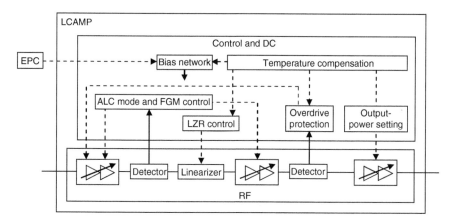

FIGURE 7.11   LCAMP unit architecture example (mostly after Khilla et al. (2002)).

alignment between the LCAMP output and TWT input (Thales, 2012) and (2) to compensate for TWT aging (Khilla et al., 2002). The control-and-DC module creates the biases for the active elements including the three VGA sections. It temperature-compensates the bias network, the linearizer, the overdrive protection, and the output power.

7.4.4.2 CAMP Specification   Table 7.2 gives an example of the parameters in a CAMP specification. For a LCAMP, the nonlinear performance of the linearizer is not specified with the CAMP but together with the TWTA.

TABLE 7.2   Example of Parameters in a CAMP Specification

Mode	Parameter in CAMP Specification	Unit
Both FGM and ALC mode	Operational frequency range	GHz
	RF input-drive level to saturate TWTA, min and max	dBm
	Noise figure—at max gain, min gain	dB
	Return loss—input, output	dB
	Operational temperature range	°C
FGM	Commandable gain range and step size	dB
	Gain variation—over 36 MHz, over full band	dB pk–pk
	Gain stability at any frequency—over 15°, over operational temp range; from aging and radiation	dB pk–pk
ALC mode	Commandable output-level range and step size	dB
	Output power variation—over 36 MHz, over full band	dB pk–pk
	Output power stability at any frequency—over 15°; from aging and radiation	dB pk–pk
	ALC time constant (Section 12.10.2)	ms

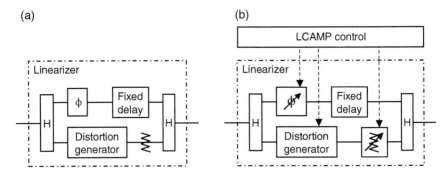

FIGURE 7.12 Linearizer architecture examples with (a) fixed tuning, (b) commandable tuning (part (b) after Zhang and Yuen (1998)).

7.4.5 Linearizer

7.4.5.1 Linearizer Architecture and Technology
Since at least 2010, most TWTAs are linearized (Menninger, 2016).

A linearizer may have an architecture similar to that shown in Figure 7.12a. This linearizer has a *bridge* structure, in that it has linear and nonlinear arms flanked at input and output with hybrid couplers (Khilla, 2011c). The linear arm contains a phase shifter and a delay line to equalize the delay in the two arms. The nonlinear arm contains a distortion generator and an attenuator. The distortion generator generates gain and phase predistortion versus input RF-drive level. The attenuator equalizes the signal level in the two arms. Such a linearizer at L-band is described in Khilla et al. (2002), where the distortion generator is a MESFET. This linearizer has NPR better than 14.5 dB even in overdrive (15 dB being a typical NPR requirement for a backed-off LTWTA). The linearizer must be tuned during manufacture to match the particular nonlinear characteristics of the TWTA(s) it will be used with.

At Ku-band, a fixed linearizer does not compensate an especially wideband TWTA well (Zhang and Yuen, 1998). If for example the linearizer may be used for two different channels that are separated by a very wide band, then the architecture may be similar to that of Figure 7.12a but with commandable control by the LCAMP's control module, as shown in Figure 7.12b. The linearizer's gain-expansion and phase-advance curves can be shifted on-orbit by modification of the phase shifter and attenuator biases, and the functions' curvatures can be adjusted by modification of the distortion-generator bias (Zhang and Yuen, 1998). This particular linearizer, whose distortion generator uses Schottky diodes, can be used for channels within a 30% bandwidth (Yuen et al., 1999). A similar linearizer was described in Villemazet et al. (2010).

At K-band frequencies and above, a fixed linearizer does not compensate well over a TWTA's usual bandwidth (Nicol, 2019a). One truly wideband linearizer has a gain-and-phase equalizer in each arm, allowing the linearizer to have varying

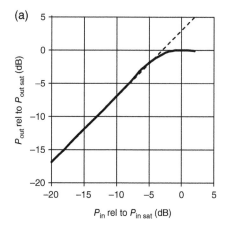
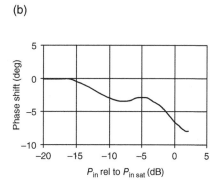

FIGURE 7.13   Example of LTWTA's (a) P_{out} versus P_{in} and (b) phase shift versus P_{in}. (Credit: Tesat-Spacecom).

nonlinear behavior across frequency (Khilia [*sic*] et al., 2013). This linearizer has a bandwidth of 2 GHz at K-band (Khilla, 2011b).

7.4.5.2   LTWTA Nonlinear Performance   An example of a TWTA's nonlinear performance was given in Figure 7.1. Here we show, in Figure 7.13, the P_{out} versus P_{in} and phase shift versus P_{in} curves for a LTWTA, where the TWTA's behavior before linearization is very much like the TWTA's of Figure 7.1. The LTWTA is close to being a clipper, which is a theoretical device which is linear up to a certain RF input drive and then saturated beyond. A small area of minor over-compensation of the TWT's gain compression by the linearizer can be seen in Figure 7.13a, at between 8 and 5 dB of IBO. This is typical. The phase shift from the LTWTA is almost an order of magnitude smaller than that of the TWTA; in this example, the maximum phase shift difference between the small-signal regime and 2-dB over-drive is only about 7.5°. In 2011, state-of-the-art LTWTAs produced by Tesat-Spacecom showed a phase shift at saturation within ±9° of the small-signal phase shift (Khilla, 2011a). The nonlinear performance of the LTWTAs of L3Harris is similar. For frequencies from C-band through K-band, linearizer performance is virtually the same for all frequencies (Menninger, 2016).

For multi-carrier use, the linearizer typically allows the TWTA to be operated at 3 dB higher output power, for a NPR of 20 dB (see Figure 7.14).

7.4.6   Electronic Power Conditioner

The TWTA consists of the TWT and the EPC. The EPC performs many functions for the TWT and (L)CAMP. The EPC and the TWT are designed together for best performance (L-3, 2009). Among other things, the EPC provides DC voltages to the cathode heater, cathode, anode, focusing electrode, helix, and the collector stages

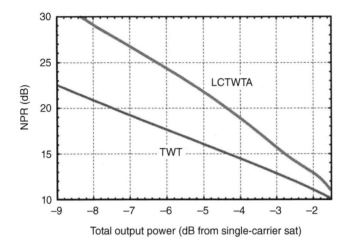

FIGURE 7.14 Typical NPR improvement from linearizing TWTA. (©2016 IEEE. Reprinted, with permission, from Menninger (2016)).

(see below). A common S-band and C-band cathode voltage is 4 kV, Ku-band is 6 kV, and Ka-band and above is 8–14 kV (Barker et al., 2005). The EPC also provides DC power to the (L)CAMP, a small number of Watts, for example 3 W (Thales, 2012).

The two leading global TWTA suppliers reported their EPC power efficiency as 94% (L-3, 2012) and 95% (Braetz, 2011a), respectively, for a regulated DC voltage input to the EPC.

7.4.7 Traveling-Wave Tube

7.4.7.1 *TWT Behavior*   Perhaps all commercial TWTs for L-band through K-band are of the helix type. The nonlinear behavior of helix TWTs is very similar for all frequency bands from C-band through V-band (Nicol, 2019a). A typical example of this behavior was given in Figure 7.1.

All space TWTs are capable of about 10% intrinsic bandwidth (Nicol, 2019a). This is the device's basic bandwidth as it exists without extra measures to extend the bandwidth (Nicol, 2019b).

All TWTs have small, aperiodic variations in their small-signal gain and phase responses versus frequency. The size of the gain variation is less than about 0.5 dB. There are many of these variations across the band. The variations exist in the saturated sweeps, too, but are much smaller. They are caused by small imperfections formed along the length of the helix that cause multipath (Nicol, 2019c).

A few individual TWTs have ripple in their gain and phase responses versus frequency, of period roughly 2% of the center frequency. Its existence depends on the TWT model and if it has any characteristic mismatch locations on either end of the high-gain circuits, causing multipath (Menninger, 2019).

7.4.7.2 *TWT Architecture and Technology*

The TWT is a complex and touchy device to design, is extremely demanding to build, and takes many months to manufacture. Typically the procurement of the TWTAs drives the payload schedule, so the satellite manufacturer places the order at the very beginning of the satellite program, sometimes earlier, even before their output powers are known exactly. A TWTA is the most expensive unit of a non-processing payload. For these reasons, some payload engineers and spacecraft customers become involved in discussions with the TWT manufacturer, for which it is very helpful to have a broad understanding of the TWT, which we aim to give here.

An excellent exposition on the TWT is given in Barker et al. (2005). A shorter but confusingly organized exposition is L-3 (2009). The latter does contain, however, a description of the full set of TWTA performance parameters.

Figure 7.15 is a drawing of the vacuum assembly of a typical helix TWT (Feicht et al., 2007). A three-dimensional drawing of a similar TWT vacuum assembly is given in Amstrong (2015). There are three connected cylindrical sections with the thinnest one in the middle, the **tube** containing the **helix**. In brief, how the TWTA works is that the electron *gun* of the first cylindrical section shoots a high-power beam of electrons down the tube, through the center of the helix. The RF signal is launched into the helix. The signal travels at the speed of light down the helix wire but, because of the helix windings, its speed down the tube's axis is slow enough to match that of the electron beam. The RF electric field and the beam interact, causing the RF signal to pick up power from the beam. The amplified RF signal is coupled out of the helix. The *collector* of the third cylindrical section collects the *spent electrons* (Thales, 2001).

The TWT vacuum assembly is enclosed in a package that serves several purposes: mechanical support, thermal path (including the baseplate) for conduction of waste heat, electromagnetic-interference shield, and protection for the magnets and high-voltage connections (L-3, 2009).

We now describe the TWT in more detail, first the electron-beam aspects (gun, tube, and collector) then the RF aspects, with thermal considerations of both. Last is a description of how the RF signal and electron beam interact to produce the amplification.

The gun has a cathode, which ejects electrons. The flow of electrons is the *cathode current*. The gun's anode has a higher voltage potential than the cathode so it strongly attracts the electrons, which speed up and pass through the middle of the anode. A diagram of the relative voltages of gun electrodes and helix is given in Figure 7.16. The focus electrode aids in proper formation of the electron beam (L-3, 2009). In even more detail, the cathode comes in two types, the M-type and the MM-type (Thales, 2001). The M-type is the more common (Barker et al., 2005). Both cathodes are made of a porous tungsten matrix (making them *dispenser cathodes* (L-3, 2009)) impregnated with barium oxide (Thales, 2001) or a compound of

Principal components of a helix TWT

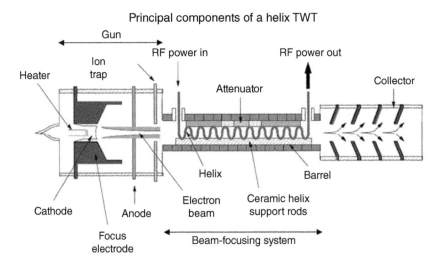

FIGURE 7.15   Drawing of typical helix-TWT vacuum assembly. (Image courtesy of US Air Force, from Feicht et al. (2007)).

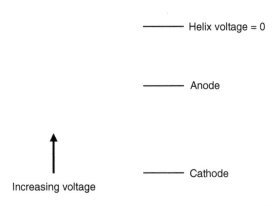

FIGURE 7.16   Diagram of relative voltages of gun electrodes and helix (not to scale).

other metal (L-3, 2009). The barium migrates to the emitting surface of the cathode, and the cathode's high temperature causes the barium to emit electrons (Thales, 2001). The M-type (metal-coated) cathode is coated with osmium, while the MM-type (mixed-metal) is coated with tungsten and osmium. The coating lowers the temperature required for the cathode to emit electrons to "only" about 1000 °C (Thales, 2001). The gun sets the TWTA's noise figure (Limburg, 1997), which is typically 25–35 dB for 50–150 W TWTAs (Barker et al., 2005).

Changing a TWT's *anode voltage* changes the amount of *beam current* and thus its saturated output power. Thomson-CSF, a precursor to Thales, reported in 1977 on the advantage of a dual anode over a single anode, where the first anode of the

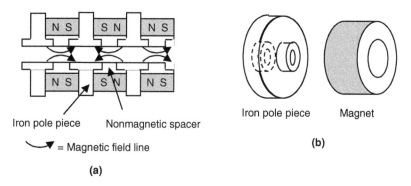

FIGURE 7.17 TWT's periodic-permanent-magnet stack: (a) part of stack showing alternating magnet orientation and magnetic field lines; (b) pole piece and magnet.

dual anode controls the beam current to any value between zero and the cathode current. Thomson-CSF used the dual anode to create a TWT with three possible saturated output power levels (Henry et al., 1977). They designed the TWT's electron gun to obtain good focusing over a large range of beam current (Strauss and Owens, 1981). L3Harris has been using a dual anode in all their TWTs launched since 1997 (Dibb et al., 2011). Their second anode of the dual anode, farther from the cathode than the first one, has a slightly positive voltage relative to the helix. It prevents positive ions from reaching the cathode and damaging the surface coating (Menninger et al., 2005; Barker et al., 2005).

The electrons are held in a tight beam down the length of the tube, in order that they go through the middle of the helix, ideally without touching it. This *beam focusing* is accomplished by a series of annular (that is, ring-shaped) magnets that run the length of the tube. The magnets are made of samarium cobalt almost exclusively (L-3, 2009). They are assembled with alternating magnetic orientation, as shown in Figure 7.17a, producing *periodic permanent magnetic (PPM) focusing*. The structure that holds the magnets is made of iron *pole pieces* and non-magnetic *spacers* brazed together (Karsten and Wertman, 1994; Thales, 2001). The inner surface of the tube is the *barrel* and forms part of the *vacuum envelope* of the vacuum assembly. The helix does not touch the barrel. Figure 7.17b shows a detail of an iron pole piece and a magnet. The outside of the PPM structure has cooling fins (not shown) to conduct heat to the baseplate.

When the RF input power is backed off, the beam power remains the same even though the RF output level is lower, so the TWTA's power efficiency is reduced (Katz et al., 2001).

The spent electrons, which still have most of their energy, are gathered in the collector. The collector first reduces the speed of the electron beam, thus its energy, causing less energy to be converted to heat in the collector. Since this means less waste energy, the TWT is more power-efficient (Section 7.2.4) (L-3, 2009). The electrons are slowed down by the negative voltage of the collector compared to the helix, the latter being at ground potential; such a collector is *depressed*. The space

TWT's collector has typically four stages (L-3, 2009), where the first stage slows down and collects the lowest-energy electrons, the second stage the next lowest-energy electrons, and so on. The second collector stage has a lower voltage than the first stage does, the third lower than the second stage, and so on as shown in Figure 7.18. The repelling force among the electrons spreads them out as they approach the collector walls (L-3, 2009).

The greatest part of the heat in the TWT to be dissipated is in the collector. There are two ways to dissipate the heat. One is to thermally couple the collector to a baseplate, which is in contact with a spacecraft heat pipe, which conducts away the heat. When the heat is entirely dissipated by conduction, the TWT is **conduction-cooled (CC)**. The other way to dissipate heat is to couple the collector to large attached fins or stacked cups, which radiate the heat into space (Thales, 2001). When the heat is partly dissipated this way, the TWT is **direct-radiation-cooled (DRC)**. DRC TWTs are located as usual on the inside surface of a spacecraft panel but the radiators protrude beyond the panel into space. An example of a 110W C-band TWTA shows that if the TWT is DRC, only half as much heat has to be conducted away as when it is CC (Barker et al., 2005). Photographs of a TWT in two different cooling versions are given in Figure 7.19.

If the TWTA is to be operated backed off, as for multiple carriers, the heat to be dissipated is less, so the cooling requirements are lower and the TWTA can be made with lighter weight (Katz et al., 2001).

Now for the RF part of the TWT. The RF is coupled into the helix wire. The helix windings slow down the axial (that is, down the length of the tube) phase velocity of the RF wave sufficiently to establish synchronism with the electron beam, making the helix a *slow-wave structure*. The helix slow-wave structure is inherently wideband (L-3, 2009). At the end of the helix, the amplified RF is coupled off. The helix is supported by three ceramic rods that run the length of the tube. The rods hold the helix in the middle of the barrel (Thales, 2001). They also serve to dissipate to the barrel the heat generated in the helix from RF losses and electron impacts (L-3, 2009). An alternative but much less common slow-wave structure is the coupled cavity, described in L-3 (2009). The coupled-cavity TWT has a narrower bandwidth but higher power than the helix TWT does. It will not be further discussed.

If there are at least two impedance mismatches along the RF path, for example, at the RF input and output couplers, an oscillation can arise. The most common means to prevent this is to introduce a *sever*, that is, cutter, about halfway down the length of the helix (L-3, 2009). Most TWTs have one sever, but some have two or none. The sever attenuates the RF signal entirely; the information of the RF signal is carried across the cut by the electron beam. The sever itself is also a potential impedance discontinuity. In some TWTs the sever is distributed, consisting of carbon deposited on the ceramic rods, with the amount of carbon being the largest about in the middle of the tube (L-3, 2009). TWTs at 30GHz and above can experience *backward-wave oscillations* (Barker et al., 2005), which occur out-of-band but rob power from the intended output signal. A distributed sever is more effective than a discrete one for suppressing this (Nusinovich et al., 1998).

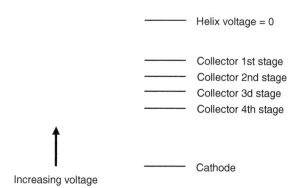

FIGURE 7.18   Relative voltages of collector stages, helix, and cathode (not to scale).

(a)   (b)

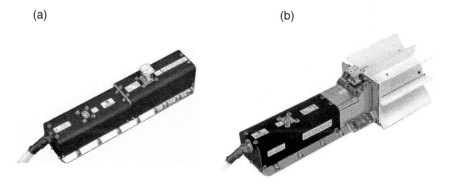

FIGURE 7.19   Photos of L3Harris's K-band model 9250 TWT in two versions:
(a) conduction-cooled, (b) radiation-cooled. (© 2016 IEEE. Reprinted, with permission,
from Robbins et al. (2016)).

The electron beam interacts with the axial component of the RF electric field
(Barker et al., 2005). In the first region of interaction, the electrons *bunch*. When
individual electrons enter the electric field, depending on the phase of the field
polarity they are either sped up or slowed down. They thus form bunches. In the
second region of interaction, the bunches induce a modified wave in the helix. The
decelerating regions of the wave's electron field line up with the bunches, which
slow down and give kinetic energy to the induced wave. The induced RF signal gets
amplified exponentially with distance down the tube. At some point, the bunches
start to come apart or they can no longer be kept in synchronism with the RF wave.
This is where the helix ends. The area of synchronism is often extended by *tapering*
the end of the helix, that is, decreasing the helix pitch to further slow down the RF
wave. When the TWT is operated at saturation, the electron bunches at helix output
have a 30–50° phase delay relative to when the TWT is operated at small-signal.
The bunches *pull* the RF wave back by this same amount (Barker et al., 2005).

Figure 7.1b showed that the TWT decreases the signal phase of high-power input, that is, delays it, relative to low-power input.

If the TWTA is to be used for multi-carrier operation, where the operating point will be below single-carrier saturation, the TWT can be tuned to maximize the PAE at the operating point. This can improve the PAE by up to 5% points beyond the PAE improvement provided by the linearizer (Katz et al., 2001).

7.4.8 TWTA Subsystem Specification

Table 7.3 gives an example of the specification parameters for a (L)TWTA. This particular (L)TWTA would be used over a range of IBOs, so its nonlinearity is specified by a mask as a function of IBO. A (L)TWTA to be used only for

TABLE 7.3   Example of Parameters in a (L)TWTA Specification

Parameter in (L)TWTA Specification	Unit
Operational frequency range	GHz
CW-saturation output power, minimum across frequency range at BOL	W
CW-saturation output-power stability—over 15°, over full operational temperature range; over life	dB
RF input-drive level for CW to saturate—BOL, EOL	dBm
OBO versus IBO curve	Mask
Phase shift versus IBO curve	Mask
Gain flatness—on 36 MHz, on full band; at CW sat, at small signal	dB pk–pk
Gain slope—at CW sat, at small-signal	dB/MHz
Phase deviation from linear versus frequency, at small-signal	° pk–pk
Second- and third-harmonic output power at CW saturation	dBc
Spurious outputs, inband, minimum gain, in any 4 KHz bandwidth	dBc
Spurious outputs, inband, minimum gain, due to AC heater and EPC	dBc
Spurious outputs, out-of-band, noncoherent, in any 4 KHz resolution bandwidth	dBc
Out-of-band rejection relative to inband gain, over uplink band (LTWTA only)	dB
Noise figure	dB
PAE at operating point	%
Return loss—input, output	dB
Operational temperature range	°C

Notes on the table:
- Resolution bandwidth is the bandwidth of the RF chain in a spectrum analyzer before the power measurement device.
- The International Telecommunications Union (ITU) specifies out-of-band spurious signals for all space services with reference bandwidth 4 KHz for frequencies below 15 GHz and 1 MHz for frequencies above 15 GHz. The spectrum analyzer's resolution bandwidth ideally equals the reference bandwidth (ITU-R, 2015).
- The out-of-band spurious outputs come from both EPC spurs and TWTA noise (Nicol, 2019d).

multi-carrier operation could instead be specified on NPR, typically 15 dB (see Section 7.2.2 for discussion on nonlinearity specification).

7.4.9 TWTA Subsystem Performance

Table 7.4 summarizes the current performance of TWTA subsystems made by Tesat-Spacecom for commercial payloads. Tesat-Spacecom's TWT supplier is Thales. Subsystem performance is given in terms of the key parameters output power, bandwidth, and subsystem power efficiency. The latter is calculated from the formula in Section 7.2.4. The numbers all apply for CW saturation. L3Harris's performance is similar as far as can be verified (L-3, 2016a, 2016b). In addition, L3Harris has a 300 W K-band TWTA of 1 GHz bandwidth and subsystem efficiency of 61%.

Particularly at Ku- and K-bands, wideband or broadband TWTAs and linearizers can be of interest. We define this category as those with instantaneous bandwidth of at least 1 GHz. Some such TWTAs from both suppliers are listed in Table 7.5, where the emphasis is on high power. Some entries are repeated from Table 7.4. A wideband linearizer, of 2 GHz bandwidth at K-band (Khilla, 2011b), is described in Khilia [sic] et al., (2013).

7.4.10 Flexible TWTA Subsystem

We describe a **flexible TWTA** subsystem which can alter $P_{\text{out sat}}$ by telecommand, so that as on-orbit the required P_{out} changes, the TWTA can be kept operating at its most power-efficient (Khilla et al., 2005; Khilla, 2008). The approach relies on the fact that changing a TWT's anode voltage causes $P_{\text{out sat}}$ to change in the same direction. As $P_{\text{out sat}}$ is reduced by 1 dB, the TWT's gain reduces by about 5 dB and gain slope versus frequency arises. The LCAMP commands the EPC to change the cathode current and at the same time puts out a different power level to the TWT and compensates the gain slope. The LCAMP has an additional, medium-power amplifier in its output amplification section. $P_{\text{out sat}}$ can be set to within 0.1 dB over a range of 4 dB. For a 4 dB reduction in $P_{\text{out sat}}$, the DC power consumption is reduced by 24% and heat to be dissipated by 30% relative to no $P_{\text{out sat}}$ reduction, as shown in Figures 7.20 and 7.21. In the figures, "IOA setting" refers to the commanded decrease in $P_{\text{out sat}}$. Such K-band subsystems fly on Hylas-1 (Phys.org, 2009), launched in 2010 (Gunter, 2017).

L3Harris uses the second anode of the dual anode of their TWTs (Section 7.4.7.2) to enable a wide range of adjustable $P_{\text{out sat}}$ (Menninger, 2015).

7.4.11 TWTA Subsystem Environmental

The (L)CAMP, the TWT, and the EPC change over life, and the (L)CAMP is designed to compensate the changes in all three (Khilla et al., 2002). The categories

TABLE 7.4   Summary of One Supplier's TWTA Subsystem Performance[a] for CW

Frequency Band	High-Power			Low-Power			Reference for TWT Efficiency
	P_{sat} (W)	Bandwidth (MHz)	Typical Power Efficiency (%)	P_{sat} (W)	Bandwidth (MHz)	Typical Power Efficiency (%)	
L-band	280	100	61	70	50	56	Thales (2013a)
S-band	500	150	64	70	100	60	Thales (2013a)
C-band	125	350	65	20	350	56	Thales (2013b)
X-band	160	500	62	12	500	44	Thales (2013c)
Ku-band	300	2050	61	25	2050	59	Hanika et al. (2015) and Thales (2013d)
Ku-band	300	1000	63				Hanika et al. (2015)
K-band	250	2500	59	15	2500	50	Thales (2013e)

[a] Assuming 95% EPC efficiency from Braetz (2011a) and 2 W DC power consumption for (L)CAMP.

TABLE 7.5   Performance of Some Wideband TWTAs^a for CW

Frequency Band	P_{sat} (W)	Bandwidth (MHz)	Typical Power Efficiency (%)	Reference
C-band	150	800	67	Dürr et al. (2015)
Ku-band	300	2050	61	Hanika et al. (2015)
Ku-band	300	1000	63	Hanika et al. (2015)
K-band	250	2500	59	Thales (2013e)
K-band	170	2500	62	Eze and Menninger (2017)
K-band	300	1000	61	Robbins et al. (2016)

^{a} Assuming 95% EPC efficiency from Braetz (2011a) and 2 W DC power consumption for (L)CAMP.

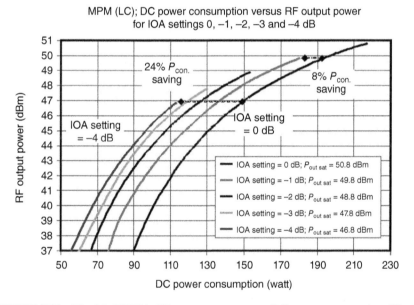

FIGURE 7.20 Flexible TWTA's RF output power versus DC power consumption. (Used with permission of M Khilla and the European Space Agency, from Khilla (2008)).

of environmental influence are temperature, radiation, and aging. Aging and radiation are sometimes combined to form the category "life."

How the TWTA deals with the heat it generates was covered in Section 7.4.7.2.

7.4.11.1 Temperature The (L)CAMP would have decreasing gain with higher baseplate temperature, without temperature compensation. However, the unit has a temperature sensor, whose readings the control module uses to adjust both the input and output amplifier sections (see Figure 7.11) (Khilla et al., 2002).

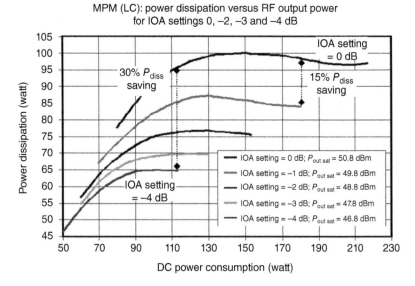

FIGURE 7.21 Flexible TWTA's power dissipation versus DC power consumption. (Used with permission of M Khilla and the European Space Agency, from Khilla (2008)).

For a constant RF drive level into the (L)CAMP, as the TWTA temperature rises, the (L)CAMP increases its output power to the TWTA to compensate the TWTA (Khilla et al., 2002).

7.4.11.2 Radiation The (L)CAMP's gain decreases over life, due to its component amplifiers' exposure to radiation. This is compensated by adjustment of the output amplifier section (Khilla et al., 2002).

The TWT is intrinsically radiation-hard, while the solid-state power-conditioning circuits in the EPC require radiation protection (Barker et al., 2005).

7.4.11.3 Aging The gain of the (L)CAMP's component amplifiers decreases over life, but the LCAMP fully compensates for this.

Tesat-Spacecom believes that its EPCs fully compensate TWT aging (Braetz, 2011b).

7.4.12 TWTA Reliability

TWTA reliability has already been partially addressed in Section 7.3.1.

The TWTA failure rate in Space Systems Loral (SSL) satellites significantly decreased every year from 2007 to 2016 in all frequency bands. This reflected a drop in both TWT failure rate and EPC failure rate (Nicol et al., 2016a). Seventy-one percent of their satellites had no TWTA failures and only 8.4% had two or more failures by 2016 (Nicol et al., 2016b).

A 2018 study compared and combined the TWTA reliability statistics of both SSL and Boeing. The failure statistics seen by the two companies were similar enough to be combined intelligently. More than 90% of satellites experienced no TWTA failure. About one third of TWT and EPC failures occurred in the first year of operation. The rate for other years was much lower. Ku-band TWTAs had the highest failure rate of any frequency band. This was most likely because these TWTAs generally put out much higher RF power and have higher DC power input than C-band TWTAs do. The EPC had a higher failure rate than the TWT. The dual EPC failure rate was less than twice that of the single EPC (Nicol and Robison, 2018). There is a rare failure in the dual EPC that brings down both TWTs, but most of the EPC failures bring down only one TWT (Phelps, 2008).

The study made the point that the number of on-orbit operating hours of TWTAs by these two manufacturers is now sufficient to form a value for the TWTA *failures-in-time (FIT) rate*, which is the number of failures in 10^9 hours. The value is so good that a strong case can be made for reducing the level of redundancy (Nicol and Robison, 2018).

7.5 SOLID-STATE POWER AMPLIFIER

When the frequency band is L-, S-, or C-band and the transmit antenna is a phased array or a reflector-based antenna with a phased-array feed, SSPAs are used. An exception is that the OneWeb low-earth-orbit satellites (LEOs) use Ku-band SSPAs on the user downlinks, one per beam, and K-band SSPAs on the feeder downlinks (Barnett, 2016). The deciding factor was the small size of the SSPA compared to the TWTA.

The SSPA is different from the TWTA subsystem in that the SSPA unit contains the preamplifier, linearizer if any, and DC power supply. (An exception is the first-generation Globalstar SSPAs that feed a multi-beam phased array for the user downlinks: the EPC is separate and serves all 91 SSPAs (Metzen, 2000).)

7.5.1 SSPAs in Payload

SSPAs are used in active and semiactive phased arrays at L- and S-bands (Section 11.12). They find a few other uses, such as powering Inmarsat-4's C-band global transmit horn (Section 20.3.6). Unusually, the OneWeb satellites use them at Ku-band, to power passive phased arrays that transmit to users (Section 11.10.3).

7.5.2 SSPA Unit Architecture and Technology

The SSPA unit architecture depends slightly on which amplifier technology is used in the SSPA's last section(s). There are two technologies of choice, GaAs pseudo high-electron-mobility transistor (pHEMT) and GaN high-electron-mobility transistor (HEMT) (Katz and Franco, 2010) (refer to Section 6.5.1 for description of

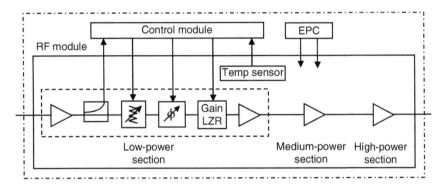

FIGURE 7.22   SSPA unit architecture, Inmarsat-4 L-band example (after Seymour (2000)).

HEMT). The standard technology and the most widespread is the GaAs pHEMT, which is built on or grown on top of a GaAs substrate. Such devices are used in Inmarsat-4's L-band user-link SSPAs (Seymour, 2000) and C-band gateway-link SSPAs (Kiyohara et al., 2003). The new technology is the GaN HEMT, which has a GaN channel and a SiC (silicon carbide) substrate. Flight models were shipped in 2014 (Hirano et al., 2014).

Figure 7.22 gives an example of a GaAs SSPA's unit architecture, from Inmarsat-4 (Seymour, 2000). This SSPA adjusts its internal attenuation based on the RF input power it measures; thus, it has no ALC, which bases adjustment on output power. However, an ALC is common among SSPAs. Like most SSPAs, this SSPA has low-power, medium-power, and high-power amplification sections. The low-power section corresponds most closely to a TWTA's CAMP, and the medium-power and high-power amplification sections together correspond most closely to the TWTA itself, in terms of function and the amount of gain each has. The SSPA's high-power section consists of GaAs component-amplifiers usually in parallel whose outputs are combined. A newer, GaN SSPA has similar architecture but with ALC fed by the unit's output power (NEC, 2017a). GaN component-amplifiers are used in the SSPA's amplification stages after the first (Nakade et al., 2010; Kido et al., 2016). Because of the higher output-power capability of GaN transistors, there is no need for power-combining of component-amplifiers, that is, there is only one device in each GaN stage (Nakade et al., 2010).

Much research and development has been performed on the GaN HEMT-based SSPA, culminating by 2017 in multiple contracts for flight SSPAs (Airbus, 2016) and even shipments (Hirano et al., 2014; Mitsubishi, 2015a; NEC, 2017d). The GaN HEMT has many advantages over the GaAs:

- Higher output power per transistor (Microsemi, 2017)
- Higher PAE (Microsemi, 2017)
- Higher power density, so smaller transistor size (Microsemi, 2017)

- Tolerance of higher temperature, so smaller SSPA and cooling surfaces (Nakade et al., 2010)
- Excellent temperature stability (Microsemi, 2017)
- Higher reliability (Damian, 2014)
- Wider bandwidth (Ishida, 2011)
- Radiation hardness (Waltereit et al., 2013).

The GaN HEMT is believed to be suited for high-power application up to at least 100 GHz. GaN HEMT power-MMIC (monolithic microwave integrated circuit) has been qualified on frequencies through 22 GHz (Quay et al., 2013).

7.5.3 GaAs SSPA Behavior

7.5.3.1 Nonlinearized Behavior A GaAs SSPA has different behavior near saturation from a TWTA. An example of SSPA nonlinear behavior is shown in Figure 7.23, for the Inmarsat-3 L/C-band upconverter/SSPA unit (Khilla and Leucht, 1996). The C-band SSPA was used on the fixed-terminal links. The upconverter was operated in its linear region, so the nonlinear behavior was that of the SSPA. The P_{out} versus P_{in} curve was nearly linear for P_{in} up to within just a few dB of the 1.6-dB compression point at the right edge of the graph. The SSPA's phase shift at this point was much smaller in magnitude than a TWTA's.

The **2-dB compression point (P2dB)** is referred to almost exclusively for a GaAs SSPA rather than the saturation point. The reason is that P2dB should not be

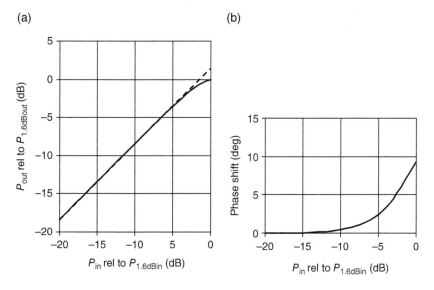

FIGURE 7.23 Example of SSPA's (a) P_{out} versus P_{in} curve and (b) phase shift versus P_{in} curve (after Khilla and Leucht (1996)).

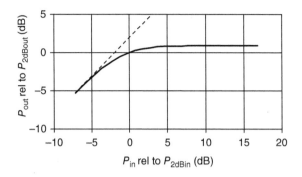

FIGURE 7.24 Example of small transistor's P_{out} versus P_{in} curve (after Khilla (2011d)).

exceeded by much, because otherwise the GaAs transistor's drain will suffer break-down and the transistor will be damaged (Khilla, 2011d). Sometimes in the litera-ture, P2dB or thereabouts is even called the "saturation point."

Figure 7.24 shows the over-driven behavior of a small transistor, which is what a SSPA's would be if it were overdriven (Khilla, 2011d). True saturation is reached, with P_{out} about 1 dB greater at saturation than at P2dB. The P_{out} versus P_{in} curve does not turn down as it does for a TWTA. The phase shift (not shown) keeps grow-ing beyond P2dB, approximately linearly with logarithmic P_{in}.

In some payload applications, the nonlinear performance of an unlinearized SSPA is sufficient. In the example above, at the 1.6 dB compression point, the phase shift is 9° (Khilla and Leucht, 1996). In general, the phase shift of an unlinearized GaAs SSPA at P2dB and the phase shift of a LTWTA at saturation have about the same magnitude (Khilla, 2011a).

7.5.3.2 *Linearized Behavior*   In other applications, linearization is needed to meet high requirements, and the high linearity allows the GaAs SSPA to oper-ate with higher efficiency. Linearization is implemented in the low-power ampli-fication part of the SSPA. Examples in GaAs are the first-generation Globalstar SSPAs for the C-band gateway links (Ono et al., 1996), the Inmarsat-3 SSPAs for the C-band fixed-terminal links (Khilla and Leucht, 1996), and the Inmarsat-4 SSPAs for the C-band gateway links (AIAA JSFC, 2003). All these SSPAs amplify multiple carriers.

Figure 7.25 shows the linearized performance of the same GaAs SSPA whose nonlinearized performance was shown in Figure 7.23. Over the range of 20 to 0 dB IBO, where 0 dB IBO is the P_{in} at which unlinearized gain compression is 1.6 dB, the linearized phase shift is within ±1.5° and the gain compression is within ±0.25 dB. The multi-carrier NPR specification was 23 dB; linearization allowed the SSPA to be operated 2 dB closer to the 1.6 dB compression point (Khilla and Leucht, 1996).

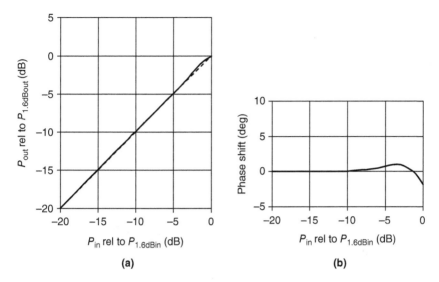

FIGURE 7.25 Example of linearized GaAs SSPA's (a) P_{out} versus P_{in} curve and (b) phase shift versus P_{in} curve (after Khilla and Leucht (1996)).

7.5.4 GaN SSPA Behavior

7.5.4.1 Nonlinearized Behavior The GaN SSPA may be driven to saturation, unlike the GaAs SSPA, so there is no need to talk about the P2dB. The point of 0 dB OBO can be defined as the point at which PAE has its peak. This happens also to match the way the TWTA behaves (Nicol et al., 2013b).

At C-band, the nonlinearized SSPA's plot of $P_{out}/P_{out\,sat}$ versus $P_{in}/P_{in\,sat}$ for CW is very similar to the plot of the nonlinearized LTWTA. The same holds true for the noise-loaded curve (Nicol et al., 2013b). Noise-loading mimics the input of multiple carriers.

7.5.4.2 Linearized Behavior Figures 7.26 and 7.27 show the nonlinear performance of a linearized GaN SSPA at C-band (Kido et al., 2016). This SSPA has 100 W output power and a bandwidth of 300 MHz. The linearity is remarkable, since the 1-dB compression point is at only about 2 dB backoff from saturation. NPR is 15 dB at about 4.5 dB backoff from saturation. The plot of gain flatness over frequency, shown in Figure 7.28, confirms the stated 300 MHz bandwidth.

7.5.5 Flexible SSPA

Besides the flexible ways in which the payload architecture can use SSPAs (Section 11.12), a **flexible SSPA** can also be designed and built. A payload that implements both kinds of flexibility is Globalstar-2, in its S-band architecture with multiport amplifier (MPA) (Darbandi et al., 2008).

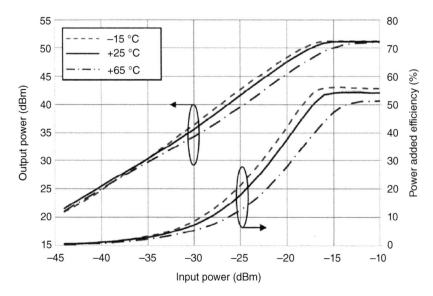

FIGURE 7.26   Example of P_{out} and PAE versus P_{in} for a linearized, C-band GaN SSPA in vacuum. (© 2016 IEEE. Reprinted, with permission, from Kido et al. (2016)).

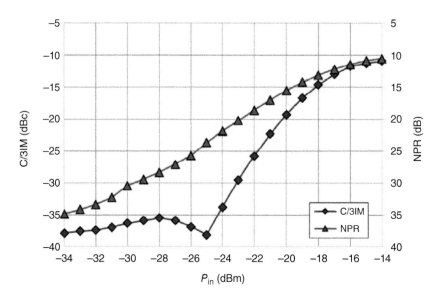

FIGURE 7.27   Example of C/3IM and NPR versus P_{in} for a linearized, C-band GaN SSPA. (© 2016 IEEE. Reprinted, with permission, from Kido et al. (2016)).

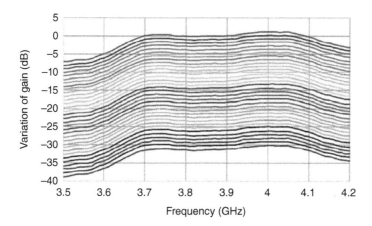

FIGURE 7.28   Example of dynamic range of variable gain for a linearized, C-band GaN SSPA. (© 2016 IEEE. Reprinted, with permission, from Kido et al. (2016)).

The SSPA can have two different kinds of flexibility designed into it. Some SSPAs have both. Here we do not count an ALC capability as a flexibility element, because it is so common among SSPAs. The two kinds of flexibility are as follows:

- Gain and phase compensation over input RF-drive level, accomplished in the low-power-amplification section of the SSPA. Examples of this are Inmarsat-4's L-band user links (Seymour, 2000) and C-band gateway links (Kiyohara et al., 2003).
- Automatic re-biasing of the SSPA in order to change its output power in response to detected input RF-drive level. This maintains good power efficiency even though the traffic level varies. There is also gain and phase compensation to keep the gain and phase constant. Examples of this are the first-generation Globalstar S-band user links (Metzen, 2000) and C-band gateway links (Ono et al., 1996), as well as Globalstar-2's S-band user links (Darbandi et al., 2008). The improvement in power efficiency obtained in the latter case is shown in Figure 7.29.

7.5.6   SSPA Environmental

The categories of environmental influence are temperature, radiation, and aging.

As the SSPA's temperature increases, its gain drops and its noise figure increases (Microwave Encyclopedia, 2010). To limit this, the SSPA is installed on a heat pipe (Section 2.2.1.6) to carry away the heat. At low temperatures, a heater may be necessary when the SSPA is amplifying to a low power level, for example the Iridium L-band (Schuss et al., 1999). The SSPA in an active antenna array employs temperature compensation of gain and usually phase, to keep all the SSPAs tracking. The beam-forming network (BFN) needs this, as well as any MPA or multimatrix amplifier (MMA).

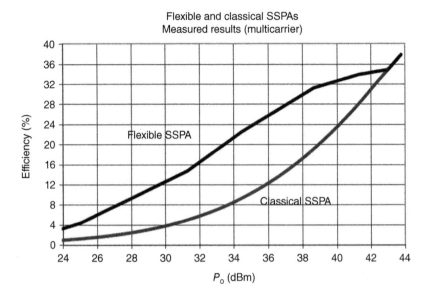

FIGURE 7.29   Example of efficiency comparison of flexible and classical SSPAs. (© 2008 IEEE. Reprinted, with permission, from Darbandi et al. (2008)).

The GaN HEMT has a much higher operating temperature than the GaAs pHEMT does, so it is expected that the thermal-control subsystem can be much reduced (Muraro et al., 2010). This SSPA requires temperature compensation to maintain stable output power and gain (Nicol et al., 2013b).

GaAs devices must be shielded against radiation, unlike GaN devices, which are radiation-hard (Barnes et al., 2005).

Over life, there are at least two different aging effects. The GaAs device's gain gradually decreases over life even with low RF drive (Chou et al., 2004). For such devices run long-term at saturation, there is also another degradation of gain as well as of P_{out} and power efficiency. For example, a saturating-RF lifetest of a 20-GHz, two-stage MMIC power amplifier resulted in the entire P_{out} versus P_{in} curve shifting down by the same amount, while power efficiency decreased by almost 2 percentage points (Chou et al., 2005). A similar result regarding P_{out} versus P_{in} was observed in pHEMT MMIC power amplifiers designed for the band from 6 to 18 GHz (Anderson et al., 2000).

7.5.7   SSPA Specification

Table 7.6 gives an example of the specification parameters for a SSPA that is not part of an active transmit phased array, that is, where no phase or gain matching is required. It is for multi-carrier use since NPR is specified. A typical NPR

TABLE 7.6   Example of Parameters in a SSPA Specification

Mode	Parameter in SSPA Specification	Unit
Both FGM and ALC mode	Operational frequency range	GHz
	NPR	dB
	Noise figure—at max gain, min gain	dB
	Return loss—input, output	dB
	Spurious outputs—inband and noncoherent out-of-band	dBc
	Out-of-band rejection relative to inband gain, at uplink frequencies	dB
	Power efficiency for multiple carriers	%
	Operational temperature range	°C
FGM	Commandable gain range and step size	dB
	Gain flatness at small signal—over 36 MHz, over full band	dB pk–pk
	Gain stability at any frequency—over temperature; from aging and radiation	dB pk–pk
ALC mode	Commandable output-level range and step size	dB
	Output power variation at small signal—over 36 MHz, over full band	dB pk–pk
	Output power stability at any frequency—over temperature; from aging and radiation	dB pk–pk
	ALC time constant (Section 12.10.2)	ms

TABLE 7.7   Performance of Some GaAs SSPAs

Frequency	P_{ref} (W)	Power Efficiency at P_{ref} (%)	P_{ref}	Note	Reference
L-band	40	50	P2dB		Thales (2012)
L-band	55	—	P_{sat}	C/3IM 18 dB	NEC (2017b)
L-band	110	45	P2dB		Thales (2012)
S-band	40	50	P2dB		Thales (2012)
S-band	110	45	P2dB		Thales (2012)
C-band	14	—	P_{rated}	C/3IM > 30 dB	Mitsubishi (2015b)
C-band	40	36	P2dB		Thales (2012)
X-band	20	—	P_{rated}	C/3IM > 25 dB	Mitsubishi (2015c)
X-band	30	25	P2dB		Thales (2012)
Ku-band	15	15	P2dB		Thales (2012)
K-band	3	10	P2dB		Thales (2012)

requirement for multi-carrier use is 15 dB. This SSPA has both FGM and ALC mode. A SSPA has a very much lower noise figure than a TWTA does (Barnes et al., 2005). The out-of-band rejection parameter has the purpose of minimizing any power that could potentially sneak back into the receiver.

TABLE 7.8   Performance of Some GaN SSPAs

Frequency	P_{ref} (W)	Power Efficiency at P_{ref} (%)	P_{ref}	Note	Reference
L-band	65	55	P_{nom}	NPR 17.5 dB at 3 dB OBO relative to P_{nom}	NEC (2017a)
L-band	120	—	P_{sat}		Tesat-Spacecom (2017)
S-band	65	45	P_{nom}	NPR 17.5 dB at 3 dB OBO relative to P_{nom}	NEC (2017a)
S-band	220	—	P_{sat}		Tesat-Spacecom (2017)
C-band	60	—	P_{sat}		Tesat-Spacecom (2017)
C-band	100	50	P_{sat}	NPR 17.5 dB at 6 dB IBO relative to CW $P_{in\,sat}$; BW 300 MHz; mass and footprint much smaller than of TWTA	Kido et al. (2016)
X-band	30	—	P_{sat}		Tesat-Spacecom (2017)
Ku-band	30	—	P_{sat}		Tesat-Spacecom (2017)
K-band	10	—	P_{sat}		Tesat-Spacecom (2017)

7.5.8 SSPA Performance

Table 7.7 summarizes the performance of some current SSPAs in terms of the key parameters output power and power efficiency. Note that a SSPA's stated power corresponds to a different reference point according to manufacturer. The Mitsubishi C- and X-band data points are appropriate for multi-carrier use.

Table 7.8 summarizes the performance of some GaN SSPAs. The data points where NPR = 17.5 dB are appropriate for multi-carrier use. In addition to those SSPAs listed in the table, Mitsubishi Electric offers a C-band GaN SSPA (Mitsubishi, 2015a). Tesat-Spacecom qualified GaN SSPAs at various frequency bands up to 220 W and delivered them to a customer in 2018 (Siegbert, 2019).

REFERENCES

Agilent Technologies (2000). Agilent AN 1287-4, Network analyzer measurements: filter and amplifier examples. Application note. On cp.literature.agilent.com/litweb/pdf/5965-7710.pdf. Accessed Nov. 12, 2010.

AIAA Japan Forum on Satellite Communications (2003). Capital products & review: C-band solid state power amplifer (SSPA). *Space Japan Review*; no 27; Feb./Mar. On satcom.nict.go.jp/English/english2-3/index.html. Accessed June 27, 2011.

Airbus Defence and Space (2016). Airbus Defence and Space awarded third contract in 18 months for advanced gallium nitride (GaN) satellite amplifiers. Press release. On airbusdefenceandspace.com/newsroom/news-and-features/. Accessed May 26, 2017.

Anderson WT, Roussos JA, and Mittereder JA (2000). Life testing and failure analysis of PHEMT MMICs. *Proceedings, GaAs Reliability Workshop*; Nov. 5; 45–52.

Armstrong CM (2015). The quest for the ultimate vacuum tube. *IEEE Spectrum*; 52 (12) (Dec.); 28–51.

Barker RJ, Booske JH, Luhmann Jr NC, and Nusinovich GS, editors, (2005). *Modern Microwave and Millimeter-Wave Power Electronics*. New Jersey: IEEE Press and John Wiley & Sons.

Barnes AR, Boetti A, Marchand L, and Hopkins J (2005). An overview of microwave component requirements for future space applications. *European Gallium Arsenide and Other Semiconductor Applications Symposium*; Oct. 3–4; 5–12.

Barnett RJ on behalf of WorldVu Satellites (2016). File no. SAT-LOI-20160428-00041 of FCC International Bureau. Application to FCC for access to US market. Apr. 28.

Bosch E, Jaeger A, Seppelfeld E, Monsees T, and Nunn RA (2003). TWTA dominance, C-band traveling wave TWTs versus solid state amplifiers. *AIAA International Communications Satellite Systems Conference*; Apr. 15–19; 1–10.

Braetz M, Tesat-Spacecom, Backnang, Germany (2011a). Private communication. July 19.

Braetz M, Tesat-Spacecom, Backnang, Germany (2011b). Private communication. Nov. 25.

Briskman RD (2010). Attachment A, FM-6, technical description. Part of application to FCC for authority to launch and operate the FM-6 satellite. Apr. 9. On licensing.fcc.gov/myibfs/download.do?attachment_key=81002. Accessed Aug. 9, 2017.

Chou YC, Grundbacher R, Leung D, Lai R, Liu PH, Kan Q, Biedenbender M, Wojtowicz M, Eng D, and Oki A (2004). Physical identification of gate metal interdiffusion in GaAs PHEMTs. *IEEE Electron Device Letters*; 25 (Feb.); 64–66.

Chou Y-C, Grundbacher R, Lai R, Allen BR, Osgood B, Sharma A, Kan Q, Leung D, Eng D, Chin P, Block T, and Oki A (2005). Hot carrier effect on power performance in GaAs PHEMT MMIC power amplifiers. *IEEE MTT-S International Microwave Symposium Digest*; June 12–17; 165–168.

Damian C (2014). Reliability of GaN based SSPAs, a major technological breakthrough. White paper. On www.advantechwireless.com/wp-content/uploads/2015/02/WP-Reliability-GaN-based-SSPAs-15033.pdf. Accessed Jan. 21, 2017.

Darbandi A, Zoyo M, Touchais JY, and Butel Y (2008). Flexible S-band SSPA for space application. *NASA/ESA Conference on Adaptive Hardware and Systems*; June 22–25; 70–76.

Dibb DR, Aldana-Gutierrez S, Benton RT, McGeary WL, Menninger WL, and Zhai X (2011). High-efficiency, production 40-130 W K-band traveling-wave tubes for satellite communications downlinks. *IEEE International Vacuum Electronics Conference*; Feb. 21–24; 79–80

Dürr W, Ehret P, and Bosch E (2014). 500W S-band traveling wave tube. *IEEE International Vacuum Electronics Conference*; Apr. 22–24.

Dürr W, Dürr C, Ehret P, and Bosch E (2015). Thales 150 W C-band radiation cooled travelling wave tube. *IEEE International Vacuum Electronics Conference*; Apr. 27–29.

Eze DC and Menninger WL (2017). 170-W radiation-cooled, space K-band TWT. *IEEE International Vacuum Electronics Conference*; Apr. 24–26.

Feicht JR, Martin RH, and Williams BC (L-3 Communications-Electron Technologies) (2007). (Congressional-Microwave Vacuum Electronics Power Res. Ini.) TWT coatings improvement investigation–TWT gain growth. Final technical report. Feb 1. Arlington, VA: USAF, AFRL AF Office of Scientific Research. Report no AFRL-SR-AR-TR-07-0079. Contract no FA9550-05-C-0173. On www.dtic.mil/cgi-bin/GetTRDoc?AD=ADA463632&Location=U2&doc=GetTRDoc.pdf. Accessed July 6, 2011.

Gunter's Space Page (2017). Hylas 1. June 2. On space.skyrocket.de/doc_sdat/hylas.htm. Accessed Aug. 21, 2017.

Hanika J, Dietrich C, and Birtel P (2015). Thales 300 Watt Ku-band radiation and conduction cooled travelling wave tube. *IEEE International Vacuum Electronics Conference*; Apr. 27–29.

Henry D, Pelletier A, and Strauss R (1977). A triple-power-mode advanced 11-GHz TWT. *European Microwave Conference*; Sep. 5–8; 231–236.

Hirano T, Shibuya A, Kawabata T, Kido M, Yamada K, Seino K, Ichikawa A, and Kamikokura A (2014). 70W C-band GaN solid state power amplifier for satellite use. *Proceedings, Asia-Pacific Microwave Conference*; Nov. 4–7; 783–785.

Ishida T (2011). GaN HEMT technologies for space and radio applications. *Microwave Journal*; 54 (8) (Aug.); 56–66.

ITU-R (2015). Recommendation SM.1541-6. Unwanted emissions in the out-of-band domain. Aug. Geneva: International Telecommunications Union, Radio Communication Sector.

Jaumann G (2015). Reliability of TWTAs and MPMs in orbit—update 2014. *IEEE International Vacuum Electronics Conference*; Apr. 27–29.

Kaliski M (2009). Evaluation of the next steps in satellite high power amplifier technology: flexible TWTAs and GaN SSPAs. *IEEE International Vacuum Electronics Conference*; Apr. 28–30; 211–212.

Karsten KS and Wertman RC, inventors; ITT Corp, assignee (1994). Interlocking periodic permanent magnet assembly for electron tubes and method of making same. US patent 5,334,910. Aug. 2.

Katz A, Pallas G, Gray R, and Nicol E (2001). An integrated Ku-band linearizer and TWTA for satellite applications. *AIAA International Communications Satellite Systems Conference*; Apr. 17–20.

Katz A and Franco M (2010). GaN comes of age. *IEEE Microwave Magazine*; 11 (Dec.); S24–S34.

Khilla M (2008). In-orbit adjustable microwave power modules MPMs. *ESA Workshop on Advanced Flexible Telecom Payloads*; Nov. 18–20; 1–11.

Khilla A-M, Tesat-Spacecom, Backnang, Germany (2011a). Private communication. Feb. 16.

Khilla A-M, Tesat-Spacecom, Backnang, Germany (2011b). Private communication. Aug. 19.

Khilla A-M, Tesat-Spacecom, Backnang, Germany (2011c). Private communication. Aug. 22.

Khilla A-M, Tesat-Spacecom, Backnang, Germany (2011d). Private communication. Sep. 7.

Khilla A-M and Leucht D (1996). Linearized L/C-band SSPA/upconverter for mobile communication satellite. *Technical Papers, AIAA International Communications Satellite Sysems Conference*; pt 1; Feb. 25–29; 86–93.

Khilla M, Gross W, Schreiber H, and Leucht D (2005). Flexible Ka-band LCAMP for in-orbit output power adjustable MPM. *AIAA International Communications Satellite Systems Conference*; Sep. 25–28; 1–12.

Khilla AM, Scharlewsky D, and Niederbaeumer J (2002). Advanced linearized channel amplifier for L-band-MPM. *AIAA International Communications Satellite Systems Conference*; May 12–15; 1–10.

Khilia [sic] A-M, Leucht D, Gross W, Jutzi W, Schreiber H, Inventors (2013). Predistoration [*sic*] linearizer with bridge topology having an equalizer stage for each bridge arm. US patent 8493143 B2. July 23.

Kido M, Kawasaki S, Shibuya A, Yamada K, Ogasawara T, Suzuki T, Tamura S, Seino K, Ichikawa A, and Tsuchiko A (2016). 100W C-band GaN solid state power amplifier with 50% PAE for satellite use. *Proceedings, Asia-Pacific Microwave Conference*; Dec. 5–9.

Kiyohara A, Kazekami Y, Seino K, Tanaka K, Shirasaki K, Fukazawa S, Iwano N, Kittaka Y, and Gill R (2003). Superior tracking performance of C-band solid state power amplifier for Inmarsat-4. *AIAA International Communications Satellite Systems Conference*; Apr. 15–19; 1–9.

Kosinski B and Dodson K (2018). Key attributes to achieving > 99.99 satellite availability. *IEEE International Reliability Physics Symposium*; Mar. 11–15.

Kubasek SE, Goebel DM, Menninger WL, and Schneider AC (2003). Power combining characteristics of backed-off traveling wave tubes for communications applications. *IEEE Transactions on Electron Devices*; 50 (6) (June); 1537–1542.

Limburg H, Hughes Electron Dynamics (now L-3 Communications Electron Technologies, Inc), Torrance, CA (1997). Private communication. Oct. 21.

L-3 Communications Electron Technologies, Inc. (2009). *TWT/TWTA Handbook*, 13th ed., Torrance, CA: L-3 Communications Electron Technologies, Inc.

L-3 Communications Electron Technologies, Inc. (2012). Space LTWTA products, space qualified EPCs product listing; Feb. 1. On www.l-3com.com/eti/downloads/summarytable_space.pdf. Accessed June 23, 2017.

L-3 Electron Devices (2016a). Ku-band space traveling wave tube (TWT). On www2.l3t.com/edd/pdfs/datasheets/TWT_Ku-band%20datasheet.pdf. Accessed May 8, 2017.

L-3 Electron Devices (2016b). K-band space traveling wave tube (TWT). On www2.l3t.com/edd/pdfs/datasheets/TWT_K-band%20datasheet.pdf. Accessed May 8, 2017.

Menninger WL, Benton RT, Choi MS, Feicht JR, Hallsten UR, Limburg HC, McGeary WL, and Zhai XL (2005). 70% efficient Ku-band and C-band TWTs for satellite downlinks. *IEEE Transactions on Electron Devices*; 52; 5; May.

Menninger WL (2015). High-efficiency, qualification-tested, next-generation 50-130 W K-band TWTs for satellite communications downlinks. Viewgraph presentation. *IEEE International Vacuum Electronics Conference*; Apr. 27–29.

Menninger WL (2016). Fifteen years of linearized traveling-wave tube amplifiers for space communications. *IEEE International Vacuum Electronics Conference*; Apr. 19–21.

Menninger WL, L-3 Communications (2017). Private communication. June 28.

Menninger WL, L-3 Communications (2019). Private communication. June 4.

Metzen PL (2000). Globalstar satellite phased array antennas. *Proceedings, IEEE International Conference on Phased Array Systems and Technology*; May 21–25; 207–210.

Microsemi Corp (2017). Gallium nitride (GaN) technology. On www.microsemi.com/design-support/gallium-nitride-gan-technology#overview. Accessed July 27, 2017.

Microwave Encyclopedia (2010). Power amplifiers. Oct. 15. On www.microwaves101.com. Accessed July 1, 2011.

Mitsubishi Electric Corp (2015a). C-band GaN solid state power amplifier. Product sheet. On www.mitsubishielectric.com/bu/space/satelite_components/rf_equipment/index.html. Accessed June 25, 2017.

Mitsubishi Electric Corp (2015b). C-band solid state power amplifier. Product sheet. On www.mitsubishielectric.com/bu/space/satelite_components/rf_equipment/index.html. Accessed June 25, 2017.

Mitsubishi Electric Corp (2015c). X-band solid state power amplifier. Product sheet. On www.mitsubishielectric.com/bu/space/satelite_components/rf_equipment/index.html. Accessed June 25, 2017.

Muraro J-L, Nicolas G, Nhut DM, Forestier S, Rochette S, Vendier O, Langrez D, Cazaux J-L, and Feudale M (2010). GaN for space application: almost ready for flight. *International Journal of Microwave and Wireless Technologies*; 2; 121–133.

Nakade K, Seino K, Tsuchiko A, and Kanaya J (2010). Development of 150W S-band GaN solid state power amplifier for satellite use. *Proceedings, Asia-Pacific Microwave Conference*; Dec. 7–10; 127–130.

NEC Space Technologies Ltd. (2017a). L,S-band GaN SSPA (S30 series). Product sheet. On www.necsace.co.jp/en/products/indexhtml#payload. Accessed July 17, 2017.

NEC Space Technologies Ltd. (2017b). L-band solid state power amplifier (SSPA). Product sheet. On www.necsace.co.jp/en/products/indexhtml#payload. Accessed July 17, 2017.

NEC Space Technologies Ltd (2017c). S-band solid state power amplifier (SSPA). Product sheet. On www.necsace.co.jp/en/products/indexhtml#payload. Accessed 2017 July 17.

NEC Space Technologies Ltd. (2017d). X-band solid state power amplifier (SSPA). Product sheet. On www.necsace.co.jp/en/products/indexhtml#payload. Accessed July 17, 2017.

Nicol EF, Mangus BJ, Grebliunas JR, Woolrich K, and Schirmer JR (2013a). TWTA versus SSPA: a comparison update of the Boeing satellite fleet on-orbit reliability. *IEEE International Vacuum Electronics Conference*; May 21–23.

Nicol EF, Mangus BJ, Grebliunas JR, Woolrich K, and Schirmer JR (2013b). TWTA versus SSPA: on-orbit reliability of the Boeing satellite fleet. Viewgraph presentation. *IEEE International Vacuum Electronics Conference*; May 21–23.

Nicol EF, Robison J, Ortland R, Ayala A, and Saechao GS (2016a). TWTA on-orbit reliability for Space Systems Loral satellite fleet. *IEEE International Vacuum Electronics Conference*; Apr. 19–21.

Nicol EF, Robison J, Huang W, Ortland R, Saechao GS, and Ayala A (2016b). TWTA on-orbit reliability of the SSL satellite fleet. Viewgraph presentation. *IEEE International Vacuum Electronics Conference*; Apr. 19–21.

Nicol EF and Robison JM (2018). TWTA on-orbit reliability for satellite industry. *IEEE Transacations on Electron Devices*; 65 (6) (June); 2366–2370.

Nicol E (2019a). Private communication. May 17.

Nicol E (2019b). Private communication. June 2.

Nicol E (2019c). Private communication. June 4.

Nicol E (2019d). Private communication. Nov. 19.

Nusinovich GS, Walter M, and Zhao J (1998). Excitation of backward waves in forward wave amplifiers. *Physical Review E*; 58 (Nov.); 6594–6605.

Ono T, Ozawa T, Kamikokura A, Hayashi R, Seino K, and Hirose H (1996). Linearized C-band SSPA incorporating dynamic bias operation for Globalstar. *Technical Papers, AIAA International Communications Satellite Systems Conference*; pt 1; Feb. 25–29; 123–130.

Phelps TK (2008). Reliability of dual TWTAs—spacecraft system considerations. *IEEE International Vacuum Electronics Conference*; Apr. 22–24; 173–174.

Phys.org (2009). Hylas payload shipped to India. Nov. 6. On phys.org/news/2009-11-hylas-payload-shipped-to-india.html. Accessed Aug. 21, 2017.

Quay R, Waltereit P, Kühn J, Brückner P, van Heijningen M, Jukkala P, Hirche K, and Ambacher O (2013). Submicron-AlGaN/GaN MMICs for space applications. *Digest, IEEE MTT-S International Microwave Symposium*; June 2–7.

Robbins NR, Menninger WL, Zhai XL, and Lewis DE (2016). Space qualified, 150-300-Watt K-band TWTA. *IEEE International Vacuum Electronics Conference*; Apr. 19–21.

Schuss JJ, Upton J, Myers B, Sikina T, Rohwer A, Makridakas P, Francois R, Wardle L, and Smith R (1999). The Iridium main mission antenna concept. *IEEE Transactions on Antennas and Propagation*; 47 (Mar.); 416–424.

Seymour D (2000). L band power amplifier solutions for the Inmarsat space segment. *IEE Seminar on Microwave and RF Power Amplifiers*; Dec. 7; 6/1–6/6.

Siegbert M, Tesat-Spacecom CTO (2019). Private communication. Aug. 16.

Strauss R and Owens JR (1981). Past and present Intelsat TWTA life performance. *AIAA Journal of Spacecraft and Rockets*; 18 (6) (Nov.–Dec.); 491–498.

Tesat-Spacecom GmbH (2016). Amplifier products. Product information. On tesat.de/images/downloads/AP_Brochure_2016.pdf. Accessed Apr. 26, 2019.

Tesat-Spacecom GmbH & Co (2017). Satellite 2017. Brochure. On www.tesat.de/en/media-center/downloads. Accessed June 23, 2017.

Thales Alenia Space (2012). Channel amplifier/linearizer. Product datasheet. May. On www.thalesgroup.com/sites/default/files/asset/document/CAMP_SSPA_2012.pdf. Accessed July 14, 2017.

Thales Electron Devices (2001). Space, advanced technologies for peak performance. Product brochure. May. Received from TED Ulm in 2006 May.

Thales Electron Devices (2013a). L & S-band space TWTs. Product literature. On www.thalesgroup.com/sites/default/files/asset/document/thales_space_l_s_band.pdf. Accessed July 31, 2017.

Thales Electron Devices (2013b). C-band space TWTs. Product literature. On ~/thales_space_c_band.pdf. Accessed July 31, 2017.

Thales Electron Devices (2013c). X-band space TWTs. Product literature. On ~/thales_space_x_band.pdf. Accessed July 31, 2017.

Thales Electron Devices (2013d). Ku-band space TWTs. Product literature. On ~/thales_space_ku.pdf. Accessed July 31, 2017.

Thales Electron Devices (2013e). K & Ka-band space TWTs. Product literature. On ~/thales_space_k_ka_band.pdf. Accessed July 31, 2017.

Villemazet J-F, Yahi H, Lopez D, Perrel M, Maynard J, and Cazaux J-L (2010). High accuracy wide band analog predistortion linearizer for telecom satellite transmit section. *IEEE MTT-S International Microwave Symposium Digest*; May 23–28; 660–663.

Waltereit P, Bronner W, Quay R, Dammann M, Cäsar M, Müller S, Reiner R, Brückner P, Kiefer R, van Raay F, Kühn J, Musser M, Haupt C, Mikulla M, and Ambacher O (2013). GaN HEMTs and MMICs for space applications. *Semiconductor Science and Technology*; 28 (July); 074010.

Weekley JM and Mangus BJ (2004). TWTA versus SSPA; a comparison of on-orbit reliability data. *IEEE International Vacuum Electronics Conference*; Apr. 27–29; 263.

Weekley JM and Mangus BJ (2005). TWTA versus SSPA: a comparison of on-orbit reliability data. *IEEE Transactions on Electron Devices*; 52 (May); 650–652.

Will K (2016). High-power Ku- and Ka-band MPMs for satellite communications. *IEEE International Vacuum Electronics Conference*; Apr. 19–21.

Yuen CH, Yang SS, Adams MD, Laursen KG, inventors; Space Systems/Loral, Inc., assignee (1999). Broadband linearizer for power amplifiers. US patent 5,966,049. Oct. 12.

Zhang W-M and Yuen C (1998). A broadband linearizer for Ka-band satellite communication. *IEEE MTT-S Microwave Symposium Digest*; vol. 3; June 7–12; 1203–1206.

CHAPTER 8

PAYLOAD'S ANALOG COMMUNICATIONS PARAMETERS

8.1 INTRODUCTION

This chapter presents the analog communications parameters found in the top-level payload requirements document. The payload-level communications parameters particular to a digital processing payload are discussed in Section 10.5. In the present chapter, we intend to fully explain the following for each analog parameter:

- What the parameter is
- How the parameter's application at unit level relates to its application at payload level
- How the parameter value is verified to be in compliance.

Figure 8.1 gives the simplified transponder diagram used in the sections below, for marking the payload elements that contribute to a parameter value. ("Element" means either a unit or an element used to integrate the payload, such as a piece of waveguide.) Waveguide is shown as the transmission line up to the frequency converter and beyond the high-power amplifier (HPA), with coaxial cable in between, in order to have both waveguide and coax in the figure since they have different characteristics.

Satellite Communications Payload and System, Second Edition. Teresa M. Braun and Walter R. Braun.
© 2021 John Wiley & Sons, Inc. Published 2021 by John Wiley & Sons, Inc.

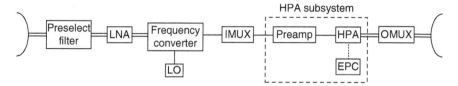

FIGURE 8.1 Simplified transponder diagram for showing main contributors to each communications parameter.

Each payload element enhances the reception and retransmission of the signal in some way but also inevitably distorts the signal in some minor way. The elements can be divided into two classes: those that cause basically linear distortion and those that cause nonlinear distortion. The effect of linear distortion on communications is easier to analyze. The elements that are significantly nonlinear are the following:

- Frequency converter
- HPA, especially
- Spacecraft edges and corners that cause passive intermodulation products (PIMs).

Some elements' distortion properties depend on bandwidth, which is characterized in this book by the percentage it represents of the carrier frequency: **narrowband**, on the order of 0.5%; **wideband**, on the order of 3%; and **very wideband**, on the order of 15%. We consider no wider bandwidths since they are uncommon.

Not all payload parameters represent distortions, but for those that do a recommendation is made in each section on how to combine the signal distortion as measured or estimated for every payload element into a distortion value at payload level. We emphasize the combining upward to payload level instead of the allocation downward from payload level to the elements, because typically the payload engineer starts at the beginning of a satellite program with a collection of preliminary element specifications from an earlier similar payload and must check if they will collectively fit together for the new payload. During the course of the program, the payload engineer continually refines his calculations to show that he continues to expect to meet the payload-level requirements. For some types of distortion, there is no way to do the combining that makes perfect sense. In that case, a proven sensible way that is readily computable is given. There are special cases, not addressed in this chapter, for which the payload engineer must take another approach to distortion-combining that he himself must develop. The recommended method for combining unit-level distortions sometimes depends on the channel bandwidth (Section 8.4).

Parameters must meet their specifications in all on-orbit conditions:

- Over the payload life on orbit, that is,
  - Over temperature—specifically, for each element over its **operating temperature range**. For units on baseplates, this is the baseplate temperature range

- ○ From aging while operated
- ○ From radiation exposure.
- Over the range of operating signal levels
- Over the parameter's **specification bandwidth** if the parameter is defined over a frequency band
- In all specified directions toward coverage area if the parameter is related to power level received or transmitted.

There are various reasons that some parameters change value over life. Some units' performance is temperature-sensitive. The temperature of most of the payload equipment generally rises over on-orbit life in a known way. The lifetime parameter variation from temperature can be calculated from the temperatures over life and the factory-measured unit temperature-sensitivity. Other parameter values change because a unit's or component's performance slowly changes simply from being operated. The payload manufacturer relies on long-term test results on similar devices from the unit or component manufacturer. Other parameter values change because a unit or component is sensitive to accumulated radiation exposure on orbit, and here again the data comes from the manufacturer. If the sensitive device is not a unit but is incorporated into a unit, the unit engineer uses the data to make a prediction for the unit. For more general information on effects, see Section 2.3; for more information specific to elements, see Chapters 3 through 7.

For every communications parameter, the **verification** method will be briefly presented. Verification shows that a specification is met. The customer and the payload engineer must come to agreement on the **verification matrix**, which is a table that shows the verification method proposed for every parameter in the payload requirements document. The most common method and regarded as the most robust, although it has its own problems, is direct test or measurement. However, for some parameters, test is not necessary: parameters that were measured at a lower level of integration and that are truly not affected by the higher level(s) of integration. In this case, the parameter value is *carried up* from the lower level of test. For some parameters, a direct measurement is not possible, in which case the method *analysis* is applied. The analysis is based on other parameters that were directly measured. The other two verification methods, *inspection* and *by design*, do not apply to the communications parameters we discuss in this book.

8.2   GAIN VARIATION WITH FREQUENCY

8.2.1   What Gain Variation with Frequency Is

Every two-port device can be considered as a filter in that it can be characterized by a transfer function $H(f)$, where f is frequency in Hz (Section 5.2.2). For an active device, the transfer function depends on the radio-frequency (RF) drive level input to it and the component-amplifier biases. As we have seen, the transfer function has two parts, the gain response $G(f)$ and the phase response $\varphi(f)$:

$$H(f) = 10^{G(f)/20}\, e^{j\varphi(f)}$$

where $G(f)$ is in dB and $\varphi(f)$ is in radians.

Payload or unit **gain variation with frequency** has a few descriptive names but the main ones are **gain flatness**, **gain tilt**, and **gain ripple**. Putting values to them is best done with the qualifier *pk-pk*, *pk-to-pk*, or *p-p*, meaning *peak-to-peak*, that is, highest relative to lowest in dB, which is unambiguous. (Examples of ambiguous qualifiers are "maximum" and no qualifier at all.) Gain flatness is simply the highest gain in dB over the specified band minus the lowest gain over the band, as illustrated in Figure 8.2. When gain flatness is specified, no additional gain-variation parameters are specified. An alternative characterization of gain variation is gain tilt and gain ripple together. Gain tilt is the delta gain over the band of the best-fit line over the band; it can be positive or negative. Gain ripple is then the variation in the gain over the band that is left after the tilt is taken out. Gain tilt and ripple are illustrated in Figure 8.3. This example is in fact the decomposition of the simple gain flatness of Figure 8.2.

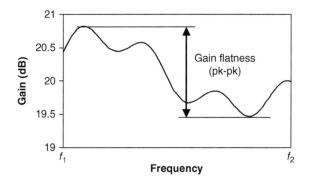

FIGURE 8.2 Definition of gain flatness.

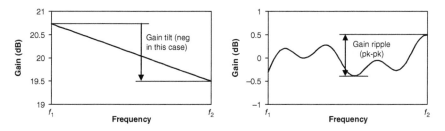

FIGURE 8.3 Definitions of gain tilt and gain ripple.

When the band has ripple that is more or less a sinewave over frequency, the ripple slope may additionally be specified. The slope together with the ripple amplitude gives you the ripple frequency period (= π times pk-pk ripple in deg divided by ripple slope in deg/Hz). The sinewave may be in any phase.

How gain variation with frequency degrades the signal is addressed in Section 12.15.1.

8.2.2   Where Gain Variation with Frequency Comes from

We look at the gain variation characteristics of the payload units each over their own specified band. The following is meant to be a guide to expectations of gain-variation specification and test data results but may not be accurate in all cases.

Most units have so little gain variation that they are specified on simply gain flatness, as shown in Figure 8.4. This is true for the filters (preselect, input multiplexer (IMUX), and output multiplexer (OMUX)) because they are equiripple or nearly equiripple (Section 5.2.4). It is true of the frequency converter and the solid-state power amplifier (SSPA).

The complementary set of payload elements has gain tilt, as shown in Figure 8.5. First are the antennas. The most common types of communications-satellite antennas for the GHz frequencies are aperture antennas (Section 3.3.2). Their apparent area measured in square wavelengths goes up with the square of frequency. A reflector antenna's gain ideally goes up linearly in dB with frequency. A phased array's gain goes up somewhat less than linearly because the common use of phase shifters instead of delay elements causes a scanned beam to point a little off-axis for frequencies above and below mid-band (beam squinting) (Section 11.7.2.2). Besides the antennas, also waveguide, coax, the (L)CAMP, and the traveling-wave tube (TWT) commonly present gain tilt.

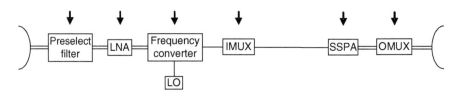

FIGURE 8.4   Payload elements whose gain variation is specified on simply gain flatness.

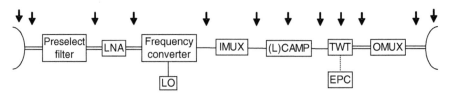

FIGURE 8.5   Payload elements with gain tilt.

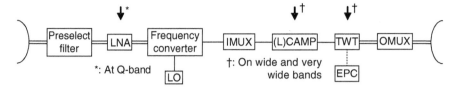

FIGURE 8.6   Payload elements with quadratic gain component.

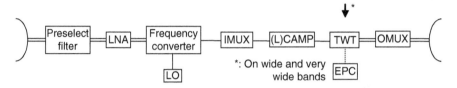

FIGURE 8.7   A few individual TWTs have sinewave-ripple gain component.

The (L)CAMP and the TWT have a quadratic component of gain variation in addition to tilt, on wide and very wide bands. Additionally, a Q-band low-noise amplifier (LNA) has been seen to have a quadratic component (Grundbacher et al., 2004). These three units are marked in Figure 8.6.

The TWT, marked in Figure 8.7, is the only unit that may have a sinewave-ripple versus frequency component of gain variation. A few individual TWTs have ripple in their gain and phase responses versus frequency, of period roughly 2% of the center frequency. The TWT gain variation is measured at both small-signal and saturation. The ripple amplitude at small-signal is several times what it is at saturation (Section 7.4.7.1). If a communications channel has bandwidth about half of the ripple period, it may be necessary to take the gain slope into account in estimating channel performance. If a modulated signal has bandwidth equal to a ripple period or more, it may be necessary to apply a frequency-varying model of the traveling-wave tube amplifier (TWTA) in performance estimation (Section 16.3.5).

8.2.3   How Gain Variations with Frequency at Unit Level Carry to Payload Level

If the payload carries more than one channel, then at the payload level the gain variation specifications will be on two layers. One layer is by the channel, to insure minimal distortion of the channel. Ideally, all the channels have the same specification if they will all carry the same kind of modulation; however, one or two channels may need different specifications if they are near another signal such as a command channel and need to be protected from it and to protect it. The second layer of payload specifications will typically be over a bandwidth wide enough to include all the channels, to insure that the channels receive approximately equal treatment.

For the payload elements whose gain variation is specified simply on pk-pk gain flatness, it is difficult to say how to combine them into a payload-level estimate.

If the gain variation plot of each element across the band is similar to plots with jumpy measurement error, then it is probably best to rss together the rms values of the elements. **Rss** is the square root of a sum of squared values, and **rms** is the square root of the average of a set of squared values. Taking the rms value across a band minimizes the impact of any large but narrow deviation, which is a legitimate approach if the signal going through the band will fill the band. Taking the rss as a way to combine all the elements' values is a good idea if the gain variations from element to element look unrelated, for example do not all have a peak at the same frequency. The last step would be to multiply the rss value by $\sqrt{2}$ to create an approximation of a pk-pk value. There are many cases in which this is *not* a good approach. The payload engineer must inspect the gain variation plots of the major contributing elements to be able to form an idea of what is a good way to combine. For example, if two elements have a large variation at about the same frequency, then the best thing to do is to form the algebraic sum of the large variations and add it to $\sqrt{2}$ times the rss of the rms values of the other elements. For another example, if two elements have a large variation at different frequencies, then the best thing to do is to take the larger of the two variations and add it to $\sqrt{2}$ times the rss of the rms values of the other elements.

For payload elements whose gain variation specs are on tilt, if consistently signed the gain tilts over a band can be added to obtain the total tilt through the payload on that band. The same goes for quadratic gains.

8.2.4   How to Verify Gain Variation with Frequency

A scalar network analyzer is used to measure the payload's end-to-end gain variation with frequency.

When the HPA's specification bandwidth is at least wide, two measurements of the HPA must be made. One is the **small-signal frequency-sweep**, for which the sweeping continuous wave (CW) is at a constant level, low enough for the HPA to be operating in its quasi-linear region across the full band. The other is the **large-signal or saturated frequency sweep**, for which the sweeping CW is also at a constant level, high enough to saturate the TWTA across the full band. Both sweeps are required in order to provide enough information if the HPA will be modeled in a simulation (Section 16.3.7).

8.3   PHASE VARIATION WITH FREQUENCY

8.3.1   What Phase Variation with Frequency Is

Phase response $\varphi(f)$, in radians, is a frequency-domain characterization. Phase response is more useful than group-delay response for digital communications because its effect on the signal phasors is easier to visualize. If group delay is provided, the phase can be estimated by an integration (Section 5.2.2).

Payload or unit **phase variation with frequency** has a few descriptive names, but the main ones are **phase linearity**, **quadratic phase**, and **phase ripple**. Putting values to them is best done with the qualifier "pk-pk" because it is unambiguous. The phase response is assessed after its slope is removed, since the slope corresponds only to the average delay through the payload or unit. Phase linearity is actually deviation from linear.

How phase variation with frequency degrades the signal is addressed in Section 12.15.1.

8.3.2   Where Phase Variation with Frequency Comes from

We look at the phase variation characteristics of the payload units each over their own specified band. The following is meant to be a guide to expectations of phase-variation specification and test data results but may not be accurate in all cases.

Most units have so little phase variation that they are specified on simply phase linearity, as shown in Figure 8.8. This is true for the filters (preselect, IMUX, and OMUX), the frequency converter, and the SSPA.

Most of the rest of the payload elements, but not the antennas, have quadratic phase, as shown in Figure 8.9. The only units that do are the (L)CAMP and the TWT. Besides them, waveguide does.

The (L)CAMP and the TWT have a cubic component of phase variation in addition to quadratic, on wide and very wide bands, as shown in Figure 8.10.

The TWT, marked in Figure 8.11, is the only unit that may have a sinewave-ripple versus frequency component of phase variation. It arises for the same reason that periodic gain ripple does and has the same characteristics. Because the ripple source is passive, the phase ripple is directly related to the gain ripple: there is

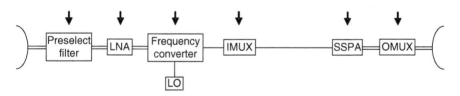

FIGURE 8.8   Payload elements whose phase variation is specified on simply phase linearity.

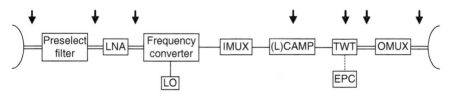

FIGURE 8.9   Payload elements with quadratic phase component.

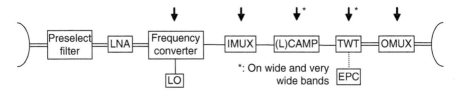

FIGURE 8.10   Payload elements with cubic phase component.

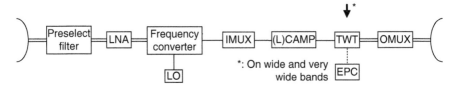

FIGURE 8.11   A few individual TWTs have sinewave-ripple phase component.

7° pk-pk of phase ripple for every 1 dB of pk-pk gain ripple, and the phase and gain ripples are 1/4 cycle apart in the frequency domain. This is shown in the Appendix 8.A.1 of this chapter.

8.3.3   How Phase Variations with Frequency at Unit Level Carry to Payload Level

Phase variations with frequency at unit level carry to payload level in the same way that gain variations do. If consistently signed the quadratic components are summable and the cubic components are summable.

8.3.4   How to Verify Phase Variation with Frequency

If it were not for the repeater's frequency converter, a vector network analyzer (VNA) could measure the repeater's phase response versus frequency. A VNA measures the S-parameters of a two-port device (Section 4.7.2). However, the VNA requires that the input and output signals be at the same frequency.

One way to get around this problem requires either that the frequency converter's local oscillator (LO) signal be accessible or that its signal be replaceable by an external LO. The same LO must feed both the repeater and a reference mixer. The technique also requires a reciprocal calibration mixer, which has the same gain and phase response in both up- and down-conversion (Agilent Technologies, 2003).

A newer technique measures group delay by the definition of derivative. Recall that group delay τ_g is defined as

$$\tau_g(f) = -\frac{1}{2\pi}\frac{d\varphi(f)}{df}$$

This can be approximated by

$$\tau_g(f) \doteq -\frac{1}{2\pi}\frac{\Delta\varphi(f)}{\Delta f}$$

where
$\varphi(f) = \angle$(repeater transfer function including frequency translation)
f = frequency in frequency band of repeater input signal.

The technique inputs a pair of tones that are near each other and centered at f_1, say, into the repeater. A VNA in a special setup measures the phase difference of the tones at input and at output. The difference between the two-phase differences, divided by the delta frequency and scaled, is an approximation of $\tau_g(f_1)$. The measurement is made across the input frequency band. When done properly the phase and frequency instabilities of the frequency converter's LO are canceled out. The phase response across the band is calculated by integration and scaling (Rohde & Schwarz, 2012).

When the HPA's specification bandwidth is at least wide, two phase measurements of the HPA must be made, same as for gain response.

8.4 CHANNEL BANDWIDTH

The term **channel bandwidth** in this chapter means the bandwidth of a channel on its total path through the payload. If a specific channel bandwidth is a requirement, it is a replacement requirement for gain and phase responses. A typical type of channel bandwidth specified is the 3-dB bandwidth (Section 5.2.3).

The units that primarily determine a channel's bandwidth are marked with arrows in Figure 8.12. They are the IMUX and to a lesser extent the OMUX.

The gain response over the channel's frequency band, through the whole payload, must be characterized at cold operating temperature and again at hot operating temperature. The intersection, in the sense of set theory, of the two responses is taken to show specification compliance. When the filter is made of waveguide, the filter response shifts from air to vacuum, so measurements in air must be adjusted accordingly (Section 5.4.1).

8.5 PHASE NOISE

Phase noise is an unwanted variation in the signal phase that comes from the LOs used in frequency conversion (Section 6.6.6). Section 12.10.2 gives tells what part of the phase noise spectrum should have specifications on it and how phase noise affects the signal.

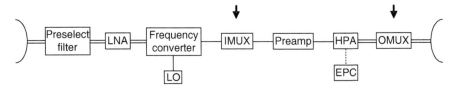

FIGURE 8.12   Sources of channel-bandwidth limitation.

There are three types of phase noise test set currently in use. The simplest one is a spectrum analyzer augmented with phase noise-testing software. This type does not differentiate between phase noise and amplitude noise but combines them. However, apparently, amplitude noise is minimal in an oscillator (see for example (Anritsu Corp, 2008)). Such a test set measures the single-sideband phase noise $\mathcal{L}$ down to 10 Hz at the lowest extreme. A more complex and better performing type of test set is the signal source analyzer (see for example (Agilent Technologies, 2007)). It is also based on a spectrum analyzer but is much more. It is able to separate out phase noise from amplitude noise. It measures phase noise in two or three ways including a cross-correlation technique. It can measure $\mathcal{L}$ down to at least as low as 1 Hz (you do not need it this low, but this helps you to identify the kind of test set you have). A third type of test set is all digital and can measure phase noise at frequencies down to 0.1 mHz but currently only for oscillators with frequency up to 30 MHz, so they are only usable for the payload's master reference oscillator.

With whichever type of test set is used, the engineer must make certain that if the test set itself adds phase noise, its spectrum at all frequencies of interest is much smaller than that of the device under test.

8.6 FREQUENCY STABILITY

The **frequency stability** of the LO, if it derives from a reference oscillator, is the same as that of the reference oscillator in parts per million (ppm).

An oscillator is normally specified on both its long-term and short-term stability. **Long-term stability** or **drift** characterizes frequency variations over intervals from days to years. For communications, the interval of interest is usually a year. This is systematic drift due to aging and is stated in ppm per the time interval. The engineer does not test this but relies on the numbers from the manufacturer. **Short-term stability** is stability over an interval of a few seconds at most. The oscillator's phase noise spectrum (Section 6.6.6) describes the short-term stability on intervals of 1 second or less (1 Hz on up). For communications, stability over a few seconds (i.e., down to 0.1 Hz) does not matter if it is not extreme.

8.7   SPURIOUS SIGNALS FROM FREQUENCY CONVERTER

The frequency converter can create five kinds of **spurious signals** (also called *spurious outputs, spurious,* or just *spurs*) in its output. (Other units create spurious signals, too, as we see below.) The mixer in the frequency converter can create four kinds of spurious signals:

- Oscillator harmonics
- Harmonics of the input signal
- Cross-products of the input signal and LO carrier besides the intended cross-product
- Leakage of input signal and LO directly through to the output.

The fifth can arise in the LO and is transferred to the signal by the mixer. It arises most often in synthesized LOs (Hitch, 2019):

- Discrete spurs close to carrier.

The first three are intermodulation products (IMPs).

What the payload engineer can do to mitigate the spurious signals is the following:

- Ensure that the satellite's frequency plan keeps the spurious signals of significant power out of the channels
- Specify realistically low levels of IMPs for the frequency converter
- Add filters where required.

The spurious signals specification depends on signal level because there are some spurious signals, namely those with a contribution from the signal, whose level is different at a different signal level.

When any signal input to the payload is only narrowband or wideband but not very wideband, there is usually no difficulty in dealing with the spurious outputs. That is, the spurious signals are so far out of the band of the downconverted signals that the only concern is interference to signals from other systems being received on the ground. However, it may happen with a very wideband signal that the spurious outputs would be a problem with only one frequency conversion on the satellite, in which case a dual conversion must be performed.

If the level of any of the spurious signals would be too high at the payload output given the filtering provided by the active units and the passive filtering already foreseen in payload design, then a passive filter must be added, carrying a penalty in RF loss, mass, and spacecraft-panel mounting area.

The spurious signals are measured with a spectrum analyzer by applying a CW as converter input signal and sweeping its frequency. A CW is used because it is easier to see the mixer outputs than with a modulated signal. The type of spurious signal seen can be partially determined by inspection of whether and how it moves as the CW is swept, since the mixer output (m,n) will move m times as fast as the CW is moving and in the direction given by the sign of m.

The International Telecommunications Union (ITU) specifies out-of-band spurious signals for all space services with reference bandwidth 4 KHz for frequencies below 15 GHz and with 1 MHz for frequencies above 15 GHz. The spectrum analyzer's resolution bandwidth ideally equals the reference bandwidth (International Telecommunication Union, Radiocommunication Sector (ITU-R), 2015).

8.8 HPA NONLINEARITY

What the HPA nonlinearity is was discussed in Section 7.2. The four ways in which the nonlinearity can be specified were also discussed in that section and are as follows:

- P_{out} versus P_{in} curve and phase shift φ versus P_{in} curve
- Derivative of P_{out} versus P_{in} curve and derivative of phase shift versus P_{in} curve
- C/3IM
- NPR.

Verification of the HPA nonlinearity depends on the way it is specified. To measure P_{out} versus P_{in} and φ versus P_{in}, the VNA is used, but instead of sweeping the frequency, you sweep the signal level. The CW stays at one frequency during the power sweep. If the HPA is to be used on a very wide band, the measurements should be taken at about three frequencies, at midband and at about 1/6 and 5/6 of the way across the band. Verification for the other specification methods was described in Section 7.2.2.

The nonlinearity is measured at HPA level and carried up to payload level.

8.9 NEAR-CARRIER SPURIOUS SIGNALS FROM HPA SUBSYSTEM

8.9.1 What HPA-Subsystem Spurious Signals Are

A HPA puts out spurious signals near the carrier. They are better described in the literature for the TWTA than for the SSPA.

For a TWTA, some of them exist whether or not a signal is present, and some exist only when there is a signal (L-3, 2009). Their frequencies are usually between 100 Hz and 500 KHz from the carrier (L-3, 2009). These spurious signals are so close to the carrier that they cannot be filtered out. A TWTA's spurious signals arise from both spurious phase modulation (PM) and **spurious amplitude modulation (AM)**. A spurious PM that has a given frequency represents a small sinewave component of the signal phase. A spurious AM that has a given frequency represents a small sinewave component of the signal amplitude. On a spectrum analyzer, one of either type would be seen to have many harmonics, symmetric in spacing and power about the carrier. On a phase-noise test set, spurious PM would be seen as a spike on the phase noise spectrum.

There are some spurious signals, namely those with a contribution from the signal, whose level is different at a different signal level.

8.9.2   Where HPA-Subsystem Spurious Signals Come from

For the TWTA, both the TWT and the EPC create spurious signals, and for a SSPA it may be the same, as illustrated in Figure 8.13.

For a SSPA, there are in-band spurious and EPC switching-noise spurious (AIAA Japan Forum on Satellite Communications, 2003).

The following information on spurious signals for a TWTA comes from (L-3, 2009). The cause of the spurious signals that exist even without a carrier present is usually ripple on power supply voltages:

- The cathode heater, usually a filament, causes PM if it is connected to an alternating-current (AC) power supply.
- Cathode voltage fluctuation in the form of ripple causes PM and AM, but the spurious signals due to AM are nearly always 10 dB lower than those due to PM. The spurs are usually 100–500 KHz from the carrier. The ripple has two sources: AC components at the fundamental and harmonics of the DC-to-AC converter's chopper (which precedes the voltage transformer and the AC-to-DC converter in the power supply), and AC ripple or transients on the power supply's input bus voltage. A change in cathode voltage affects the speed of the electrons in the beam, which affects the phase of the RF output signal.

The cause of the spurious signals that exist only when a carrier is present is interactions within the TWT itself:

- Cathode voltage fluctuation at a rate equal to the modulation-symbol rate R_s causes PM. Since the RF signal being amplified has variations in its signal level, the amount of electron current that undesirably hits the TWT's helix also varies, which causes the cathode voltage to fluctuate.

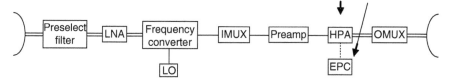

FIGURE 8.13   Sources of HPA-subsystem spurious signals.

8.9.3 How HPA-Subsystem Spurious Signals Carry to Payload Level

Spurious that could interfere into other channels may be removed by the OMUX. For the TWTA, since the maximum frequency of these spurious signals is 500 KHz, for the removal to occur the carrier spacing would have to be less than 1 MHz. Spurious signals due to PM do not harm the signal once the signal is received on the ground (Section 12.10.2). The spurious that are due to AM will degrade the bit error rate (BER) unless the modulation symbol rate is much less than 500 KHz. Also, if the antenna beam that carries this signal impinges on another beam, this signal will cause a varying level of interference that must be accounted for in the self-interference calculation (Section 8.14). Spurious PM due to signal-amplitude fluctuation will be mostly or entirely removed (that is, averaged out) by the detection filter on the ground (Section 12.10.1) so will cause little or no harm.

8.9.4 How to Verify HPA-Subsystem Spurious Signals

At both the unit level and payload level, these spurious signals can be measured with a spectrum analyzer with a CW signal. However, if it is desired to determine which spurious signals are due to PM and which are due to AM, two special test setups are required (L-3, 2009).

8.10 STABILITY OF GAIN AND POWER-OUT

8.10.1 What Gain Stability and Power-Out Stability Are

Stability of gain and **stability of power-out** are characteristics of the payload gain and power-out at the input to the transmit antenna. Stability is the change in overall level of the parameter over the operating temperature range and life, with all other conditions being equal. The stability may be specified at just the mid-channel frequency or at a few frequencies.

A particular payload may have either one of these parameters specified or both, depending on the operating modes of the preamp. If the unit has a fixed-gain mode (FGM), then the unit has the capability to amplify the signal by a commanded fixed gain. If the unit has an automatic level-control (ALC) mode, then it has the capability to amplify the signal (plus interference and noise) to a commanded level. The commanded level corresponds to a particular HPA input backoff for noise-free single-carrier operation (Section 7.2.1). When the unit is in FGM, then the applicable specification is gain stability; when it is in ALC mode, the applicable specification is power-out stability.

8.10.2 Where Gain Stability and Power-Out Instability Come from

The stability of the gain and power-out depend primarily on the preamp and the HPA, as depicted in Figure 8.14. The only other possibility is the OMUX, but as long as it is on heat pipes its effect is probably negligible.

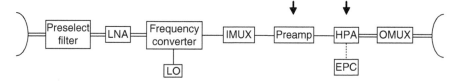

FIGURE 8.14   Sources of gain and power-out instability.

TWTA performance changes over temperature and life are well characterized, and the (L)CAMP compensates them. The (L)CAMP also compensates its own changes (Section 7.4.11).

The SSPA compensates its own performance change over temperature. With aging, its gain decreases for a fixed RF drive level (Section 7.5.6).

For both types of HPA, gain and power-out instability are only residual effects from imperfect compensation.

8.10.3   How Gain Stability and Power-Out Stability Carry to Payload Level

These parameters are only defined at the payload level.

8.10.4   How to Verify Gain Stability and Power-Out Stability

The contributing factors to gain stability and power-out stability, exclusive of aging and radiation, are tested at the unit level. The parameters themselves, except for variation from aging and radiation, are tested at spacecraft system test. The tendencies that the units have over life, due to aging and radiation, are known from earlier testing of such units by their manufacturers, and these effects are factored in by analysis.

8.11   EQUIVALENT ISOTROPICALLY RADIATED POWER

Equivalent isotropically radiated power (EIRP) (Section 3.3.3.1) is often the first payload parameter that comes to mind. Its specification tells how much power the payload must radiate toward the transmit coverage area, which is the combined area on the earth over which the payload's transmission must meet minimum-power and other specifications. EIRP is the product of the power into the transmitting antenna's terminal and the antenna gain in a given direction.

The main contributors to EIRP in the positive sense are the preamp, the HPA, and the transmit antenna. The preamp sets the RF drive level into the HPA, thereby determining the HPA's output power. The negative contributors are the OMUX and the post-HPA RF lines. Figure 8.15 illustrates this.

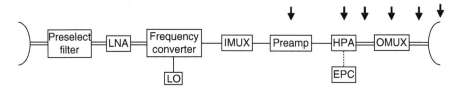

FIGURE 8.15 Payload elements that determine EIRP.

EIRP is a payload-level requirement. Actually, it is a spacecraft-level requirement because it depends on the antenna's on-orbit pointing error, which depends on the spacecraft bus's ability to control it. For a spot-beam payload at geostationary orbit (GEO) the spacecaft's pointing error can be a critical factor in the EIRP.

EIRP without consideration of antenna pointing error is verified in spacecraft-level testing at the compact antenna test range (CATR) (see Sections 3.12 and 11.15 on antenna testing). The effect of pointing error on the EIRP is folded in analytically.

8.12 FIGURE OF MERIT G/T_s

8.12.1 What G/T_s Is

Part of the payload's function is to serve as a receive terminal for uplink signal(s) from the ground station(s) and, for some payloads, from user terminals. The receive aspect of the payload is commonly characterized by the terminal figure of merit G/T_s (Section 3.3.3.1). (It is usually written "G/T" but we want to emphasize what particular T it is and to differentiate it from T used elsewhere in this book.) G is the gain of the receive antenna and is stated in dBi, namely dB relative to the gain of an ideal isotropic antenna (with gain 0 dB). T_s is the **system noise temperature** and is stated in Kelvin, where $0\,K = -273\,°C$. G/T_s is stated in units of dBi/K, which means G in dBi minus T_s in the form of $10 \log_{10}T_s$. Both G and T_s must be **referenced** to, that is, apply at, the same point in the payload, usually the antenna terminal. (Any point can be the reference point, as is shown in the Appendix 8.A.2 of this chapter, and the answer comes out the same.) The G/T_s definition is illustrated in Figure 8.16.

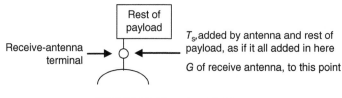

FIGURE 8.16 G/T_s definition.

G/T_s is used in the calculation of the uplink signal-to-noise ratio (SNR), if the signal level S into the antenna is known:

$$\text{Uplink SNR} = \frac{S}{N}$$

where

$$S = P_{\text{den}} \left(\frac{c}{f_c} \right)^2 \frac{1}{4\pi} G$$

P_{den} = signal flux density at input to antenna in units of W/m^2
c = speed of light = 2.998E8 m/s
$N = T_s \kappa B$
κ = Boltzmann's constant = 1.379E-23 W/(Kelvin-Hz)
B = equivalent noise bandwidth on which noise is measured

The terms G and T_s require further explanation.

G is the receive antenna gain in the direction of the signal, referenced to the antenna terminal, taking the pointing error into account in some way agreed with the customer.

The uplink system noise temperature T_s is more complex. It consists of three additive inband components:

- Antenna noise temperature T_a—the noise temperature that the antenna receives from where it looks
- **Thermal noise temperature T_e** due to rest of payload—the noise generated by all the payload elements together, whether active or passive, besides the antennas
- **Ringaround noise temperature T_r** due to payload transmitted signals being received by payload (means against it are addressed in Section 4.4.5).

There is an excellent treatment on the noise components in (Pritchard and Sciulli, 1986). We have

$$T_s = T_a + T_e + T_r$$

Antenna temperature T_a is not actually due to the antenna but to what the antenna receives along with the uplink signal (Section 14.5).

Thermal noise temperature T_e is due to the rest of the payload beyond the antenna terminal. The repeater noise figure can be estimated before spacecraft system test by combining the noise figures (Section 9.2) of all the payload elements past the G/T_s reference point. If F is the payload noise figure, the thermal noise temperature is given by

$$T_e = (F - 1)290 \text{ in Kelvin}$$

The payload receives a very weak signal which will pass through many elements in the payload. Most of these elements cause a loss to the signal level, and some like the HPA have very large noise figures. The payload noise figure must be **set** in the payload's **front end**, which extends from the receive antenna terminal through the downconverter. Shortly past the antenna, there must be a unit which primarily determines the noise figure for the entire payload. This unit is the LNA. It must greatly amplify the signal while adding little noise. If the payload is required to handle a large range (often 30 dB) of input signal level, the LNA may not be able to set the noise figure for all signal levels. The preamp is the unit that must compensate, fully or partially, the large range of payload input-signal levels. It may do this by amplifying the signal by many dB and then attenuating by the amount required. In this case, the preamp will have such a large noise figure that it contributes noticeably to the overall payload noise figure but is still not the main determinant of it. Alternatively, this unit may switch in the amplifier stages that it needs and then do only a small amount of attenuation. In this case, its noise figure may not be so bad. The bottom line is that the uplink SNR is satisfactory for all input signal levels because the larger noise figure only occurs when the signal level is high. For the estimation of payload noise figure before measurement, the calculation must be made for at least two cases, namely when the preamp is configured for minimum attenuation and for maximum attenuation.

Ringaround noise is not actually noise but unwanted signal. It is uncorrelated with the desired receive signal so it can be well modeled as noise.

8.12.2   How to Verify G/T$_s$

G/T_s is verified more than once.

The antenna gain is measured for the first time on the near-field range (Section 3.12).

The antenna temperature is a value that must be agreed upon with the customer. It is typically the worst case that the antenna will see on orbit.

The noise figure of the payload minus antennas is measured during thermal-vacuum testing so the worst case over temperature can be obtained. Noise figure is measured with a spectrum analyzer augmented for noise-figure measurements, including a noise source as input. The preamp must be in FGM. The noise figure must be measured in at least two cases, that is, when the preamp is configured for minimum attenuation and for maximum attenuation. The noise figure must be measured across the full frequency band, as it is not flat over the band. The way the noise figure is measured is the following. The gain G_p of the payload minus antennas must be already known from a prior measurement. The noise source puts out two possible noise temperatures, T_{off} and T_{on}, where $T_{on} > T_{off}$. The noise power out of the transmit antenna terminal in a noise bandwidth B is measured with both temperatures as input. The two points are plotted with input temperature as the x-axis. A straight line is drawn between the two points and is extrapolated to input temperature of 0 K. The y-value at that point equals $(F-1)G_p$ 290 B, so F can be calculated from it. Such a plot is illustrated in Figure 8.17.

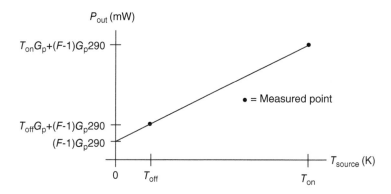

FIGURE 8.17   Plot for calculating payload noise figure.

Then from the antenna gain, the agreed-upon antenna temperature, and the payload noise figure, G/T_s can be calculated.

Antenna gain is measured again in the CATR. There also G/T_s is measured.

8.13   SATURATION FLUX DENSITY

Saturation flux density (SFD) is the power flux density at the spacecraft antenna in the boresight antenna direction, required to saturate the HPA of the selected channel. It depends on the HPA's preamplifier settings. It is most often stated in units of $(dBW)/m^2$.

8.14   SELF-INTERFERENCE

8.14.1   What Self-Interference Is

Self-interference is the payload interfering into itself, on transmit. It may be spot beams interfering into other spot beams or into a regional beam or a regional beam interfering into spot beams. **Carrier-to-interference ratio (C/I)** is the usual measure of self-interference. Since *C/I* is a function of coverage location, practical computation limits the number of ground locations that it can be calculated for.

8.14.2   Where Self-Interference Comes from

C/I has many independent dimensions, all of which must be taken into account:

- Modulation on signals in channels. This is a function of the uplink ground station, not the spacecraft, but *C/I* computations require knowledge of signal power levels in the channels and their modulation-symbol rates and modulation schemes.

- Channel spacing in frequency domain. This depends on the size of the guard band between channels. If the payload transmits on two polarizations, the channel assignments on both must be known.
- HPA nonlinearity characteristics, as well as HPA operating point. This knowledge is required for computation of the third-order IMPs of the HPA, which can interfere into other signals.
- OMUX characteristics. They tell how much the OMUX will reject the third-order IMPs of the HPA.
- EIRP.
- Antenna beam patterns. Spot beams do not necessarily all have the same pattern. The way the pattern rolls off, its sidelobes, and the beam spacing are key to the computation of I.
- Coverage requirements.
- Antenna polarization isolation. This determines how much of the complementary polarization is transmitted as interference. This is usually excellent.
- Antenna cross-polarization. This partially determines how much of one polarization will interfere into the other upon reception on the ground. This is usually excellent
- Assignment of channels and polarizations to the set of beams.

The C/I that is computed in this way is only one component of the total communication system-level C/I, which also includes interference from other uplink signals, downlink signals from other satellite systems, and signals from this satellite but on the other polarization. Section 14.6 provides further discussion on this topic.

Calculation of C/I variation requires, first of all, knowledge of the nominal C and I. There are a few extra, simultaneous dimensions which must be taken into account separately for C and for I variations:

- Gain stability and power-out stability
- Antenna pointing error. If the spot-beam antennas are fixed on a plate or in a structure, they will move together. On the other hand, two antennas may be separately steered. Antenna pointing error is a probabilistic phenomenon
- Antenna beam-pattern edge. The beam pattern sharply decreases/increases at the beam edge, which is where I comes from.

Chapter 15 provides a probabilistic treatment of the C/I variation which is not as conservative as the usual combining of worst cases.

8.14.3 How Self-Interference Carries to Payload Level

Self-interference is a payload-level characteristic.

8.14.4 How to Verify Self-Interference

Self-interference levels are verified from a combination of testing and calculation. The individual antenna beams are tested for pattern and polarization isolation, or they may be tested simultaneously so the C/I contribution from the transmit antenna can be measured (Section 3.12). The payload channels are normally tested with CW(s). Their output is taken at the test couplers immediately before the antenna terminals, for the power-out and frequency responses. The rest of the verification is by analysis.

8.15 PASSIVE INTERMODULATION PRODUCTS

PIMs are a phenomenon that occurs when transmitted signals hit edges or corners of spacecraft features, which perform a nonlinear combining of the signals, and then an IMP happens to feedback into a receive antenna. The transponder elements that can cause PIMs are shown in Figure 8.18. If it happens that the PIM is at about the same frequency as a signal on that channel, the PIM becomes a distortion on that signal and is from then on transmitted with that signal.

When a communications satellite has many antennas and/or transmits on several frequency bands, PIMs are a significant concern. One example of a satellite for which they were a big concern is the Optus-C1 satellite, which had UHF and X-, Ku-, K-, and Ka-band payloads and 16 antennas (Singh and Hunsaker, 2004).

Satellite manufacturers have done a lot of work with PIMs, which are difficult to analyze with any accuracy so a worst-case approach is taken. The approach by one satellite manufacturer is multi-stage (Singh, 2008):

- First, perform analysis. It takes into account near-field antenna analysis, spacecraft structure scattering, multipath, and antenna subsystem layout
- Then assess the risk that the PIMs created will be a concern
- Design the spacecraft to avoid the occurrence of harmful PIMs. Design measures can be taken on the antennas, spacecraft thruster, TWTA radiators, and spacecraft thermal blankets
- Last, test at two levels to ensure that the potentially harmful PIMs are weak: at the antenna-subsystem level with a spacecraft mock-up and flight blankets and at spacecraft level including payload receivers.

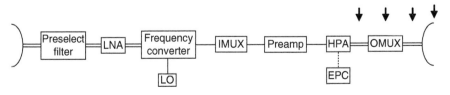

FIGURE 8.18   Sources of PIMs, besides other exposed spacecraft surfaces.

Nothing further on PIMs is included in this book because they do not lend themselves to accurate analysis, aside from the possible frequencies where they may occur. For further reading, see for example (Singh and Hunsaker, 2004).

8.A   APPENDICES

8.A.1   Relation of Gain and Phase Ripple

Multipath, for example from an impedance mismatch, always causes a ripple in both gain and phase. For every 1 dB of pk-pk gain ripple, the phase ripple is 6.6° pk-pk, and the phase and gain ripple are 1/4 cycle apart in the frequency domain. Suppose that the impulse response (Section 5.2.2) of the multipath channel is this:

$$h(t) = (1-\varepsilon)\delta(t) + \varepsilon e^{j\lambda}\delta(t-\tau)$$

where
  ε = real number of amplitude much less than 1
  λ = rotation angle in radians
  τ = delay of small signal relative to main signal

Then the gain response $G(f)$ and the phase response $\varphi(f)$ of the transfer function $H(f)$ are approximately given by the following:

$$G(f) \doteq \frac{20}{ln10}\varepsilon\left[\cos(2\pi f\tau)-1\right]$$

$$\text{Ripple part of } G(f) \doteq \frac{20}{ln10}\varepsilon\cos(-\lambda+2\pi f\tau)$$

$$\varphi(f) \doteq -\varepsilon\sin(-\lambda+2\pi f\tau)$$

The ratio of the magnitude of the phase ripple to the magnitude of the gain ripple is then 6.6 deg/dB.

8.A.2   Independence of G/T_s on Reference Location

In Section 8.12.1, we discussed G/T_s, the figure of merit for the payload when the payload is viewed as a receive terminal. We stated that both G and T_s must be referenced to the same point, where G is the gain up to that point and T_s is the system noise temperature after that point. The receive-antenna terminal is usually taken as the reference point. We show here that in fact any point can be chosen without altering the value of G/T_s.

Figure 8.19 shows two cases of the reference point location, one at the antenna terminal and the other somewhere farther down the payload hardware chain. In both

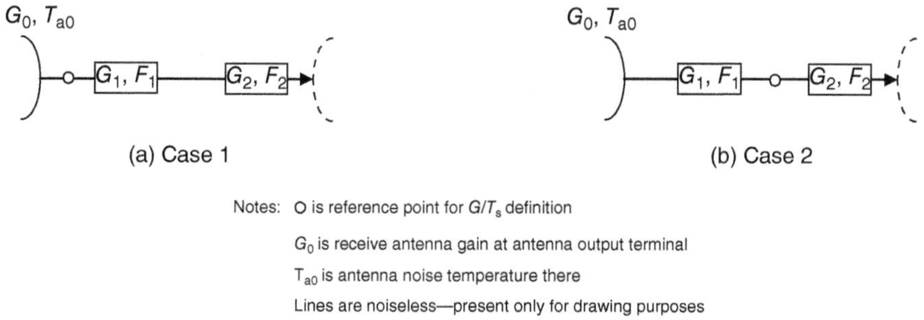

G0, Ta0
G1, F1
G2, F2
(a) Case 1
G0, Ta0
G1, F1
G2, F2
(b) Case 2
Notes: O is reference point for G/Ts definition
G0 is receive antenna gain at antenna output terminal
Ta0 is antenna noise temperature there
Lines are noiseless—present only for drawing purposes

FIGURE 8.19 Two cases of reference point for G/T_s.

cases, the entire payload electronics (minus the antennas) chain is represented by the cascade of two electronic elements, where each element would actually be the cascade of any number of consecutive elements in the signal path.

In the first of the two cases, where the reference point is at the antenna terminal, we have

Gain to reference point $G = G_0$

Electronics past reference point has combine noise figure $F = F_1 + \dfrac{F_2 - 1}{G_1}$

Therefore,

$$G/T_s = \frac{G_0}{T_{a0} + T_{r0} + (F-1)290}$$

In the second case, we have

$$G = G_0 G_1$$

$$T_a + T_r = T_{a0} G_1 + T_{r0} G_1 + (F_1 - 1) G_1 290$$

$$F = F_2 \text{ and } T_e = (F_2 - 1)290$$

Therefore,

$$G/T_s = \frac{G_0 G_1}{T_{a0} G_1 + T_{r0} G_1 + (F_1 - 1) G_1 290 + (F_2 - 1)290}$$

$$= \frac{G_0}{T_{a0} + T_{r0} + \left[F_1 - 1 + \dfrac{F_2 - 1}{G_1} \right] 290}$$

The results are the same at both locations.

REFERENCES

Agilent Technologies (2003). Agilent PNA microwave network analyzers, mixer transmission measurements using the frequency converter application. Application note 1408-1; May.

Agilent Technologies (2007). Agilent E5052B signal source analyzer, advanced phase noise and transient measurement techniques. Application note.

AIAA Japan Forum on Satellite Communications (2003). Capital products & review: C-band solid state power amplifer (SSPA). *Space Japan Review;* no 27; Feb/Mar. On satcom.nict. go.jp/English/english2-3/index.html. Accessed June 27, 2011.

Anritsu Corp (2008). Using Anritsu's Spectrum Master™ and Economy Bench Spectrum Analyzers to measure SSB noise and jitter. Application note 11410-00461, rev A.

Grundbacher R, Chou Y-C, Lai R, Ip K, Kam S, Barsky M, Hayashibara G, Leung D, Eng D, Tsai R, Nishimoto M, Block T, Liu P-H, and Oki A (2004). High performance and high reliability InP HEMT low noise amplifiers for phased-array applications. *IEEE MTT-S International Microwave Symposium Digest;* Vol. 1; June 6–11; pp 157–160.

Hitch B (2019). Private communication. July 25.

International Telecommunication Union, Radiocommunication Sector (ITU-R) (2015). Recommendation SM.1541-6. Unwanted emissions in the out-of-band domain. Aug. Version in force on 2020 Sep. 21.

L-3 Communications Electron Technologies, Inc. (2009). *TWT/TWTA Handbook*, 13th ed., Torrance.

Pritchard WL and Sciulli JA (1986). *Satellite Communication Systems Engineering*, Englewood Cliffs, NJ: Prentice-Hall, Inc.

Rohde & Schwarz (2012). Group delay and phase measurement on frequency converters. Application note 1EZ60_1E. On www.rohde-schwarz.com/us/search_63466.html?term= application%20notes. Accessed June 10, 2019.

Singh R (2008). Passive intermodulation (PIM) requirements for communications satellites. *International Workshop on Multipactor, Corona and Passive Intermodulation in Space RF Hardware*; Sep. 24–26.

Singh R and Hunsaker E (2004). PIM risk assessment and mitigation in communications satellites. *AIAA International Communications Satellite Systems Conferences*; May 9–14; pp 1–17.

CHAPTER 9

MORE ANALYSES FOR PAYLOAD DEVELOPMENT

9.1 INTRODUCTION

This chapter presents various analyses, additional to those in Chapter 8, which are required in all stages of payload development. Each analysis can be useful alone or as part of a higher-level analysis. Sections 9.2 and 9.3 deal with payload performance budgets and, thus, with payload design and evaluation. Sections 9.4 and 9.5 cover topics that have to do with the high-power amplifier (HPA), additional to what has already been discussed in Section 7.2. Section 9.5 pertains to other nonlinearities in the payload besides the HPA, such as a component amplifier driven too close to saturation or an analog-to-digital converter. Section 9.6 gives advice on payload analysis via simulation.

9.2 HOW TO DEAL WITH NOISE FIGURE

Noise figure (NF) was defined in Section 6.4. That section also lists the other sections in the book that address noise. The present section describes the mechanics of NF.

Recall that a two-port electronic element can be characterized by its gain G and NF F, where both are power ratios not in dB.

Satellite Communications Payload and System, Second Edition. Teresa M. Braun and Walter R. Braun.
© 2021 John Wiley & Sons, Inc. Published 2021 by John Wiley & Sons, Inc.

9.2.1 Noise at Output of Passive Element

A passive element has gain $G < 1$; its NF is $1/G$. If the input-noise temperature is 290 K, as in a NF measurement, then the output-noise temperature is also 290 K. This is because the input noise sees the element gain, so its temperature at output is $G290$ K, and the internally generated noise temperature at output is $(F-1)G290$ K $= (1-G)290$ K, and their sum is 290 K. More generally, for any input-noise temperature of T_0, the output-noise temperature of a passive element is $GT_0 + (1-G)290$ K.

A passive element's output-noise temperature can be lower than the input-noise temperature; this happens whenever $T_0 > 290$ K. There is a floor, that is, a minimum, to the output-noise temperature, namely 290 K, since when 290 K is input, 290 K is also output.

9.2.2 Gain and Noise Figure of Two-Element Cascade

Key to estimating the NF of the repeater by calculation (Section 8.12.1) is the formula for combining the gains and NFs of two electronic elements in cascade, yielding the gain and NF of the cascade. The formula is given in Figure 9.1. It is to be noted that although the gain and NF of the second element, before combining, are referenced to the input of that element, the gain and NF of the cascade are referenced to the cascade input.

It is especially easy to calculate the NF of a two-element cascade when the first element is passive. Suppose that the passive element has gain $G_1 < 1$; then its NF is $F_1 = 1/G_1$. Suppose that the active element has NF F_2. Then the cascade's NF is simply $(1/G_1)F_2$, as illustrated in Figure 9.2. Its gain is the product of the gains.

9.2.3 Playing Off Gains and Attenuations

Component amplifiers can be purchased with a limited selection of fixed gain values. On the other hand, attenuators come in many varieties: fixed selectable from a fine-grained set of values, variable but with value set in payload integration,

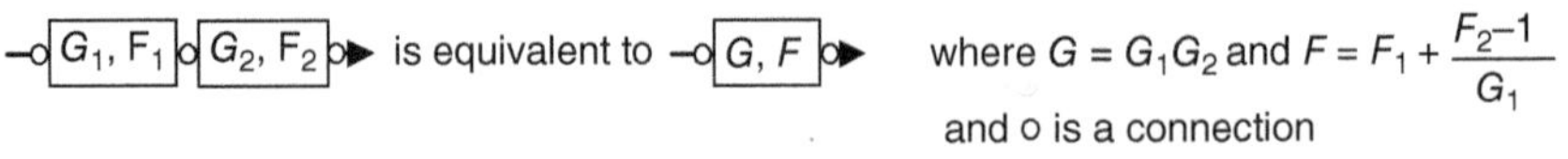

where $G = G_1 G_2$ and $F = F_1 + \dfrac{F_2 - 1}{G_1}$ and o is a connection

FIGURE 9.1 Gain and noise figure of cascade of two general electronic elements.

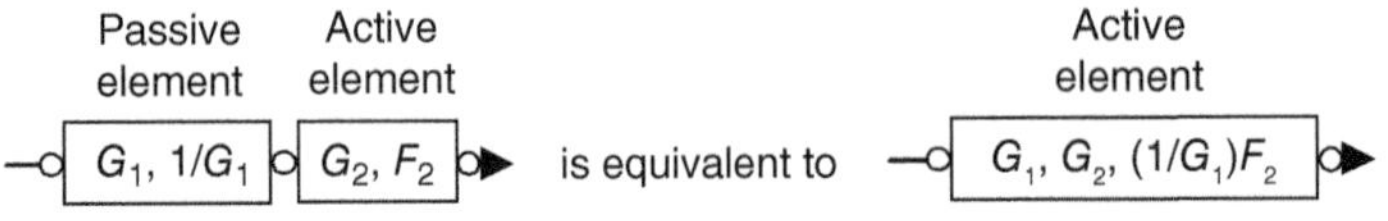

FIGURE 9.2 Gain and noise figure of passive element followed by active element.

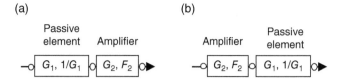

FIGURE 9.3 Same gain attained by means of (a) attenuator followed by amplifier and (b) amplifier followed by attenuator.

variable set autonomously onboard in an automatic level-control (ALC) unit, and variable commandable from the ground. Usually, when an amplifier is needed, it is not quite the right value, so an attenuator must be paired with it.

In this section, we address two questions in unit design that affect overall NF. Since NF is always to be minimized, at least down to a certain level, good design does that.

The first question is which is better unit design, to place a component amplifier in front of or after an attenuator. The two possibilities are depicted in Figure 9.3. The passive element's gain is G_1 in both cases. In the first case, the combined NF F is given by

$$F = \frac{1}{G_1} F_2$$

In the second case, it is given by

$$F = F_2 + \frac{\frac{1}{G_1} - 1}{G_2}$$

The second NF is always smaller, since for an amplifier $F_2 G_2 > 1$ (actually, each term is greater than 1). Thus, the second design is better. As an example, let us assume that $G_1 = 1/\sqrt{10}$ and $G_2 = 10$. In the first case $F \doteq 3.2\, F_2$, and in the second case $F \doteq F_2 + 0.2$, which is much smaller.

The second question regards the HPA's preamp design. Recall that the preamp usually must be able to amplify over a large range, often about 30 dB. The preamp could be designed to amplify by many dB and then attenuate as much as required. Or it could separate the amplification into two or more stages which would be switched in when required and then have lower values of attenuation. (The decision of which way to go depends on other factors, too, such as mass and volume, but we look only at the NF here.) An example of two such cases is shown in Figure 9.4a,b. In the first case, two identical component amplifiers are followed by an attenuator which undoes the gain of the second amplifier. In the second case, there is just one of these amplifiers and no attenuator.

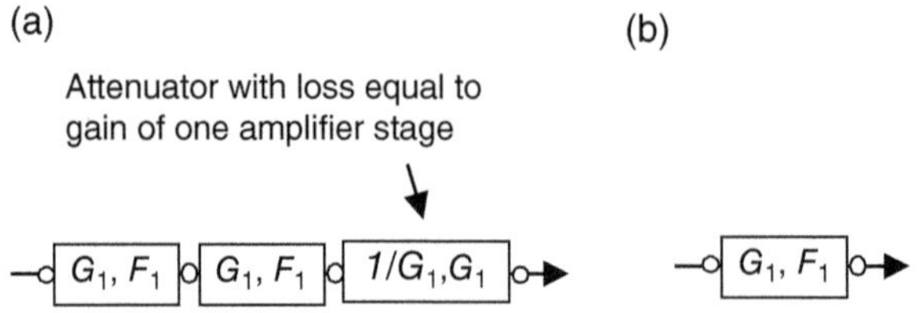

(a)   (b)

Attenuator with loss equal to
gain of one amplifier stage

$\downarrow$

| G_1, F_1 | G_1, F_1 | $1/G_1, G_1$ | | G_1, F_1 |

FIGURE 9.4   Same gain attained by means of (a) two-stage amplifier followed by attenuator and (b) single-stage amplifier and no attenuator.

In the first simplified case, the combined NF F is given by

$$F = F_1 + \frac{F_1 - \dfrac{1}{G_1}}{G_1}$$

while in the second simplified case it is simply

$$F = F_1$$

F in the second case is smaller than in the first case, so the second design is better. For the reasonable number of $G_1 = 10$, in the first simplified case $F \doteq 1.1\,F_1$, while in the second case $F = F_1$. It may be good enough to use the simpler design of the first case.

9.3   HOW TO MAKE AND MAINTAIN PAYLOAD PERFORMANCE BUDGETS

9.3.1   Payload Requirements Analysis

Fundamentally, the payload systems engineer's job is to make sure that the payload meets its requirements from the customer. A bent-pipe payload typically has at least a hundred pages of customer requirements. The payload must be able to do what it needs to do, for as long as specified, with proven reliability. The payload requirements are part of the overall spacecraft requirements, which are part of the end-to-end system requirements. These comprise requirements for the spacecraft, ground segment, user segment, and interface requirements.

Analysis of these requirements is a large effort that starts even before the satellite-build program and goes on through to the end of the program. Before contract award, the analysis is developed in enough detail that the customer sees he will get what he wants and the satellite manufacturer is confident of building it on schedule and making a profit. After award, the payload units, the payload layout on the spacecraft, and the payload's response to the on-orbit environment must be understood in detail. The first analysis phase during the program must show in detail that

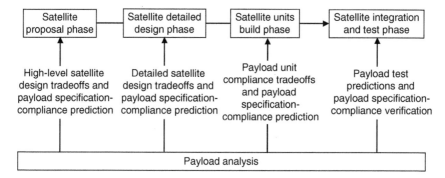

FIGURE 9.5 Contributions of payload analysis through all stages of satellite development.

the customer's payload requirements are expected to be met, with margin. Every requirement must be met in a number of cases ranging from one to hundreds.

The payload engineer performs continuing analysis of the payload requirements all during the program to show that at the end all of them will be met. When it looks like one will not be, it is his job to work with the unit engineers to put things back in line. As units get built, better and more detailed numbers become available, and the analysis incorporates them and carries a decreasing risk margin (Section 9.3.5). Finally, after verification, all the requirements are shown to hold and the customer accepts the satellite. This process is illustrated in Figure 9.5.

If a customer requirement cannot quite be met without a big negative impact on the satellite program, communications analysis of the end-to-end communications system (Chapter 16) can be decisive on how to proceed. It may persuade the payload engineer, the program manager, and the customer that the requirement can be loosened without significantly degrading communications performance.

9.3.2 Example Budget without Uncertainty: Signal and Noise Levels

We start our discussion of budgets with the construction of one particular budget, a key one. We will go so far as to compute the nominal values of the line items and the bottom line but will delay the crucial but complicated topic of uncertainty to the next section.

The **signal-and noise-level budget** tracks the signal level P and the noise level N_0 (one-sided RF noise power spectral density) through the payload. Each line in the budget corresponds to one payload-level element, namely either a payload unit or a payload-integration element such as waveguide, pad attenuators, and switches. This budget is important because among the payload requirements there will effectively be some requirements on G/T_s and some on equivalent isotropically radiated power (EIRP), which are to first order the most important payload parameters. We write "effectively" because the requirements may take slightly different forms from

G/T_s and EIRP, respectively; for example, instead of G/T_s it may be receive gain and repeater NF, and instead of EIRP it may be downlink availability. G/T_s sets the uplink signal-to-noise ratio (SNR) and EIRP the downlink SNR (Section 14.7), as far as the payload can affect these quantities.

Even in the design of active units, there is a need to check input P and N_0. For example, the noise level entering a component amplifier must not be so high that signal plus noise power drives the amplifier near compression (Section 6.3.2). For another example, the noise level entering an ALC circuit must not be so high that the noise power has a significant effect on where the ALC circuit sets its gain.

In the budget, P and N_0 are both computed at the output of every element. Also, the composite gain and NF to that point but referenced to the beginning, are tracked. Additionally but not done in the example below, at any line or lines a bandwidth B could be declared that is of concern or interest there; then the SNR in B at that line, with value $P/(N_0 B)$ (Section 12.16), would also be calculated and shown. The budget is normally executed in dB or dB-type units. This particular budget normally has to be constructed in four cases for every signal path: two cases of signal level into the payload (lowest and highest) times two cases of preamp mode (ALC and fixed-gain).

The steps in making the budget for the payload are the following:

1. Define the reference point for the repeater's NF calculation (Section 8.12.1). It is the point chosen for the G and T_s calculations. At every line item, the composite NF through to the output of that item will be referenced to the same initial point.
2. Define T_0 at input to the reference point as the antenna noise temperature.
3. Obtain P_0, the power into the reference point, from the payload specification.
4. Enter into the spreadsheet all the gains and NFs, in order, of all the payload elements in the signal path to be analyzed.
5. Top the spreadsheet, that is, right above the first element, with the start values of gain of 1 (0 dB) and NF of 1 (0 dB). These stand for the gain and NF through to the output of the reference point and are meaningless except as the necessary starting point of the recursion. Also, top the spreadsheet with T_0 and P_0 into the reference point; these are also necessary for the recursion.
6. Do the following at each succeeding element from first to last. Form the composite gain G through to the output of this element, from the current element's gain and the composite gain of all elements through to the element before the current one. Form the composite NF F from the current element's NF and the composite gain and NF of all elements through to the element before the current one. Compute system noise temperature T_s, always applicable at the reference point, for the composite of the elements through to this element by applying the formula below:

$$T_s = T_0 + (F - 1)290$$

Example of signal-and noise-level budget without uncertainty

To 300.00 deg K   kappadB –198.60 dB(mW/(K-Hz))
PodB –30.00 dBm

Element	Elt gain (dB)	Elt NF (dB)	Elt gain (not dB)	Elt NF (not dB)	G (dB)	G (not dB)	F (not dB)	F (dB)	Ts (K)	Ts (dBK)	N0 (dB(mW/Hz))	P (dBm)	Source of elt gain and NF	Date
Reference point					0.00	1.00	1.00	0.00	300	24.77	–173.83	–30.00		
Element 1	–0.15	0.15	0.97	1.04	–0.15	0.97	1.04	0.15	310	24.92	–173.84	–30.15	Supplier X spec	xx/xx/xx
Element 2	–1.50	1.50	0.71	1.41	–1.65	0.68	1.46	1.65	434	26.38	–173.88	–31.65	Supplier X spec	xx/xx/xx
Element 3	21.00	5.50	125.89	3.55	19.35	86.10	5.19	7.15	1515	31.80	–147.45	–10.65	Supplier Y spec	yy/yy/yy
Element 4	–2.00	2.00	0.63	1.58	17.35	54.33	5.19	7.16	1516	31.81	–149.45	–12.65	Supplier Y spec	yy/yy/yy
Element 5	–3.65	3.65	0.43	2.32	13.70	23.44	5.22	7.18	1524	31.83	–153.08	–16.30	J. Fisher memo	yy/yy/yy
Element 6	–3.65	3.65	0.43	2.32	10.05	10.12	5.28	7.22	1540	31.87	–156.68	–19.95	J. Fisher memo	yy/yy/yy
Element 7	–0.05	0.05	0.99	1.01	10.00	10.00	5.28	7.22	1540	31.88	–156.73	–20.00	Measured in unit test	xx/xx/xx
Element 8	–4.20	4.20	0.38	2.63	5.80	3.80	5.44	7.36	1587	32.01	–160.80	–24.20	Measured in unit test	xx/xx/xx
Element 9	–2.00	2.00	0.63	1.58	3.80	2.40	5.59	7.48	1632	32.13	–162.68	–26.20	Est. based on length	xx/xx/xx
Element 10	–0.15	0.15	0.97	1.04	3.65	2.32	5.61	7.49	1636	32.14	–162.82	–26.35	Supplier X spec	yy/yy/yy

FIGURE 9.6   Example of signal- and noise-level budget without uncertainty.

Compute noise power spectral density N_0 at output of this element by applying the formula below:

$$N_0 = \kappa G T_s$$

where $\kappa =$ Boltzmann's constant $= 1.379 \times 10^{-20}$ mW/(K-Hz).
Compute P at output of this element as follows:

$$P = GP_0$$

Figure 9.6 gives an example of a signal- and noise-level budget for a simplified signal path of 10 elements, without consideration of uncertainty. The signal power and noise temperature at the input, respectively T_0 and P_0, would be entered in the boxes in the upper left corner. The other items enclosed in thick boxes, near the left side, would be entered into the spreadsheet to describe the individual elements. Note that the source and date of every one of these items are to be entered in the spreadsheet. The middle columns in the table are cumulative.

Variations on this form of the signal- and noise-level budget are sometimes of interest:

- Same form but with $T_0 = 290$, for computing the NF of the repeater when the NF is specified.
- Similar form as shown but on just a section of the signal path, as a payload development aid. T_0 is then T_s at output of the part of the signal path that precedes this section.
- Same form as just above but with $T_0 = 290$, for computing the NF of a repeater section.

It is important to understand what bandwidth the noise level applies to. For the purpose of computing uplink P/N_0 as a term that will go into an end-to-end link budget (Section 14.8), the noise that counts is the noise that is inside the detection filter's bandwidth (Section 12.10.1). For the purpose of computing how close to saturation an amplifier will be driven, the noise that counts is the noise that enters the amplifier, which may have a greater bandwidth than the detection filter. In either case, if the noise inside the applicable bandwidth does not have a uniform value of N_0, the average N_0 over the bandwidth is to be used.

9.3.3 Dealing with Impedance Mismatch

In the payload, any place where are two elements are connected, there is an impedance mismatch, however small (Section 4.7.1). The payload engineer is responsible for accounting for the mismatches between two units, between a unit and an integration element, and between two integration elements. The signal-and-noise-level budgets include a mismatch loss for each such location. Both signal and noise are attenuated by the mismatch, but the mismatch does not add noise. The mismatch reflects a small portion of the signal and noise backward, which gets absorbed by a termination on an isolator, a circulator, or a switch. If the reflected signal and noise hit another mismatch, a small portion of that gets reflected forward. It may get further reduced by a pad attenuator on its way. In any case, it is negligible when the payload design is good and nothing is broken.

9.3.4 Dealing with Uncertainty in Budgets

9.3.4.1 *Two General Ways of Dealing with Uncertainty*   A performance budget contains the computation of the *nominal value* of the parameter involved and then degrades the nominal value by a number that reflects uncertainty. The bottom-line result of the budget is then the degraded nominal value, representing a sort of worst case, which has a high probability of being achieved over the specified range of on-orbit conditions.

There are two general ways to degrade the nominal value of the parameter value. The most common way is to compute the nominal value's current standard deviation or σ value (book appendix Section A.3.3) in the budget; then at the bottom of the budget 2σ or some other multiple of σ is added to the nominal value to create the bottom-line result, but it is subtracted if the difference is a worse case than the sum. This way of obtaining uncertainty is for a budget whose line items all have known or rather well-estimated σ. It is discussed in detail in the rest of this section. The other way of degrading the nominal value is not to deal with σ values at all but instead to employ a large design margin (Section 9.3.5). This way will not be further discussed as it is based on a rule of thumb or a "gut feeling."

The actual value of a parameter will be greater than nominal minus 2σ with a probability of 97.7% (book Appendix Section A.3.5) if the uncertainty has a Gaussian probability distribution. In fact, the uncertainty does not have a Gaussian probability distribution because it consists of the sum of some Gaussian- and some non-Gaussian-distributed component uncertainties. The actual probability distribution is unknowable, since even the assumed distributions of the component uncertainties are only approximations. In summary, the computed combined σ value is a good estimate of the actual combined σ, but the probabilities are only approximations, and the greater the factor applied to σ, the less accurate the Gaussian approximation. Even though the uncertainty may not be Gaussian, given enough elements in the budget, Gaussian is still a useful approximation (book Appendix Section A.3.9)

What exactly to use for the nominal value is a nontrivial consideration that is discussed near the end of this section on uncertainty.

9.3.4.2 *Types of Line-Item Uncertainty* In a budget where the σ approach to uncertainty is taken, the σ for the parameter is some combination of σ values of the budget's line items. We must explore what types of line-item uncertainty there are, so we can see how to combine them into the bottom-line uncertainty. To some extent, the types are differentiated by whether the line item itself has a nonzero or zero mean. Line items that have a nonzero mean reflect intentional parts of the design, such as presence of integration elements between units (for example, waveguide, switches, and hybrids). Some line items that have a zero mean reflect intangibles or unknowables such as unit design immaturity, error in computer modeling, and manufacturing tolerance. These items have uncertainties that drop out at some point in time during the program. Another group of zero-mean line items apply even on orbit: performance variation over temperature; performance variation from aging and radiation total dose; and error in power measurement, if required. The examples of line-item uncertainty are listed in Table 9.1 with the payload-build stages when they must be taken into account.

The good thing is that the various types of uncertainty are mutually uncorrelated (book Appendix Section A.3.3), so a budget's combined σ is the root-sum-square (rss) of the σs of the various types. As we will see, the hard part is finding how to compute the σ for performance variation over temperature on orbit.

9.3.4.3 *Easy Dealing with Some Uncertainty Types* The σs for some types of uncertainty are straightforward to deal with, so let us get them out of the way first. Again we reference our discussion to the uncertainty types in Table 9.1, which may not be all types but give the general ideas. Uncertainty in loss in the payload-integration elements at room temperature is of two kinds, one from the uncertainty in waveguide and coax lengths and one from the uncertainty in loss in the other integration elements. The spacecraft engineer who lays out the spacecraft estimates the lengths of waveguide and coax runs, while the payload engineer estimates the error in loss from the other payload-integration elements. Uncertainty from unit design immaturity, computer-modeling error, and manufacturing tolerance are estimated by the unit engineers and passed on to the payload engineer. Variation over temperature, that is, the temperature-sensitivity coefficients, is provided for the units by the unit engineers and by the payload engineer for the integration elements.

So there are a few σs the payload engineer must estimate himself, from uncertainties provided by the element manufacturers. If the manufacturer just gives an unsigned number, it is best to ask the manufacturer whether it is 1σ, $\pm$ bound with

TABLE 9.1 Some Types of Budget-Item Uncertainty and When to Take Them into Account

	Payload and Unit Design Stage	Payload-Detailed-Design-and-Unit-Build Stage	Spacecraft Integration-and-Test Stage	On Orbit
Error in estimated loss in payload-integration elements at room temperature	√	√		
Error from unit design immaturity	√	√		
Error in computer model	√	√		
Error from manufacturing tolerance	√	√		
Performance variation over temperature	√	√	√	√
Performance variation from aging and total radiation dose	√	√	√	√
Error in power measurement, if required	√	√	√	√

a uniform probability density function (pdf) in between (book Appendix Section A.3.6), or something else. If no answer is forthcoming, the number is probably 1σ (and if not, this is a conservative assumption). If the manufacturer just gives a ± number, it is best to ask whether it is $\pm 1\sigma$, $\pm 2\sigma$, ± bound with a uniform pdf in between, or something else. If the manufacturer writes "worst-case ± *number*" then it is probably $\pm 2\sigma$ (and if not, this is a conservative assumption). The payload engineer then derives the 1σ number from what he is given or his guess. Book Appendix A.3 provides some help on how to do this.

The other fairly straightforward uncertainty type is measurement error, discussed next.

9.3.4.4 Dealing with Uncertainty in Power Measurement

Whether and how measurement error is taken into account in payload-level power measurements may be driven either by the customer or the manufacturer. There are different ways to take measurement error into account. One way is to require subtraction of a certain amount of measurement error, for example 2σ, from the measured value. Another way is to ask for the power level above which, with for example 84% probability, the actual power will lie. Suppose that the measured power is X and the measurement has a rms error of σ. (Uncertainty in a power meter measurement is internationally stated in terms of a rms value (Agilent Technologies, Inc, 2009).) The question is then, X minus how many multiples n of σ is the right value to give to the customer. It turns out that n is the same number as if a similar question had been asked about P, as illustrated in Figure 9.7:

$$\Pr\left(P \geq X - n\sigma\right) = \text{given probability} \Leftrightarrow \Pr\left(X \leq P + n\sigma\right) = \text{given probability}$$

In taking payload measurements used for payload sign-off, the engineer may decide to take more than one measurement if he thinks that the first one yielded a somewhat unrepresentative answer. If, say, he wants to make three measurements altogether and report the average, the engineer must conscientiously include the very next two measurements after the first one. He may only leave a very bad one out if he knows the probable cause and the customer agrees. Another possible reason for taking several measurements and using the average is to reduce measurement error. If the measurements can be considered to be independent (although they should be taken on the same test equipment), then the reduced measurement error is as follows:

$$\text{Std.dev.of error of } n \text{ averaged measurements} = \frac{1}{\sqrt{n}} \text{std.dev.of error of}$$
$$\text{one measurement}$$

9.3.4.5 Specifying Environment in Lifetime on Which Payload Performance Must Be Met

Now for the hard part in dealing with uncertainties in payload performance budgets: computing the σ for the type of uncertainty due to payload performance variation over temperature on orbit over life. Similar in some ways but much simpler to deal with is uncertainty due to aging and radiation. We first consider the spacecraft-level specifications which drive these types of uncertainty and in the following sections translate them into what it means for the payload.

In the spacecraft requirements document, there are two high-level conditions on which the payload performance requirements must be met. The first such condition is the environment: the spacecraft orbit (altitude, inclination, station-keeping or

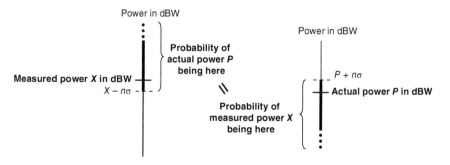

FIGURE 9.7   Probabilities related to output-power measurement.

lack thereof) determines the thermal and radiation environment that the spacecraft will see. The second high-level condition regards the spacecraft lifetime and is variously stated, perhaps dependent on the customer's business model. An example is at end of life (EOL), that is, that performance requirements must be met at the stated EOL with x probability. Another example of lifetime condition is **average over life**, which means for example, that EIRP must be met $y\%$ of the time averaged from beginning of life (BOL) to EOL. The latter example could even mean that a particular rain model is taken into account, useful for Ku-band and higher frequencies (Section 14.4.4.1).

9.3.4.6 *Dealing with Uncertainty from Aging and Radiation*   Some units, for example, an oscillator, change characteristics with aging and radiation. Unit engineers have estimates for the pdfs of these drifts (see examples in Chapters 4 through 7). If the drift from BOL to any later year on orbit may be positive or negative with equal probability, then the drift has average or nominal value of zero but an uncertainty that increases with time. If the drift is more likely to be in one direction, it may be compensated so that the residual drift is equally likely to be up or down with the years. If such a drift is uncompensated, at whatever year in the life the budget is to be made, the drift's average value for the year will be nonzero (Section 9.3.4.9 below discusses what to use for nominal value in such a situation).

9.3.4.7 *Converting Thermal Environment in Lifetime into Unit Temperature Variations*   So now we have only temperature variation to contend with. The manufacturer understands the required spacecraft orbit as implying a whole set of sun direction and sun distance conditions over the year and possibly varying from year to year, for example, when there is no station-keeping. Thermal engineers translate the set of sun direction and sun distance conditions into a set of unit temperatures, more specifically, a set of temperatures of mounting plates (unit thermal interfaces to the spacecraft bus). (Actually, this is an iterative process between payload and bus.) Even if the orbit stays the same over the years, the unit-interface temperatures of units on heat pipes will gradually rise because the thermal subsystem's effectiveness slowly degrades.

So the information on mounting-plate temperatures is available to the payload engineer to use in budgets. For a particular signal path through the payload, the particular units used are known, so their locations on panels or earth deck and therefore temperatures can be determined. For comparison of two signal paths through the payload, as is done in C/I analysis, the task is twice as much work. (Examples of temperature dependence on spacecraft location are given in Chapter 15.)

Let us see how the thermal environment varies with time so we can begin to think about how to deal with it. Figure 2.10 showed how a geostationary orbit (GEO) satellite looks to the sun over the course of the summer solstice day in the northern hemisphere. The satellite always has its earth deck pointed down to the earth and its thruster pointed away from the earth. Let us suppose that on the earth deck are the low-noise amplifiers (LNAs), on the north and south panels are the traveling-wave tube amplifiers (TWTAs) and input multiplexers (IMUXes), and on the east and west panels are the output multiplexers (OMUXes). The spacecraft drawing at the bottom of the figure represents how the spacecraft looks to the sun at the satellite's local noon: the sun sees the thrusters and the north panel at a sharp angle (23.44°). The TWTAs and IMUXes on the north panel get a little sun—actually, they get the same amount of sun at all times that day. The spacecraft drawing on the right side of the figure is 6 hours later, and now the sun is almost perpendicular to the west panel, actually at 66.56° to the panel, so the OMUXes on the west panel get a lot of sun. Another 6 hours later, at midnight, and the sun is at 66.56° to the earth deck, so the LNAs get a lot of sun. Another 6 hours later, and the sun is at 66.56° to the east panel, so the OMUXes there get a lot of sun. So even just over a day the units on the earth deck and the east and west panels have a wide variation in illumination.

At other times of the year, the situation is similar. At the winter solstice, the south panel instead of the north panel gets a little sun all day; the earth deck and east and west panels have the same situation as in summer (Figure 2.11). At the equinoxes, the north and south panels get no sun at all and the earth deck and east and west panels get direct sun at times; both equinoxes are the same in this regard (Figure 2.9).

As the years go by, a unit's daily temperature swing on any particular day of the year will have an increasingly high average temperature.

The thermal situation for a medium-earth-orbit (MEO) or a low-earth-orbit (LEO) spacecraft is more complex than for a GEO (Section 2.2.1.3) and cannot be generalized since the orbits between and even sometimes within constellations are so various.

9.3.4.8 Dealing with Performance Variation with On-Orbit Temperature

Therefore, the ideal way to incorporate the variation in payload performance with temperature into the sensitive budgets is to make a version of the budget at every minute of the spacecraft life. If the payload requirements document says the performance requirements have to apply at EOL, only the budgets over the last year would be needed.

This way is of course totally infeasible. So we come to the second-to-the-last condition on making budgets, and that is that the chosen set of budgets must be computationally feasible.

Now, it is almost as bad to have conservative budgets as optimistic ones. Having optimistic budgets is clearly bad—they can lead to payload design mistakes that have to be corrected at a late stage or to payload performance that does not meet customer requirements. However, having conservative budgets can mean that spacecraft capacity is being wasted that could have been used to offer higher performance or additional capabilities to the customer, or it could mean that the extra mass drives the launch vehicle to be unnecessarily large, or it could mean that the extra resources need never have been implemented to begin with and the manufacturer could have saved money. So the last condition on making budgets is that they must be conservative if they cannot be accurate, but as little conservative as possible.

Therefore, some simplifications must be made. There are some budgets for which a simplification is easy to formulate. Some payload parameters have variations whose worst case seems readily accommodated by the unit engineers. G/T_s is a somewhat similar parameter except that it spans many units. G of a passive antenna will not change over life. Since T_s is set by relatively low-power devices, it is often not a tough design decision to add the relatively few "extra" resources to accommodate its lifetime temperature variation. For these parameters, there is no need to use a σ value; in the one version of the budget, the line items are entered at their worst temperature condition. Another case with easy simplification is when the EOL temperature variation σ has negligible effect on the rss of all the σs in the budget.

The difficult budget simplifications are for EIRP. Payload performance is generally worse at hot than at cold. Amplifiers perform worse at hot. Some units or other payload elements may need to be temperature-compensated to maintain a nearly constant performance over temperature. A HPA changes characteristics with temperature (also with aging and radiation) (Chapter 7). If the payload transmits non-interfering beams, it may be sufficient to just "throw HPA power" at the problem. However, if bus power is tight or the payload transmits overlapping spot beams, some means of compensation by the preamp must be implemented. In summary, with temperature and the years, the payload changes performance and possibly also the way it functions.

So for a GEO's EIRP budget, to make good simplifications, we have to consider more closely the thermal environment variation over the day and year, with the help of Figures 2.9 through 2.11. One feasible simplification is to choose the worst day of the year in terms of payload performance and let it stand in for the whole year. What day this is depends on the spacecraft layout and the payload technology. What is left then are the diurnal variation and the lifetime requirement (that is, whether the payload specifications apply at EOL, at a few years earlier, or to some average or percentage of the time over the years).

Let us assume that in the signal path under consideration, only one unit plus associated integration waveguide is on the earth deck or east or west panel; such is the case in the examples in Chapter 15. For simplicity's sake, let us look now just at EOL. The temperature variation of that unit and waveguide over the worst day will have a pdf like that given in book Appendix Section A.3.7, where for half the day the temperature is at its daily minimum and for the other half of the day the pdf increases with temperature to the maximum. The mean unit temperature is about 0.32 of the way from minimum to maximum, and the temperature's standard deviation σ is about 0.39 of the maximum minus the minimum. The nominal budget should be computed with the temperature-varying unit and waveguide at their mean temperature over the day. The budget would be computed again at 0.71 of the way from minimum to maximum temperature, and the difference between the budgets would be the σ due to temperature variation. The second budget would not be further used; only the σ would be entered into the budget at the mean temperature. All the other units and integration elements will have constant temperature.

9.3.4.9 *Nominal Value*

We have defined variation as being about a nominal value, and exactly what the nominal value should be is a nontrivial consideration.

When the lifetime requirement is "at EOL," clearly the nominal value should be for the EOL. Most of the types of uncertainty in performance of a unit or payload-integration element have the same value whether the performance is being predicted for EOL or BOL. For those types, the nominal performance value should be the expected value or current best estimate, not a conservative estimate, since the conservatism is taken care of in the addition or subtraction of 2σ in the budget's bottom line. The nominal antenna gain in a particular direction is the gain averaged over antenna-pointing error. There are two types of performance uncertainty that sometimes do have a different estimate for EOL from that for BOL: that due to radiation and aging and that due to unit temperature rising over life from the thermal subsystem aging. In our payload example in Chapter 14, radiation and aging cause a different expectation with the years but rising temperature causes only a negligible difference.

When the lifetime requirement is "average over life," the nominal value should be the expected performance in the middle of life, assuming that drifts are linear over life. A suggestion of how to handle this in the bottom line of a budget, where normally 2σ is either subtracted or added, is to use $|\delta| + 2\sigma$ or $|\delta| - 2\sigma$ instead, where δ is the drift expected to occur from the middle of life to EOL. When the payload specification has a link availability requirement instead of EIRP requirements and the frequency band is 10 GHz or higher, |δ| up to about 0.1 dB can be ignored because availability at EOL and at BOL will effectively average out to availability in the middle of life.

9.3.4.10 *Combining Line-Item Uncertainties*

Recall that the various types of uncertainty are mutually uncorrelated, so a budget's combined σ is the rss of the σs of the various types. Since the budget items are in dB, it is convenient if also the uncertainties are in dB. Some uncertainties are clearly available in dB, for example, waveguide loss per foot and coupler loss. If, however, an uncertainty is not in dB it can easily be turned into dB. If an uncertainty is given as a power ratio X, $+u$, $-v$, first turn this into the fractions $+u/X$, $-v/X$. Then let ε equal either one of these fractions. The uncertainty represented by the fraction is approximately equal to $(10/\ln 10)\varepsilon$ dB or, more accurately, $(10/\ln 10)(\varepsilon - 0.5\varepsilon^2)$ dB.

Figure 9.8 shows an example of a simplified signal-level budget carried through three units and their integration hardware. Normally the noise level would also be carried in such a budget but for the sake of clarity we leave this out. Both high and low signal levels are carried. A nominal signal level is carried, the low case is decreased further by the 2σ of uncertainties, and the high case is augmented by the addition of this amount. The nominal levels at the output of unit 1 form the

Unit 1 and associated integration elements

Element	Elt gain (dB)	Cum G (dB)	σ (dB)	Low signal Nominal P (dBm)	Low P (dBm)	High signal Nominal P (dBm)	High P (dBm)	Source	Date
Reference point		0.00	0.00	−60.00		−30.00			
Connector	−0.15	−0.15	0.05	−60.15		−30.15		Supplier X spec	xx/xx/xx
LNA	10.00	9.85	0.00	−50.15		−20.15		K. Timm memo	tt/tt/tt
Design immaturity	0.00	9.85	0.40	−50.15		−20.15		K. Timm memo	tt/tt/tt
Measurement error	0.00	9.85	0.10	−50.15		−20.15		K. Timm memo	tt/tt/tt
Cum rss of σ's			0.42						
Signal level				−50.15	−50.98	−20.15	−19.32		

Unit 2 and associated integration elements

Element	Elt gain (dB)	Cum G (dB)	σ (dB)	Low signal Nominal P (dBm)	Low P (dBm)	High signal Nominal P (dBm)	High P (dBm)	Source	Date
Reference point		0.00	0.42	−50.15		−20.15			
Connector	−0.15	−0.15	0.05	−60.15		−20.30		Supplier X spec	xx/xx/xx
Waveguide	−1.50	−1.65	0.20	−61.65		−21.80		J. Frank memo	yy/yy/yy
HPA	30.00	28.35	0.40	−31.65		8.20		Supplier Z spec	zz/zz/zz
Aging and radiation	0.00	28.35	0.25	−31.65		8.20		Supplier Z spec	zz/zz/zz
Measurement error	0.00	28.35	0.20	−31.65		8.20		Supplier Z spec	zz/zz/zz
Cum rssof σ's			0.69						
Signal level				−31.65	−33.03	8.20	9.58		

Unit 3 and associated integration elements

Element	Elt gain (dB)	Cum G (dB)	σ (dB)	Low signal Nominal P (dBm)	Low P (dBm)	High signal Nominal P (dBm)	High P (dBm)	Source	Date
Reference point		0.00	0.69	−31.65		8.20			
Waveguide	−1.20	−1.20	0.05	−32.85		7.00		J. Frank memo	yy/yy/yy
OMUX	−0.50	−1.70	0.05	−33.35		6.50		C. Hendrix PDR	zz/zz/zz
Over temp in life	0.00	−1.70	0.07	−33.35		6.50		C. Hendrix PDR	zz/zz/zz
Measurement error	0.00	−1.70	0.10	−33.35		6.50		C. Hendrix PDR	zz/zz/zz
Cum rss of σ's			0.71						
Signal level				−33.35	−34.76	6.50	7.91		

FIGURE 9.8   Example of simplified signal-level budget (with uncertainty).

reference values at input to unit 2 and so on. The combined 2σ increases as we progress through the units. All the uncertainties are uncorrelated, so they are rss'ed together. Unit 3 is on the earth deck or east or west panel. The values entered in the left section of the table are not meant to be representative.

9.3.5 Keeping Margin in Budgets

Performance budgets usually carry one or more kinds of **margin**. Margin is headroom. The various kinds of margin are carried for different purposes. Most budgets carry only one or two, depending on the purpose of the budget.

- *Risk margin*: This margin is carried in certain budgets throughout the program by the satellite manufacturer. It is a measure of the uncertainty that the engineer feels in his calculations at the point in time. It is usually a percentage of the total budget. Based on his experience on other programs, the engineer knows about how much his current numbers could grow from that time on. (Numbers can also decrease, but when a decrease is benign, as is almost always the case, the engineer does not need to protect himself.) The risk margin decreases over the course of the program; when the unit or payload is completed and measured, this margin equals 0 dB.

- *Design margin in high-error budgets*: This kind of margin is only used in the few budgets in which the individual terms cannot be well estimated or measured. An example is passive intermodulation products (PIMs) (Section 8.15).
- *Customer-required margin*: This kind of margin is not commonly required. If required, it would be for a clear and definite purpose for the customer, since requiring the payload to carry "extra" margin is an expensive proposition.
- *Bottom-line margin*: This is the excess margin after taking into account uncertainties in the budget items and any of the above other margins. The payload engineer must ensure that this is nonnegative at all times.

Choosing the margins other than the calculated bottom-line margin is a delicate balancing act. If they are too small, in the end, the payload may not meet its specifications. On the other hand, it is almost as bad to carry too much margin. The margins must be chosen with a reasonable amount of risk in mind.

The manner in which risk margins are held at both payload and unit levels should be consciously decided on. The payload engineer must either know how the unit engineers are doing it or tell them how they should do it; in either case, he must depend on their openness in the matter. It is wasteful if the unit engineers carry their own risk margin and the payload engineer carries an additional margin for those units for the same purpose.

9.3.6 Maintaining Budget Integrity

The payload budgets must be kept current and accurate, to reflect the current state of knowledge of the payload. The reason is that many decisions must be made throughout the program on the basis of the budgets: decisions on unit specifications and acceptance, on spacecraft layout, on what kind and how much waveguide to buy and so on.

The payload budgets need not be updated every day, but ideally, every time that better numbers are available for entries. The first set of numbers are the result of the proposal analysis; new numbers become available at detailed payload design, detailed unit design, unit test, integration, final test. The budgets must at least all be updated for preliminary design review (PDR), critical design review (CDR), and final acceptance review.

The individual elements' gain and NFs would be entered into the spreadsheet. The source and date of every one of these items are to be entered in the spreadsheet. Practical experience shows that this is absolutely essential at payload level to avoid a lot of head-scratching and repeated work. There are so many payload performance budgets, there are so many entries in each budget, and the entries must be kept up to date as the satellite program progresses that without notes it is impossible to remember whether or not individual elements are due for an update or not.

9.4   HPA TOPICS

9.4.1   How to Know If HPA Nonlinearity Should Be Specified on C/3IM or NPR

In Section 7.2.2 we listed the parameters variously used for the specification of the HPA nonlinearity. In this section, we discuss under what conditions C/3IM or noise-power ratio (NPR) is appropriate. One or the other is appropriate when the operational signals into the HPA will be much like the corresponding characterization-test signals, but otherwise, these characterizations are not useful. The explanation of this was deferred to here since it rests on probability theory (book Appendix A.2).

The C/3IM test signals are two equal-power tones of different frequencies, and the NPR test signals are a large number of equal-power tones equally spaced in frequency. The NPR test signals are an approximation to white Gaussian noise that has been brickwall-filtered (Section 12.4). The pdf of the instantaneous total power of the combined test signals is shown in Figure 9.9 for both characterizations. Both plots in the figure are scaled so that the long-term average total power is 1. For C/3IM the pdf curve is fairly flat over the middle of the interval [0,2] but rises to infinity at 0 and 2 and is zero beyond 2. For NPR the pdf curve falls exponentially over the interval from 0 to infinity. These are quite different.

The plots in Figure 9.9 tell us how the HPA will be exercised in these two tests. Recall that the HPA operating point has as its P_{in} the long-term average input power; thus, the operating point corresponds to the x-value of 1 in the plots of Figure 9.9. The HPA will be tested on a wide range of operating points. For C/3IM the HPA will be driven only up to 3 dB beyond each operating point. For NPR the HPA will be driven as far beyond the operating point as is feasible for the HPA, indeed at least 3 dB above the operating point about one-third of the time.

Thus, C/3IM and NPR characterize the HPA at any given operating point in very different ways. Care must be taken in deciding if the operational signals into the HPA will be sufficiently close to either set of test signals for C/3IM or NPR to justify choosing one of these characterizations for the HPA nonlinearity.

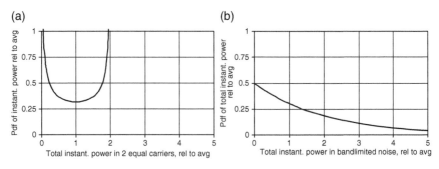

(a)   (b)

FIGURE 9.9   Probability density functions of instantaneous total power in (a) two equal-power carriers and (b) bandlimited Gaussian noise.

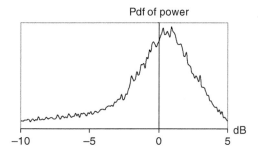

FIGURE 9.10   Pdf of power of QPSK signal with RRC pulse of roll-off factor $\alpha = 0.2$.

Both C/3IM and NPR are bad characterizations when the operational input to the HPA will be one modulated carrier. The pdf of the instantaneous power in one modulated carrier is shown in Figure 9.10 for quaternary phase-shift keying (QPSK) modulation with root raised-cosine (RRC) pulse of roll-off factor $\alpha = 0.2$ (Section 12.8.4.3). The pdf is heavily weighted around 1, the long-term average power. The same holds more or less true for other values of α. The pdf is clearly very different from either pdf in Figure 9.9, meaning that the modulated signal would exercise the HPA very differently from either set of test signals, so neither test would characterize the HPA usefully.

9.4.2   How to Ease Payload Integration of Combined TWTAs

In a transponder where the signal is split, fed to two or more TWTAs, and combined in phase and where the post-TWTA RF line is waveguide, the lengths of the wave-guides to the combiner must be closely matched. If the lengths are not matched, the path delays are different. The TWTAs themselves can be made phase-matched to within a few degrees, even though the absolute phase shift through a TWTA is thousands of degrees. The situation is illustrated in Figure 9.11. If the channel band is wide (on the order of 3% of the carrier frequency), then the phase shift at the upper end of the channel band relative to the phase shift at the lower end of the channel band may be many degrees different between the two waveguide runs. This would cause serious degradation in the combined signal. A good solution is to adjust the

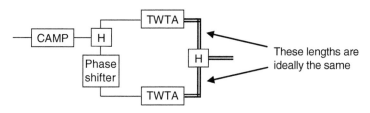

FIGURE 9.11   In-phase combining of TWTA outputs.

relative length of the two pieces of pre-TWTA coaxial cable to provide a compensating differential delay in the waveguide.

The technique is as follows. We first figure out how many degrees of phase shift the lower and upper channel frequencies will see in the extra length of waveguide, and we take the difference. By "extra length" we mean the length of the longer piece of waveguide minus the length of the shorter piece. Then we find the length of coax that will provide the same difference between the frequencies. (The absolute phase shift at each of the two end frequencies will be different in coax from what they are in the waveguide.) Between the two end frequencies, the match will not be perfect but almost.

In detail, the calculation is as follows. In rectangular waveguide excited by the usual TE_{10} mode, the wavelength λ_g at propagating frequency f above the waveguide's cutoff frequency f_c is as follows (Ramo et al., 1984):

$$\lambda_g = \frac{c/f}{\sqrt{1-\left(f_c/f\right)^2}} \quad \text{where } c \text{ is the speed of light in free space}$$

If d is the extra length of waveguide, then the phase shift at the channel band's upper frequency f_2 minus the phase shift at the lower frequency f_1 is the following:

$$\Delta\varphi_{wg} = \left(\frac{d}{\lambda_{g2}} - \frac{d}{\lambda_{g1}}\right)360 \text{ in deg}$$

$$\text{where } \lambda_{gi} = \text{guided wavelength at frequency } f_i, i = 1,2$$

Now, in coax, the guide wavelength is simply proportional to the free-space wavelength. In coax with dielectric constant ε_r excited by the usual transverse electromagnetic (TEM) mode, the guide wavelength at frequency f is the following:

$$\lambda_g = \frac{c/\sqrt{\varepsilon_r}}{f} \text{ or } \frac{1}{\lambda_g} = \frac{f}{c/\sqrt{\varepsilon_r}}$$

In a piece of coax of unknown length m, the phase shift at f_2 minus the phase shift at f_1 is the following:

$$\Delta\varphi_{coax} = m\left(\frac{1}{\lambda_{g2}} - \frac{1}{\lambda_{g1}}\right)360 \text{ in deg where now } \lambda_{gi} \text{ applies to the coax}$$

So to make $\Delta\varphi_{coax} = \Delta\varphi_{wg}$ we simply use the length m of coax given by

$$m = \frac{\Delta\varphi_{wg}/360}{\left(\dfrac{f_2 - f_1}{c/\sqrt{\varepsilon_r}}\right)}$$

For example, a channel band is the frequency band between 12.2 and 12.7 GHz. WR75 waveguide has the cutoff frequency 7.868 GHz and a 141 coax has a dielectric constant of 2.04. An 8-cm extra length of waveguide can be compensated by a 7.23-cm extra length of coax. The phase shift over the channel band, relative to the phase shift at 12.2 GHz, is shown in Figure 9.12 for both the waveguide and the coax. Both the waveguide and the coax see a phase shift of about 62° across the band. The difference across the band of the relative phase-shift plots is shown in Figure 9.13. The maximum difference across the band, that is, the error in the compensation, is only about 0.21°.

A 360°-capable phase shifter is required in one of the arms of the combiner for several reasons: (1) the waveguide and coax pieces can never be cut exactly right; (2) the phase shift in waveguide and coax bends cannot be accurately predicted, to the author's knowledge; and (3) the guide wavelengths will shift a little when the

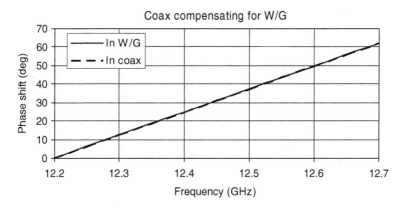

FIGURE 9.12   Relative phase shift over the band in a length of waveguide and in compensating coax length.

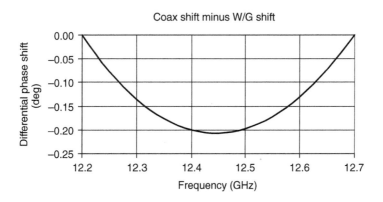

FIGURE 9.13   Compensation error over the band.

waveguide goes from air to vacuum (Section 5.4.1). If the use of a 360°-capable phase shifter is not practical, then two of the 180°-capable ones can be used.

9.5 WHAT NONLINEARITY DOES TO MODULATED SIGNAL

When the HPA is operated in a nonlinear region, besides amplifying the signal it also creates intermodulation products (IMPs) (Section 6.3.1). One must know what this causes, in order to make a good payload design, for example, in regard to the post-HPA filtering.

The HPA is a prime example of a nonlinearity, so the following subsections refer to HPA as the nonlinearity. In fact, though, there are other examples, such as a component amplifier that has to be operated undesirably closer to saturation or an analog-to-digital converter (Section 10.2.1).

These IMPs are mostly inband or near-out-of-band. Since the inband ones cannot be removed, they must be low enough not to be too big of a problem. Near-out-of-band means outside the channel but in nearby channels.

9.5.1 In Terms of Intermodulation Products

9.5.1.1 Case 1: HPA Amplifies One Signal
If the HPA amplifies only one modulated signal, the HPA outputs the main signal plus odd-number IMPs, for all positive real numbers, centered at the carrier frequency f_c. Usually, the 3rd- and 5th-order IMPs are the only ones large enough to matter. The IMPs become less powerful with increasing order. The shape of the IMPs' spectra depends on the modulation pulse (e.g., RRC with $\alpha = 0.2$). For the simple-to-compute case of the rectangular pulse, the 3rd- and 5th-order IMPs have the same shape spectrum as the original signal, as illustrated in Figure 9.14.

9.5.1.2 Case 2: HPA Amplifies Two Signals of Equal Power and Same Spectrum
The case of two signals is a more complicated. Again, it is the 3rd-order IMPs and some of the IMPs of each odd higher order that overlap the signals. The signals and their 3rd-order IMPs for RRC pulses are shown in Figure 9.15. There are four 3rd-order IMPs. A 3rd-order IMP overlays each signal and is almost the same as that signal but of opposite sign; it is effectively a power reduction in the

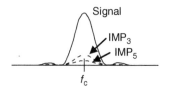

FIGURE 9.14   Example of signal and its 3rd- and 5th-order IMPs.

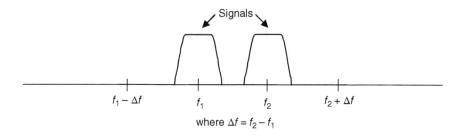

where $\Delta f = f_2 - f_1$

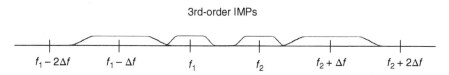

FIGURE 9.15 Example of two equal signals and their inband and near out-of-band 3rd-order IMPs.

signal, so it is not separately detectable. The two other 3rd-order IMPs are offset, and each has twice the power as one of the other kind. The spectrum of each is nearly the same as that of the other kind of IMP but twice as wide (latter assuming that the signals were modulated with independent data clocks).

9.5.1.3 *Case 3: HPA Amplifies Two Unequal Signals*   When two signals are unequal in power and are amplified together, the HPA will suppress the smaller signal relative to the larger one. The HPA cannot be operated at saturation when there are two signals. The input backoff typically must be at least 3 dB, where still the suppression must be taken into account in calculations of output-power specification compliance for the weaker signal. The "missing" power of the weaker signal goes into the IMPs. The suppression of the weaker signal by the stronger is known as power robbing and is further addressed in Section 9.5.3 below.

Figure 9.16 shows two unequal modulated signals and their 3rd-order IMPs. The IMPs now have unequal power. The power ratio of the two IMPs that overlap the output signals is that of the output signals, but, as before, they are not noticeable. The IMPs on the sides are intermediate in size, with the one next to the larger signal being larger than the one next to the smaller signal.

9.5.1.4 *Case 4: HPA Amplifies a Few Signals*   When there are more than two signals, say three or four, it is best if they are unequally spaced across the band, to minimize the number of 3rd- and 5th-order IMPs overlapping the signals. Reference to Figure 9.16 and some thought show the danger of spacing the signals equally, especially if they are of different power levels: if three signals are evenly spaced and one is much smaller than the other two, a 3rd-order IMP overlapping the

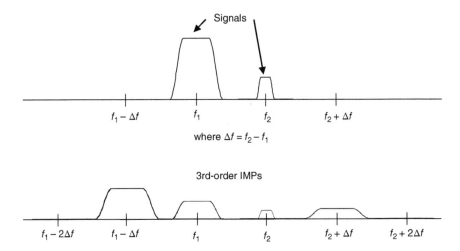

FIGURE 9.16   Example of two unequal signals and their 3rd-order IMPs.

smallest signal could have size on the same order as the smallest signal. If the signals have to be evenly spaced or randomly spaced, it is best if the ratio of powers of the various signals to each other be within a limited range, even if some signals are thereby over-powered. Operating the HPA well backed off is an alternative solution. The HPA that will always be run well backed off is optimized by the HPA manufacturer differently from the HPA that will always be run at saturation or just a few dB backed off.

9.5.1.5 Case 5: HPA Amplifies Many Evenly Spaced, Independent Signals of Equal Power and Same Spectrum

Another simple case is a very large number of evenly spaced, independent signals of the same power and bandwidth. Figure 9.17 shows the spectra of the intended signals and the outlines of the spectra of their 3rd- and 5th-order IMPs. Since the signals are independent, the overall shape of the kth IMP of the group of signals is that of the original group convolved with itself $k-1$ times. In practice, the HPA must be well backed off in this case.

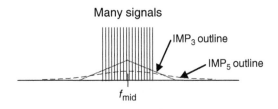

FIGURE 9.17   Many signals and outlines of their 3rd- and 5th-order IMPs.

9.5.1.6   Case 6: HPA Amplifies Many Unequal Signals   What the 3rd- and 5th-order IMPs look like for any other signal-set case can roughly be evaluated from the other cases presented here but a good evaluation requires simulation.

9.5.2   In Terms of Spectrum-Spreading

We saw in case 1 above that the HPA creates odd-order IMPs that overlay the input signal's spectrum and are wider. Thus, the output signal's spectrum is altogether wider than that of the input signal. The HPA has caused the signal's spectrum to spread. An example is shown in Figure 9.18 for the RRC pulse with $\alpha = 0.35$. The loss of power in the intended signal is visible, and this power has gone into IMPs.

In the case where there was an original signal spectrum that was narrowed by filtering in front of the HPA, for example by the IMUX, the HPA would cause some of the outer edges of the signal spectrum to come back, at a lower level. This is **spectral regrowth**.

The spreading of the spectrum is a concern for the adjacent channels (adjacent-channel interference or ACI) so must be controlled.

9.5.3   In Terms of Power-Robbing

Power-robbing is a term that reflects the fact that the HPA suppresses a weaker signal relative to a stronger signal when two unequal-power signals are input to the HPA. It appears that the larger signal has stolen power from the weaker one. Power-robbing occurs when the HPA operates in its nonlinear-amplification regime: the closer to saturation the HPA is, the more the suppression is.

Case 3 of Section 9.5.1 represented the situation of two modulated signals of different power with little noise.

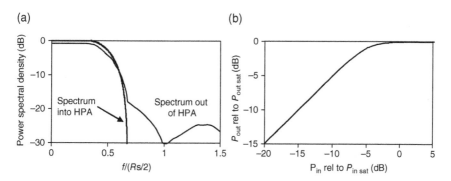

FIGURE 9.18   Spectral regrowth: (a) one side of signal spectrum, into and out of HPA; (b) assumed P_{out} versus P_{in} of HPA.

Consider the *ideal bandpass hard limiter*, which is an ideal hard limiter followed by a bandpass filter. The ideal hard limiter sets a bandpass signal to the instantaneous sign of the signal. The ideal bandpass hard limiter causes different suppression depending on the input signals:

- When Gaussian noise and a much weaker tone are input, the limiter degrades the SNR by 3 dB
- When one tone and another much weaker tone without noise are input, at output the smaller tone is 6 dB lower in power compared to the larger tone than it was at input
- When two tones and Gaussian noise are input:
    o If the noise is much stronger than the two tones, the limiter degrades the SNRs of the tones by about 1 dB
    o If one tone is much stronger than both the other tone and the noise, the limiter increases the SNR of the strong tone by 3 dB and decreases the SNR of the weak tone by 3 dB (Jones, 1963).

The HPA is not an ideal bandpass hard limiter so it suppresses less. However, linearized HPAs operated near saturation come rather close to being such a hardlimiter.

When one modulated signal is much stronger than both a second modulated signal and the Gaussian noise input to the HPA, the SNR of the stronger signal actually gets enhanced. However, for phase-shift-keying (PSK) modulations this creates no advantage, since all that has happened is that the radial component of the noise is suppressed, but this component has nothing to do with symbol decision (Section 12.10.6). The radial and tangential components of noise are defined relative to the instantaneous signal vector.

To get a good idea of the power-robbing in a particular case, it is necessary to perform a simulation or measurement (Chapter 16).

9.6   SIMULATING PAYLOAD PERFORMANCE AS A FUNCTION OF GAUSSIAN RANDOM VARIABLES

Gauss–Hermite integration is a tool that can replace most Monte Carlo simulations over independent Gaussian random variables with a much quicker evaluation. What is being evaluated in either method is a function whose independent variables have independent Gaussian probability distributions. Gauss–Hermite integration is a weighted sum of the function evaluated at a few points, so the computational burden is much less. This is especially a benefit when the function is being evaluated over many random variables including the Gaussian ones or when the function is evaluated as just part of a longer calculation. Gauss–Hermite integration is described in book Appendix A.4. Following are some examples of payload calculations where Gauss–Hermite integration is a possibility:

- EIRP as a function of antenna pointing error
- EIRP as a function of repeater-caused uncertainty
- Loss in combining HPA outputs aligned in phase, as a function of phase misalignment
- HPA output power or phase shift, averaged over the input noise or over the other signals into the HPA which together look like noise.

REFERENCES

Agilent Technologies, Inc (2009). Fundamentals of RF and microwave power measurements, part 3. Application note 1449-3. June 5.

Jones JJ (1963). Hard-limiting of two signals in random noise. *IEEE Trans on Information Theory*; 9; 34–42.

Ramo S, Whinnery JR, and Van Duzer T (1984). *Fields and Waves in Communication Electronics*, 2nd ed., New York: John Wiley & Sons, Inc.

CHAPTER 10

PROCESSING PAYLOAD AND FLEXIBLE PAYLOAD

10.1 INTRODUCTION

10.1.1 Why a Processing Payload

The capabilities of satellite payloads have been increasing in complication and scale since the beginning. The technology keeps improving, and the worldwide demand for data keeps increasing. In the beginning, a payload provided only a few beams and all communications went through a ground station, so all connections between up- and downlink were few and fixed and the payload was a bent-pipe. Then came small-scale time-domain switching among beams, in 1989 (Intelsat VI). Then came a payload with a small-scale digital processor (NASA's Advanced Communications Technology Satellite—ACTS), in 1993. Next came payloads with multiple spot beams, in 1996 (Inmarsat-3), and hundreds of spot beams as well as user-to-user communications, in 2003 (Thuraya). After that came the first small-scale regenerative payload, in 2004 (Amazonas 1, now known as Hispasat 55W-1) and then a large-scale regenerative payload (Spaceway 3), in 2007. (We omit those with an analog processor from this history because there is no standard definition of what that is.) The emphasis today is on flexible non-regenerative payloads that can deal with the rapid evolution of modulations, protocols, and formats (Angeletti et al., 2008).

Satellite Communications Payload and System, Second Edition. Teresa M. Braun and Walter R. Braun.
© 2021 John Wiley & Sons, Inc. Published 2021 by John Wiley & Sons, Inc.

Data traffic from mobile users, some of which goes over satellite and some of which uses Internet Protocol (IP), has been dramatically increasing. It was reported to be 4000 times greater in 2016 than in 2006 and almost 400 million times greater in 2016 than in 2001. It was expected to grow eightfold between 2015 and 2020 (Cisco, 2016).

End-to-end satellite systems are converging toward seamless broadband IP-access networks with flexible and on-demand data transfers between satellite and terrestrial nodes (Gupta, 2016). It was expected in 2020 that the number of devices connected to IP networks (not all of which go over satellite) would be more than three times the global population by 2023, with over 70% of the population having Internet access (Cisco, 2020).

10.1.2 System Connectivity

All the processing payloads have different processing for forward-link signals and return-link signals. The forward link is a signal path from the ground station to the user terminal through the satellite, and the return link is a signal path from the user terminal to the ground station through the satellite, as shown in Figure 1.1. A hop is a one-way link through the payload, so communication from one user to another takes two hops if it has to go through a ground station.

There are two different connectivity topologies that a processing payload can have with its user terminals and ground stations. A satellite system for which all communications between users go through a ground station is a **star network**. A satellite system for which users can communicate directly, through the satellite, is a **mesh network**.

The standards for communications on the forward and return links may be different. The standard for a link direction is called the **air interface**. It is not just signal format but also the means of multiple access (Section 13.2.1).

In a mesh network, since there is no ground station involved, the meanings of forward link and return link are different from in a star network. The signal from a user is on the return link on the uplink but on the forward link on the downlink to the destination user, as illustrated in Figure 10.1. The payload must change the

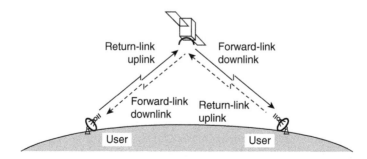

FIGURE 10.1   Definition of forward and return links for user-to-user communications.

signal format from the return-link standard to the forward-link standard, if they are different. The user-to-user links are one-hop. Thus, the payload processes user-to-user signals differently from signals that go through a ground station.

10.1.3 Terminology

A processing payload possesses flexibility in most or all of the following capabilities: bandwidth allocation, signal gain, radio frequency (RF) power allocation, and frequency/beam/timeslot routing. Some of the reasons for deciding on a processing payload over a non-processing one are the following:

- Infeasibility of performing the quantity of desired processing with traditional hardware
- Need to accommodate daily variation in traffic patterns
- Need to accommodate long-term traffic changes
- Need to accommodate satellite mission change, for example, orbital slot change
- Provision of direct user-to-user connectivity.

The two types of processing payload are **regenerative** and **non-regenerative**, where a regenerative one processes the signals more deeply than a non-regenerative one does. A regenerative payload performs at least the operations of demodulation and remodulation. These days it also performs decoding and re-encoding (Chapter 12) since all signals today are coded, and it performs conversion of time-frame formatting. Non-regenerative processing payloads are sometimes called **transparent processing** or **bent-pipe processing** payloads in the literature.

A processing payload has an **onboard processor (OBP)**. (This terminology is not universal, as in some papers the term means strictly a regenerative processor.) The main processor is most often digital, but the payload may have analog processors as well that complement the digital processor.

The term **flexible payload** is more generic than the term "processing payload." A flexible payload may be processing or it may be lesser, possessing only one or two of the capabilities of a processing payload and it may even be without a processor.

The choice of non-regenerative versus regenerative payload is a choice for the end-to-end system architecture. So is the choice of flexible payload versus fixed payload. However, the choice among an analog, a digital, and a mixed payload is an implementation choice (Chie, 2020).

10.1.4 Regenerative versus Non-Regenerative Processing Payload

Regenerative payloads have real advantages over non-regenerative ones:

- Possibility of onboard routing based on message headers in signals
- Possibility of employing time-domain multiplexing (TDM), in which signals are assigned time slots in one modulated signal, so if traveling-wave tube amplifiers (TWTAs) are used (and not solid-state power amplifiers—SSPAs) they can be operated at their most power-efficient, maximizing use of satellite resources.

If the regenerative payload does further processing, it can provide another advantage:

- Error-correction decoding of uplink and re-encoding for downlink, improving uplink signal-to-noise ratio (SNR) that is not strong, which reduces the capabilities required of the user transmitter.

Regenerative payloads used to have the disadvantage that the uplink and downlink coding and modulation schemes were frozen (Chie, 2010). However, today Iridium Next has overcome that by using field-programmable gate arrays (FPGAs) in its processors, which can be reprogrammed to accommodate new coding and modulation schemes (Section 19.2).

Nonetheless, most processors today (2019) are non-regenerative. They have many advantages over regenerative ones, including the OBP being much less complex, direct current (DC) power-hungry, costly, and heavy (García et al., 2013). Since non-regenerative payloads manipulate only frequency bands, not time slots, it makes no difference to the payload what the signal formats are. Various coding and modulation schemes can be in use at any given time, or they can evolve over time as standards change.

10.1.5   Flexible Payload

The term "flexible payload" clearly encompasses processing payloads but is more general. One can look at the areas in which a payload needs to be flexible and then select the technologies that provide the required degrees of flexibility (Chan, 2011). The areas are as follows, with the solutions ordered by increasing cost and complexity:

- Beam coverage flexibility, provided potentially by
  - Steerable and/or rotatable spot beam antenna
  - Mechanically reconfigurable antenna
  - Phased-array antenna, either direct-radiating or forming the feeds for a reflector; with electronic beam-steering or digital beam-forming (Chan, 2011)
  - Beam-hopping; performed by digital processor (Alberti et al., 2010).
- Frequency plan and channel-to-beam connection flexibility, provided potentially by
  - Analog processor, possibly with agile frequency converters and variable-bandwidth surface acoustic-wave (SAW) filters
  - Digital non-regenerative processor
  - Digital regenerative processor, which can additionally add flexibility in timeslot connection.
- RF power-distribution flexibility, provided potentially by
  - Flexible TWTAs or flexible SSPAs
  - Multiport or multimatrix amplifiers
  - Transmit phased array with coefficients that represent not just delay or phase shift but also gain (Chan, 2011).

All of these technologies are described in this book, some in this chapter but most in Chapters 3 through 7 and 11.

10.1.6 Digital versus Analog Processors

There are both analog and digital processors. In the past, some spacecraft had exclusively analog OBPs. An **analog processor** performs all the operations of frequency-conversion, signal-dividing, channel-filtering, routing, and signal-combining with analog hardware. The channel-filter technology is SAW (Section 10.2.4). Routing is performed by switch matrices, which limits the number of possible connections. One example of spacecraft with an analog OBP was the Inmarsat-3 series, with first launch in 1996. Each spacecraft had a forward-link processor and a return-link processor. The processors provided two-way connectivity for seven L-band spot beams and one L-band global beam, through the ground station. Frequency conversion was by mixers, some fed by fixed local oscillators (LOs) and some by selectable LOs. Channel bandwidths varied within a factor of 10. Any channel could be routed to any beam (Pelton et al., 1998). A second example was the Anik F2 spacecraft, launched in 2004. Each Beam*Link processor handled the return link of eight spot beams to one ground station. Before entering the processor, the signals were converted from Ka-band to L-band. Each spot beam used a different channel from the others, out of 12 possible. The processor took whatever eight channels the spot beams were currently transmitting and converted them to almost any subset of eight of the possible 12 downlink channels and combined them to form the ground-station signal (Lee et al., 2002).

Today most processing payloads have digital OBPs, possibly supplemented by analog OBPs. There are two ways in which a **digital processor** can be used: (1) to perform some large subset of the operations demultiplexing, frequency conversion, multiplexing, and routing (and more operations by a regenerative processor), and (2) to perform digital beam-forming. A payload may include processors that do both. Digital processing has the advantage of increasing the payload capability, to a scale not possible with analog technology. For example, the payload can handle more spot beams, hundreds of them as on Spaceway 3, allowing smaller beams and thus higher frequency reuse, which means increased capacity. A general advantage of digital processing is lack of sensitivity to temperature changes (Craig et al., 1992).

One analog processor that a payload may have along with a digital processor usually precedes the digital processor in the signal path. An analog pre-processor could contain a switch matrix for routing, multiple downconverters fed by LOs of programmable frequency, and/or SAW bandpass filters with variable bandwidth. There may be another analog processor after the digital processor in the signal path. For example, the Inmarsat-4 series of spacecraft (Section 20.3) have both a pre-processor and a post-processor. The L-band pre-processor performs downconversion, filtering, LO distribution, and redundancy switching. The post-processor performs similar but inverse functions (Kongsberg Defence Systems, 2005). The two

Inmarsat-6 satellites, the first of which is expected in 2020 to launch in 2021, will have analog pre- and post-processors that perform frequency conversion from L-band to the digital processor and vice versa. The analog processors will contain SAW filters (Kongsberg Norspace, 2016).

10.1.7 Payload Architecture

At the highest level, the architecture of a non-regenerative processing payload is the same as that of a bent-pipe payload, but the names of some unit banks are different: the IMUXes and OMUXes are called the **channelizers**, and the channel and beam-routing switch banks are called the **routers**. The output channelizer is sometimes called the *synthesizer*. The router provides the connections of input beams and channels to output channels and beams (Cherkaoui and Glavac, 2008). There may be hundreds of channel and beam combinations. In some cases, the functions of channelizer and router are intertwined so the two units are combined into one.

The following three figures show some simplified architectures for processing payloads. Figure 10.2 gives a diagram of a non-regenerative payload that has no phased array. There may be no need for downconverters and upconverters if the receive and transmit frequencies are in the lower C-band or even lower (Sections 10.2.1 and 10.2.2). If the signal is multiplexed in the time domain, as for Thuraya (Section 20.2), the processor also demultiplexes and remultiplexes in the time domain.

Figure 10.3 shows a simplified diagram of non-regenerative payload that has phased arrays for both receive and transmit. The possibilities for the architectures of the first and third blocks are given in Chapter 11. A non-regenerative payload may also have a phased array on only one end.

Figure 10.4 gives a simplified diagram of a regenerative payload that has no phased array. Again, there may be no need for downconverters and upconverters if the receive and transmit frequencies are in the lower C-band or below. If the signal is multiplexed in the time domain, the processor also demultiplexes and remultiplexes in the time domain. The order of the listed functions of the digital processor may not be the order in which the operations are performed. A simplified diagram for a regenerative payload with one or more phased arrays can be deduced from this diagram and the one above.

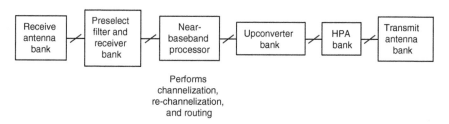

FIGURE 10.2 Simplified diagram of non-regenerative processing payload that has no phased array.

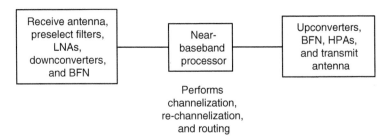

FIGURE 10.3 Simplified diagram of non-regenerative payload that has both receive and transmit phased arrays.

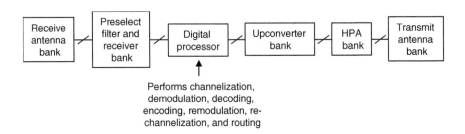

FIGURE 10.4 Simplified diagram of regenerative payload that has no phased array.

10.2 PROCESSING OPERATIONS

The following subsections each describe one processing operation, aside from the subsection on digital filtering, which explains multiple operations: filtering, channelization, and frequency conversion.

The channelizers and routers have to provide so much more flexibility than in a bent-pipe non-processing payload that they require different architectures and technologies.

There are other new elements in a processing payload that we have not seen so far in this book, for example, signal converters between analog and digital. A digital beam-former for a phased array is also considered a processor.

The operations that are unique to a regenerative payload, such as demodulation and decoding, are described in Chapter 12.

10.2.1 Analog-to-Digital Conversion

A digital processor in the signal path must be preceded by an **analog-to-digital converter** (**ADC**, *A-to-D*, or *A/D*).

The greatest factor in determining the processing capability of a processor is the bandwidth per processor port. Increasing the bandwidth means more can be processed without increasing the pre- and post-processing hardware chains (Brown

(a)

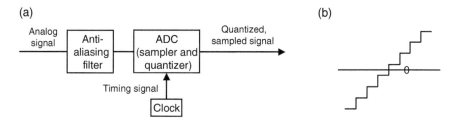

(b)

FIGURE 10.5 Analog-to-digital converter: (a) ADC with associated elements and (b) levels of 3-bit quantizer.

et al., 2014). The input port bandwidth equals the ADC's bandwidth (Cornfield, 2019b).

A port bandwidth of 500 MHz, incorporated into processors by about 2015 by major manufacturers, matches Ka-band broadband systems well because 500 MHz is the spectrum allocation for user links at that band (Brown et al., 2014).

A block diagram of an ADC and associated elements is shown in Figure 10.5, along with an example of quantizer levels. The number of levels in an actual ADC would be greater.

The ADC must sample at least as fast as the **Nyquist rate** of the signal, that is, twice the highest frequency of the signal. This is because the sampling will cause any frequency components beyond ±1/2 sampling rate to overlay the components within ±1/2 the sampling rate (Section 12.10.1.2). So, any unwanted high-frequency components must be filtered out before the sampling or they will distort the sampler output. The filter for this is the *anti-aliasing* filter. This filter and the clock are often not shown in diagrams but they are understood to be present.

The next element in the ADC is the sampler-and-quantizer. A clock provides the timing signal to this. The sampler outputs a weighted average of its input over a period of time. The quantizer has a stated number of levels which is a power of two, where the power is the number of bits. The quantizer will clip any value that lies outside its min/max.

A SNR can be defined for the ADC, the ratio of signal power to power of the added noise within the channel. The ADC introduces noise to the input signal, as analog components do, but the main noise sources are different. One type of noise is *quantization noise*, which is uniformly distributed over ±1/2 quantization step size. The other main source of noise, for sampling rates from about 2 Ms/s to about 4 Gs/s, is *aperture jitter*. This is a timing jitter: the sampling epoch is characterized by a mean and a standard deviation. For even higher sampling rates, the main noise source besides quantization noise is *comparator ambiguity*, which comes from the comparator responding too slowly to a small voltage difference from last sample to this. The occurrence is related to the comparator's regeneration time (Walden, 1999a, 1999b).

An effective number of bits can be defined, which is less than the stated number of bits by a reduction that accounts for ADC noise other than quantization error.

The reduction is about 1.5 bits for all sampling rates. A universal measure of ADC performance is the product of its effective number of quantization levels and its sampling rate (Walden, 1999b).

In regard to the signal power, if clipping rarely comes into play then the input and output signal power are the same. However, if the input signal either has a lot of noise on it already or is noise-like itself (for example, consists of many equal-magnitude carriers), the ADC will clip and thus create intermodulation products (IMPs) (Section 6.3.1). Some signal power is lost to the IMPs, which can be treated as an additional output noise term in analysis (Section 14.9). The IMPs are minimized if the quantizer levels are centered about zero as shown in Figure 10.5b (Taggart et al., 2007).

An example in 2019 of an ADC for communications satellites performs 12-bit quantization. It has dual channels, each sampling at 1.6 Gsps. The channels can be operated in phase or in opposition, when they can be combined to provide one channel at 3.2 Gsps. The 3-dB input bandwidth is up to 4.3 GHz, which allows direct digitization of signals up to lower C-band (Teledyne e2v, 2019a). Another ADC for space also performs 12-bit quantization. It has four channels, each sampling at 1.6 Gsps. Besides being able to operate the four channels independently, it can also operate as two channels, each at 3.2 Gsps, or as one channel at 6.4 Gsps. The instantaneous bandwidth is up to 3.2 GHz (Teledyne e2v, 2019c).

10.2.2 Digital-to-Analog Conversion

A digital processor in the signal path must be followed by a **digital-to-analog converter** (**DAC**, *D-to-A*, or *D/A*). The DAC turns the digital signal back into an analog signal, while using the timing clock to do so. The analog signal has unwanted high-frequency components that are removed by the following *reconstruction filter* (Salim et al., 2004). This process is illustrated in Figure 10.6. The reconstruction filter adds noise to the signal as all analog elements do. This filter and the clock are often not shown in diagrams but they are understood to be present.

In 2019, one space-grade DAC operates on a 12-bit quantized digital signal. Its four channels each operate at a guaranteed 3 Gsps. Each two channels can be multiplexed in parallel or all four channels can. The output bandwidth is 7 GHz, allowing direct RF output at up to lower C-band. It provides gain adjustment (Teledyne e2v, 2019b).

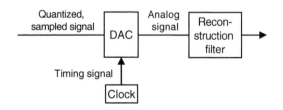

FIGURE 10.6 Digital-to-analog converter with associated elements.

10.2.3 Digital Filtering

Digital filtering is a fundamental operation of a digital processor. Applications include filtering, demultiplexing, multiplexing, and frequency conversion, potentially to and from baseband.

A digital filter is either a **finite impulse-response (FIR) filter** or an **infinite impulse-response (IIR) filter**. A FIR filter has an impulse response (Section 5.2.2) of finite length. In the time domain, the filter output is formed from the current input and a finite number of earlier inputs. In contrast, an IIR filter has an infinite-length impulse response. The filter is recursive in the time domain, so that its output is formed from the current and previous inputs and previous outputs. A FIR filter is guaranteed stable and can be designed to have linear phase response, while an IIR cannot (Oppenheim and Schafer, 1975). However, an IIR filter that meets gain specifications in the frequency domain is usually of a lower order than a FIR filter that meets the same specifications, meaning less computation (Wikipedia, 2019a). In addition, the IIR filter can be designed to have nearly linear phase in the passband (Andersen, 1996).

Alternatively, filtering can be accomplished in the frequency domain using the fast Fourier transform (FFT) and the inverse FFT (IFFT). For a long-enough time-domain impulse response, frequency-domain filtering is faster than time-domain. The dividing line has been described as being between 16 and 80 points, depending on computing method and resources (Borgerding, 2006; Smith, 2007).

10.2.3.1 *In Time Domain*   Two simplifying techniques are widely used in time-domain digital filtering: multirate filtering and half-band filtering.

Multirate filtering significantly improves the efficiency of long FIR filters, making them very desirable. It can also be applied to IIR filters, increasing their efficiency as well. In many cases, multirate filtering is the only practical way to implement a FIR or IIR filter with steep roll-off. Multirate filtering, when applied to a narrowband low-pass filter, breaks the filtering down into three main operations: decimation of the input signal down to a lower sample rate, filtering at the lower sample rate (by the kernel filter), and interpolation of the signal back to the original sample rate (Milić et al., 2006).

Decimation divides the sampling rate of a signal. The usual assumption is that the sampling rate is higher than it needs to be to capture the frequency components present in the signal. Decimation has two steps, as illustrated in Figure 10.7a. Suppose that the sampling rate is to be divided by an integer M. The first step is to

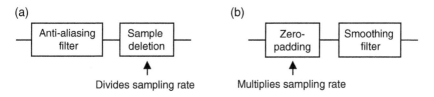

FIGURE 10.7   Two digital processing operations: (a) decimation and (b) interpolation.

low-pass filter the signal, with an anti-aliasing filter. The second step is to delete $M-1$ samples out of every M (Purcell, 2010).

The *kernel filter* performs the filtering operation that was desired in the first place, but on a signal with a lower sampling rate, making for a lower processing load.

Interpolation multiplies the sampling rate of a signal. It must work in a way so as not to add frequency components to the signal that were not present to begin with. Interpolation has two steps, as illustrated in Figure 10.7b. Suppose that the sampling rate is to be multiplied by an integer L. The first step is to pad the original samples with $L-1$ zeros after each original sample. This creates a signal with very sharp transitions, namely very high frequencies. The second step is to low-pass filter the signal. This *smoothing filter* performs smoothing that is consistent with the frequency components present in the original signal (Purcell, 2010).

The anti-aliasing and decimation are efficiently implemented together as a *polyphase filter*, which is a parallel-processing implementation (Fowler, 2007). For a linear-phase FIR filter, the number of arithmetic operations is reduced by the decimation factor (Milić et al., 2006). Similarly, the interpolation and smoothing are also implemented together as a polyphase filter.

The two polyphase filters and the kernel filter have significantly relaxed specifications compared to the original narrowband filter that they replace (Milić, 2009).

A further increase in the efficiency of the multirate filter comes from a *multistage* design, in which the original narrowband low-pass filter is broken down into a cascade of a few such, and each is implemented with the multirate technique. Multirate filtering is applied to a narrowband bandpass filter by first frequency-shifting the input signal from intermediate frequency (IF) down to baseband, applying the multirate implementation of the corresponding narrowband low-pass filter, then frequency-shifting the signal back to IF (Purcell, 2010). A good example of multirate filtering is given in Section 20.3.6.4 on the Inmarsat-4 and Alphasat OBPs.

Half-band filtering is another technique to simplify digital filtering. Such filters have a particular shape that allows for implementation with little more than half as many arithmetic operations as a general digital filter requires. These filters are particularly useful as channel filters. They can be either FIR or IIR filters. Half of a half-band filter response is shown in Figure 10.8, where f_s is the sampling rate. What makes it a half-band filter is that it is antisymmetric about $f_s/4$. Every other value of the impulse response equals zero, so the filter can be implemented efficiently as a polyphase filter (Mathworks, 2019). In cases where the phase response does not have to be exactly linear, an IIR implementation instead of the usual FIR can bring reduction of up to 35% in terms of circuit area and power (Coskun et al., 2013).

Frequency conversions are performed digitally in the following manner.

- To convert a complex-baseband sequence in the time domain to a sequence centered about the frequency f_1, start with the following:

$$x_n + jy_n \text{ where } x_n \text{ and } y_n \text{ are real-valued baseband sequences}$$

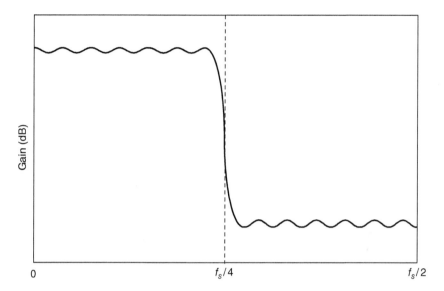

FIGURE 10.8   Example of half-band filter. (After Mathworks (2019)).

Form

$$x_n \cos\left(2\pi f_1 n\Delta t\right) + y_n \sin\left(2\pi f_1 n\Delta t\right)$$

where $\Delta t = 1/F_{per}$ and F_{per} is the frequency period of the sequence, equal to the sampling rate.

- To shift a sequence centered at f_1 down to complex baseband, multiply it by $2\cos(2\pi f_1 n\Delta t)$ and take the baseband part to obtain x_n and multiply it by $-2\sin(2\pi f_1 n\Delta t)$ and take the baseband part to obtain y_n.
- To shift a sequence centered at f_1 down by the frequency $f_0 < f_1$, multiply it by $2\cos(2\pi f_0 n\Delta t)$ and take the part centered at $f_1 - f_0$.

10.2.3.2   *In Frequency Domain*   Filtering in the frequency domain using the FFT and IFFT is sometimes called *fast convolution* or *FFT convolution*.

In summary, the filtering is performed as follows. The signal is divided into time segments. Each segment is filled out at its end with zeros. The transform is taken, it is multiplied by the filter's frequency response, and it is inversely transformed back to the time domain. The filtered segment is longer than the original, so consecutive filtered segments have to be overlapped in time and added (Smith, 2007).

An example of a channelizer and a synthesizer allows for channel bandwidths and center frequencies to be programmed from the ground. The FFT and IFFT use 4096 samples. The FFT is implemented by decimation in time and the IFFT by decimation in frequency (Cherkaoui and Glavac, 2008).

10.2.4 Analog Filtering

In an analog processor, a channel filter is built in **SAW** technology. A SAW filter has two transducers, for input and output, which convert between voltage and mechanical stress. These conversions are the *piezoelectric effect*, and the transducers lie on a *piezoelectric substrate*. Each transducer is an *interdigital filter*, which looks like two combs facing each other with interleaved teeth. Between the transducers travels a surface acoustic wave (API Technologies, 2019). This is illustrated in Figure 10.9. The acoustic wavelength is 100,000 times shorter than electro-magnetic signals of the same frequency. Thus, SAW filters are highly miniaturized. The percentage bandwidth of SAW bandpass filters ranges from 0.01 to above 50% (Microsemi, 2018). SAW products are used primarily up to 2 GHz (API Technologies, 2019).

There are several types of SAW filter for space. One important characteristic of a SAW filter is its *shape factor*, which is the ratio of the 40-dB bandwidth to the 3-dB bandwidth. The smaller the shape factor is, the sharper is the filter's roll-off. The type of SAW filter for channelizing is the *transversal filter*, which can have a shape factor as low as 1.1, inband ripple as low as 0.2 dB pk–pk, and stopband rejection as high as 60 dB, but an insertion loss on the order of 20 dB. Other SAW filter types have much lower insertion loss of a few dB but higher shape factor, so they are suitable for use as a wideband filter. These types have shape factors of 2 or 3, greater ripple than the transversal type, and stopband rejection of 40 to 45 dB (Com Dev International, 2014).

As we have seen above, the channel filters on Inmarsat-3 and Anik F2 were SAW filters. Inmarsat-3 used the sharp-roll-off property of these filters in a technique called *bandwidth-switchable SAW filtering*, to implement flexible bandwidth and guard-band recovery (Peach et al., 1994). Normally, there is a guard band between adjacent channel bands, with bandwidth about 10% of channel spacing, unused except to allow for a small overlap of the signal spectra and the detection filter responses. The SAW filters in the bank can be used singly or in combinations, where a combination passes the combined passbands including the guard bands (Shinonaga and Ito, 1992).

Thuraya (Section 20.2) uses SAW filters for input and output filtering in its digital mobile-to-mobile switch.

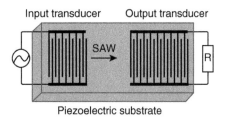

FIGURE 10.9 Diagram of a typical SAW filter (Wikipedia (2019b)) (Credit: M. Buchmeier 2007).

10.2.5 Routing

There are at least three kinds of routing. One kind, in non-regenerative payloads, flexibly connects input beams to output beams or, at a lower level, frequency bands on any input beam to frequency bands on a multitude of output beams. This kind of router is equivalent to a switch matrix, which is how it is implemented in an analog processor. It can alternatively be implemented in a digital processor. In regenerative payloads, a second kind of router performs packet-switching, as on Spaceway 3, and a third kind performs circuit-switching, as on Amazonas 2 (Section 10.4).

10.2.6 Digital Beam-Forming

The forming of reconfigurable beams by a phased array can be done either in an analog fashion, by ground command of analog hardware, or digitally. We consider only digital beam-forming to be processing. (Electronic beam-steering is addressed in Section 11.9.3.)

Digital beam-forming is currently done only by non-regenerative payloads. On receive the beam-former samples, digitizes, and stores, the signals from all the array elements. To form each beam, it delays or phases the signals, weights them, and combines them. On transmit for each beam, it samples and digitizes the corresponding signal from the repeater. For each array element, it delays and weights the signal properly, and it accumulates these numbers over all the beams (Godara, 1997). The use of delay instead of phase shift makes the antenna more wideband, as scanning causes no squint (Section 11.7.2.2). Beam-forming in general includes the function of beam-to-channel assignment on receive and channel-to-beam assignment on transmit (Craig et al., 1992).

All phased arrays known to the author are flat and have the radiating elements in a regular grid pattern. Uniform spacing allows one to take full advantage of signal processing techniques (Dudgeon, 1977).

Digital beam-forming does not have to be executed directly at RF. The RF signals could be downconverted to an IF, decimated, processed at IF, interpolated, and upconverted back to RF (Bailleul, 2016).

Digital beam-forming is described further in Section 11.9.3.

10.3 NON-REGENERATIVE PROCESSING PAYLOADS

10.3.1 Summary of Some Current Payloads

Table 10.1 summarizes the features of some current non-regenerative processing payloads, the ones the authors could find sufficient information on. The satellites are ordered by year of launch or, for a series of spacecraft, year of first launch. All the satellites are geostationary-orbit (GEO) satellites. The table shows the features of the part of each payload that has processing; this part includes any channel or beam switching that is outside of the processor but used with the processor. The quoted

TABLE 10.1 Capabilities of Some Non-Regenerative Processing Payloads

Satellite(s)/Property	Thuraya Series	Inmarsat-4 Series	SES-12 and 14	SES-17
Processor or payload name	—	—	Digital Transparent Processor of Airbus-UK	Spaceflex VHTS of Thales Alenia
Successful launches	2003, 2008	2005–2008	Both 2018	2021 (expected, in 2020)
Frequency band(s)	L- and C-bands	L- and C-bands	Ku-band	Ka/K-band
Downlink BW or bit rate per satellite	1 GHz	272 MHz	Up to 14 GHz and up to 12 GHz, respectively	Unknown
User connectivity	User-ground-user, user–user	User-ground-user, user–user	User-ground-user	User-ground-user, user–user
Frequency at which processing occurs	IF	IF	Baseband	Unknown
Flexible bandwidth allocation	Yes	Yes	Yes	No
Flexible gain	Yes	Yes	Yes	Unknown
Flexible uplink-to-downlink channel mapping	Yes	Yes	Yes	Yes
Flexible channel-to-beam assignment	Yes	Yes	Yes	Yes
Channel filters and routing technology	Digital	Digital	Digital	Digital
Flexible RF power allocation	Yes	Yes	Unknown	Unknown
Digital beam-forming	Yes	Yes	No	Yes
Sources	Thuraya Telecommunications Co (2009a,b), Sunderland et al. (2002)	Farrugia (2006), EADS Astrium (2009), Martin et al. (2007)	Kongsberg Norspace (2015), ESA (2011), SES (2017a), Brown et al. (2014), Emiliani (2019)	Venet (2019), Nichols (2019)

downlink bandwidth or bit rate includes the multiplying effect of any frequency reuse. Recall that in this book frequencies around 20 GHz are termed K-band, not Ka-band, so the uplink/downlink frequency combination of 30/20 GHz is Ka/K-band.

Flexible bandwidth allocation is sometimes called *bandwidth on demand* (*BOD*), while flexible power allocation is sometimes called *power on demand* (*POD*). These capabilities are currently all managed on the ground, and the payload is commanded.

Other common characteristics of the payloads in the table are flexibility in bandwidth allocation, gain, uplink-to-downlink channel mapping, and channel-to-beam assignment. All have a digital processor performing channelization and routing.

The satellite systems of the spacecraft in the table are described in this book: Thuraya and Inmarsat-4 in Chapter 20 in detail, and the SES systems below in less detail.

10.3.2 SES-12 and 14

SES-12 and SES-14 carry SES's first processing payloads. SES-12 was launched in 2018. Its high-throughput Ku-band payload provides broadband data communications via 72 spot beams (SES, 2018b). SES-14 was launched before SES-12, also in 2018. Its processing payload serves the aeronautical and maritime markets and can provide cellular backhaul and broadband services (SES, 2018a). It creates 40 Ku-band spot beams. SES-12's antenna subsystem has eight multi-beam reflector-based antennas (Aerospace Technology, 2018), SES-14 fewer. Both antenna subsystems have the single-feed-per-beam architecture (Section 11.6) (Emiliani, 2019). The digital processor is the same on both satellites, the Digital Transparent Processor or the Channelizer by Airbus-UK (Morelli and Mainguet, 2018). Complementing the processor on both satellites are analog pre- and post-processors that take the signals from Ku-band to baseband and vice versa and that perform filtering with SAW filters (Kongsberg Norspace, 2015).

Table 10.2 summarizes the development of Airbus-UK's digital processors since the early 2000's. (The table is provided partly as a decoding tool for anyone reading articles about the processors, as the collection of their names is bewildering.) Airbus-UK's generation-3 processor, whose development was completed in about the year 2008, flies on Alphasat (Section 20.3). Its port bandwidth is 250 MHz, and it processes at baseband. The Next Generation Processor (NGP) was a further development performed in cooperation with the European Space Agency (ESA) (Cornfield et al., 2012). The processors onboard SES-12 and SES-14 are the NGP (Cornfield, 2019a). Further processor improvements, significant but not dramatic, were made by about the year 2015. These included replacing FIR filters with IIR filters and using multirate filtering. The generation-4 processor was qualified in 2015 and provides a dramatic improvement over the generation-3. The doubling of the port bandwidth allows a five-fold increase in processing capability and a reduction by half of the processor mass. It processes at IF. It can support almost any scale of mission (Brown et al., 2014). (References for this paragraph, aside from the previous sentence, are to be found in the table.)

TABLE 10.2 Development History of Airbus-UK's Digital Processor

Processor Name(s)	Approximate Year of Completion of Processor Development	Onboard Which Satellite(s)	Characteristics
Generation 3; Digital Transparent processor; Digital Sub-Channelizer	2008, year after Alphasat contract award (ESA, 2007)	Alphasat (Brown et al., 2014)	Ports of 250 MHz, can be used either for channel routing and gain setting or as digital BFN (Brown et al., 2014)
Next Generation Processor (NGP); Digital Transparent Processor; Channelizer	2012 (Coskun et al., 2013)	SES-12 and 14 (Morelli and Mainguet, 2018; Cornfield, 2019a)	Goal of port bandwidth of up to 320 MHz (ESA, 2011), processes at baseband (Kongsberg Norspace, 2015)
Unnamed, but an improvement over the NGP	2015 (Coskun et al., 2016)	Unknown if any	Uses IIR filters instead of FIR and uses multirate filtering (Coskun et al., 2016)
Generation 4; Small Modular Processor	2015 (Thomas et al., 2015)	Two Inmarsat-6 spacecraft to be launched in 2020	Five-fold increase in processing capability compared to gen 3, can be used either for channel routing and gain setting or as digital BFN, same architecture as gen 3 (Brown et al., 2014), inputs at IF and port BW of 500 MHz (Thomas et al., 2015)

Other characteristics of the NGP and thus of the SES-12 and SES-14 processors are the following. Each single processor has 14 input and 14 output ports. More than one processor can be used, to extend the entire capacity. The ADCs and DACs operate at almost ten times the sampling rate of those on Inmarsat-4, which directly links to processor's port bandwidth. The satellite operator can have the processor measure the RF power of individual uplink and downlink signals and set the gain of individual signals (Cornfield et al., 2012).

10.3.3 SES-17

SES-17 will also have a Ka/K-band payload with a digital processor. In 2020, the launch is expected in 2021. The payload's main role will be to provide Internet connectivity for aircraft, with the first large anchor-customer being Thales operating its FlytLIVE. The payload will provide close to 200 spot beams of various sizes (SES, 2016).

The digital processor is the Spaceflex VHTS (very high-throughput satellite) of Thales Alenia Space France. It will allow mesh, broadcast, and multicast network configurations (SES, 2017b) and will perform digital beam-forming (Nichols, 2019). It is the fifth generation of the Spaceflex processor. Some of its properties are as follows (Venet, 2019):

- Star and mesh, multicasting, and broadcasting
- Full connectivity between input and output ports
- I/O ports of 480 MHz
- Up to 160/160 I/O ports
- Total capacity or useful bandwidth 480 GHz
- Channel bandwidth 3.5 MHz
- Gain control (automatic level control [ALC] or fixed-gain mode [FGM]) selectable per channel
- Mass about 0.8 Kg/GHz
- DC power consumption about 8.5 W/GHz.

The 2.5th-generation processor, named Spaceflex 5, has been flying on the Inmarsat-S-EAN spacecraft since 2017 (Nicolas, 2020).

Both the mass and power consumption of the Spaceflex VHTS are drastically reduced compared to the Spaceflex 5, mass by a factor of more than 12 and power consumption by more than 17. Spaceflex VHTS is the first generation to use optical digital interconnects (Venet, 2019).

10.4 REGENERATIVE PAYLOADS

10.4.1 Summary of Some Current Payloads

Table 10.3 summarizes the features of three regenerative payloads, the ones the authors could find sufficient information on. The Amazonas 2 and Spaceway 3 spacecraft are GEO, while the Iridium Next constellation is LEO.

TABLE 10.3 Capabilities of Some Regenerative Payloads

Satellite(s)/Property	Spaceway 3	Amazonas 2	Iridium Next Constellation
Satellite constellation	1 GEO	1 GEO	66 LEOs
Processor(s) or payload name	–	AmerHis 2	–
Successful launches	2007	2009	2017–2019
Frequency band(s)	Ka/K-band	Ku-band	L, Ka/K, K (crosslinks)
Downlink BW or bit rate per satellite, processed	10 Gbps	216 Mbps	600 MHz per spacecraft
User connectivity	User-ground-user, user–user	User-ground-user, user–user	User-ground-user, user–user
Switching	Packet	Circuit	Both
Flexible bandwidth allocation	Yes	Yes	Yes
Flexible RF power allocation	Yes	No	Yes
Digital beam-forming	No	No	No
Decoding	Yes	Yes	Yes
Coding	Yes	Yes	Yes
Communications standards	DVB	DVB	Proprietary
Sources	Whitefield et al. (2006), Fang (2011), ETSI TS 102 188-1 v1.1.2 (2004), Wu et al. (2003)	Wittig (2003), ESA (2004), ETSI TS 102 429-1 v1.1.1 (2006), ETSI TS 102 602 v1.1.1 (2009a), Thales Alenia Space (2009)	Buntschuh (2013), Murray et al. (2012)

Since all three systems perform decoding, the processors are necessarily digital.

All three systems provide a star network as well as a mesh network (Section 10.1.2). The Iridium Next system is different, though, in that a signal will usually be relayed through more than one satellite before getting to a ground station or another user (Section 19.2).

The ground stations of all three systems have gateways to the Internet and public telephone networks.

These satellite systems are fundamentally different from those with non-regenerative satellites because these deal with not just one but three layers of the

Open Sysems Interconnect (OSI) communications model (Section 13.2.1). The communications protocols of Amazonas 2 and Spaceway 3 are described in Section 13.3.

10.4.2 Amazonas 2

Recall from Table 10.3 that the total processed data rate of Amazonas 2 is 216 Mbps. The payload provides four fully connected channels. Originally there were four beams (Wittig, 2003), of which currently only three are active. The AmerHis 2 processor does not provide flexible gain control, but the system provides the means for the uplinking terminals to adjust their output power. Figure 10.10 offers a diagram of the Amazonas 2 payload minus antennas.

Thales Alenia has developed a third generation of AmerHis for the Redsat payload on the GEO Smallsat Hispasat 36W-1 launched in 2017 (Yun et al., 2010).

The ground stations have the same kind of terminal as the users do.

10.4.3 Spaceway 3

Recall from Table 10.3 that the total processed data rate of Spaceway 3 is 10 Gbps. Spaceway 3 has a huge number of downlink cells, 784, that it connects with 112 uplink cells. Each uplink cell contains seven downlink cells. Twenty-four simultaneous downlink spot beams hop extremely rapidly among the downlink cells (Section 11.11).

Figure 10.11 shows some of the flexibility that the payload provides. The ground stations have the same kind of terminal as the users do. The multiple access methods are described in Section 13.2.2.

Spaceway 3 performs uplink power control in a closed loop with a terrestrial terminal. For each burst from the terminal, the payload measures the received SNR

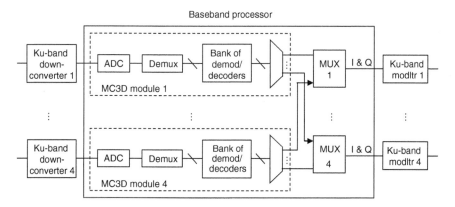

FIGURE 10.10   Amazonas 2 regenerative payload, minus antennas. (After Wittig (2003)).

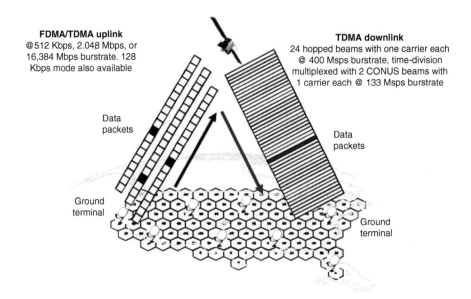

FIGURE 10.11 Spaceway 3 uplink and downlink. (© 2011 IEEE. Reprinted, with permission, from Fang (2011)).

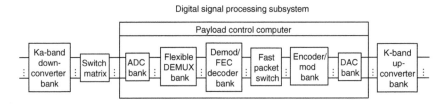

FIGURE 10.12 Spaceway 3 regenerative payload, minus antennas.

and the timing error and sends this information to the terminal. The terminal can then adjust itself.

Spaceway 3 has a reserve pool of RF power to compensate downlink rain fades in a hybrid closed- and open-loop system. In the closed-loop part of the system, a terrestrial terminal can request an increase in downlink power for its beam. In the open-loop part, the payload preemptively increases downlink power to a beam, based on rain predictions coming from ground-based radar.

Figure 10.12 shows a block diagram of the regenerative payload, minus antennas, as described in Fang (2011). The uplink antenna is a dual offset Cassegrain with multiple feeds. The downlink antenna is a direct-radiating phased array with 1500 radiating elements (Boeing, 2006). Beam-forming is not performed by the digital processor but is analog (Ramanujam and Fermelia, 2014).

FIGURE 10.13   Model of non-regenerative processing payload, minus antennas, for payload noise figure estimation.

10.5   COMMUNICATIONS PARAMETERS OF DIGITAL PROCESSING PAYLOAD

10.5.1   For Non-Regenerative Payload

For a non-regenerative payload with digital processor, there are two types of parameters that must be dealt with differently from an all-analog payload: repeater (that is, payload minus antennas) noise figure and filter performance.

For the repeater noise figure estimation, the repeater can be considered to have the block diagram shown in Figure 10.13, where each box has a gain and noise figure (Sulli et al., 2016). The gains and noise figures are combined analytically (Section 9.2.2) to yield the repeater noise figure. The ADC noise was described above in Section 10.2.1. The DAC noise comes from the reconstruction filter (Section 10.2.2) but may be negligible, as it is not mentioned in the referenced paper. The calculation of the "noise" added by the digital processor is time-consuming and complex. This "noise" comes from quantization. One evaluation method is presented in the already referenced paper and another is provided in Rocher et al. (2012).

The characterization of filtering is different in a non-regenerative processing payload in that (1) some of the filters are digital and (2) the digital filters do not have a noise figure since their error due to coefficient quantization is already accounted for in the processor "noise."

10.5.2   For Regenerative Payload

The regenerative payload breaks the forward or return link into two links which must be separately characterized in terms of communications performance. The uplink processing for demodulation, detection, and decoding is similar to what is done in the ground terminal receiving the downlink so the performance would be similarly characterized. The downlink processing for coding and modulation is similar to what is done in the ground terminal preparing the uplink transmission so the performance would be similarly characterized. The processes involved and some performance parameters are described in Chapter 12, but the rest is outside the scope of this book.

REFERENCES

Aerospace Technology (2018). SES-12 telecommunications satellite. News article. On www.aerospace-technology.com/projects/ses-12-telecommunications-satellite. Accessed Sep. 13, 2019.

Alberti X, Cebrian JM, Del Bianco A, Katona Z, Lei J, Vazquez-Castro MA, Zanus A, Gilbert L, and Alagha N (2010). System capacity optimization in time and frequency for multi-beam multi-media satellite systems. *Advanced Satellite Multimedia Systems Conference and Signal Processing for Space Communications Workshop*; Sep. 13–15.

Andersen BR (1996). Digital filter bank designs for satellite transponder payloads: implementation on VLSI circuits. *IEEE International Conference on Universal Personal Communications*; Oct. 2.

Angeletti P, De Gaudenzi R, and Lisi M (2008). From "bent pipes" to "software defined payloads": evolution and trends of satellite communications systems. *AIAA International Communications Satellite Systems Conference*; June 10–12.

API Technologies (2019). Introduction to SAW filter theory & design techniques. White paper. Oct. On www.micro.apitech.com/saw_filters.aspx. Accessed Feb. 20, 2019.

Bailleul P (2016). A new era in elemental digital beamforming for spaceborne communications phased arrays. *Proceedings of the IEEE*; 104 (3) (Mar.); 623–632.

Boeing (2006). Spaceway™ North America, Bandwidth-on-demand. Spacecraft factsheet. On web.archive.org/web/20060619122729/http://boeing.com/defense-space/space/bss/factsheets/702/spaceway/spaceway.html. Accessed Aug. 14, 2019.

Borgerding M (2006). Turning overlap-save into a multiband mixing, downsampling filter bank. *IEEE Signal Processing Magazine*; 23 (2) (Mar.); 158–161.

Brown SP, Leong CK, Cornfield PS, Bishop AM, Hughes RJF, and Bloomfield C (2014). How Moore's law is enabling a new generation of telecommunications payloads. *AIAA International Communications Satellite Systems Conference*; Aug. 4–7.

Buntschuh F of Iridium Constellation LLC (2013). Appendix 1, Iridium Next engineering statement. Part of application to FCC. Dec. 27. On licensing.fcc.gov/cgi-bin/ws.exe/prod/ib/forms/attachment_menu.hts?%20id_app_num=102703&acct=265137&id_form_num=15&filing_key=-260629.

Chan H (2011). Advanced microwave technologies for smart flexible satellite. *International Symposium of IEEE Microwave Theory and Techniques Society*; June 5–10.

Cherkaoui J and Glavac V (2008). Signal frequency channelizer/synthesizer. *International Workshop on Signal Processing for Space Applications*; Oct. 6–8.

Chie CM, retired from Boeing Satellite Center (2010). Personal communication, Mar. 7.

Chie CM (2020). Personal communication, Mar. 2.

Cisco (2016). Cisco visual network index: global mobile data traffic forecast update, 2015–2020. White paper. Feb. 3 On www.cisco.com/c/dam/m/en_in/innovation/enterprise/assets/mobile-white-paper-c11-520862.pdf. Accessed Apr. 3, 2020.

Cisco (2020). Cisco annual Internet report (2018–2023). White paper. Mar. 20. On www.cisco.com/c/en/us/solutions/executive-perspectives/annual-internet-report/index.html. Accessed Apr. 3, 2020.

Com Dev International, now Honeywell Aerospace (2014). Com Dev SAW filters. Application note 102. On www.comdev.ca/docs/com_dev_saw_filters.pdf. Accessed Jan. 27, 2014.

Cornfield P (2019a). Private communication, Nov. 28.

Cornfield P (2019b). Private communication, Dec. 23.

Cornfield P, Bishop A, Masterton R, and Weinberg S (2012). A generic on-board digital processor suitable for multiple missions. *Proceedings of the ESA Workshop on Advanced Flexible Telecom Payloads*; Apr. 17–19.

Coskun A, Kale I, Morling RCS, Hughes R, Brown S, and Angeletti P (2013). Halfband IIR filter alternatives for on-board digital channelisation. *AIAA International Communications Satellite Systems Conference;* Oct. 15–17.

Coskun A, Kale I, Morling RCS, Hughes R, Brown S, and Angeletti P (2016). Efficient digital signal processing techniques and architectures for on-board processors. *ESA Workshop on Advanced Flexible Telecom Payloads*; Mar. 21–24.

Craig AD, Leong CK, and Wishart A (1992). Digital signal processing in communications satellite payloads. *IEE Electronics and Communication Engineering Journal*; 4 (3) (June); 107–114.

Dudgeon DE (1977). Fundamentals of digital array processing. *Proceedings of the IEEE*; 65 (6); 898–904.

EADS Astrium (2009). Inmarsat-4, the very latest in communications technology. Program article. On www.astrium.eads.net. Accessed Mar. 2009.

Emiliani L of SES (2019). Private communication, Sep. 17.

ESA (2004). *AmerHis, A New Generation of Satellite Communications Systems*. Publication BR-226. The Netherlands: ESA Publications Division.

ESA (2007). ESA and Inmarsat sign innovative Aphasat satellite contract. Press release. On www.esa.int/About_Us/Business_with_ESA/ESA_and_Inmarsat_sign_innovative_Alphasat_satellite_contract. Nov. 23. Accessed Sep. 17, 2019.

ESA (2011). Next generation processor. Telecommunications and integrated applications project description. Nov. 22. On telecom.esa.int/telecom/www/object/index.cfm?fobjectid=28134. Accessed Oct. 10, 2014.

ETSI TS 102 188-1 v1.1.2 (2004). Satellite earth stations and systems (SES); regenerative satellite mesh—A (RSM-A) air interface; physical layer specification; part 1: general description.

ETSI TS 102 429-1 v1.1.1 (2006). Satellite earth stations and systems (SES); broadband satellite multimedia (BSM); regenerative satellite mesh—B (RSM-B); DVB-S/ DVB-RCS family for regenerative satellites; part 1: system overview.

ETSI TS 102 602 v1.1.1 (2009a). Satellite earth stations and systems (SES); broadband satellite multimedia; connection control protocol (C2P) for DVB-RCS; specifications.

Fang RJF (2011). Broadband IP transmission over Spaceway® satellite with on-board processing and switching. *IEEE Global Telecommunications Conference*; Dec. 5–9.

Farrugia L, EADS Astrium UK (2006). Astrium view of future needs for interconnect complexity of telecommunications satellite on-board digital signal processors. Presentation at meeting at European Space Research and Technology Centre; Feb. 9. On escies.org/GetFile?rsrcid=1643. Accessed Feb. 1, 2010.

Fowler M (2007). Polyphase filters. Notes for course EE521 Digital Signal Processing, lecture IV-04. On ws.binghamton.edu/fowler/fowler%20personal%20page/EE521.htm. Accessed Aug. 3, 2019.

García AY, Rodriguez-Bejarano JM, Jimenez I, and Prat J (2013). Future opportunities for next generation OBP enhanced satellite payloads. *AIAA International Communications Satellite Systems Conference*; Oct. 14–17.

Godara LC (1997). Application of antenna arrays to mobile communications, part II: beam-forming and direction-of-arrival considerations. *Proceedings of the IEEE*; 85 (8); 1195–1245.

Gupta RK (2016). Communications satellite RF payload technologies evolution: a system perspective. *Asia-Pacific Microwave Conference*; Dec. 5–9.

Kongsberg Defence Systems (2005). Inmarsat-4 F1 launched. News release. Apr. 14. On www.kongsberg.com/en/kds/kns/newsarchive. Accessed Nov. 29, 2014.

Kongsberg Norspace (2015). Kongsberg Norspace wins orders with Airbus Defence and Space. Nov. 24. Press release. On www.kongsberg.com/norspace/news-and-media/news-archive. Accessed Sep. 13, 2019.

Kongsberg Norspace (2016). Kongsberg Norspace wins orders for deliveries to Inmarsat 6. Press release. July 8. On www.kongsberg.com/news-and-media/news-archive/2016/kongsberg-norspace-wins-orders-for-deliveries-to-inmarsat-6. Accessed Nov. 1, 2019.

Lee M, Wright S, Dorey J, King J, and Miyakawa RH (2002). Advanced Beam*Link® processor for commercial communication satellite payload application. *AIAA International Communications Satellite Systems Conference*; May 12–15.

Martin DH, Anderson PR, and Bartamian L (2007). *Communications Satellites*, 5th ed. El Segundo (CA): The Aerospace Press; and Reston (VA): American Institute of Aeronautics and Astronautics, Inc.

Mathworks (2019). FIR halfband filter design. MATLAB article. On www.mathworks.com/help/dsp/examples/fir-halfband-filter-design.html. Accessed Aug. 3, 2019.

Microsemi (2018). SAW products. Product brochure. May. On www.microsemi.com/product-directory/bandpass-filters/4878-surface-acoustic-wave-filters. Accessed Oct. 31, 2019.

Milić L (2009). *Multirate Filtering for Digital Signal Processing: MATLAB Applications*. Pennsylvania: Information Science Reference.

Milić L, Saramäki T, and Bregović R (2006). Multirate filters: an overview. *IEEE Asia Pacific Conference on Circuits and Systems*; Dec. 4–7.

Morelli G and Mainguet A (2018). Automated operations of large GEO telecom satellites with Digital Transparent processor (DTP): challenges and lessons learned. *SpaceOps Conference*; May 28–June 1.

Murray P, Randolph T, Van Buren D, Anderson D, and Troxel I (2012). High performance, high volume reconfigurable processor architecture. *IEEE Aerospace Conference*; Mar. 3–10.

Nichols S (2019). AIX: SES sees a future with smarter MEO/GEO satellites. News article. Apr. 5. On www.getconnected.aero/2019/04/aix-ses-sees-a-future-with-smarter-meo-geo-satellites. Accessed Oct. 26, 2019.

Nicolas C of Thales Alenia Space (2020). LinkedIn page. Accessed Oct. 22, 2020.

Oppenheim AV and Schafer RW (1975). *Digital Signal Processing*. New Jersey: Prentice-Hall.

Peach RC, Lee YM, Miller ND, van Osch B, Veenstra A, Kenyon P, and Swarup A (1994). The design and implementation of the Inmarsat 3 L-band processor. *AIAA International Communications Satellite Systems Conference*; Feb.–Mar. 3.

Pelton JN, Mac Rae AU, Bhasin KB, Bostian CW, Brandon WT, Evans JV, Helm NR, Mahle CE, and Townes SA (1998). Global satellite communications technology and systems. WTEC panel report. Dec. On wtec.org/loyola/pdf/welcome.htm#satcom2. Accessed Apr. 14, 2010.

Purcell JE, Momentum Data Systems Inc (2010). Multirate filter design—an introduction. Application note. On www.mds.com. Accessed Feb. 2010.

Ramanujam P and Fermelia LR (2014). Recent developments on multi-beam antennas at Boeing. *European Conference on Antennas and Propagation*; Apr. 6–11.

Rocher R, Menard D, Scalart P, and Sentieys O (2012). Analytical approach for numerical accuracy estimation of fixed-point systems based on smooth operations. *IEEE Transactions on Circuits and Systems*; 59 (10) (Oct.); 2326–2339.

Salim T, Devlin J, and Whittington J (2004). Analog conversion for FPGA implementation of the TIGER transmitter using a 14 bit DAC. *IEEE International Workshop on Electronic Design, Test and Applications*; Jan. 28–30.

SES (2016). SES orders high throughput satellite from Thales with first secured anchor customer for inflight connectivity. Press release. Sep. 12. On www.ses.com/press-release/

ses-orders-high-throughput-satellite-thales-first-secured-anchor-customer-inflight. Accessed Oct. 22, 2019.

SES (2017a). Investor day 2017. Presentation. June 28. On www.ses.com/investors/presentations. Accessed Oct. 28, 2020.

SES (2017b). SES and Thales unveil next-generation capabilities onboard SES-17. Press release. Apr. 4. On www.ses.com/press-release/ses-and-thales-unveil-next-generation-capabilities-onboard-ses-17. Accessed Aug. 2, 2019.

SES (2018a). SES-14, redefining broadcasting and connectivity across the Americas. On www.ses.com/sites/default/files/2018-01/SES-14%20Factsheet_0.pdf. Accessed Sep. 14, 2019.

SES (2018b). Fact sheet: SES-12. On www.ses.com/sites/default/files/2018-05/SES-12_Fact%20Sheet_May_0.pdf. Accessed Sep. 14, 2019.

Shinonaga H and Ito Y (1992). Microwave SAW bandpass filters for spacecraft applications. *IEEE Transactions on Microwave Theory and Techniques*; 40 (6) (June); 1110–1116.

Smith SW (2007). FFT convolution and the overlap-add method. *Electronic Engineering Times*. On eetimes.com/fft-convolution-and-the-overlap-add-method. Accessed Apr. 2, 2020.

Sulli V, Giancristofaro D, Santucci F, and Faccio M (2016). An analytical method for performance evaluation of digital transparent satellite processors. *IEEE Global Communications Conference*; Dec. 4–8.

Sunderland DA, Duncan GL, Rasmussen BJ, Nichols HE, Kain DT, Lee LC, Clebowicz BA, and Hollis IV RW (2002). Megagate ASICs for the Thuraya satellite digital signal processor. *IEEE International Symposium on Quality Electronic Design*; Mar. 18–21.

Taggart D, Kumar R, Krikorian Y, Goo G, Chen J, Martinez R, Tam T, and Serhal E (2007). Analog-to-digital converter loading analysis considerations for satellite communications systems. *IEEE Aerospace Conference*; Mar. 3–10.

Teledyne e2v (2019a). EV12AD550B dual 12-bit 1.6 Gsps ADC, space grade. Datasheet. Feb. On teledyne-e2v.com/products/semiconductors/adc. Accessed Oct. 26, 2019.

Teledyne e2v (2019b). EV12DS130AG, EV12DS130BG, low power 12-bit 3 Gsps digital to analog converter with 4/2:1 multiplexer. Datasheet DS1080. June On teledyne-e2v.com/products/semiconductors/dac. Accessed Oct. 26, 2019.

Teledyne e2v (2019c). EV12AQ600 quad 12-bit 1.6 Gsps ADC with embedded cross-point switch, digitizing up to 6.4 Gsps. Datasheet. Sep. On teledyne-e2v.com/products/semiconductors/adc. Accessed Oct. 26, 2019.

Thales Alenia Space (2009). Amazonas-2 satellite to embark Thales Alenia Space's Amerhis-2 system. Press release. Sep. 22. On www.thalesgroup.com/Pages/PressRelease.aspx?id=10303. Accessed Jan. 26, 2010.

Thomas G, Jacquey N, Trier M, and Jung-Mougin P (2015). Optimising cost per bit: enabling technologies for flexible HTS payloads. *AIAA International Communications Satelite Systems Conference*; Sep. 7–10.

Thuraya Telecommunications Co (2009a). Space segment. On www.thuraya.com. Accessed Jan. 27, 2010.

Thuraya Telecommunications Co (2009b). Thuraya satellite. On www.thuraya.com. Accessed Mar. 19, 2009.

Venet N (2019). Spaceflex onboard digital transparent processor: a new generation of DTP with optical digital interconnects. Paper and slide presentation. *International Conference on Space Optics*; Oct. 9–12.

Walden RH (1999a). Analog-to-digital converter survey and analysis. *IEEE Journal on Selected Areas in Communications*; 17 (4); 539–550.

Walden RH (1999b). Performance trends for analog-to-digital converters. *IEEE Communications Magazine*; 37 (2) (Feb.); 96–101.

Whitefield D, Gopal R, and Arnold S (2006). Spaceway now and in the future: on-board IP packet switching satellite communication network. *IEEE Military Communications Conference*; Oct. 23–25.

Wikipedia (2019a). Infinite impulse response. Article. Jan. 17.

Wikipedia (2019b). Surface acoustic wave. Article. Oct. 25.

Wittig M, European Space Research and Technology Centre (2003). Telecommunikation satellites: the actual situation and potential future developments. Presentation; Mar. On www.dlr.de/rd/Portaldata/28/Resources/dokumente/RK/wittig-multimedia_systeme.pdf. Accessed Jan. 28, 2010.

Wu YA, Chang RY, and Li RK (2003). Precision beacon-assisted attitude control for Spaceway. *AIAA Guidance, Navigation, and Control Conference*; Aug. 11–14.

Yun A, Casas O, de la Cuesta B, Moreno I, Solano A, Rodriguez JM, Salas C, Jimenez I, Rodriguez E, and Jalon A (2010). AmerHis next generation global IP services in the space. *Advanced Satellite Multimedia Systems Conference and Signal Processing for Space Communications Workshop*; Sep. 13–15.

CHAPTER 11

MULTI-BEAM ANTENNA AND PHASED ARRAY

11.1 INTRODUCTION

This chapter deals with two topics that have much overlap, the multi-beam antenna (MBA) and the phased array (PA). MBAs were developed in the 1990s and 2000s to provide a large number of higher-gain, smaller beams, in order to raise the capacity in the overall coverage area. Frequency and polarization combinations are reused in the beams, sometimes by a large factor. There are several general types of MBA. PAs are used commercially most often as part of a MBA either to provide feeds to a geostationary-orbit (GEO) spacecraft's reflector for multiple beams or to directly form regional beams from low earth orbit (LEO). This chapter also treats PAs in their less common application of forming one or a few contoured beams with a reflector.

We cover all aspects of these topics including the various antenna schemes, the reflectors, the radiating elements, the beam-forming network (BFN) for a PA, PA amplification, and MBA testing.

This chapter comes last in the book's payload part because it relies on material from most of the preceding chapters.

Satellite Communications Payload and System, Second Edition. Teresa M. Braun and Walter R. Braun.
© 2021 John Wiley & Sons, Inc. Published 2021 by John Wiley & Sons, Inc.

We need to declare some terminology for this chapter. Some of the following terms are not used consistently across the industry, while others may not be common knowledge:

- **Radiating element** of antenna—may also stand for RF-receiving element. It could, in another application, be an antenna in its own right.
- **Cluster** or **array**—a collection of radiating elements, set near each other.
- **Single-feed-per-beam** (**SFPB**) antenna—a reflector-based MBA where each radiating element forms one spot beam.
- **Multiple-feed-per-beam** (**MFPB**) antenna—a reflector-based MBA where a cluster of radiating elements out of a large number of elements, forms each beam.
- **Phased array**—array of radiating elements with beams formed from multiple radiating elements by a BFN based on either phase shifts or time delays. The MFPB antenna is an example of a PA.
- **Direct-radiating array** (**DRA**)—a phased array that directly radiates, that is, without a reflector.
- **Scanned beam**—a beam that points off the perpendicular to the plane of a PA.
- **Feed** for MBA or PA
  - For SFPB, a radiating element.
  - For MFPB, a cluster of radiating elements that form a primary radiation pattern corresponding to one beam (recall that in a reflector-based antenna the feeds form the primary radiation patterns and the reflector the secondary radiation patterns).
  - For **full-PA-fed reflector-based antenna**, a primary radiation pattern formed by the full PA, corresponding to one beam.

11.2 MBA INTRODUCTION

11.2.1 Beams

MBAs are mostly used to create a large number of spot beams. In many cases, all the beams have the same size and form a hexagonal pattern, as seen from the spacecraft. In other cases, there is a mixture of beam sizes, where smaller beams are used for terrestrial areas to be more densely served, and the beams are not contiguous. Examples of both these situations are present in the spot-beam coverage of Eutelsat's Ka-Sat, launched in 2010, shown in Figure 11.1 and in color on the book cover. The part of the coverage with the hexagonal pattern is an example of the common four-color scheme, where a **color** is a frequency band/polarization combination. The four-color coverage scheme is illustrated in the middle section of Figure 11.2, along with the other popular schemes of three and seven colors. In the figure, all the beams denoted as color "1" are highlighted to make it easier to see the relative distances between them. The larger the number of colors, the farther apart are the beams of the same color.

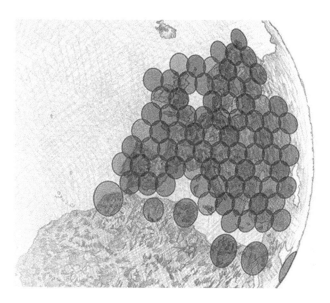

FIGURE 11.1 Spot-beam coverage of Eutelsat's Ka-Sat. (Drawing courtesy of Eutelsat).

FIGURE 11.2 Three-, four-, and seven-color beam coverage schemes. (After Rao (1999) and Guy (2009)).

Naturally, the beams are not clean circles as depicted in Figure 11.2. They are shown in the figure as not overlapping, for simplicity, but each beam actually has a mainlobe which extends outside the circles shown and sidelobes which go out far beyond. Section 14.6 discusses beam-to-beam (that is, self) interference.

Parameters that characterize hexagonal-pattern coverage are as follows. The **peak of beam** is the location (direction from the antenna) in a beam where the gain is highest, near or at the beam center. **Beam spacing** or *beam separation* is the center-to-center angular separation of adjacent beams, where the vertex of the angle is at the spacecraft antenna. When the beams are all of the same size and regularly spaced, there is additional terminology. Imagine the beams represented by circles just large enough to fully cover their coverage area, as shown in Figure 11.3 for three beams. These circles possess the **beam diameter** (Rao, 1999). The point at which

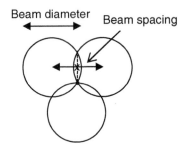

■ = Triple-beam overlap point

x = Point of greatest overlap
between adjacent beams

FIGURE 11.3   Some definitions of terms for equal-size spot-beam coverage.

three adjacent circles meet is the *triple-beam crossover point* or *triple-beam overlap*. The dashed line between the two intersection points of adjacent circles divides the coverages of the two beams. The center of the line is the point of greatest *adjacent-beam overlap*. The edge of a circle is its edge of coverage (EOC).

11.2.2   Pointing Error

Antennas on the spacecraft in orbit always have a small error in pointing. Beam pointing error is treated in calculations of EOC gain and carrier-to-interference ratio *C/I* by considering the beams to be enlarged by the pointing error. It is up to the customer or payload engineer, how many sigmas of pointing error to treat. The pointing errors must be considered for both the beam under consideration (for which the pointing error decreases *C*) and for the interfering beams. If the MBAs are reflector-based, the reflectors will have some kind of closed-loop pointing correction. If an interfering beam is on the same reflector, the beams move together, so the interfering beam never gets any closer to the beam of interest. If the interfering beam is on a different reflector, the pointing errors are independent; a worst-case analysis would assume the primary beam and the interfering beam moved in opposite directions, while a more realistic analysis would average over the two-dimensional pointing errors of the two beams. If one PA provides all the beams, the primary beam and interfering beams would move together.

11.2.3   Types

The relationship among MBA types is represented pictorially in Figure 11.4 with a Venn diagram. There are two partially overlapping categories, reflector-based MBAs and MBAs with PA. The PA-fed reflector MBA constitutes the intersection of the two categories. In this, there are two types: (1) the MFPB antenna, also known as **array-fed reflector**, *focal array-fed reflector*, *MBA with overlapping*

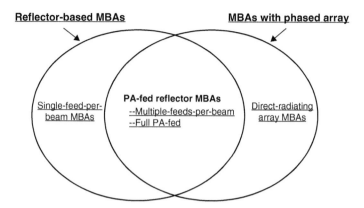

FIGURE 11.4   Venn diagram showing relationship among MBA types.

feed clusters, and *enhanced-feed MBA*; and (2) the full PA-fed reflector. One type of reflector-based MBA does not have a PA, the SFPB MBA. Similarly, one type of MBA has a PA but no reflector, the DRA.

11.2.4 Choosing Type

The choice of which type of MBA to use for a particular application has many drivers. Some of them are the following, where the first two are in general the main ones:

- Orbit
- Carrier frequency
- Beam size
- Capability and flexibility
- Accommodation on spacecraft
- Performance in terms of antenna efficiency, edge-of-coverage gain rolloff, and C/I
- Cost
- Mass.

The orbit determines whether the antenna is reflector-based or not. For a GEO, high gain is achieved by means of a reflector-based antenna. At low GHz frequencies, the reflector must be huge and there is only room for one, which requires a PA to form the feeds, either MFPB or full PA. At C-band and above, multiple reflectors with SFPB are possible. To a LEO the earth appears large, so high scanning capability is required, which can only be offered by a DRA.

The second main driver is frequency band. For GEOs, the most commonly used MBA types by frequency band are given in Table 11.1. For each band, the MBA types are listed in order of decreasing popularity. At L- and S-bands, the PA-fed solution is preferred; the reflector is an unfurlable mesh type since the reflector must be very large and a solid reflector cannot be so large. At the higher frequency

TABLE 11.1   Commonly Used MBAs for GEO, by Frequency (Amyotte et al., 2011, 2014)

Frequency Band	Commonly Used MBAs for GEO
L and S	MFPB, unfurlable mesh reflector, digital beam-forming on spacecraft, and on-ground beam-forming calculations
C	SFPB, 4 solid reflectors, diplexed Tx/Rx feed horns MFPB with 1 solid reflector for receive and 1 for transmit
Ku	For large number of small spot beams: SFPB, 4 non-oversized solid reflectors For up to 20 or 30 beams of diameter >1°: SFPB, 1 oversized shaped solid reflector
Ka/K	For large number of small spot beams: SFPB, 4 non-oversized solid reflectors For up to 20 or 30 beams of diameter >1°, possibly contoured: SFPB, 1 oversized shaped solid reflector

bands, SFPB is the most common because of its low cost, low mass, highest antenna efficiency, and simplicity; the reflector is solid (Amyotte et al., 2014). The term "oversized" in the table means that the reflector is larger than the normal size dictated by the carrier frequency.

For GEOs, some tradeoffs by frequency band are as follows. At L- and S-bands, digital beam-forming onboard provides flexibility in power allocation and beam size but is only possible if the number of array elements is relatively small, in contrast to analog beam-forming (Ramanujam and Fermelia, 2014). At all bands, the SFPB with four reflectors would offer the best performance, but only at C-band and above are this many reflectors feasible. The MFPB needs only two reflectors, one for receive and one for transmit at C-band and above, but needs more radiating elements than SFPB, more power amplifiers on transmit, a BFN, and a more complex payload architecture (Amyotte et al., 2014). At Ku- and Ka/K-bands, if only up to 20 or 30 beams need to be formed larger than 1° in diameter, a MFPB solution with one oversized reflector is feasible, where the reflector is at least twice the diameter of a non-oversized one (Amyotte et al., 2011, 2014). At Ka/K-band, the single oversized-reflector solution provides performance equal to the four-reflector SFPB but only for larger beams (Amyotte et al., 2011). In all SFPB cases with non-oversized reflector, using three reflectors works with degraded performance compared to four reflectors (Amyotte et al., 2011).

11.3 REFLECTOR FOR MBA OR CONTOURED BEAM AND CONFIGURATION OF FEEDS

On commercial GEOs currently (2020), almost all MBAs are reflector-based without subreflector and are offset-fed. The one exception appears to be Viasat-1, which is discussed in Chapter 18. Most MBA reflectors are either solid or unfurlable mesh. For the frequencies where both technologies are available, the choice between them is driven primarily by a difference in mass, available size, and deployment risk. Let us recall that, for non-oversized reflectors, the size of the reflector aperture, projected perpendicular to the paraboloid's axis, determines the size of spot beams, to first order.

When there are multiple reflectors, all beams of any one color are on the same reflector so that these beams have the same antenna-pointing error. This limits the worst self-interference *C/I* since the worst interfering beams are usually those of the same color as the beam of interest (Section 14.6.3). There can be more colors than reflectors, which increases *C/I*.

An unshaped reflector must be used for spot beams. A defocused reflector (Section 11.5.3) that has been shaped to have an irregular coverage area cannot also be used for spot beams. An exception is if the reflector is shaped for an irregular coverage area at one frequency and the spot beams are at a much lower frequency, perhaps by a factor of 10, so the shaping appears negligible at the lower frequency (Aliamus, 2020).

The feeds are in the near field of the reflector.

11.3.1   For C-, Ku-, and Ka/K-Bands

First, we describe the MBA reflector technology then the current reflector/feed configurations.

11.3.1.1   Reflector Technology   The original MBA reflectors for C-band and higher frequencies were solid, made of graphite. An example of a Ka-band shaped solid reflector has a diameter of 1.2 m. The reflecting surface is made of carbon fiber-reinforced polymer fabric. The reflector under-side has 1.3 cm-thick ribs that support the reflector shell. The shell is a honeycomb sandwich of fabric face-sheets and a core (Amyotte et al., 2006).

L3Harris, formerly Harris Corporation, introduced five alternatives made of mesh for C-band and above:

- In 2012, the fixed-mesh reflector. Its mass can be lower than that of a solid reflector by 40%. This reflector is qualified in sizes from 1.5 to 3.5 m and for frequencies from UHF to V-band. It has a compact stowed configuration (Harris Corp, 2018a).
- In 2016, the unfurlable Ka-band reflector, also known as the radial-rib (Harris Corp, 2019b). The Viasat-2 spacecraft, launched in 2017, carries four of these. Two each of the 5-m reflectors are deployed off the east and west panels. The reflector has eight ribs and unfurls like an umbrella (Farrar, 2016, 2018).
- In 2018, the high-compaction-ratio reflector antenna for small satellites. The name means that it folds into an unusually small volume for launch. Its size ranges from 1 to 5 m in diameter and operates at up to 40 GHz (Harris Corp, 2018b).

- In 2019, the Ka-band perimeter truss reflector. It has 50% lower mass than a solid reflector. Its size ranges from 3 to 22 m in diameter (Harris Corp, 2019a).
- In 2020, the Ka-band smallsat perimeter truss reflector. It also has 50% lower mass than a solid reflector. Its diameter is up to 4 m. Its design was optimized for high production rates (Harris Corp, 2020).

All of the reflectors for C-band and above have their radiating elements in the focal plane. An array of radiating elements in the focal plane is called a **focal array**, *focal-plane array*, or *focused array*.

11.3.1.2 *Eight Reflectors with SFPB*

The oldest on-orbit MBA implementation for C-band and above is eight solid reflectors mounted off the east and west panels of a GEO spacecraft. Each panel has four reflectors, of which two are for receive and two are for transmit. Each reflector is sized for its frequency band. The feeds are SFPB. Every receive reflector creates the beams for one color, as illustrated in Figure 11.5. The diagram is the same for the four transmit reflectors.

The reason for four reflectors for each direction is that, to obtain proper aperture illumination and the beam attributes that are usually required, the feeds have to have diameters about twice the required feed spacing in the focal plane for two adjacent beams (Burr, 2013).

The reflector's focal length divided by projected diameter, f/D (Section 3.4.1), is about 1 for these reflectors, allowing feed spacing of about 2–3 times the wavelength. If the feeds illuminate the reflector optimally with a taper of 10–15 dB, the antenna efficiency is about 78% and the sidelobe levels are about 25 dB down (Rao, 2015a).

The beams produced are Gaussian-shaped in their mainlobe (Amyotte et al., 2011; Rao, 2003).

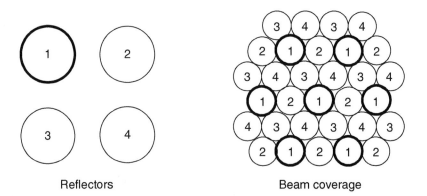

Reflectors   Beam coverage

FIGURE 11.5 Four-reflector SFPB MBA layout and beam coverage, where the numbers represent colors. (©1999 IEEE. Reprinted, with permission, from Rao (1999)).

11.3.1.3 *Four Reflectors with SFPB*   The most common MBA implementation today for C-band and above is four solid reflectors mounted off the east and west sides of GEO spacecraft, two on each side. The feeds are SFPB. All four reflectors are used for both receive and transmit. This implementation is in general the one that offers the combination of high performance, low risk, and reasonable price (Amyotte et al., 2014).

The feed horns have high-efficiency apertures (Section 11.4). Each reflector is sized for the lower frequency, the transmit frequency. A high-efficiency horn helps to improve peak and EOC gains in the transmit band, while it reduces the peak gain, broadens the beam, and improves the EOC gain in the receive band because of the reflector being oversized for the receive band. At the same time, if desired, the MBA can handle dual polarization on both receive and transmit if the feeds can (Rao, 2015a).

The transmit beams are Gaussian-shaped in their mainlobe (Amyotte et al., 2011; Rao, 2003).

11.3.1.4 *Three Reflectors with SFPB*   A slight variation on the four-reflector MBA is the three-reflector MBA with SFPB. The corresponding beams have three (or more) colors. The reflector configuration and the beam coverage are illustrated in Figure 11.6.

The beams of any one color are closer together than they are in the four-reflector configuration. The directivity is degraded by 0.5–1 dB and the self-interference C/I by 1–2 dB for receive and 3–4 dB for transmit (Amyotte et al., 2014). (See also Section 11.4.3 for further discussion.)

An example of a three-reflector Ku-band MBA is on Ciel 2, launched in 2008, which provides 54 circularly polarized beams (Lepeltier et al., 2012).

The transmit beams produced are Gaussian-shaped in their mainlobe (Rao, 2015a).

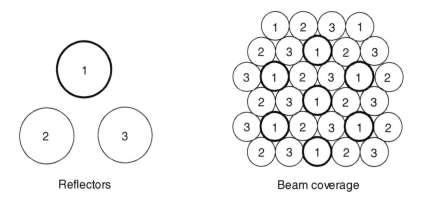

Reflectors   Beam coverage

FIGURE 11.6   Three-reflector SFPB MBA layout and beam coverage, where the numbers represent colors. (©1999 IEEE. Reprinted, with permission, from Rao (1999)).

11.3.1.5 Two Oversized Reflectors with SFPB A different SFPB MBA implementation for the C-, Ku-, and Ka/K-bands has two oversized reflectors, one for receive and one for transmit. The reflector and the feeds have at least double the diameter of those in the four-reflector antennas. Such a reflector may be more feasible than multiple reflectors when the spacecraft has other antennas to accommodate. The beams form the same hexagonal lattice as in the four-reflector implementation with four colors. Without reflector shaping, the self-interference at the triple crossover point would be higher than in the four-reflector implementation, so the reflector is shaped to widen the beams (Balling et al., 2006). The beams have nearly flat gain and can have improved EOC gain (Amyotte et al., 2011).

11.3.1.6 Different Two Oversized Reflectors with SFPB A different SFPB MBA implementation for the C-, Ku-, and Ka/K-bands has two oversized reflectors for either transmit or receive. An example of this is the two reflectors for K-band transmit on the Inmarsat-5 series of spacecraft (Section 20.6.7.2).

11.3.1.7 One Oversized Reflector with SFPB A similar MBA implementation has only one oversized reflector for simultaneous receive and transmit. The reflector has more than twice the diameter of a reflector in the four-reflector configuration. The reflector is shaped to widen and flatten the beams. This implementation is best suited for 20 or 30 beams at most. If implemented with a solid reflector of about 3-m diameter, at Ka-band the beam diameters would be greater than about 1° (Amyotte et al., 2011, 2014).

11.3.1.8 Two Reflectors with MFPB A compact design for Ka/K-band has just two reflectors, but only four radiating elements go to form each beam (usually MPFB uses at least seven elements (Rao, 2015a)). The antenna provides beams in four colors, as shown in Figure 11.7. Each reflector can handle both receive and transmit

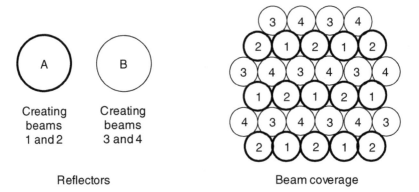

FIGURE 11.7 Two-reflector MFPB MBA layout and beam coverage, where the numbers represent colors.

and dual polarization. Each reflector provides the beams in one row, and consecutive beams have different polarizations (Lafond et al., 2014). The spacing of beams of the same color is the same as in the four-reflector case illustrated in Figure 11.5.

11.3.1.9 Two Oversized Reflectors with MFPB

Another MBA implementation has two oversized reflectors, one for receive and one for transmit. The feeds are MFPB. A long focal length is usually required, meaning an extra-long boom or a subreflector (Lepeltier et al., 2012). If the number of beams is large but the polarization plan is not favorable, the BFN is complex; if the number of beams is large but the coverage is sparse, the number of radiating elements is very large (Amyotte et al., 2014).

11.3.1.10 Contoured Beam Created by One Shaped Reflector, Full PA-Fed

There is an antenna configuration which does not provide multiple beams but does use a reflector and full PA to create a reconfigurable contoured beam. An example at Ku-band is one shaped reflector for transmit in dual polarization, fed by a full PA of 25 elements that probably employs scanning. The element coefficients are only phases, and the BFN is analog (Nadarassin et al., 2011). The coverage areas of the two polarizations can be different (Voisin et al., 2010).

11.3.2 For L- and S-Bands

First, we describe the MBA reflector technology then the current reflector/feed configurations.

11.3.2.1 Reflector Technology

The most common MBA reflector implementation for L- or S-band is one huge, offset-fed, unfurlable, mesh reflector that hangs off the spacecraft at the end of a long boom when deployed. Such a reflector, on Inmarsat's Alphasat, is illustrated in Figure 20.8. A very impressive and entertaining video of the reflector deployment of the ICO G1 (now known as EchoStar G1) spacecraft is available on YouTube (Space Systems Loral, 2012).

The reflector mesh is knitted metal wire. Tungsten and molybdenum are used for the wire because of their thermal stability. The wire is gold-plated. The reflector is tensioned with cables and/or nets which divide the surface into planar, usually triangular faces (Migliorelli et al., 2013).

Northrop Grumman introduced the first unfurlable mesh reflector on a commercial satellite at these frequencies. It was a 12.25-m L-band reflector on the Thuraya satellite, launched in 2000 (Thomson, 2002). An image of it is given in Figure 20.2. These reflectors are made of gold-plated molybdenum mesh, stretched across the convex side of a front net (Thomson, 2002).

L3Harris is the primary manufacturer of most of these reflectors. They come in size from one to 25 m (Harris Corp, 2018c). A photo of Harris's L-band, 22-m SkyTerra 1 reflector in the factory is given in Figure 20.28.

In general, the reflector size is driven by the beam size and the self-interference requirement (Rahmat-Samii et al., 2000).

11.3.2.2 One Reflector with MFPB The spacecraft which produce a large number of beams at L- and S-band are part of a mobile satellite service (MSS), to which Chapter 20 is dedicated. The current MBAs at these frequencies can handle both receive and transmit if desired and dual polarization if desired, as, for example, the EchoStar XXI (formerly known as TerreStar-2 and launched in 2017) antenna does.

The f/D is not large, about 0.5, since the boom length is limited (Gallinaro et al., 2012). A different source says about 0.6–0.7 (Rao, 2015a). An example is Inmarsat-4 with 0.53 (Guy, 2009). The first satellite of the Inmarsat-4 series was launched in 2005 (Section 20.3). A small f/D means that the outer beams in a MBA would have high phase errors across the beam, which are corrected by using PAs, either MFPB or the full PA (Gallinaro et al., 2012).

Some of the MSS antennas have focused arrays, and some have defocused arrays (Section 11.5.3).

11.3.2.3 Contoured Beams Created by One Reflector, Full PA-Fed Defocusing can be used to continually reconfigure a single contoured beam over a geosynchronous-satellite orbit (GSO) or highly elliptical orbit (HEO). The flexible reflector is fed by a small, full PA. This is further described in Section 11.5.3.

11.4 HORN AND FEED ASSEMBLY FOR GEO

This section addresses the horns and horn-based feed assemblies used for MBAs or PAs in GEO application (see Section 11.8 for further discussion on radiating elements specifically for PAs, which may be in GEO or non-GEO application). The horn is by far the most common radiating element in MBAs.

A horn with dielectric material in it is not suitable for multi-beam space antennas because the material induces electro-static discharge (Chan and Rao, 2008).

A feed assembly includes not just the horn but also the polarizer and orthomode transducer (OMT), if any, and the diplexer, if any (Section 3.8).

11.4.1 Potter Horn

The Potter horn was described in Section 3.7.1. It is a circular horn often used as a reference horn to compare other horns to. Its efficiency is about 74% (Rao, 2003). This horn was designed to have identical aperture distributions along the E- and H-planes (Section 4.4.1) to minimize cross-polarization. The result, though, is that the aperture distribution is highly tapered in both planes, so the horn cannot be highly efficient (Bhattacharyya and Goyette, 2004).

The Potter horn is optimal for use in the three-reflector SFPB scheme of three colors. The mainlobe of its primary beam falls off faster than a high-efficiency horn's does, and in a three-color SFPB scheme, the main source of self-interference

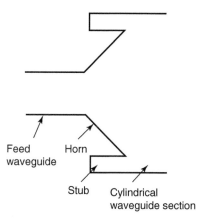

FIGURE 11.8   Diagram of Airbus's high-efficiency SCRIMP horn. (After Muhammad et al. (2011)).

comes from the sides of the mainlobes of interfering beams (Section 14.6.3). The Potter horn is not to be used in a four-color SFPB scheme (Rao, 2003).

11.4.2   SCRIMP Horn of Wolf and Sommer

The short, circular, ring-loaded horn with minimized cross-polarization (SCRIMP horn) of Airbus (formerly Astrium) is a circular multimode horn (Section 3.7.3) with an aperture efficiency of at least 77%, low cross-polarization, and low voltage standing-wave ratio (VSWR) over a bandwidth up to 19%. The horn is illustrated in Figure 11.8. The stub generates the higher-order modes necessary for high aperture efficiency, in contrast to the multiple locations required in the stepped and corrugated horns (Section 3.7.1). It has compact size (aperture diameter of 1.1–2 wavelengths), low mass, and low volume. It provides dual circular polarization (CP) and was initially developed for C-band (Wolf and Sommer, 1988). It currently flies on the Intelsat IX series of satellites as radiating elements of PAs to form C-band hemi and zone beams in reflector-based antennas. Its limited bandwidth means that transmit and receive antennas are separate (Hartmann et al., 2002a).

An L-band CP TX/RX module for up to 120 radiating elements that includes SCRIMP horns and 8×8 Butler matrices (Section 11.12) for distribution of the high-power amplification, was a further development of the C-band horns (Hartmann et al., 2002a).

11.4.3   High-Efficiency Horn in General

Guy Goyette was the first person to recognize the importance of high aperture efficiency in feed horns for reflector-based MBAs. Goyette worked then at Hughes Space and Communications (now part of Boeing Satellite Development Center).

Today, high-efficiency horns are used in almost all multibeam satellites and in some PAs (Bhattacharyya, 2019a).

In SFPB antennas where the feed horns are tightly packed together, low horn efficiency means higher spillover of the radiation into nearby horns, making for mutual coupling (Section 11.14) and higher sidelobes (Amyotte et al., 2006). A high-efficiency horn creates a narrower primary beam than a lower-efficiency horn does, so it can illuminate the reflector with a greater edge taper. This makes the secondary beam flatter in its mainlobe, lower in its peak gain, and lower in its sidelobes (Rao, 2003).

The following two figures show a comparison of three horns over the band 18–30 GHz: a high-efficiency horn, an ideal Potter horn, and a corrugated horn. A true Potter horn could not cover this bandwidth, so the comparison is with an ideal Potter horn. A corrugated horn is used in the comparison as another horn that can support the full band. In fact, a corrugated horn has thick walls that reduce the aperture by about 15% for the same horn spacing, so it is not a good choice for overlapping beams. The particular high-efficiency horn is not well described (Rao and Tang, 2006a; Rao et al., 2005), and current horns have better performance. But still the comparison is informative.

Figure 11.9 gives a comparison of the aperture efficiencies. The high-efficiency horn has an average aperture efficiency across the band of about 85%, the ideal Potter about 69%, and the corrugated about 53% (Rao and Tang, 2006a).

Figure 11.10 shows the greater edge taper of the high-efficiency horn compared to the two others.

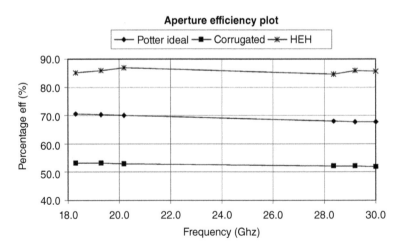

FIGURE 11.9 Aperture efficiency comparison of a high-efficiency horn, an ideal Potter, and a corrugated horn. (©2006 IEEE. Reprinted, with permission, from Rao and Tang (2006a)).

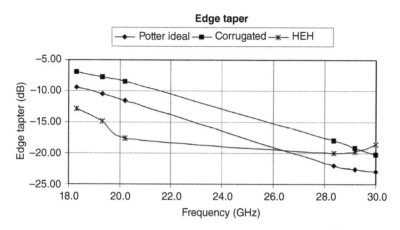

FIGURE 11.10   Reflector-edge illumination taper comparison of a high-efficiency horn, an ideal Potter, and a corrugated horn. (©2006 IEEE. Reprinted, with permission, from Rao and Tang (2006a)).

High efficiency requires generation of multiple propagating modes in order to yield nearly uniform field distributions in the aperture in the **E**- and **H**-planes. This contrasts with the field distributions in the Potter horn (Rao, 2003).

Only transverse electric (TE) modes (Section 4.4.5) should be excited on the horn aperture. The various modes should have particular relative amplitudes and phases determined by the aperture geometry (Bhattacharyya and Goyette, 2004).

The high-efficiency horn is optimal for a four-color SFPB scheme but is not to be used for a three-color one (Rao, 2003).

For a given multibeam coverage, the three-reflector SFPB scheme has smaller feed horns than a four-reflector design. From a gain perspective, high feed-aperture efficiency is even more critical for three reflectors than for four reflectors. However, for three reflectors, increasing horn efficiency degrades C/I (Amyotte, 2019c).

All of the high-efficiency horns described below have excellent return loss and cross-polarization.

11.4.4   High-Efficiency Horns of Bhattacharyya

What is believed to be the first patent for a high-efficiency horn was by Arun Bhattacharyya (2019a). It is for a smooth-wall horn with multiple sections of various flare angles and some small steps. (Inward steps were depicted but apparently the patent encompassed both inward and outward.). The step discontinuities generate the desired transverse magnetic (TM) modes in the correct proportions. Its aperture efficiency is 85% over a 10% bandwidth when there were two steps. Its rather large size makes it suitable for DRA applications but not reflector-based MBAs (Bhattacharyya and Goyette, 2004).

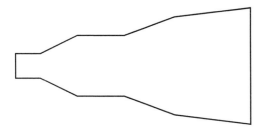

FIGURE 11.11 Diagram of Bhattacharyya's high-efficiency horn without steps (Bhattacharyya and Sor, 2007).

Such a circular horn, with one outward step and one inward step and diameter triple the wavelength, was shown to have aperture efficiency above 90% over a 15% bandwidth and above 85% over a 20% bandwidth. In its square form, the performance was just a little worse (Bhattacharyya and Goyette, 2004).

A second patent is for the first dual-band high-efficiency horn (Bhattacharyya, 2019a). It has multiple sections with outward steps and various flare angles. Each step generally excites one more TE mode. One of the steps additionally generates a TM mode, and a later step generates the same mode but with opposite phase, canceling out the mode. The horn generated the same TE modes in both frequency bands. At the 12- and 14-GHz bands, the efficiency was better than 85% (Bhattacharyya and Roper, 2003).

A third patent, believed to be for the first for a high-efficiency horn without steps (Bhattacharyya, 2019b), was designed for ease of manufacturing. It is illustrated in Figure 11.11. The horn has multiple sections of various flare angles. A change in the flare angle creates multiple waveguide modes. By adjustment of the section lengths, the undesired modes can be made to phase-cancel and the desired modes to constructively intensify. Straight sections allow selected unwanted modes to attenuate. A horn with four sections has better than 92% aperture efficiency over 10% bandwidth at 20 GHz (Bhattacharyya and Sor, 2007).

11.4.5 Dual-Band High-Efficiency Horns of Rao

Sudhakar Rao and others reported on a circular, high-efficiency horn developed for the receive band at 30 GHz and the transmit band at 20 GHz simultaneously (Rao et al., 2005; Rao and Tang, 2006a). The horn was not patented so details are not available. It was said to be thin-walled, smooth-walled with slope discontinuities (Rao and Tang, 2006a), and smooth-profiled (Chan and Rao, 2008). The feed horn has four slope discontinuities and is designed for a four-reflector SFPB scheme at Ka/K-bands. The reflectors were sized for the transmit frequency but shaped for the receive frequency in order to broaden and flatten the receive beams to match the size of the transmit beams. The horn could provide dual CP (Rao and Tang, 2006a). The transmit primary beam has a Gaussian shape in most of its mainlobe (Rao, 2003).

A feed assembly was developed for this Ka/K-band horn (Rao and Tang, 2006a).

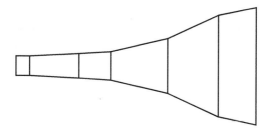

FIGURE 11.12   Diagram of Rao's 60/20-GHz high-efficiency feed horn (Rao et al., 2013).

Rao and others later patented a different horn that could simultaneously handle 60 and 20 GHz. The horn is circular and its internal surface has a piecewise linear slope, as illustrated in Figure 11.12. The horn shown has five slope discontinuities and no steps. At 20 GHz, the feed has TE modes and only a few TM modes at its aperture. At 60 GHz, it has more TM modes, to reduce the aperture efficiency and consequently widen the beam to match the beamwidth at 20 GHz (Rao et al., 2013).

11.4.6   Dual-Band High-Efficiency Horn of MDA

Eric Amyotte and others of MDA patented early a general high-efficiency circular horn that embodies discontinuities of surface and/or of surface slope to generate higher-order TE or even TM modes. The patent does not say how to design the horn's discontinuities and dimensions (Amyotte et al., 2002). The patent was meant to apply to horns operating on any number of bands simultaneously (Amyotte, 2019b).

MDA has used and continues to use many of these smooth-wall horns in MBAs and other antennas (Amyotte, 2019b). Most applications have been for dual band. For a single band, the aperture efficiency can be higher; an example is receive at 30 GHz where the efficiency is 94%. The designer will typically trade aperture efficiency for cross-polarization and bandwidth (Amyotte, 2019a).

MDA also makes feed assemblies with their horns.

11.4.7   Dual-Band, High-Efficiency, Spline-Profile Horns of Thales Alenia

Thales Alenia has developed a square, spline-profile, dual-band, high-efficiency multimode horn for use as radiating elements in a full PA-fed reflector-antenna. (A spline is smooth and is defined piecewise by polynomials.) An array of seven horns, used for mutual-coupling testing, is illustrated in Figure 11.13. Spline profiles with smooth wall are generally a good compromise among aperture efficiency, cross-polarization levels, and VSWR (Lepeltier et al., 2012). The spline-profile horn, in a circular form for a single band, was developed to allow a considerably shorter high-efficiency, low-cross-polarization horn (Deguchi et al., 2001).

The horn was developed for only transmit at Ku-band (10.7–12.75 GHz). It can handle dual linear polarization. These horns were developed as parts of feed assemblies.

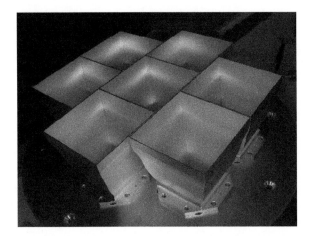

FIGURE 11.13   Septet of Thales Alenia's high-efficiency horns. (Used with permission of E. Vourch and Thales Alenia Space, from Nadarassin et al. (2011)).

Thales Alenia has since then developed the KISS horn, a circular, spline-profile, dual-band Ka/K-band feed horn for a lattice of at least 1.9 times the transmit wavelength. The horn has a low mass and is suited for use with a reflector of reduced focal length or for small beams. The aperture efficiency is unstated. The corresponding feed assembly has also been developed (Lepeltier et al., 2012). Thales Alenia proposed this equipment for use in a two-reflector, four-color MFPB scheme (Section 11.3.1.8) (Lafond et al., 2014).

11.5   LOCATION OF RADIATING ELEMENTS IN OFFSET-FED REFLECTOR MBA

For all the reflector-based antennas treated in this chapter, most if not all of the primary radiating elements are obviously not at the reflector's focal point. In one situation, one element is at the focal point. In another situation, none of the elements is at the focal point but all are closer to the reflector, in the defocused cases mentioned above in Section 11.3.2.2.

A number of questions arise about the locations of the radiating elements in such an antenna. We address these for the offset-fed antenna since it is more common than the center-fed:

- What is the nominal geometry of the offset-fed single reflector with multiple radiating elements?
- What happens with the radiating elements that are sideways displaced?
- What happens when the array of radiating elements is displaced along the offset axis?

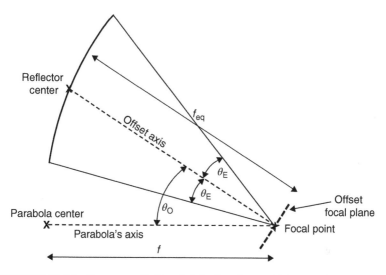

FIGURE 11.14 Geometry of offset-fed single reflector. (After Milligan (2005)).

11.5.1 Nominal Location of Radiating Elements

The geometry of an offset-fed single reflector (that is, without subreflector) is illustrated in Figure 11.14. The reflector is only part of a paraboloid. The focal point of the reflector is the same as that of the full paraboloid. Imagine the angle that the reflector subtends from the focal point. The **offset axis** goes through the focal point and divides the subtended angle in half, yielding the angle θ_E on either side. We call the point on the reflector that the offset axis goes through the *reflector center*. Another defining angle is the **offset angle** θ_O at the focal point between the offset axis and the full paraboloid's axis. From the focal distance f and the angles, the distance f_{eq} from the focal point to the reflector center can be calculated. The **offset focal plane,** or just **focal plane**, is the plane that contains the focal point and is perpendicular to the offset axis.

When there are multiple primary radiating elements, they are arranged so that their phase centers lie in a plane. That plane is nominally the focal plane. The antenna is then said to have a focused array or a focal-plane array. The optimal location of the multitude of phase centers is not exactly a plane but a curved surface, where the focal plane is tangent to the curved surface (Mittra et al., 1979).

The focal plane is in the near field of the reflector.

11.5.2 Laterally Displaced Radiating Elements

The radiating elements on the focal plane but not at the center are said to be **laterally displaced** from the focal point. The question is what this does to the beams.

The first-order analysis of how the displacement distance relates to the beam angular separation from the center beam is based on the assumption that all the feeds point to the center of the reflector. Figure 11.15 shows the geometry for a

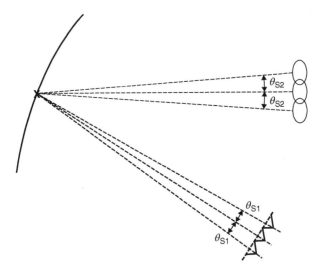

FIGURE 11.15 Geometry of laterally displaced radiating elements and their beams. (After Palacin and Deplancq (2012)).

linear array of three horns. The horns have their apertures in the focal plane. Three dashed lines go from the aperture centers to the reflector center. The rays reflect off the dish and spread out. At the reflector center, the angle between consecutive rays is θ_{S1} and upon reflection the angle is θ_{S2}. Based on f_{eq}, the lateral displacements, and the angles of Figure 11.14, it is possible to compute θ_{S2}. Now, the ratio of θ_{S2} to θ_{S1} is the *beam deviation factor* (Palacin and Deplancq, 2012):

$$\text{BDF} = \frac{\theta_{S2}}{\theta_{S1}} < 1$$

It is equally possible, and more to the point, to start with the desired θ_{S2} and compute the necessary θ_{S1} and lateral displacement. The source (Ingerson and Wong, 1974) contains a plot of BDF as a function of both θ_{O} and θ_{E} for a feed edge taper of $-10\,\text{dB}$. This same reference shows how the offset BDF is calculated from the non-offset BDF. It also points out that for a given θ_{E}, as θ_{O} increases from zero the BDF gets smaller. So offset-feeding the reflector has the benefit of allowing the radiating elements to have larger lateral displacements than a center-fed antenna does.

Now, pointing the radiating elements toward the reflector center is not practical from a manufacturing point of view. So this analysis is not sufficient for antenna design but is sufficient for a first-order understanding of the situation.

11.5.3 Axially Displaced Radiating Elements

When the phase centers of the primary radiating elements do not lie in the offset focal plane but in a different plane parallel to the offset focal plane, the antenna has

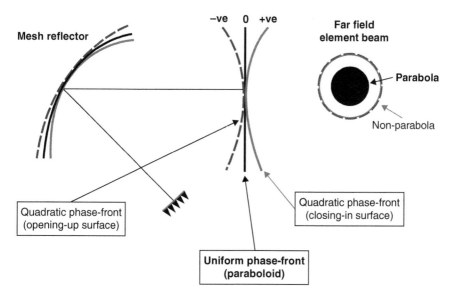

FIGURE 11.16   Non-focused reflector concept for reconfigurable contoured-beam antenna. (©2006 IEEE. Reprinted, with permission, from Rao et al. (2006b)).

a **defocused or non-focused array**. The axial displacement causes the element beams to become broader and to have lower peak gain, but the array requires fewer elements (Rao, 2015b). The effect is approximately equal for defocusing toward or away from the reflector by the same distance, but toward the reflector is advantageous because then the antenna is smaller (Ingerson and Wong, 1974). Defocusing, preferably by several wavelengths from the reflector's focal plane, provides better reconfigurability of the beams (Ramanujam et al., 1999). When the PA is not in the focal plane, a wavefront arriving from the reflector at the PA has a parabolic phase across the PA. The phase shifters of the radiating elements are used to correct this. Inmarsat-4 provides an example of a defocused PA that has to reconfigure throughout the day (Section 20.3).

Defocusing can be used to continually reconfigure a single contoured beam over a GSO or HEO orbit by means of a flexible reflector. A small, full PA feeds the reflector. The PA is fixed in location but the reflector is flexed so that sometimes its focal point is closer and sometimes farther than the PA. The defocusing broadens the element beams significantly (Rao et al., 2006b). The concept is illustrated in Figure 11.16.

11.6   SINGLE-FEED-PER-BEAM MBA

It is important to know that the pattern of the feeds is directly mapped to the pattern of the beams on the ground (Amyotte and Camelo, 2012).

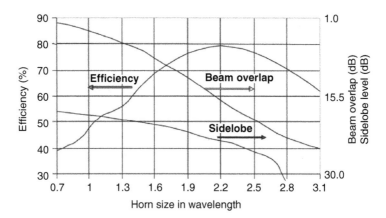

FIGURE 11.17   Antenna performance as a function of feed-horn size, for one offset-fed SFPB antenna. (© 2015 IEEE. Reprinted, with permission, from Rao (2015a)).

For some SFPB MBAs, there exist typical values of gain at adjacent-beam overlap and triple-beam overlap points (defined in Section 11.2.1). If a reflector is used only for receive or only for transmit, the gain at the adjacent-beam overlap point is about 3 dB down from the peak gain of the beam and the gain at the triple-beam overlap point is about 4 dB down (Rao, 1999). It is typical nowadays for the same reflectors to be used for both receive and transmit when both directions are required. The reflector is sized for the lower, transmit frequency and the 3 dB and the 4 dB apply to those beams. The beams for the higher frequency are flattened by the feed horns to widen them, so these numbers do not apply.

Here are some examples of SFPB antennas on satellites. An example of eight reflectors for separate receive and transmit is the Ka/K-band Anik F2 spacecraft launched in 2004 (Amyotte et al., 2006). An example of four reflectors for both receive and transmit is the Ka/K-band Ka-Sat (Amyotte et al., 2010). An example of three reflectors is the Ku-band Ciel 2 spacecraft (Lepeltier et al., 2012). It provides high-definition TV (HDTV), so most beams are only transmit.

There is a reason why SFPB works for C-band and above but does not work for L- and S-bands. When there are four reflectors as at C-band and above, the feed size can be rather large in units of wavelength, which can provide the best performance. Figure 11.17 shows various antenna performance parameters as a function of horn diameter, for the SFPB case where the feeds are horns. A horn diameter value of 2.2 λ provides optimal antenna efficiency, 80%, and sidelobe level of about 23 dB down from peak gain (Rao, 2015a). (It is not stated what values are assumed for the rest of the components of antenna efficiency.) For a four-color spot-beam scheme, the worst contributors to self-interference are most likely the sidelobes of the nearest same-color beams. In this case, the sidelobe level in dB down from beam peak gain must be at least as great as the required C/I plus about 8 dB. For example, if the C/I required is 15 dB, the sidelobe must be at least 23 dB down. Thus, for this case, we have confirmation again that horn size should be roughly 2.2 λ. A larger horn would cause some decrease in antenna efficiency. For more than four colors, higher sidelobes are tolerable so the horns could be smaller with some loss in antenna efficiency. However, if instead of four reflectors only one non-oversized one can be used as at L- and S-bands, then the horn diameter must be halved. But this puts the antenna efficiency into the area of only 52%. So at L- and S-bands, the only feasible option is MFPB or full PA-fed (Rao, 2015a).

11.7 PHASED ARRAY INTRODUCTION

A PA is an array of radiating elements all of the same kind, laid out in a regular grid, that forms beams by means of a BFN. The BFN forms each primary beam on receive by combining the signals of radiating elements in either particular phase or particular delay relationships. Both are known as PAs. On transmit, the BFN does something similar. Some BFNs also provide different amplification to the various element signals.

The PA has a few forms. One form is the MFPB, which uses a cluster of only a few elements to make each beam, and each cluster overlaps with nearby clusters for nearby beams. The PA where all the elements go to form each beam is used either to form the feeds for a reflector-based antenna or as a DRA without reflector. In commercial satellites, the DRA is used only in non-GEO satellites currently (2020). PAs have been flown on commercial spacecraft only in L- and S-bands, as far as the author knows, except for the simple PAs for OneWeb user links (Section 11.10.3).

11.7.1 Phasing Principle

The basic principle of a PA is illustrated in Figure 11.18a, which shows a wavefront arriving in the direction of the arrow at the last two elements in a linear array (the linear array being used here only for explanatory purposes). The direction represents an angle θ off boresight. The dashed line represents the wavefront hitting the middle of the second-leftmost array element a short time ago, and the solid line represents the wavefront hitting the middle of the leftmost element now. From θ and the distance between array elements, the distance d that the wavefront traveled

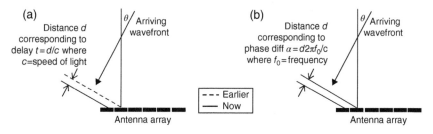

FIGURE 11.18 Principle of a receive linear PA in terms of (a) time delay and (b) phase difference.

between the two times can be computed. The delay t can be computed from d and the speed of light. Note that this delay is independent of frequency. If the payload hardware delays the earlier signal received at the second-leftmost element by t, it is now lined up in time with the signal being received at the leftmost element, so the two can be added. The signals being received at the other elements can be correspondingly delayed and added in. Similarly, a signal could be transmitted in the direction θ off boresight by delaying the individual signal copies going to the various array elements.

A similar effect can be achieved by the use of phases instead of delays, as illustrated in Figure 11.18b. We think about the two different wavefronts that are hitting the middle of the same two array elements, now. The one hitting the second-leftmost element is phase α ahead of the one hitting the leftmost element. Phase difference α can be computed from d, the speed of light, and the center frequency f_0 of the signal being received. Note that α is not quite correct for any other frequency from f_0. Shifting the signal from the second-leftmost element by $-\alpha$ puts it in phase with the signal hitting the leftmost element, so the two can be added in phase. The signals being received at the other elements can be correspondingly phase-shifted and added in. An antenna array that uses phases instead of delays is inherently band-limited, but most PAs do use phases.

11.7.2   When Element Signals Have Equal Amplitudes

We consider a PA whose element signals have equal amplitudes, then in the next section address unequal amplitudes.

A PA can be thought of as the spatial convolution of one radiating element's aperture with an array of equal-size delta-functions. Recall that the antenna pattern of a general antenna is related to the two-dimensional transform of the antenna's aperture. For small off-boresight angles, the antenna pattern nearly equals the transform. Thus, the antenna pattern of the PA is related to the product of one radiating element's antenna pattern and the antenna pattern of the array of delta-functions. The antenna pattern of one element is wide and equals zero for boresight angles beyond ±90°. The antenna pattern of the array of delta-functions has a mainlobe and sidelobes. The product of the antenna patterns is zero beyond ±90°. Inside of ±90° is the *visible region*.

11.7.2.1   *Broadside Antenna Pattern*   The PA creates a *broadside* antenna pattern when the beam's boresight is perpendicular to the plane of the array. All the elements have the same phase. Let us start with an array whose elements are in a rectangular grid, for simplicity. Then the antenna is the spatial convolution of one radiating element's aperture, a linear array of evenly spaced delta-functions, and another such linear array orthogonal to the first one.

Each linear array has a two-dimensional Fourier transform that repeats along the orthogonal direction to the array's orientation. For element spacing $d \le \lambda/2$, the repeat is not in the visible region. As d increases beyond $\lambda/2$, the antenna gain starts to rise at large off-boresight angles. For $d = \lambda$, two more gain peaks appear at the off-boresight angles of +90° and −90°. As d increases beyond λ, the secondary peaks become closer to the main peak. If the separation gets large enough, a second pair of gain peaks appears beyond the first pair, and so on. These other peaks in the pattern, besides the desired one, are called **grating lobes** (Chang, 1989).

The antenna pattern of the spatial convolution of the linear arrays of delta-functions is related to the product of the antenna patterns of the linear arrays. A rectangular array produces a rectangular grating lobe pattern (Milligan, 2005).

The last factor in the array's antenna pattern, the antenna pattern of one radiating element, has a wide mainlobe. The multiplication of this pattern with the rest decreases the size of the grating lobes. In particular, for $d = \lambda$ the pair of secondary peaks at +90° and −90° are taken to zero.

Grating lobes are not necessarily harmful. The off-boresight angle at which they occur and/or their magnitude may result in them not interfering into other beams of the payload or into other satellite systems.

The case of a non-rectangular PA is more complicated. An example of antenna pattern is given in (Milligan, 2005). This case is fully addressed in Chang (1989).

The widths of the antenna pattern's mainlobe and sidelobes depend on the array dimensions, but the gain depends on the composite area of the radiating apertures.

11.7.2.2 *Scanned Antenna Pattern*

A scanned beam of a PA is one with boresight not perpendicular to the array's plane. The PA achieves scanning by altering the phases of the radiating elements in a regular progression, as already discussed.

Scanned beams have different properties from non-scanned beams. For one thing, the antenna gain is less since gain is proportional to the area of the array projected on a plane perpendicular to the beam direction. For another, CP is no longer circular but elliptic—just imagine a circle being viewed off axis. Linear polarizations are no longer orthogonal; the exception is if the scan is about an axis parallel to one of the linear polarizations. A third effect of scanning, for arrays with phases, is that for all the signal frequencies except the nominal frequency, the beam has squint, that is, points not quite in the intended direction, as illustrated in Figure 11.19.

FIGURE 11.19 Wavefronts added in phase, for various frequencies, in a PA scanned beam: (a) at nominal frequency, (b) at higher frequency, (c) at lower frequency.

We now look at the antenna pattern. Let us assume that the PA is rectangular (the non-rectangular case is too complicated for this book but is addressed in Chang (1989)). When the array is scanned along one of the array axes, the scanned array is the spatial convolution of one radiating element's tilted aperture with one linear array of evenly spaced delta-functions, with another such linear array that is orthogonal to the first one. The spacing of one linear array of delta functions has shrunk compared to when the array is unscanned.

The tilted radiating element is now oblong, so its pattern is oblong and oriented orthogonally.

The linear array that has closer delta-functions has grating lobes that are now farther apart.

The gain pattern of the spatial convolution of the two linear arrays of delta-functions now has peaks in a rectangular pattern with different spacing in the two dimensions, if the apparent element spacing is more than λ in both directions. One peak is the desired one and the others are grating lobes. The problem is that one or more grating lobes can be within the field of view of the PA. For example, for $d = \lambda$ and 30° scanning angle, there is a grating lobe at −30°; for 10° scanning angle, at −56°; for 60° scanning angle, at −8° (Chang, 1989).

There is a criterion for the maximum element spacing for a given scan angle to make the grating lobe occur at −90°. This leads to the requirement that element spacing be not much more than $\lambda/2$ for wide scan angles (Chang, 1989). We find this partially confirmed by the choices of DRA element spacing made by Iridium Next and Globalstar-2. These LEO satellites need large fields of view: ±60° for Iridium Next and ±54° for Globalstar-2. The element spacings used on the two programs are between 0.55 and 0.8λ, depending on the satellite altitude (Lafond et al., 2014).

11.7.3   When Element Signals Have Unequal Amplitudes

Just as a reflector antenna has lower illumination near its edges to decrease the discontinuity at the edges, in order to create lower side lobes in the antenna pattern a PA is made to have lower gain in its outer elements. The case where all element signals have equal amplitude has the greatest boresight gain but high sidelobes. There is another extreme pattern of element amplitudes which gives no sidelobes but a wide mainlobe. There is a family of optimal distributions of amplitudes for a PA. For a given sidelobe level, it provides the narrowest mainlobe; for a given mainlobe width, it provides the lowest sidelobes (Kraus and Marhefka, 2003).

11.8   RADIATING ELEMENT OF PHASED ARRAY

Table 11.2 shows current types of radiating elements in PAs and their applications and properties. The elements either are in currently on-orbit satellites or have been developed by a major spacecraft manufacturer. The examples are at all bands. The three types of application are MFPB reflector, full PA feeding reflector, and DRA.

TABLE 11.2 Types of Radiating Elements in Phased Array and Their Properties

Radiating Element	Frequency Band	Antenna Application	Program/Manufacturer	TX/RX or Just One Direction	Polarization	Reference
Horn	S-band	MFPB, making two contoured beams	ABS 4 (formerly known as MBSat)	One	Single LP	Smith et al. (2004)
	Ku-band	MFPB	Boeing	One	Dual LP	Bhattacharyya and Goyette (2004)
	Ku-band	Full PA-fed	Thales Alenia	Transmit	Dual LP	Nadarassin et al. (2011)
	Ka/K-band	MFPB	Thales Alenia	Both	Dual CP?	Lafond et al. (2014)
Cup-horn	L-band	MFPB	MDA	Both	Dual CP	Richard et al. (2007)
Helix	L-band	MFPB	Inmarsat-3	Both	Single CP	Perrott and Griffin (1991), Angeletti and Lisi (2008b)
	L-band	MFPB	Inmarsat-4 and Alphasat	Both	Single CP	Guy et al. (2003), Dallaire et al. (2009)
Patch radiator	L-band	DRA	Iridium	Both but not simultaneously	RHCP	Schuss et al. (1999)
	L-band	DRA	Globalstar-2 transmit	One	LHCP	Croq et al. (2009), Huynh (2016)
	S-band	DRA	Globalstar-2 receive	One	LHCP	Croq et al. (2009), Huynh (2016)
Patch-excited cup	L-band	MFPB	Thuraya	Both	LHCP	Roederer (2005), Ruag Space (2016)
	L-band	MFPB	SkyTerra	Both	Dual CP	LightSquared (2006), Ruag Space (2016)
	S-band	MFPB	EchoStar XXI (formerly known as TerreStar-2)	Both	LHCP	Semler et al. (2010), Simon (2007)

11.8.1 Horn

For a given horn size and a desired PA gain, high-efficiency horns significantly reduce the number of horns needed in the PA. For a scan angle up to some limit determined by the horn size, a PA of high-efficiency horns has higher gain than if it had non-high-efficiency horns (Bhattacharyya and Goyette, 2004).

A high-efficiency horn that can be made in either circular or square cross-section is the Bhattacharyya horn (Section 11.4.4). Its large aperture diameter in wavelengths is suitable for application in a MFPB, a full PA-fed reflector, or a DRA (Bhattacharyya and Goyette, 2004). The square version is used on the ABS 4 (Mobisat) spacecraft, formerly known as MBSat 1 (Smith et al., 2004).

Thales Alenia has built a Ka/K-band, dual-polarization horn for use in MFPB antennas. It is a further development of the KISS horn (Section 11.4.7).

Space Engineering has qualified an engineering qualification model of a horn for use in the Ku-band receive DRA of Eutelsat's Quantum GEO satellite, to be launched in 2021. The horn is a spline horn. The feed chain covers the entire Ku-band frequency range for the fixed satellite service (Chapter 17). The active DRA, located on the earth deck, will have about 100 radiating elements and create eight beams. The beams will be electronically steerable and the BFN will control both gain and phase. It will be capable of creating a null toward interference (Pascale et al., 2019; Airbus, 2019).

11.8.2 Cup-Horn

The circular cup-horn was developed for L-band MSS satellites by MDA. It provides RX/TX in dual CP for MFPB application. A breadboard septet of them is shown in Figure 11.20. The radiating element consists of circular waveguide that radiates through a cup and splash plates. It has a gain of 10–11 dBi (Richard et al., 2007).

FIGURE 11.20 Cup-horn septet breadboard. (Used with permission of MDA Corporation, from Richard et al. (2007)).

FIGURE 11.21   Part of L-band helix array of Inmarsat-4 in assembly. (©2006 IEEE. Reprinted, with permission, from Stirland and Brain (2006)).

11.8.3   Helix

The helix is in use in L- and S-bands for the MFPB reflector and the DRA. It can provide simultaneous receive and transmit because it is inherently wideband, but only in one polarization.

The helix is used on the Inmarsat-4 spacecraft and Alphasat in an L-band RX/TX full-PA-fed application. It has a gain between 10 and 12 dBi (Guy et al., 2003; Dallaire et al., 2009). A photo of a corner of the Inmarsat-4 array in assembly is given in Figure 11.21. The helix is set into a cup to isolate the element from the surrounding helices in the array (Guy et al., 2003).

11.8.4   Patch or Ring-Slot Radiator

A **patch or ring-slot radiator** is one of the most useful radiating elements for PAs because of its wide scan capability. It consists of a flat piece of metal backed by a larger flat piece of metal which forms the ground plane. Dielectric lies between. An annular slot, cut out of the top piece of metal, radiates. A cavity underneath the slot has metal walls that stretch between the two pieces of metal. The boresight gain is 5.4 dBi (Bhattacharyya et al., 2013).

Thales Alenia developed three DRAs for the LEO Globalstar-2, as shown in Figure 11.22. Two DRAs, conical with the tops cut off, consisting of patch radiators are for L-band left-hand circular polarization (LHCP) receive. One consists of nine panels of 16 radiators each and the other consists of six panels of 8 radiators each.

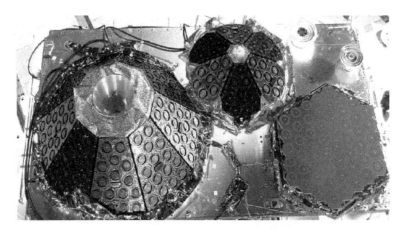

FIGURE 11.22   Globalstar-2 DRAs on earth deck, transmit DRAs on left and center and receive DRA on right. (Used with permission of Thales Alenia Space France, from Croq et al. (2009)).

Each panel creates one fixed beam. A flat DRA with patch-in-cavity radiators is for S-band LHCP transmit. Each circular patch lies in a very shallow cavity. This DRA forms 16 fixed beams (Croq et al., 2009).

11.8.5   Patch-Excited Cup

The patch-excited cup was developed by Saab Ericsson Space (now Ruag Space) and deployed on Thuraya and SkyTerra in L-band RX/TX MFPB applications at L-band (Roederer, 2005; LightSquared, 2006). Thuraya employs only LHCP while SkyTerra employs dual CP. A drawing of the patch-excited cup is given in Figure 11.23. It consists of a cylindrical cup and a patch tower comprising two stacked circular exciter disks and a reflector disc. The gain is 10–12 dBi (Ruag Space, 2016).

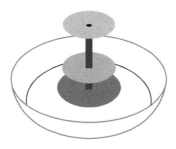

FIGURE 11.23   L-band patch-excited cup. (After Ruag Space (2016)).

EchoStar XXI uses something similar at S-band. The element is described as a "stacked, cupped, microstrip disk element fed by coaxial probes" (Simon, 2007). The application is a MFPB feeding one reflector, RX/TX, LHCP (Semler et al., 2010).

11.9   BEAM-FORMING NETWORK

The **BFN** is the part of a PA that creates the beams. On receive, a BFN collects the signals from the various array elements, phases or delays them correctly so as to form a beam, and adds them. Amplfication or the element signals is optional. On transmit, it divides a beam signal, phases or delays the resultant signals appropriately, and feeds them to the array elements. The beam can be contoured by applying different phases or delays, other than those required to point, and possibly different gains to different array elements. The BFN coefficients may have fixed values for unchanging beams or be resettable for reconfigurable beams. Figure 11.24 shows the concept of a BFN for two beam signals and eight radiating elements used for both beams.

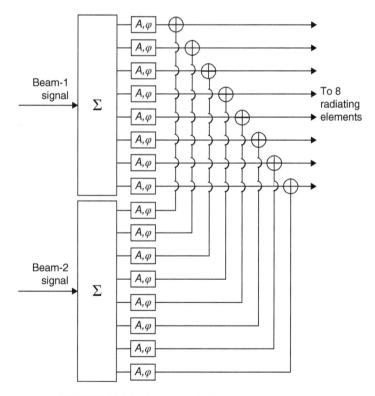

FIGURE 11.24   Example of BFN concept on transmit.

All the signal paths must closely match in gain and phase and must maintain this matching, that is, track, over temperature, power, and life. The high-power amplifiers (HPAs) alone must also exhibit this. When the paths do not match, there are residual amounts of the other signals in the outputs. It is relatively easy to maintain tracking at L- and S-bands and difficult at Ku-band and above.

A PA can create a contoured beam through just phase control or both gain and phase control of the BFN coefficients (Shishlov et al., 2016). A PA can form a null toward an interferer through just phase control, just gain control, or both together (Hejres, 2004).

For the beam-forming architecture, there are a few choices to be made:

- Phases or delays
- Onboard or on the ground
- Analog or digital
- Relation between BFN and amplification (addressed in Section 11.12.1).

11.9.1 Phases or Delays

The general idea of beam-forming on receive was seen in Figure 11.18, in terms of either phase-shifting or delaying the elements' signals appropriately to create each receive beam. Phasing is less costly than delaying. A passive or semiactive BFN (Section 11.12.1) with phases can only be used at low GHz frequencies where a wavelength is long, but an active BFN with phases has been qualified at Ku-band for use on Quantum (Pascale et al., 2019). The elements' signals can conceptually be used any number of times to create any number of beams. Transmit beams are formed in an inverse fashion to receive beams.

Forming the beams directly as just described is called *element-space processing*. Other types of processing are beam-space and frequency-domain (Godara, 1997) but they seem to be less common.

11.9.2 Onboard or on the Ground

The BFN can be either on the spacecraft, where it is called **onboard beam-forming (OBBF)**, or on the ground, where it is called **ground-based beam-forming (GBBF)**. GBBF has the advantages of (1) requiring less onboard computing capacity, prime power, volume, and mass and (2) having the potential for modified frequency plan and number of beams (Angeletti et al., 2010). OBBF has the advantage of lower required feeder-links bandwidth. For both GBBF and OBBF, the computation of the coefficients is done on the ground.

The first GBBF came into use in NASA's Tracking and Data Relay Satellite System (TDRSS) upon on-orbit arrival of TDRS-1 in 1984. (The second-generation satellites performed OBBF but the third and current generation again uses GBBF.) The S-band multiple-access antenna has 30 radiating elements. On receive, the satellite puts the elementary signals onto separate carriers and transmits them to the ground. The signals are downconverted to near baseband and sampled. Originally, there were five BFNs, each receiving a full copy of the 30 downconverted elementary signals. Each BFN combines the signals with proper phases and gains to create one beam. Currently, 32 beams can be formed. Transmit is similar (Zillig et al., 1998; Hogie et al., 2015).

Two commercial examples of GBBF are MSS spacecraft (Chapter 20). One is EchoStar XXI, whose BFN can form more than 500 beams (Epstein et al., 2010). OBBF would be prohibitive. Another example is SkyTerra with thousands of beams (Koduru et al., 2011).

Examples of OBBF are Thuraya, Inmarsat-4, and Alphasat.

11.9.3   Onboard Analog or Digital

OBBF has several varieties.

Fixed beams are sometimes created with **analog beam-forming** using fixed phase shifters or fixed delay elements in the BFN. The BFN is often implemented with a power divider/combiner and an analog phase shifter at each element for each beam. This is cumbersome for a large number of beams (Bailleul, 2016). Analog typically creates only about 32 beams or fewer, while digital can create hundreds. Analog is generally considered efficient when the number of beams is smaller than the number of radiating elements; otherwise, digital is more efficient (Thomas et al., 2015). High-level transmit beam-forming (Section 11.12.1) always uses analog beam-forming (Amyotte, 2020b).

Reconfigurable beams are created by either electronic beam-steering or digital beam-forming commanded from the ground.

In **electronic beam-steering**, typically there are phase shifters and sometimes gain controls, and these are electronically commandable to either a fixed, limited set of values or to a continuous set of values within the available resolution. An example of a commandable phase shifter is one at Ku-band developed by Thales Alenia for high-power applications (Nadarassin et al., 2011). Another, developed by Airbus, will be used on Eutelsat's Quantum satellite (Airbus, 2019).

In **digital beam-forming**, all computations are performed digitally. The payload is considered to be a processing payload (Section 10.2.6). Only phase shifts have been used in the beam-forming, not delays as would be required for broadband operation, to the author's knowledge. (Some GBBF use delays.) On receive, each element's signal is converted from analog to digital and then copied to the processing paths for the beams, so there is no noise increase as in analog dividing. There is no need for frequency down-conversion for frequencies up through Ku-band; that is, on receive the elements' radio-frequency (RF) signals can be sampled directly. The enabler for this is the dramatic improvement in analog-to-digital converters (ADCs) in the last few years (Bailleul, 2016). Of course, older on-orbit spacecraft may have downconverters before the ADCs.

A proposed digital BFN for one beam on receive is illustrated in Figure 11.25. The low-noise amplifier (LNA) on each radiating element is not part of the BFN.

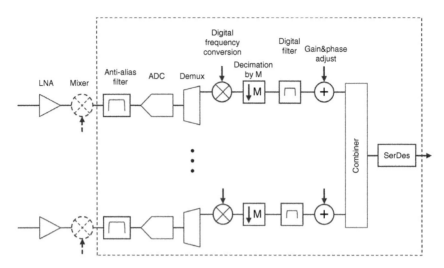

FIGURE 11.25   Block diagram of proposed digital BFN for one receive beam. (©2016 IEEE. Reprinted, with permission, from Bailleul (2016)).

Neither is the downconverter, needed for Ka/K-band. The anti-aliasing filter before the ADC is for removing the frequencies beyond half the sampling rate. The digital frequency conversion takes the signal to complex baseband. Decimation removes most of the samples, only keeping enough to capture the signal bandwidth. The digital filter then isolates the desired subband. Gain and phase adjustments are performed. Time delays would then be applied (not shown) for a broadband signal. Finally, the complex-baseband signals from all the elements are combined and read off the ASIC by the "SerDes" element (Bailleul, 2016).

Digital beam-forming is currently performed at the following frequencies:

- At L-band for PAs feeding reflectors on GEOs (Section 10.3.1).
- At Ku-band for DRAs on the LEO Starlink satellites and at Ka/K-band perhaps in 2020 (Section 19.6).
- At Ka/K-band for DRAs on SES-17, to be launched in 2020.
- At Ka/K-band for DRA on Eutelsat's GEO Quantum satellite.

11.10   APPLICATIONS OF PHASED ARRAY

11.10.1   Multiple-Feed-Per-Beam MBA

The MFPB schemes are all reflector-based by definition. They all use multiple radiating elements in a cluster for each beam, and most elements are shared among clusters. The PA does not scan. Figure 11.26 shows the three most common schemes, with respectively three, four, and seven radiating elements used per beam,

(a) (b) (c)

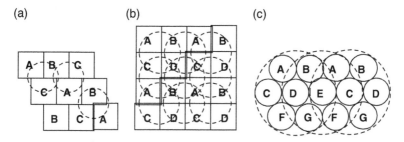

FIGURE 11.26 MFPB radiating-element usage schemes: (a) 3 elements per beam, (b) 4 elements per beam, and (c) 7 elements per beam. (©1990 IEEE. Reprinted, with permission, from Roederer and Sabbadini (1990)).

with seven being most common (Lafond et al., 2014). Nineteen have also been used (Amyotte et al., 2011). A BFN forms the beams. Just as for SFPB, the pattern of the feeds on the array is directly mapped to the pattern of the beams on the ground. Effectively, the multiple radiating elements per beam makes for a larger feed than if only one radiating element was used. The cluster size is roughly the same as in a SFPB implementation with three or four reflectors (Amyotte and Camelo, 2012).

Two examples of three-color beams are the Ciel 2 Ku-band payload (Lepeltier et al., 2012) and the Eutelsat W2A (now 10A) S-band payload as designed (Arcidiacono et al., 2010).

There is a tradeoff between MFPB and SFPB. MFPB has the advantage that the aperture area of the MFPB cluster for one beam is greater than is possible for a horn in an SFPB scheme, so the MFPB antenna can create closer beams with one reflector (Lepeltier et al., 2012). MFPB has the drawback that there must be more radiating elements than beams by a factor of 2 to 8 (Amyotte et al., 2011, 2014). This makes it unsuitable for hundreds of beams. However, for a small number of beams, the need for only one or two reflectors for receive and transmit gives MFPB an advantage. The EOC directivity suffers by less than 1 dB (Tomura et al., 2016).

One example of MFPB is the Inmarsat-4 series of spacecraft. Sixteen to 20 elements are used for each of the 193 spot beams, more for the regional beams, and pretty much all of them for the global beam (Vilaça, 2020). Another example is the Express-AMU1/Eutelsat 36C satellite, launched at the end of 2015. It offers 18 user spot beams at Ku-band, with one reflector for receive and one for transmit (Fenech et al., 2013).

MFPB lends itself to amplification by multiport amplifiers (MPAs) and multi-matrix amplifiers (MMAs), which sometimes have advantages over other forms of amplification. This topic is explained and discussed below in Section 11.12.

11.10.2 Full PA-Fed Reflector Antenna

A reflector fed by a full PA is an uncommon application of a PA. It creates a few contoured beams.

One difficulty of using a PA to form the feeds for a reflector is that the scan capability is limited. A 1° scan of the full antenna corresponds to an $x°$ scan of the array, where x is the ratio of the reflector aperture diameter to the array diameter (Jamnejad, 1992). So decreased gain and inferior CP of the scanned array come into play even at small antenna scan angles. This is much less of an issue at GEO than at LEO since the GEO field of view is much smaller.

A satellite manufacturer has developed a small Ku-band focal array of horns for use with a shaped reflector, that can create different contoured receive beams in each of the two linear polarizations (Voisin et al., 2010; Nadarassin et al., 2011).

An example of what is probably a full PA intended to form six contoured beams is the Eutelsat 10A (formerly W2A) S-band PA of 21 dual-CP elements and one reflector (Lepeltier et al., 2007; Arcidiacono et al., 2010).

11.10.3 Direct-Radiating Array

The only commercial usage today (2020) of the DRA is on LEOs. However, SES plans to put DRAs on the next generation of O3b satellites, which have medium earth orbits (MEOs) (Todd, 2017), and the Eutelsat Quantum satellite will have a Ku-band receive DRA.

The DRA is used on satellites in non-GEO orbits because of the wide scanning angles the antenna must provide. A DRA provides better off-boresight performance (that is, lower scan loss) than a PA-fed reflector would (Lisi, 2000).

A DRA on a non-GEO could potentially see a large temperature swing over the orbital period. The spacecraft will go in and out of sun, and the amount of transmit traffic can have high variation, meaning the amount of waste-heat generated by the HPAs could have high variation. To maintain phase tracking (Section 11.9), thermal control must be optimized so that (1) the simultaneously active hardware chains are kept within a mutual range of typically 10° and (2) the temperature range of each unit is minimized (Lafond et al., 2014).

The three DRAs of a Globalstar-2 satellite were shown in Figure 11.22. The field of view is ±54° (compared to ±8.7° from GEO). Iridium Next has an L-band DRA as its main mission antenna that performs both receive and transmit. It is shown in Figure 11.27. Its field of view is ±60° (Lafond et al., 2014) (Chapter 19 provides further discussion of these satellite systems, including beam patterns).

A Ku-band user antenna of a LEO OneWeb satellite has been called a "Venetian blind antenna." A qualification model is shown in Figure 11.28. The antenna is made up of 16 linear PAs, each of 32 radiating elements, called "blades." The blades are mutually rotated so as to aim at the correct areas on the earth. The pattern of each array is fixed and highly elliptical, oriented 90° from the line of the array. Together the 16 patterns aligned almost vertically, one above the other on the earth along the satellite's ground track as discussed in Section 19.6.3. The array is passive. Its radiating elements produce dual CP, one sense for transmit and the other sense for receive. The compact BFN is a power divider/combiner unit made as a single piece without flanges or fasteners. It is directly coupled to the array.

FIGURE 11.27   Iridium Next's DRA on spacecraft. (Used with permission of Iridium Satellite LLC and Thales Alenia Space, from Iridium (2019)).

FIGURE 11.28   OneWeb user antenna qualification model (Glâtre et al. 2019). (Used with permission of MDA).

The antenna design provides beams with more flexibility and higher capacity than multiple beams of other shapes would have (Glâtre et al., 2019; Amyotte, 2020a).

A 2007 study by a satellite manufacturer showed that for a GEO at Ku- or Ka/K-band, a DRA can provide better performance than a MFPB reflector antenna, with less digital processing. The drawback is that many more radiating elements are required. They proposed a scheme of overlapping clusters of 4×4-element "tiles" to produce 100 spot beams from 592 elements (Stirland and Craig, 2007).

11.11   BEAM-HOPPING

Beam-hopping allows rapid adaptation of a payload to changing traffic conditions. In beam-hopping, there are many **cells** on the ground, each of which is provided with a beam for some fraction of the time. At any given time, a subset of the cells is illuminated and that subset changes over rapidly to a different subset, and so on. A cell with a lot of traffic will be provided a beam more often than a cell with little traffic.

Here are some of the parameters of a beam-hopped system (Panthi et al., 2017):

- Period or window: repetition period of the hopping scheme
- Slot: smallest unit of time that can be allocated to a cell
- Revisit time: the length of time that a cell waits for its next slot assignment
- Time plan: transmission time-plan that allocates beams to cells.

Work has been done on how to optimize the beam assignments and dwell times (Angeletti et al., 2006; Anzalchi et al., 2009; Alberti et al., 2010). The second and third papers of these three appear to have been the result of two contracts on the same European Space Agency (ESA) study. The two papers assume the same static downlink traffic pattern with 70 cells, one fourth of the cells illuminated at once, full bandwidth and single carrier on each beam, and DVB-S2 adaptive coding and modulation (ACM) (Section 13.4.1). Beam-hopping performed better than a traditional payload but roughly as well as a flexible payload (Section 10.1.5). For both the beam-hopping payload and the flexible payload, the second paper used a four-reflector SFPB scheme while the third paper used a reflector-based PA.

When the number of cells is in the hundreds, a reflector-based PA is a necessity.

Spaceway 3, a GEO satellite with a regenerative payload launched in 2007, performs beam-hopping at Ka/K-band. There are 112 uplink cells and 784 downlink cells, where each uplink cell contains seven downlink cells. Uplink slots, downlink slots, and bandwidth can all be assigned dynamically based on demand. The designers felt that ACM was too complex and risky with limited overall payload mass and DC power, so instead the system employs user-terminal uplink power control and payload downlink power control (Fang, 2011).

Eutelsat's Quantum satellite, to be launched in 2021, will have eight beams, each of which will be able to hop among several tens of different configurations (Airbus, 2019).

11.12 AMPLIFICATION OF PHASED ARRAY

11.12.1 Passive, Active, or Semiactive Array

A PA may be active, passive, or something in between, where these terms tell us about the position of the amplification with respect to the radiating elements and, if any, the onboard BFN. Actually, the terms are a mixed bag:

- A **passive PA** has the LNAs or HPAs separate from the elements and not one-to-one with the elements. The BFN may employ either OBBF or GBBF.

- An **active PA** has a LNA and/or HPA right at each radiating element. A transmit/receive (T/R) module at each element is one way of accomplishing this (see below). The BFN may employ either OBBF or GBBF.
- A **semiactive PA** has OBBF. It is almost always a transmit PA. The amplification is directly associated with neither the beam signals nor the element signals.

One example of every PA type on transmit is shown in Figure 11.29, all with OBBF. Some of the surrounding hardware is also shown, namely frequency converter and HPA bank. The BFN itself may function at baseband, near-baseband, or RF. The converter to the transmit frequency may be before or after the BFN. The example marked "new" derives from the material of Section 11.9.3, specifically Figure 11.25 transformed into transmitting instead of receiving (Bailleul, 2016). Abbreviations are D/C for frequency downconverter, U/C for frequency upconverter, and SSPA for solid-state power amplifier.

A potential advantage of an active array is that redundant amplifiers may not be needed because the array itself has redundancy. With some high probability, no more than a small portion of amplifiers will fail over the spacecraft lifetime. The array can be designed so that there are enough elements at the beginning, so that no matter which array elements are lost the resultant amplification as well as the antenna pattern will be satisfactory with high probability.

On transmit, an active array usually has a SSPA at each element, whereas if traveling-wave tube amplifiers (TWTAs) are used the array is usually passive. Amplification by SSPAs at each element can be more linear than makes sense with TWTAs, and higher power per beam can be obtained (Maral and Bousquet, 2002).

A disadvantage of an active array is the greater number of units and the accompanying mass and mechanical issues.

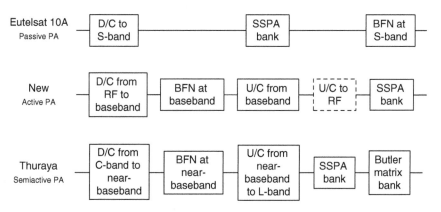

FIGURE 11.29 Architecture examples of passive, active, and semiactive BFNs on transmit.

Another terminology for the amplification placement is in terms of the signal level into the BFN. This distinction helps illuminate the choice among passive, active, or semiactive (Demers et al., 2017):

- *Low-level BFN*, where the amplifiers are one-to-one with the radiating elements. The corresponding PA is active or semiactive.
- *High-level BFN*, where the amplifiers are one-to-one with the beam signals. The BFN is between the HPAs and the PA on transmit or between the LNAs and the BFN on receive. The corresponding PA is passive.

Both low- and high-level BFNs have their pros and cons. The low-level BFN's position relative to amplification means the BFN does not degrade the payload noise figure. The signals through the BFN, though, have to be aligned in gain and phase over temperature and life. This is possible for low-GHz frequencies because the wavelength is so long. It has also been done at Ku-band. For K- and Ka-bands, this alignment is currently impossible. The low-level BFN offers more flexibility and higher performance. However, on transmit the HPAs have to be operated backed off because of the large number of signals for each radiating element, which means lower DC power efficiency. Because the amplifiers are one-to-one with the elements, for MFPB there are about two to eight times the number of amplifiers as needed for a high-level BFN.

For the high-level BFN, on transmit some RF output power is lost to the BFN and on receive the BFN increases the payload noise figure. However, the signals do not need to be aligned in gain and phase through the BFN. Amplification is on the user or ground station signals themselves and not on the signals of the radiating elements, so if it happens that the downlink signals are single-carrier, they can potentially by amplified by HPAs at saturation, increasing the DC power efficiency of the amplification. Also, far fewer amplifiers are needed because the amplifiers are one-to-one with the beams (Amyotte et al., 2011). The BFN is analog. High-level beam-forming on transmit requires hardware that can take high power. An example of a high-level BFN for a full PA-fed reflector antenna is the Ku-band BFN developed for transmit (Nadarassin et al., 2011).

The semiactive PA has a divided BFN with the amplification between the major part of the BFN and the Butler matrix (Section 11.12.3). On transmit, the smaller part of the BFN (the Butler matrix) is after the HPAs, which leads to less post-HPA loss than with the high-level BFN (passive array). However, the signals still must be aligned through the output Butler matrix. At the time of writing, there is no suitable 8×8 Butler matrix at any higher frequency than S-band.

11.12.2 Transmit/Receive Module

A **T/R module** is a unit integrated with each radiating element to make the array active on both transmit and receive. It includes a LNA, a preamplifier, and a SSPA. It connects to the BFN. It has a switch for choosing between receive and transmit at any one time. It is smaller and lighter than other means of achieving an active array.

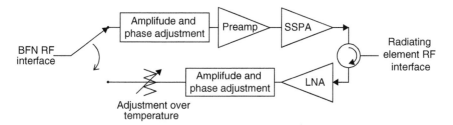

FIGURE 11.30   Simplified diagram of T/R module. (After Mannocchi et al. (2013)).

The LEO Iridium Next satellites have T/R modules at L-band. A simplified block diagram of one is given in Figure 11.30. The module's SSPA is adjustable to operating conditions, optimized for both single- and multi-carrier operations. The module employs *time-division duplex*, that is, it switches between transmit and receive, with 90-ms timeframes and 66% duty cycle for transmit (Mannocchi et al., 2013).

11.12.3   Multiport Amplifier

The **MPA** is the basic means of providing high-power amplification in a semiactive PA.

To understand how a MPA works, let us first look at a simple example of RF hybrid use. In Figure 11.31, we see two hybrids of the usual type. Each is a 0°/180° hybrid since the top signal on the left sees 0° phase shift as it goes through the hybrid and the bottom signal on the left sees 180° phase shift. The point is that with two of these hybrids, one after the other, the original signals input to the first hybrid are re-created. Note that we could have placed matched HPAs on the two lines that carry the mixed signals, without destroying the re-creation.

The next-larger such structure can be made, but this time we put HPAs in it. This is a 4×4 MPA, shown in Figure 11.32. Let us consider the box on the left defined by a dashed line. It is a 4×4 **Butler matrix** or **hybrid matrix**. Four 0°/90° hybrids properly connected make up the Butler matrix. The first Butler matrix divides each of the four input signals into four equal parts and imparts different phases to the parts. Then, it creates four output signals, each the sum of one of every divided, phase-shifted input signal. Each of the first Butler matrix's output signals has the same fraction of each input signal. We connect the four output signals to four matched HPAs. Those four outputs are connected to a second 4×4 Butler matrix.

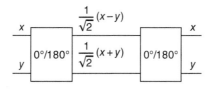

FIGURE 11.31   Signal restoration by two connected hybrids.

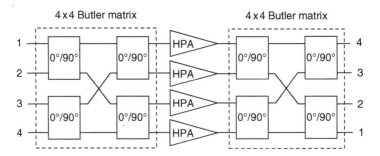

FIGURE 11.32   Four-by-four MPA, with input and output ports labeled. (After Mallet et al. (2006)).

It can be either identical to or the reverse of the first Butler matrix (Egami and Kawai, 1987). The four outputs of the second Butler matrix are the original signals, amplified.

The power relationships among the original signals are maintained at the MPA output. The benefits of the MPA are that it is possible for one signal to receive all the amplification, several signals to divide it equally, or any condition in between; the amplification is linear; and the signals may even be in the same frequency band.

MPAs generally have 2^n inputs and outputs, for some positive integer n. The largest on-orbit MPA known to the author is 8×8. Bigger than that and the output losses are too large. A satellite manufacturer has developed an 8×8 Ku-band MPA with 2 GHz bandwidth (Esteban et al., 2015).

The HPAs in a MPA could fail during the payload lifetime, which must be considered. The failure of one HPA in an 8×8 MPA causes a signal loss of 1.2 dB and degrades the isolation to 18 dB. The numbers are more benign for a larger MPA and less benign for a smaller one (Egami and Kawai, 1987). If such numbers are acceptable, the HPAs need not be redundant. The other way to deal with possible failures is to have redundant HPAs. The Eurostar generic architecture for geomobile satellite payloads employs both methods. It has redundancy rings as well as a soft response to additional failures. When the payload loses an entire active feed chain, the ground computes new beam weights and uploads them and the onboard digital processor implements them. This architecture has been applied to Inmarsat-4 (Mallison and Robson, 2001).

11.12.4   Multiport Amplifier Applications

The MPA finds its application in a transmit PA (Amyotte and Camelo, 2012). The BFN is high-level with attendant advantages and disadvantages. The outputs of a MPA are connected one-to-one with radiating elements. The amplification architectures are flexible in that they can accommodate fluctuating and unbalanced beam traffic. A potential disadvantage is that the HPAs must always be backed off to the multi-carrier region (Mallet et al., 2006).

The presence of MPAs in a payload is transparent to the rest of the payload, aside from the fact of the amplification.

The original MPA concept was apparently to have as many inputs and outputs as radiating elements (Egami and Kawai, 1987). A further development was to use several smaller MPAs instead, which is more practical (Spring and Moody, 1990). Sometimes the amplification of radiating elements with multiple MPAs is called *multimatrix amplification* but we reserve this term for something else (next subsection).

The MPA has two transmit applications. One application is to amplify multiple feed-signals of a SFPB MBA. It is especially useful if the signal to each feed is multi-carrier, since the HPAs would have to be backed off, anyway. The more common application is for a PA-fed reflector MBA. The PA is then semiactive. These two applications are shown in Figure 11.33.

A case of the second application is multiple MPAs amplifying the radiating-element signals in a MFPB antenna, where each beam is composed of one element-signal from every MPA (Spring et al., 1990). Figure 11.34 shows an example of this, with three radiating elements per beam. The radiating elements are divided into three groups, A, B, and C. One element from each group is used to form a beam. The example shows the signal paths for two beams, the one that uses elements A1, B1, and C1, and the one that uses elements A2, B2, and C2. Another beam could use, for example, A3, B1, and C2.

An example of this architecture is the Globalstar-2 S-band transmitter for user links (Darbandi et al., 2008).

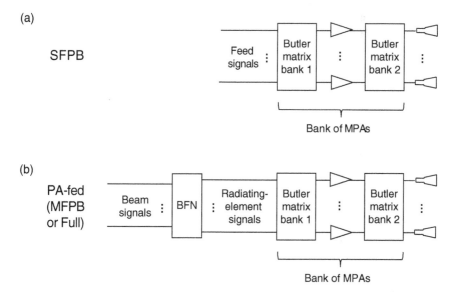

FIGURE 11.33 Two reflector-based transmit MBA architectures with MPAs: (a) SFPB and (b) PA-fed.

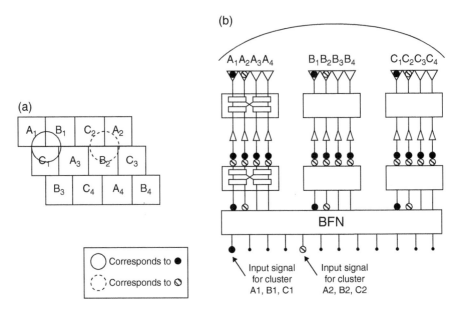

FIGURE 11.34   Example of MPAs with MFPB antenna: (a) 12 radiating elements with two three-element clusters circled; (b) BFN, MPAs, and radiating elements. (©1990 IEEE. Reprinted, with permission, from Roederer and Sabbadini (1990)).

At least at K-band and above, the MPA approach may not be the best for a MFPB antenna. The MPA approach suffers from high post-TWTA losses. A study of a transmit antenna with seven 8×8 MPAs, where the TWTAs have 10:8 redundancy, showed that an active PA with flexible TWTAs (Section 7.4.10) without redundancy performed adequately even in the case of TWTA failures and required much less hardware (Coromina and Polegre, 2004). Another study showed that for fixed beam-power allocations at Ka/K-band, the traditional application of non-flexible TWTAs out-performs MPAs (Angeletti et al., 2008a).

11.12.5 Multimatrix Amplifier and Applications

The MMA finds its application in a transmit PA (Amyotte and Camelo, 2012). The BFN is high-level.

A simplification of the MPA amplification for onboard BFN puts the first bank of Butler matrices into the BFN as shown in Figure 11.35 (Roederer and Sabbadini, 1990). It is easy to incorporate the first bank of Butler matrices into a digital BFN. The HPA bank and the second bank of Butler matrices are then called a **MMA** (although this term is also sometimes applied to the architecture with multiple MPAs). A PA amplified with one or more MMAs is called a *multimatrix antenna* (Roederer et al., 1996). This amplification architecture clearly achieves the same thing as the MPA architecture.

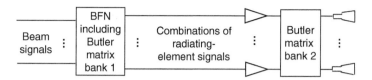

FIGURE 11.35   Transmit MBA architecture with MMAs.

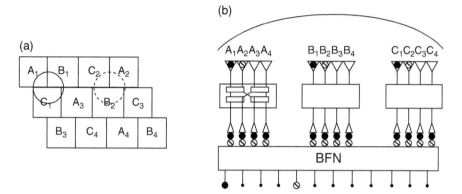

FIGURE 11.36   Example of MMAs with MFPB antenna: (a) 12 radiating elements with two three-element clusters circled; (b) BFN, MMAs, and radiating elements. (After Roederer and Sabbadini (1990)).

The MFPB architecture example of the previous subsection, but with the MPAs replaced by three 4×4 MMAs, is now a little simplified, as shown in Figure 11.36.

One example of the MMA architecture for L-band transmit amplification was on the Inmarsat-3 series of satellites. For the beams around the edge of the earth, there were five MMAs and each beam was formed from five radiating elements (Perrott and Griffin, 1991).

11.13   PHASED ARRAY POINTING ERROR

The beams formed by a PA have pointing error, as with other antennas. Part of the pointing error is due to the imperfect attitude control of the spacecraft (Section 2.2.5). PAs are fixed to the spacecraft body so cannot autotrack to reduce the error (Section 3.10). All that can be done is array recalibration or a design that does not need recalibration.

A GEO's active PA needs to have its radiating elements periodically recalibrated. There are three common ways for the ground station to determine the gain and phase of every element. On orbit, it is not possible to turn off all radiating elements except the one being tested, so the methods require some mathematics. One way involves forming a beam normally and then again with the test element's phase shifted by 180°; another is similar but involves four phases of the test element; a third way, for a very large array, requires an onboard digital processor and involves shifting the phases of special groups of elements by 180° (Bhattacharyya, 2006).

Since a non-GEO is continually moving with regard to a ground station, the above measurements cannot be taken. Thus, any calibration could only be done by the spacecraft. The following two LEO satellites do away with recalibration, by design.

The Iridium Next PA is used for both receive and transmit and is active. The T/R modules are designed not to need any adjustment in the phase of the SSPA or LNA over operating conditions. The LNA's gain is kept stable over temperature by means of a Thermopad®, which is a passive temperature-variable attenuator (Smiths Interconnect, 2020). The SSPA's gain is kept stable by means of a PIN diode attenuator (Mannocchi et al., 2013).

Globalstar-2 has taken a different approach. The transmit arrays are passive and the receive arrays are semiactive (Section 11.12.1). Thermal control for the receive radiating elements is achieved thanks to a Kapton sheet (Croq et al., 2009). Kapton is best known for its thermal management properties and is used as the outer layer of the spacecraft thermal blanket (American Durafilm, 2014).

11.14   MUTUAL COUPLING IN RADIATING-ELEMENT CLUSTER

No matter what kind of radiating elements are in a cluster, the electrical fields of the radiating elements interact with each other, that is, there is **mutual coupling** among the elements. The antenna pattern of elements in the middle of a cluster is in general different from that of the elements on the edge of the cluster (Collin, 1985).

A pictorial example of this effect is shown in Figures 11.37 and 11.38, but it is not meant to be representative of every situation. The radiating element is a horn

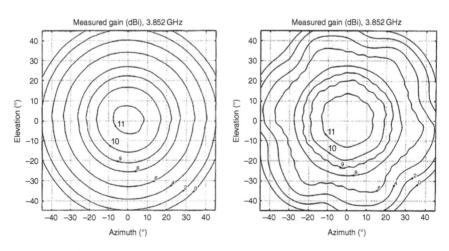

FIGURE 11.37   Gain patterns of circular horn: (a) alone, (b) in middle of 19-element cluster. (©1997 IEEE. Reprinted, with permission, from Schennum and Skiver (1997)).

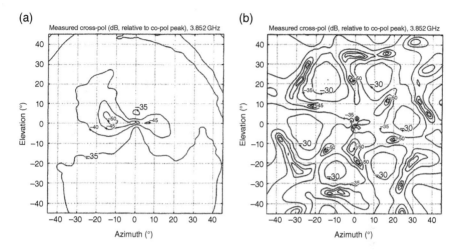

FIGURE 11.38 Cross-polarization patterns of circular horn: (a) alone, (b) in middle of 19-element cluster. (©1997 IEEE. Reprinted, with permission, from Schennum and Skiver (1997)).

with aperture diameter of 1.36 times the wavelength. The first figure shows the gain patterns both for when the horn is alone and for when it is in the middle of a 19-element cluster of horns. The second figure shows what happens to the cross-polarization pattern (Schennum and Skiver, 1997).

The higher the aperture efficiency, the less mutual coupling there is (Amyotte et al., 2006).

When the aperture is large enough in wavelengths, the mutual coupling is small (Rao, 2015a). For aperture diameter greater than three times the wavelength as in some Bhattacharyya horns (Section 11.4.4), the mutual coupling is insignificant (Bhattacharyya and Goyette, 2004).

11.15 TESTING MBA

The testing of any MBA with many beams faces potentially the same problems: (1) the huge size of the reflector if at L- or S-band, (2) dozens or hundreds of beams to test, and (3) possibly the active nature of the PA. Traditional testing methods are often infeasible.

An interesting point, before we describe test methods, is that the far-field distance for a PA-fed offset reflector may be much greater than the usual $2D^2/\lambda$ for a single-beam antenna (DiFonzo and English, 1974).

Let us recall that the antenna is tested at three levels on a spacecraft program (Section 3.12). We address here the second and third: the radiation properties of the antenna alone, then the payload beam properties. The radiation properties test is part of antenna assembly, integration, and test (AIT), and the payload-level test is part of the spacecraft system-level test. For a single-beam antenna, antenna-level testing is usually done in a near-field range (NFR) and spacecraft-level testing usually in a compact antenna test range (CATR). For a MBA, both levels of test may be different from this. Different solutions have been found, depending on the MBA, the available test ranges, and the perceived risk of not testing all the beams.

A test range for AIT of active PAs was reported in 1999. It can test in either transmit or receive mode. Since reciprocity does not exist with an active antenna, if the antenna performs both transmit and receive it must be tested for both. This test range can control thousands of T/R modules at once (Sauerman, 1999).

The AIT of the first-generation Globalstar's active DRAs was done the following way. After an antenna was all assembled except for the array itself, hard-line testing was done to check the integration of the BFN and the active modules. Then, the array was integrated with the rest and environmental testing was done. Last, NFR testing was performed (Lisi, 2000).

For Inmarsat-4, it seems that the antenna was not tested in AIT at the spacecraft manufacturer because the deployed reflector was too large. At system test, the payload including antennas was tested in a NFR. Only a limited set of patterns were measured (Stirland and Brain, 2006).

At the spacecraft level, for measuring beam properties, the standard since 1989 has been to use Airbus's compensated compact range (CCR), model CCR 75/60 (Migl et al., 2017). Since 2005, a larger version by Airbus has been available, the model CCR 120/100, for a 5-m antenna on a very large spacecraft (Section 3.12) (Hartmann et al., 2005b). This would include spacecraft with large MBAs. To test a large number of spot beams, typically more than 100, a fast controller was developed by Airbus for their CCRs. It can switch fast among frequencies, beam ports, and test polarizations (Hartmann et al., 2008). The measurement speed of the Airbus CCRs was increased even more later by faster switches. A 96-beam antenna could be completely measured in 4.5–5 hours. However, even without the faster switches, the actual data acquisition time is only about 20% of the total test-campaign time. Other components of that time are to install and uninstall the antenna, the RF set-up, the optical alignment, the data evaluation, and waiting for the customer go-ahead (Migl et al., 2013).

Another technique for increasing the measurement speed of a compensated CATR is to replace the single probe with an array of fast probes. This also reduces the required mechanical movement of the antenna (Durand et al., 2012).

REFERENCES

Airbus (2019). Airbus presents ground-breaking technology for Eutelsat Quantum. Press release. Nov 21. On www.airbus.com/newsroom/press-releases/en/2019/11/airbus-presents-groundbreaking-technology-for-eutelsat-quantum.html. Accessed Oct. 14, 2020.

Alberti X, Cebrian JM, Del Bianco A, Katona Z, Lei J, Vazquez-Castro MA, Zanus A, Gilbert L, and Alagha N (2010). System capacity optimization in time and frequency for multibeam multi-media satellite systems. *Adv Satellite Multimedia Systems Conf and Signal Processing for Space Communications Workshop;* Sep. 13–15.

Aliamus M of Space Systems Loral (2020). Private communication, Jul. 25.

American Durafilm (2014). Kapton film and the aerospace industry. Blog post. Nov 11. On market.americandurafilm.com/blog/kapton-film-and-the-aerospace-industry. Accessed Apr. 1, 2020.

Amyotte E (2019a). Private communication, Jun.14.

Amyotte E (2019b). Private communication, Jun. 17.

Amyotte E (2019c). Private communication, Nov. 18.

Amyotte E (2020a). Private communication, Mar. 28.

Amyotte E (2020b). Private communication, Oct. 12.

Amyotte E, Gimersky M, Liang A, Mok C, and Pokuls R, inventors; EMS Technologies Canada, assignee (2002). High performance multimode horn. U.S. patent 6,396,453 B2. May 28.

Amyotte E, Demers Y, Martins-Camelo L, Brand Y, Liang A, Uher J, Carrier G, and Langevin J-P (2006). High performance communications and tracking multi-beam antennas. *European Conf on Antennas and Propagation*; Nov. 6–10.

Amyotte E, Demers Y, Hildebrand L, Forest M, Riendeau S, Sierra-Garcia S, and Uher J (2010). Recent developments in Ka-band satellite antennas for broadband communications. *AIAA International Communications Satellite Systems Conf;* Aug. 30–Sep. 2.

Amyotte E, Demers Y, Dupessey V, Forest M, Hildebrand L, Liang A, Riel M, and Sierra-Garcia S (2011). A summary of recent developments in satellite antennas at MDA. *European Conf on Antennas and Propagation;* Apr. 11–15.

Amyotte E and Camelo LM (2012). Chapter 12, Antennas for satellite communications. In *Space Antenna Handbook* (Imbriale WA, Gao S and Boccia L, editors), 466–510. West Sussex, UK: John Wiley and Sons.

Amyotte E, Demers Y, Hildebrand L, Richard S, and Mousseau S (2014). A review of multi-beam antenna solutions and their applications. *European Conf on Antennas and Propagation*; Apr. 6–11.

Angeletti P, Prim DF, and Rinaldo R (2006). Beam hopping in multi-beam broadband satellite systems: system performance and payload architecture analysis. *AIAA International Communications Satellite Systems Conf*; Jun. 11–14.

Angeletti P, Colzi E, D'Addio S, Balague RO, Aloisio M, Casini E, and Coromina F (2008a). Performance assessment of output sections of satellite flexible payloads. *AIAA International Communications Satellite Systems Conf*; Jun. 10–12.

Angeletti P and Lisi M (2008b). A survey of multiport power amplifiers applications for flexible satellite antennas and payloads. *Proc, Ka and Broadband Communications Conf*; Sep. 24–26.

Angeletti P, Alagha N, and D'Addio S (2010). Space/ground beamforming techniques for satellite communications. *IEEE Antennas and Propagation Society International Symposium;* Jul. 11–17.

Anzalchi J, Couchman A, Gabellini P, Gallinaro G, D'Agristina L, Alagha N, and Angeletti P (2009). Beam hopping in multi-beam broadband satellite systems: system simulatoin and performance comparison with non-hopped systems. *IET and AIAA International Communications Satellite Systems Conf*; Jun. 1–4.

Arcidiacono A, Finocchiaro D, Grazzini S, and Pulvirenti O (2010). Perspectives on mobile satellite services in S-band. *Adv. Satellite Multimedia Systems Conf and Signal Processing for Space Communications Workshop*; Sep. 13–15.

Bailleul PK (2016). A new era in elemental digital beamforming for spaceborne communications phased arrays. *Proceedings of the IEEE*; 104 (3); 623–632.

Balling P, Mangenot C, and Roederer AG (2006). Shaped single-feed-per-beam multibeam reflector antenna. *Proc, European Conf on Antennas and Propagation*; Nov. 6–10.

Bhattacharyya AK (2006). *Phased Array Antennas: Floquet Analysis, Synthesis, BFNs, and Active Array Systems*. Hoboken (NJ): John Wiley & Sons.

Bhattacharyya AK (2019a). Private communication, Jun. 7.

Bhattacharyya AK (2019b). Private communication, Jun. 17.

Bhattacharyya AK, Roper DH, inventors; The Boeing Co, assignee (2003). High radiation efficient dual band feed horn. US patent 6,642,900 B2. Nov. 4.

Bhattacharyya and Sor J, inventors; Northrop Grumman Corp., assignee (2007). Multiple flared antenna horn with enhanced aperture efficiency. US patent 7,183,991 B2. Feb. 27.

Bhattacharyya AK and Goyette G (2004). A novel horn radiator with high aperture efficiency and low cross-polarization and applications in arrays and multibeam reflector antennas. *IEEE Transactions on Antennas and Propagation*; 52; 2850–2859.

Bhattarcharyya AK, Cherrette AR, and Bruno RD (2013). Analysis of ring-slot array antenna using hybrid matrix formulation. *IEEE Transactions on Antennas and Propagation*; 61 (4); 1642–1650.

Burr DG, inventor; Space Systems/Loral, assignee (2013). High efficiency multi-beam antenna. US patent 2013/0154874 A1. Jun. 20.

Chan KK and Rao SK (2008). Design of high-efficiency circular horn feeds for multibeam reflector applications. *IEEE Transactions on Antennas and Propagation*; 56 (1); 253–258.

Chang K, editor, (1989). *Handbook of Microwave and Optical Components, Vol. 1, Microwave Passive and Antenna Components*. New York: John Wiley & Sons.

Collin RE (1985). *Antennas and Radiowave Propagation*. New York: McGraw-Hill.

Coromina F and Polegre AM (2004). Failure robust transmit RF front end for focal array fed reflector antenna. *AIAA International Communications Satellite Systems Conf*; May 9–12.

Croq F., Vourch E, Reynaud M, Lejay B, Benoist C, Couarraze A, Soudet M, Carati P, Vicentini J, and Mannocchi G (2009). The Globalstar 2 antenna sub-system. *Proc, European Conf on Antennas and Propagation*; Mar. 23–27.

Dallaire J, Senechal G, and Richard S (2009). The Alphasat-XL antenna feed array. *Proc, European Conf on Antennas and Propagation*; Mar. 23–27; 585–588.

Darbandi A, Zoyo M, Touchais JY, and Butel Y (2008). Flexible S-band SSPA for space application. *NASA/ESA Conf on Adaptive Hardware and Systems*; Jun. 22–25.

Deguchi H, Tsuji M, and Shigesawa H (2001). A compact low-cross-polarization horn antenna with serpentine-shaped taper. *Digest, IEEE Antennas and Propagation Society International Symposium*; Jul. 8–13.

Demers Y, Amyotte É, Glatre K, Godin M-A, Hill J, Liang A, and Riel M (2017). Ka-band user antennas for VHTS GEO applications. *European Conf on Antennas and Propagation*; Mar. 19–24.

DiFonzo DF and English WJ (1974). Far-field criteria for reflectors with phased array feeds. *IEEE Antennas and Propagation Society International Symposium*. June.

Durand L, Duchesne L, Blin T, Garreau P, Iversen P, Forma G, Meisse P, Decoux E, and Paquay M (2012). Novel methods for fast multibeam satellite antenna testing. *European Conf on Antennas and Propagation*; Mar. 26–30.

Egami S and Kawai M (1987). An adaptive multiple beam system concept. *IEEE Journal on Selected Areas in Communications*; 5; 630–636.

Epstein JW and CEO of TerreStar Networks (2010). Declaration of Jeffrey W. Epstein pursuant to local bankruptcy rule 1007-2 in support of first day pleadings. To US bankruptcy court, southern district of New York. Oct 19. On www.terrastarinfo.com/pdflib/3_15446. pdf. Accessed Feb. 9, 2015.

Esteban EMG, Briand A, Soulez E, Voisin P, and Albert I (2015). Ku-band mutli-port amplifier demonstrator: measured performances over 2GHz bandwidth. *AIAA International Communications Satelllite Systems Conf*; Sep. 7–10.

Fang RJF (2011). Broadband IP transmission over Spaceway® satellite with on-board processing and switching. *IEEE Global Telecommunications Conf*; Dec. 5–9.

Farrar T (2016). 2001: a space odyssey. Blog of Telecom, Media and Finance Associates. On tmfassociates.com/blog/2016/03/09/2001-a-space-odyssey. Accessed Mar. 20, 2018.

Farrar T (2018). Viasat's curious antenna issues. Blog of Telecom, Media and Finance Associates. On tmfassociates.com/blog/2018/01/09/viasats-curious-antenna-issues. Accessed Mar. 20, 2018.

Fenech H, Tomatis A, Amos S, Soumpholphakdy V, and Merino J-L (2013). Eutelsat's evolving Ka-band missions. *Ka and Broadband Communications, Navigation and Earth Observation Conf*; Oct. 14–17.

Gallinaro G, Tirrò E, Di Cecca F, Migliorelli M, Gatti N, and Cioni S (2012). Next generation interactive S-band mobile systems. *Advanced Satellite Multimedia Systems Conf and Signal Processing for Space Communications Workshop;* Sep. 5–7.

Glâtre K, Hildebrand L, Charbonneau E, Perrin J, and Amyotte E (2019). Paving the way for higher-volume cost-effective space antennas. *IEEE Antennas & Propagation Magazine*; 61; 47–53.

Godara LC (1997). Application of antenna arrays to mobile communications, part II: beamforming and direction-of-arrival-considerations. *Proceedings of the IEEE*; 85 (8); 1195–1245.

Guy RFE (2009). Potential benefits of dynamic beam synthesis to mobile satellite communication, using the Inmarsat 4 antenna architecture as a test example. *International Journal of Antennas and Propagation*; 2009; 1–5.

Guy RFE, Wyllie CB, and Brain JR (2003). Synthesis of the Inmarsat 4 multibeam mobile antenna. *International Conf on Antennas and Propagation;* Mar. 31–Apr. 3.

Harris Corp (2018a). Fixed-mesh reflector. Spec sheet. On www.harris.com/what-we-do/space-antennas. Accessed Mar. 29, 2020.

Harris Corp (2018b). High compaction ratio reflector antenna. Spec sheet. On www.harris.com/what-we-do/space-antennas. Accessed Mar. 29, 2020.

Harris Corp (2018c). Unfurlable space reflector solutions. Product information. On www.harris.com/solution/unfurlable-mesh-reflector-antennas. Accessed Mar. 30, 2010.

Harris Corp (2019a). Harris Corporation introduces high-accuracy reflector for improved satellite comunications. Press release. On www.harris.com/press-releases. Accessed Mar. 29, 2020.

Harris Corp (2019b). Unfurlable Ka-band reflectors. Spec sheet. On www.harris.com/what-we-do/space-antennas. Accessed Mar. 29, 2020.

Harris Corp (2020). Smallsat perimeter truss reflector. Spec sheet. www.harris.com/what-we-do/space-antennas. Accessed Mar. 29, 2020.

Hartmann J, Habersack J, Steiner H-J, and Lieke M (2002a). Advanced communications satellite technologies. *Workshop on Space Borne Antennae Technologies and Measurement Techniques*; Apr. 18.

Hartmann J, Habersack J, Hartmann F, and Steiner H-J (2005b). Validation of the unique field performance of the large CCR 120/100. *Proc of the Symposium of the Antenna Measurement Techniques Association*; Oct. 30–Nov. 4.

Hartmann J, Habersack J, and Steiner H-J (2008). Improvement of efficiency for antenna and payload testing. *ESA Antenna Workshop*. May. On www.astrium.eads.net/media/document/2008-esa-workshop-improvement-of-efficiency-for-antenna-and-payload-testing.pdf. Accessed Mar. 5, 2012.

Hejres JA (2004). Null steering in phased arrays by controlling the positions of selected elements. *IEEE Transactions on Antennas and Propagation*; 52 (11); 2891–2895.

Hogie K, Criscuolo E, Dissanayake A, Flanders B, Safavi H, and Lubelczyk J (2015). TDRSS demand access system augmentation. *IEEE Aerospace Conf*; Mar. 7–14.

Huynh S (2016). Active or passive Tx/Rx Globalstar & combined GPS/Globalstar antennas: pictures, outline drawings, and specifications. Viewgraph presentation of product information. Apr. 15. On www.antcom.com/documents/catalogs/GlobalstarAntennas.pdf. Accessed Feb. 20, 2018.

Ingerson PG and Wong WC (1974). Focal region characteristics of offset fed reflectors. *IEEE Antennas and Propagation Society International Symposium*; Vol. 12; Jun.

Iridium (2019). Personal communication from Jordan Hassin, Jun. 25.

Jamnejad V (1992). Ka-band feed arrays for spacecraft reflector antennas with limited scan capability—an overview. *Digest, IEEE Aerospace Applications Conf*; Feb. 2–7.

Koduru C, Tomei B, Sichi S, Suh K, and Ha T (2011). Advanced space based network using ground based beam former. *AIAA International Communications Satellite Systems Conf*; Nov. 28–Dec. 1.

Kraus JD and Marhefka RJ (2003). *Antennas for All Applications*, 3rd ed., international ed. Singapore: McGraw-Hill Education (Asia).

Lafond JC, Vourch E, Delepaux F, Lepeltier P, Bosshard P, Dubos F, Feat C, Labourdette C, Navarre G, and Bassaler JM (2014). Thales Alenia Space multple beam antennas for telecommunication satellites. *Proc, European Conf on Antennas and Propagation*; Apr. 6–11.

Lepeltier P, Maurel J, Labourdette C, and Croq F (2007). Thales Alenia Space France antennas: recent achievements and future trends for telecommunications. *European Conf on Antennas and Propagation*; Nov. 11–16.

Lepeltier P, Bosshard P, Maurel J, Labourdette C, Navarre G, and David JF (2012). Recent achievements and future trends for multiple beam telecommunication antennas. *International Symposium on Antenna Technology and Applied Electromagnetics;* Jun. 25–28.

LightSquared (2006). Technical appendix, to FCC, seeking authority to communicate with SkyTerra 2. On licensing.fcc.gov/myibfs/download.do?attachment_key=845640. Accessed Feb. 17, 2015.

Lisi M (2000). Phased arrays for satellite communications: a system engineering approach. *Proc, IEEE International Conf on Phased Array Systems and Technology*; May 21–25.

Mallet A, Anakabe A, Sombrin J, and Rodriguez R (2006). Multiport-amplifier-based architecture versus classical architecture for space telecommunication payloads. *IEEE Transactions on Microwave Theory and Techniques*; 54 (12); 4353–4361.

Mallison MJ and Robson D (2001). Enabling technologoes for the Eurostar geomobile satellite. *AIAA International Communications Satellite Systems Conf*; Apr. 17–20.

Mannocchi G, Amici M, Del Marro M, Di Giuliomaria D, Farilla P, Macchiusi M, and Suriani A (2013). A L-band transmit/receive module for satellite communications. *Proc, European Microwave Conf*; Oct. 6–10.

Maral G and Bousquet M (2002). *Satellite Communications Systems*, 4th ed. Chichester (England): John Wiley & Sons, Ltd.

Migl J, Guelten E, Seitz W, Steiner H-J, and Meniconi E (2013). Time efficient antenna & payload technique for future multi-spot-beam antennas. *European Conf on Antennas and Propagation*; Apr. 8–12.

Migl J, Habersack J, and Steiner H-J (2017). Antenna and payload test strategy of large spacecraft's in compensated compact ranges. *European Conf on Antennas and Propagation*; Mar. 19–24.

Migliorelli M, Scialino L, Pasian M, Bozzi M, Pellegrini L, and van 't Klooster K (2013). RF performance control of mesh-based large deployable reflector. *Ka and Broadband Communications, Navigation and Earth Observation Conf;* Oct. 14–17.

Milligan TA (2005). *Modern Antenna Design*, 2nd ed. New Jersey: John Wiley & Sons and IEEE Press.

Mittra R, Rahmat-Samii Y, Galindo-Israel V, and Norman R (1979). An efficient technique for the computation of vector secondary patterns of offset paraboloid reflectors. *IEEE Trans on Antennas and Propagation*; AP-27 (3); 294–304.

Muhammad SA, Rolland A, Dahlan SH, Sauleau R, and Legay H (2011). Comparison between Scrimp horns and stacked Fabry-Perot cavity antennas with small apertures. *Proc, European Conf on Antennas and Propagation;* Apr. 11–15.

Nadarassin M, Vourch E, Girard T, Carrère JM, Soudet M, Borrell L, Rigolot H, Bouvier T, Voisin P, Onillon B, Albert I, and Taisant JP (2011). Ku-band reconfigurable compact array in dual polarization. *Proc, European Conf on Antennas and Propagation*; Apr. 11–15.

Palacin B and Deplancq X, inventors (2012). Multi-beam telecommunication antenna onboard a high-capacity satellite and related telecommunication system. U.S. patent application 2012/0075149 A1. Mar 29.

Panthi S, McLain C, King J, and Breynaert D (2017). Beam hopping—a flexible satellite communication system for mobility. *AIAA International Communications Satellite Systems Conf;* Oct. 16–19.

Pascale V, Maiarelli D, D'Agristina L, and Gatti N (2019). Design and qualification of Ku-band radiating chains for receive active array antenna of flexible telecommunication satellites. *European Conf on Antennas and Propagation*; Mar. 31 – Apr. 5.

Perrott RA and Griffin JM (1991). L-band antenna systems design. *IEE Colloquium on Inmarsat-3*; Nov. 21.

Rahmat-Samii Y, Zaghloul AI, and Williams AE (2000). Large deployable antennas for satellite communications. *IEEE Antennas and Propagation Society International Symposium*; Jul. 16–21.

Ramanujam P, Rao SK, Vaughan RE, and McClewary JC, inventors (1999). Reconfigurable multiple beam satellite reflector antenna with an array feed. U.S. patent 5,936,592. Aug 10.

Ramanujam P and Fermelia LR (2014). Recent developments on multi-beam antennas at Boeing. *European Conf on Antennas and Propagation*; Apr. 6–11.

Rao SK (1999). Design and analysis of multiple-beam reflector antennas. *IEEE Antennas and Propagation Magazine*; 41; 53–59.

Rao SK (2003). Parametric design and analysis of multiple-beam reflector antennas for satellite communications. *IEEE Antennas and Propagation Magazine*; 45 (4); 26–34.

Rao SK (2015a). Advanced antenna technologies for satellite communications payloads. *IEEE Transactions on Antennas and Propagation*; 63 (4); 1205–1217.

Rao SK (2015b). Advanced antenna systems for 21st century satellite communications payloads. Viewgraph presentation. On s3.amazonaws.com/sdieee/1820-DL_Rao_SanDiego_12Mar2015_D1.pdf. Accessed Dec. 22, 2017.

Rao, S., Chan, KK, and Tang, M. (2005). Dual-band multiple beam antenna system for satellite communications. **IEEE Antennas and Propagation Society International Symposium**; Jul. 3–8.

Rao SK and Tang MQ (2006a). Stepped-reflector antenna for dual-band multiple beam satellite communications payloads. *IEEE Transactions on Antennas and Propagation*; 54.

Rao S, Tang M, Hsu C-C, and Wang J (2006b). Advanced antenna technologies for satellite communication payloads. *Proc, European Conf on Antennas and Propagation;* Nov. 6–10.

Rao SK, Hsu C-C, and Matyas GJ, inventors; Lockheed Martin Corp, assignee (2013). Dual-band antenna using high/low efficiency feed horn for optimal radiation patterns. U.S. patent 8,514,140 B1. Aug 20.

Richard S, Demers Y, Amyotte E, Brand Y, Dupessey V, Markland P, Uher J, Liang A, Iriarte JC, Ederra I, Gonzalo R, and de Maagt P (2007). Recent satellite antenna developments at MDA. *Proc, European Conf on Antennas and Propagation*; Nov. 11–16.

Roederer AG (2005). Antennas for space: some recent European developments and trends. *International Conf on Applied Electromagnetics and Communications*; Oct. 12–14.

Roederer A and Sabbadini M (1990). A novel semi-active multi-beam antenna concept. *Digest, IEEE Antennas and Propagation Society International Symposium*; Vol. 4; May 7–11.

Roederer AG, Jensen NE, and Crone GAE (1996). Some European satellite-antenna developments and trends. *IEEE Antennas and Propagation Magazine*; 38 (2); 9–21.

Ruag Space (2016). Mobile communication antennas. Product information. On www.ruag.com/sites/default/files/2016-12/Mobile_communication_Antennas.pdf. Accessed Jul. 10, 2017.

Sauerman R (1999). A compact antenna test range built to meet the unique testing requirements for active phased array antennas. On www.nsi-mi.com/technical-papers. Accessed Jan. 2010.

Schennum GH and Skiver TM (1997). Antenna feed element for low circular cross-polarization. *Proc, IEEE Aerospace Conf*; vol. 3; Feb 1–8.

Schuss JJ, Upton J, Myers B, Sikina T, Rohwer A, Makridakas P, Francois R, Wardle L, and Smith R (1999). The Iridium main mission antennna concept. *IEEE Transactions on Antennas and Propagation*; 47; 416–424.

Semler D, Tulintseff A, Sorrell R, and Marshburn J (2010). Design, integration, and deployment of the TerreStar 18-meter reflector. *AIAA International Communications Satellite System Conf;* Aug. 30 – Sep. 2.

Seymour D (2000). L band power amplifier solutions for the Inmarsat space segment. *IEE Seminar Microwave and RF Power Amplifiers*; Dec. 7.

Shishlov AV, Krivosheev YuV, and Melnichuk VI (2016). Principal features of contour beam phased array antennas. *IEEE International Symposium on Phased Array Systems and Technology*; Oct. 18–21.

Simon PS (2007). LinkedIn page. Jan 1. Accessed Sep. 17, 2012.

Smith TM, Lee B, Semler D, and Chae D (2004). A large S-band antenna for a mobile satellite. *AIAA Space 2004 Conf*; Sep. 28–30.

Smiths Interconnect (2020). Thermopad temperature variable attenuators. Product information. On www.smithsinterconnect.com/products/rf-mw-mmw-components/resistive-components/thermopad®/. Accessed Apr. 1, 2020.

Space Systems Loral (2012). ICO G1 reflector deployment with voice over. Jun 11. Video. On www.youtube.com/watch?v=_mFnNDzxKFk. Accessed Jan. 30, 2018.

Spring KW and Moody HJ, inventors; Spar Aerospace Ltd (now MDA), assignee (1990). Divided LLBFN/HMPA transmitted architecture. U.S. patent 4,901,085. Feb 13.

Stirland SJ and Brain JR (2006). Mobile antenna developments in EADS Astrium. *European Conf on Antennas and Propagation*; Nov. 6–10.

Stirland SJ and Craig AD (2007). Phased arrays for satellite communications: recent developments at Astrium Ltd. *European Conf on Antennas and Propagation*; Nov. 11–16.

Thomas G, Jacquey N, Trier M, and Jung-Mougin P (2015). Optimising cost per bit: enabling technologies for flexible HTS payloads. *AIAA International Communications Satellite Systems Conf*; Sep. 7–10.

Thomson MW (2002). Astromesh™ deployable reflectors for Ku- and Ka-band commercial satellites. *AIAA International Communication Satellite Systems Conf;* May 12–15.

Todd D (2017). MEO is the place to go: SES orders new generation O3b mPower constellation from Boeing. News article. Sep 12. On www.seradata.com. Accessed Oct. 3, 2018.

Tomura T, Takikawa M, Inasawa Y, and Miyashita H (2016). Trade-off of multibeam reflector antenna configuration for satellite onboard application. *URSI Asia-Pacific Radio Science Conf*; Aug. 21–25.

Vilaça M (2020). Personal communication, Sep. 27.

Voisin P, Ginestet P, Tonello E, and Maillet O (2010). Payloads units for future telecommunication satellites—a Thales perspective. *European Microwave Conf*. Sep. 28–30.

Wolf H and Sommer E (1988). An advanced compact radiator element for multifeed antennas. *Proc, European Conf on Antennas and Propagation*; Sep. 12–15.

Zillig DJ, McOmber DR, and Fox N (1998). TDRSS demand access service: application of advanced technologies to enhance user operations. *SpaceOps Conf*; Jun. 1–5.

PART II

END-TO-END SATELLITE COMMUNICATIONS SYSTEM

CHAPTER 12

DIGITAL COMMUNICATIONS THEORY

12.1 INTRODUCTION

This chapter summarizes digital communications theory from encoder and modulator through demodulator and decoder. This material explains the processing in the two ground-terminal ends of a traditional satellite system, where all communications go to or from a ground station, as well as the processing in a regenerative payload. Digital communications are in layer 1 of the Open Systems Interconnect (OSI) model, namely the physical layer (Section 13.2.1).

The chapter is aimed at the engineer who has neither the time nor the inclination to read communications textbooks. It is meant to provide the following:

- The very basics of communications terminology
- Required background for performing simulations of the end-to-end satellite communications system
- Required background for making hardware measurements on the end-to-end satellite communications system or an emulator of it.

The chapter contains a large number of drawings but few equations so it can be as useful as possible to working engineers. For equations not all the validity prerequisites are stated, since in every normal case of satellite communications

Satellite Communications Payload and System, Second Edition. Teresa M. Braun and Walter R. Braun.
© 2021 John Wiley & Sons, Inc. Published 2021 by John Wiley & Sons, Inc.

the assumptions do apply. Familiarity with the Fourier transform (book Appendix A.2) is required in the sections on filtering. Most sections of the chapter are merely descriptive but references are given for those readers who want to know more.

Section 12.2 may be enough of an introduction to digital communications for those who only want to know what "I" and "Q" signal components are.

The rest of the chapter goes into more detail. Sections 12.2–12.4 present basics of signals, filters, and white Gaussian noise. Section 12.5 provides a block diagram of the satellite communications channel, showing elements that are discussed in the rest of the chapter. Section 12.6 presents the manipulations of the bit stream, such as encoding. Section 12.7 introduces baseband modulation and lists most of the modulations used on commercial satellites. Sections 12.8–12.10 present modulation, demodulation, and symbol recovery for memoryless modulations. Sections 12.11–12.13 do the same for modulation with memory, with Section 12.12 doing double duty for demodulation and convolutional decoding. Sections 12.15 and 12.16 discuss topics that apply to both types of modulation, namely inter-symbol interference (ISI) and signal-to-noise ratio (SNR), respectively.

In Sections 12.6–12.13 many choices that the Digital Video Broadcast (DVB) family of second-generation standards has made, are highlighted. Section 13.4 gives an overview of these standards.

12.2 SIGNAL REPRESENTATION

12.2.1 RF Signal Representation

A radio-frequency (RF) or **bandpass signal** is a (real-valued) signal for which there is a frequency band about 0 Hz in which it has no power. The signal's **carrier frequency** can be any of the (positive) frequencies in the signal's frequency band, chosen for convenience, but in digital communications, the carrier frequency is the center frequency. The RF signal $x(t)$ with carrier frequency f_c in Hz can be represented as follows:

$$x(t) = \sqrt{2}\,\mathrm{Re}\left\{\sqrt{P(t)}e^{j\theta(t)}e^{j2\pi f_c t}\right\} = \sqrt{2}\sqrt{P(t)}\cos\left[\theta(t) + 2\pi f_c t\right]$$

$$= \sqrt{2}\sqrt{P(t)}\cos\theta(t)\cos(2\pi f_c t) - \sqrt{2}\sqrt{P(t)}\sin\theta(t)\sin(2\pi f_c t)$$

where $\mathrm{Re}(u)$ is defined as the real part of u and f_c is much greater than the rates of change in $\sqrt{P(t)}$ and $\theta(t)$. $P(t)$ is the power of the signal at time t. Any variation in $\sqrt{P(t)}$ with t represents **amplitude modulation** (AM), and any variation in $\theta(t)$ represents **phase modulation** (PM). Digital information may be carried in amplitude or phase or both, depending on the type of modulation.

The **rotating phasor** (Couch II, 1990) representation of the RF signal is given by the complex function

$$y(t) \text{ for which } x(t) = \sqrt{2}\,\mathrm{Re}\,y(t)$$

$$y(t) = \sqrt{P(t)}e^{j\theta(t)}e^{j2\pi f_c t} = \sqrt{P(t)}\cos\left[\theta(t) + 2\pi f_c t\right] + j\sqrt{P(t)}\sin\left[\theta(t) + 2\pi f_c t\right]$$

(Actually, in Couch II (1990) the rotating phasor is defined as $\sqrt{2}$ times this, but the present author likes the rotating phasor to have the same power as the signal.) $P(t)$ is the rotating phasor's power at time t, and $\theta(t)$ as a function of time is a phase that jumps back and forth once in a while but averages to zero. Figure 12.1a shows the rotating phasor rotating around at frequency f_c Hz per second. Figure 12.1b shows an example of $\theta(t) + 2\pi f_c t$ as a function of time.

12.2.2 RF Signal's Equivalent Baseband Representation

The fact that the rotating phasor is rotating is not often of interest when we want to analyze the signal. So let us remove the average rotation from our thoughts. Now we have in mind the **phasor** (at baseband), given at time t by the complex function

$$z(t) = \sqrt{P(t)}e^{j\theta t} = \sqrt{P(t)}\cos\theta(t) + j\sqrt{P(t)}\sin\theta(t)$$

(Most authors define the phasor as $\sqrt{2}$ times this, but the present author likes the phasor to have the same power as the RF signal.) The phasor as a function of t is the **(complex) baseband equivalent** of the RF signal. (The baseband equivalent signal multiplied by $\sqrt{2}$ is called the *complex envelope* of the RF signal, but the author finds this terminology undescriptive.) The baseband equivalent of an RF signal is used in simulation (Chapter 16). Figure 12.2a shows the baseband phase versus

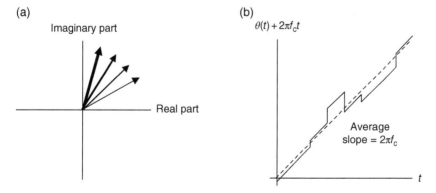

(a)

Imaginary part

Real part

(b)

$\theta(t) + 2\pi f_c t$

Average
slope = $2\pi f_c$

t

FIGURE 12.1   (a) Rotating phasor represented by arrow; previous positions are represented by thinner arrows; (b) example of phase $\theta(t) + 2\pi f_c t$ of rotating phasor versus time.

(a)   (b)

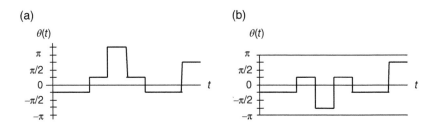

FIGURE 12.2   Same example, phase θ of signal phasor versus time: (a) original (b) original taken modulo 2π.

FIGURE 12.3   Same example of signal phasors, shown as dots, when amplitude is constant.

time corresponding to the RF phase versus time shown in Figure 12.1b. In the case where the phase jumps are multiples of $\pi/2$ and we take the phase modulo 2π, we obtain Figure 12.1b.

The real part of the phasor is called the **in-phase or I component** and the imaginary part the **quadrature-phase or Q component**. Referring to the equation we note that

$$I \text{ component} = \sqrt{P(t)}\cos\theta(t) = \frac{1}{\sqrt{2}} \text{ coefficient on } \cos\left(2\pi f_c t\right) \text{ term in RF signal}$$

$$Q \text{ component} = \sqrt{P(t)}\sin\theta(t) = \frac{1}{\sqrt{2}} \text{ coefficient on}\left(-\sin\left(2\pi f_c t\right)\right) \text{ term in RF signal}$$

When, in addition, in our example, the amplitude stays constant, the signal phasors take on only four possible values, shown in Figure 12.3. The I and Q components of the signal versus time are shown in Figure 12.4.

A **native baseband signal** is something different from the baseband equivalent of an RF signal. A native baseband signal is real-valued and has all its power in the neighborhood of 0 Hz.

12.2.3   Signal Spectrum

The signal **spectrum** is a signal's **power spectral density (psd)** function versus frequency, which has units of power per Hz. The psd is real-valued and non-negative. The integral over any frequency band tells us how much power of the signal is

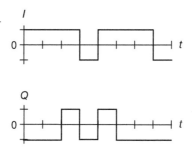

FIGURE 12.4 Same example, I and Q components of same baseband signal versus time.

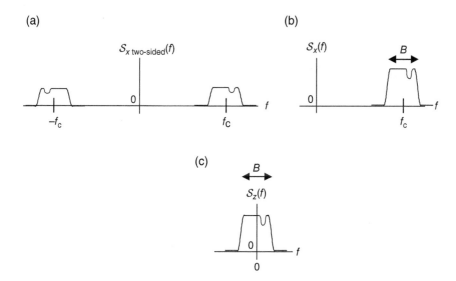

FIGURE 12.5 RF signal's spectrum representations: (a) two-sided RF; (b) one-sided RF (preferred); (c) two-sided equivalent-baseband.

within that band. The total integral of the psd is the signal power P. We write $S_u(f)$ for the psd of any signal $u(t)$. The spectrum is usually plotted with frequency on a logarithmic scale.

The psd of an RF signal has several representations. There are two RF representations, shown by example in Figure 12.5a,b. The non-preferred one, the two-sided RF spectrum, is in Figure 12.5a. It is sometimes used in textbooks. It is an even function of f, so its positive-frequency side contains full information on the whole function. The preferred one, the *(one-sided) RF spectrum*, is in Figure 12.5b. Figure 12.5c shows the psd of the baseband equivalent signal. It is the *(two-sided) equivalent-baseband spectrum*. The example spectrum has a notch for demonstration purposes. All the psds integrate to P. In the preferred RF representation and in the equivalent-baseband representation, the bandwidths B are equal (while for the non-preferred RF representation the bandwidth is double, since the psd is half as great).

The psd of a native baseband signal has two representations, shown in Figure 12.6. The non-preferred one, the two-sided RF spectrum, is in Figure 12.6a. It is sometimes used in textbooks. It is an even function of f, so its positive-frequency side contains full information on the whole function. The preferred one, the *(one-sided) spectrum*, is shown in Figure 12.6b. Both psds integrate to P. In the preferred representation the bandwidth B is one-sided (for the non-preferred representation the bandwidth is double, since the psd is half as great).

The author has been in conversations about "the" bandwidth of a spectrum and later realized that the other person and she were speaking of different things by a factor of two. Usually, the other person was inexperienced but even so performing a communications simulation (see Section 16.3.1 on the pitfalls of simulation).

Aside from the issue of whether the bandwidth is two-sided or one-sided, there are various types of spectrum bandwidth, used for different purposes. The most common for satellite communications are probably the **3-dB bandwidth** and the **mainlobe bandwidth**, both illustrated for the equivalent baseband spectrum of an RF signal in Figure 12.7. The 3-dB bandwidth is the width of that part of the spectrum

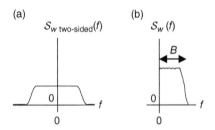

FIGURE 12.6   Native baseband signal's spectrum representations: (a) two-sided and (b) one-sided (preferred).

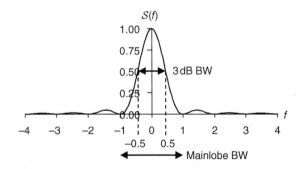

FIGURE 12.7   Definitions of two types of spectrum bandwidth, for equivalent-baseband spectrum of RF signal.

over which the spectrum is greater than $-3\,\mathrm{dB}$ or 0.5 of the peak power density. The mainlobe bandwidth is the width of the spectrum's mainlobe. Some spectra have only the mainlobe, while others have sidelobes, too, especially at output of the payload's high-power amplifier (HPA). In the latter case, the mainlobe still contains almost all the signal power.

A random process $z(t)$ is a set of random variables indexed by time (Couch II, 1990). It has an *autocorrelation function* $\mathcal{R}(t+\tau,\tau)$ defined as follows:

$$\mathcal{R}(t+\tau,\tau) \triangleq E\left[z(t+\tau)z^{*}(\tau)\right] \text{ where } E \text{ is expected value}$$

If the random process has constant mean and $\mathcal{R}(t+\tau,\tau) = \mathcal{R}(t)$ for all t and τ, then the spectrum $S(f)$ of $z(t)$ equals the Fourier transform of $\mathcal{R}(t)$:

$$S(f) = \mathcal{F}\left[\mathcal{R}(\cdot)\right](f) \text{ where } \mathcal{F} \text{ is the Fourier transform}$$

Communications signals have this property.

12.3   FILTERING IN GENERAL

In Section 5.2.2, we introduced filter representation and filter bandwidth. Here we delve further into the topic.

12.3.1   Filter Representations

Most of the time, we think of an RF filter as being centered around some RF frequency f_0. (The filter may in fact be asymmetric about f_0 and thus not have a true center, in which case f_0 is just a convenient frequency.) Figure 12.8 shows two representations of an RF transfer function, (a) the two-sided RF version, sometimes used in textbooks, and (b) the one-sided RF version, which is preferred. The heights of both representations are the same, unlike for the two RF spectrum representations. When we analyze the effect of a filter on a signal, it is easier to do this in

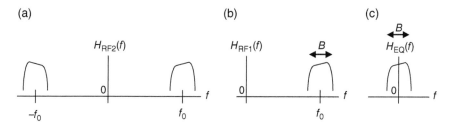

FIGURE 12.8   RF filter's transfer function: (a) two-sided RF representation, (b) one-sided RF (preferred), and (c) baseband-equivalent.

(a)   (b)

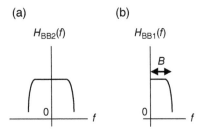

FIGURE 12.9 Native baseband filter's transfer function: (a) two-sided representation and (b) one-sided (preferred).

terms of their baseband-equivalent representations. The **baseband-equivalent transfer function** of the RF filter is shown in Figure 12.8c. It is used in simulation of an RF filter. It is constructed by sliding down the positive-frequency part of the RF transfer function to center it about $0\,\mathrm{Hz}$ instead of f_0. The baseband-equivalent impulse response is not necessarily real-valued. In the preferred RF representation and the preferred baseband-equivalent representation, the bandwidths B are equal (while for the non-preferred RF representation the bandwidth is double).

Things are different for a **native baseband filter**, which is a filter that is baseband at the beginning and stays baseband. Examples are the automatic level control (ALC) transfer function and the phase-locked loop (PLL) transfer function. Figure 12.9 shows two representations of such a transfer function, (a) the two-sided version, sometimes used in textbooks, and (b) the one-sided version, the preferred one. The impulse response is real-valued. In the preferred representation, the bandwidth B is one-sided (for the non-preferred representation the bandwidth is double). Confusion can arise about filter bandwidth in simulation.

12.3.2 Filtering a Signal

Let us start by defining the *(Dirac) delta function*, written $\delta(t)$ in the time domain. It is a useful theoretical construct. The function equals zero everywhere except where the argument equals zero, at which it equals infinity. The integral over its full argument range equals 1. The function in the time domain is always drawn as in Figure 12.10, where the arrow at zero argument reaching to the level of 1 means an infinite value with the integral of 1. The δ-function can alternatively be defined as the only function having the following property (Couch II, 1990):

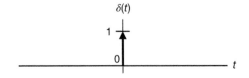

FIGURE 12.10 Dirac delta function in time domain.

$$\int_{-\infty}^{\infty} \delta(\tau) w(t-\tau) d\tau = w(t) \text{ for all functions } w(\cdot) \text{ that are continuous at } t$$

The operation of **convolution** on two functions h and u is defined as follows:

$$(h \circ u)(t) \triangleq \int_{-\infty}^{\infty} h(\tau) u(t-\tau) d\tau = \int_{-\infty}^{\infty} u(\tau) h(t-\tau) d\tau = (u \circ h)(t)$$

Thus, the δ-function convolved with a function equals the function.

Suppose that $h(t)$ is the impulse response of a filter. The filter may be at RF or baseband. $h(t)$ is the time-domain output of the filter when the input is the delta function. The transfer function $H(f)$ of the filter is the Fourier transform of $h(t)$, where f is frequency.

Suppose that $u(t)$ is the signal input to the filter. Then, the signal $v(t)$ out of the filter is given for all t by

$$v(t) = (h \circ u)(t)$$

The effect of the filter on the signal is to smear the signal in time. Filtering is a linear operation on the input signal, so it induces linear distortion.

If $U(f)$ is the Fourier transform of $u(t)$, then the Fourier transform $V(f)$ of $v(t)$ is given for all f by

$$V(f) = H(f) U(f)$$

and the spectrum $S_v(f)$ of $v(t)$ is given by

$$S_v(f) = |H(f)|^2 S_u(f)$$

What distortion filters cause to signals is discussed in Section 12.15.

12.4 WHITE GAUSSIAN NOISE

White Gaussian noise is a theoretical construct, noise with a constant-level psd over all frequencies. Figure 12.11 shows the noise psd for all three kinds of signal spectrum, each in its preferred representation. The level is N_0 in all cases, and the power is infinite.

As a function of time, white Gaussian noise is a random process. All time samples have a mean value of zero. Samples at different times are uncorrelated (book Appendix Section A.3.3). The autocorrelation function equals *No* times the δ-function.

When white Gaussian noise is filtered by a bandpass filter, the result is *colored noise*. Colored noise sampled at any time has a Gaussian probability density

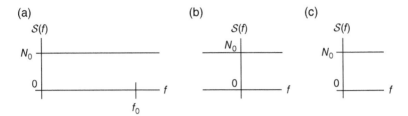

FIGURE 12.11   White noise psd in preferred representations: (a) RF, (b) baseband equivalent of RF, (c) native baseband.

function with zero mean. All time samples have the same root-mean-square (rms) value. Samples at different times are correlated, with the correlation decreasing as the time difference between the samples increases.

12.5   END-TO-END COMMUNICATIONS SYSTEM

Figure 12.12 gives a simplified diagram of an end-to-end satellite communications system with a bent-pipe payload. The diagram is broken down into ground transmitter, payload, and ground receiver. In two places, there is **additive white**

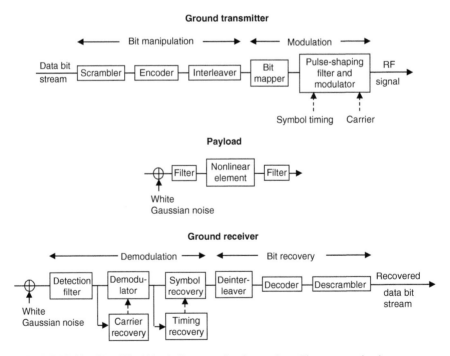

FIGURE 12.12   Simplified block diagram of end-to-end satellite communications system.

Gaussian noise (AWGN). White Gaussian noise is never actually seen since it is immediately filtered. Frequency converters, antennas, and atmospheric propagation effects are not shown in the diagram. The basic idea of a communications system is that the elements of the modulating transmitter are mirrored in the demodulating receiver by reverse elements. The diagram for a regenerative payload would have an additional demodulating receiver and an additional modulating transmitter in the payload. All the elements in the ground transmitter and ground receiver are treated in this chapter. The exposition is divided primarily into sections on bit manipulation, modulation, demodulation, and bit recovery, defined as containing the elements as marked in the figure.

12.6   BIT MANIPULATION

There are three kinds of bits in the system, of which we will have an example in Section 13.4.1. The **data bits** are the bits that contain the information. The **coded bits** are all the bits out of the encoder, where the encoder input is the data bits and some overhead bits. The **channel bits** are the coded bits plus more overhead bits. They are the bits that get mapped to symbols that will modulate the carrier.

12.6.1   Scrambler

The purpose of a *scrambler* is energy dispersal. There are two kinds, a scrambler of a bit sequence and a scrambler of a sequence of phasors out of a modulation scheme's constellation. A bit scrambler randomizes the bit sequence so that there is no long sequence of the same bit value. DVB Satellite, 2nd Generation (DVB-S2) implements one by means of a feedback shift register. DVB-S2 also has a phasor scrambler. It multiplies the phasor stream in a frame by a complex randomization sequence whose period is longer than a frame (ETSI EN 302 307-1).

12.6.2   Forward Error-Correcting Encoder

In a satellite communications system, the modulating transmitter protects the data from errors that can arise by applying a **forward error-correcting (FEC) code** to the bit stream. The signal distortions that cause the errors can arise in many places: in the ground transmitter, on the uplink propagation path, in the payload, on the downlink propagation path, and in the ground receiver. This is for a bent-pipe payload, whether processing or not; for a regenerative payload, errors arise separately on each link. The FEC code gives the demodulating receiver the ability to correct most errors in a communications system that has no reverse channel on which to request retransmission of data.

The FEC code is the *channel code* in communications theory terminology. We must introduce the term **communications-theory channel**, which is not our usual channel (Section 1.3) but the entire sequence of processes, including atmospheric propagation, that the data bit stream goes through from beginning to end. (In contrast, source encoding is a form of lossless compression (Proakis and Salehi, 2008). If there is a source encoder in the system, it would be outside of the diagram of Figure 12.12 and its output would be the data bit stream input in the diagram. Source encoding is outside the scope of this book.)

The FEC code is so effective that the required data-bit error rate (BER) (Section 12.14.2) can be achieved with a large decrease in the equivalent isotropically radiated power (EIRP) and/or the receiver sensitivity over what is necessary without coding. The drawback to the use of coding is that it increases the signal bandwidth. This is the fundamental tradeoff when it comes to introducing error control (Biglieri, 2020).

The **encoder** performs the encoding. We interchangeably use the terms "encoder" and "coder." The encoder works by adding correlations among the input bits. The idea is that if the receiver detects or guesses some coded bits wrongly, the decoder can use all the relationships to still reconstruct the original bit stream correctly.

One parameter of a code is its **code rate**. The code rate is a ratio $n/m < 1$ of integers n and m that have no common divisor, with the meaning that the number of coded bits in a time period is m/n times the number of input bits. A code with a rate close to 1 can achieve the same performance as a lower-rate code when the higher-rate encoder does more processing (and similarly the decoder does more processing). Nonetheless, normally, lower code rates are used when the channel quality is worse, for example, during heavy rain at Ku-band and higher frequencies.

A code is *systematic* if the input bit stream is embedded in the output bit stream. The DVB-S2 codes are systematic.

There are four types of channel code used over satellite, where the last two are newer than the first two:

1. Classical block code
2. Convolutional code
3. Turbo code
4. Low-density parity-check (LDPC) block code.

A **block code** takes a fixed number of bits in a block and computes a larger block of coded bits, called the *code word.*

The most common of the classical block code are *Bose, Chaudhuri, and Hocquenghem (BCH)* and *Reed–Solomon (RS)*, described, respectively, in Lin and Costello Jr. (1983) and Couch II (1990). The RS code is actually a subclass of BCH code. A binary BCH code deals with bits and its decoder corrects bit errors, while the RS code deals with multiple-bit "symbols" and its decoder corrects "symbol" errors (not the same as modulation symbols). One of the key features of BCH codes is that the code design provides precise control over the number of "symbol" errors correctable by the code. Another is that they are easy to decode (Wikipedia, 2020a).

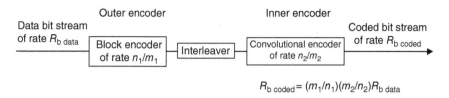

$$R_{b\ coded} = (m_1/n_1)(m_2/n_2)R_{b\ data}$$

FIGURE 12.13 Concatenated block and convolutional encoders with interleaver in between.

Convolutional coding, on the other hand, is a continuous process. It first divides the input bit stream into k parallel streams, where $k \geq 1$. When a new bit is shifted into each of the k input streams, $i > k$ linear combinations of new and old bits are formed and output. Such a code has rate equal to k/i or k/i in its reduced fractional form. If some bits are then removed in a regular pattern, that is, the code is *punctured*, then the rate becomes higher than k/i.

A block code and a convolutional code are sometimes **concatenated**, in which one code is applied on top of another in order to achieve even more of a drop in required transmitter power and/or receiver sensitivity for the same BER. The process is shown in Figure 12.13. The block code is the first one applied, called the **outer encoder**. The convolutional code is second, called the **inner encoder**, and there is an interleaver in between (Section 12.6.3). It is easier to describe the purpose of this construction in terms of the decoding. The first decoder, for the convolutional code, corrects many of the errors. When it makes a mistake its output is a bit stream with a burst of errors, that is, correlated. The second decoder, for the block code, needs as input a bit stream which has uncorrelated errors in it. The deinterleaver breaks up and widely separates the errors, making them uncorrelated as far as the block decoder is concerned. Then, the block decoder can remove those errors.

Turbo coding is also a concatenated coding scheme, newer than block and convolutional. It comes close to achieving the **Shannon bound or limit**. Shannon showed that every noisy communications-theory channel has a *channel capacity*, and for any data-bit rate arbitrarily close to it, coding schemes exist that, together with maximum-likelihood (ML) decoding (Section 12.12), lead to an arbitrarily small decoding-error probability (Lin and Costello Jr., 1983). In general, a code approaches the Shannon bound at the price of a considerable increase in its redundancy, block length, and complexity. Exceptions are turbo codes and LDPC codes (below), which explains their success (Biglieri, 2020). The Shannon bound is plotted in Figure 13.9 in Section 13.4.2.

A basic sketch of a turbo coder is given in Figure 12.14. It shows that the encoder has two convolutional encoders and a block interleaver. Each convolutional encoder is recursive, which means that its output bits are not just a combination of current and earlier input bits but also of earlier output bits (Proakis and Salehi, 2008). Each constituent encoder encodes the entire data-bit stream. Turbo codes perform better than block and convolutional codes but are more complicated to decode.

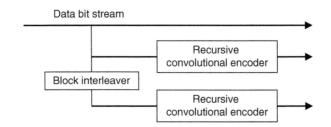

Data bit stream

FIGURE 12.14   Turbo encoder (after Proakis and Salehi (2008)).

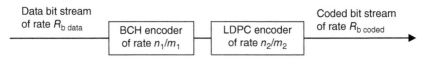

$$R_{b\ coded} = (m_1/n_1)(m_2/n_2)R_{b\ data}$$

FIGURE 12.15   Concatenated BCH and LDPC encoders.

LDPC coding is the type of code most recently applied to satellite. It functions well with code rate close to 1. It was chosen over turbo coding to be part of the DVB-S2 standard. Its structure has some similarity to that of the turbo coder. Instead of two constituent encoders, though, there are more. They are typically simple accumulators, and each encodes only a small part of the input frame. Each puts out a parity bit and the parity bits, together with the original bit stream, form the encoder output stream. LDPCs perform better than turbo codes for higher code rates, while turbo codes are better for lower code rates (Wikipedia, 2020b). The code word is very long, typically thousands of bits (Proakis and Salehi, 2008).

The LDPC encoder needs to be followed by a bit interleaver for memoryless modulations of higher order than four. This is because demodulation causes correlation among bits. A deinterleaver must decorrelate them, otherwise the performance of the LDPC decoder would be degraded (Kienle and Wehn, 2008).

A BCH coder and a LDPC coder are concatenated in DVB-S2, as shown in Figure 12.15. There is no need for an interleaver in between because the LDPC encoder, when it makes a mistake, does not put out a burst of bit errors.

12.6.3   Interleaver

The *interleaver* reorders bits on a large scale. The interleaver is there so that in the receiver the deinterleaver will decorrelate nearby received bits. The decoder requires as input an uncorrelated bit stream to work optimally.

There are two types of interleaver, block and convolutional. Data are serially written into a block interleaver column-wise, and they are serially read out row-wise.

A convolutional interleaver is better matched for use with convolutional codes (Proakis and Salehi, 2008). The bits are serially read into a collection of shift registers, each with a fixed delay. Typically, the delays are integer multiples of a fixed integer. Then, the data are serially read out of the shift registers (Unnikuttan et al., 2014).

12.7 MODULATION INTRODUCTION

The modulations described in this chapter are the ones commonly used over satellite. Some are memoryless and some have memory. They include a wide variety, as shown in the examples of Table 12.1. These examples come from DVB standards and user terminal manufacturers. A very small-aperture terminal (VSAT) is an example of a user terminal, a type mostly used by businesses (Section 17.4.2).

Whether the modulation is memoryless or has memory, at regular intervals of duration T, the baseband modulator uses the next bit(s) from the bit stream to choose the next **symbol** to incorporate into its output. The size M of the symbol alphabet available to choose from at each T is the **modulation order**. The greater M is, the greater the **spectral efficiency** of the modulation scheme considered alone, which equals $\log_2 M$ bits/Hz. (See Section 13.4.1 for a definition of spectral efficiency in a different context.)

The **symbol rate R_s** is the inverse of T. The **channel bit rate R_b** is given by

$$R_b = \log_2 M * R_s$$

The data bit rate and the coded bit rate are also written as R_b, so one must notice the context.

TABLE 12.1 Some Modulation Schemes Used in Standards and User Terminals

Where Used	Modulation Schemes	References
Inmarsat's BGAN	π/4-QPSK	Howell (2010)
DVB-S2 standard	QPSK, 8PSK, 16APSK, 32APSK	(ETSI EN 302 307-1)
DVB-S2X standard, additional to DVB-S2	8APSK, 64APSK, 128APSK, 256APSK	(ETSI EN 302 307-2)
DVB-RCS2 standard	π/2-BPSK, QPSK, 8PSK, 16QAM, CPM	(ETSI EN 301 545-2)
Hughes VSAT	OQPSK, 8PSK	Hughes (2015)
Gilat VSAT	MSK, GMSK, QPSK, 8PSK	Gilat (2013, 2014)
Satmode	Programmable phase filter for binary or quaternary CPM	Wikipedia (2018)

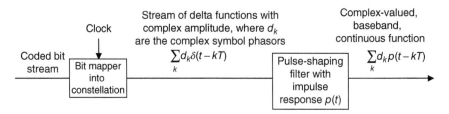

FIGURE 12.16   Memoryless modulation process (baseband part).

12.8   MEMORYLESS MODULATION

Memoryless modulation is more common than modulation with memory (Section 12.11). It is also known as *linear modulation*. The modulation schemes are (1) **phase-shift keying (PSK)**, where the phasors are different only in their phase; (2) **amplitude-and-phase-shift-keying (APSK)**, where the phasors are different in the combination of their amplitude and phase; and (3) **quadrature amplitude modulation (QAM)**, a special case of APSK, where the constellation phasors are in a regular square grid aligned with the real and imaginary axes. In memoryless modulation, each symbol is a phasor in the modulation scheme's constellation.

This section first discusses the baseband part of memoryless modulation, pictured in Figure 12.16. This is different for modulation with memory, as we will see in Section 12.11. This process is followed by the modulation of the carrier, which is discussed at the end of this section and pictured there.

12.8.1   Modulation Schemes

A modulation scheme is a constellation of baseband phasors from which symbols are selected. They are selected with equal probability. Constellations with phasors of all equal magnitude are the best for when the payload's HPA must be operated at or close to saturation, because all phasors will undergo the same distortion from the HPA nonlinearity. Second best is a collection of phasors without "corners," which are a few phasors that have amplitude greater than the others. The more amplitude differences there are in the constellation, the farther backed-off the HPA must be, generally. However, user-terminal manufacturers employ proprietary means of nonlinear predistortion of the constellations, to partially compensate for the HPA nonlinearity. It is impossible to completely compensate because the transitions from symbol to symbol are not instantaneous, so the amplitude variation in the signal into the HPA depends not just on the constellation but also on the pulse-shaping filter and the payload's input-multiplexer (IMUX) filtering.

Another consideration for the higher-order modulations is that system phase noise has more of an effect, as the phasors are closer together. The phase noise specification may need to be tighter for such modulations.

Null transition also allowed

FIGURE 12.17   BPSK constellation.

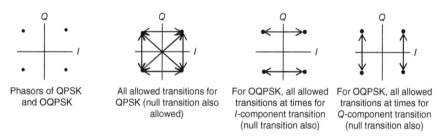

| Phasors of QPSK and OQPSK | All allowed transitions for QPSK (null transition also allowed) | For OQPSK, all allowed transitions at times for *I*-component transition (null transition also) | For OQPSK, all allowed transitions at times for *Q*-component transition (null transition also) |

FIGURE 12.18   QPSK and OQPSK.

Binary phase-shift keying (BPSK) has only two phasors in its constellation, as shown in Figure 12.17. Because BPSK is less spectrally efficient than quaternary phase-shift keying (QPSK) (that is, for the same bandwidth you only get half the bit rate), it is used less often than QPSK. It does, however, require only half the transmit power of QPSK for the same BER.

QPSK and its slight variation **offset quaternary phase-shift keying (OQPSK)** may be the modulation schemes most commonly used in satellite communications (Elbert, 2004). OQPSK is also known as *staggered quaternary phase-shift keying (SQPSK)*. The reasons they are common are that they are power-efficient and, ideally, anyway, of constant amplitude so suffer little degradation from the HPA's nonlinearity. They are also moderately bandwidth-efficient and simple to implement. Their four phasors are depicted in the leftmost drawing of Figure 12.18. The next drawing to the right depicts, for QPSK, all allowed transitions between phasors. In OQPSK, the *I* and *Q* phasor components transition sequentially instead of at the same time, as depicted in the rightmost two drawings of the figure. QPSK produces a larger amplitude variation than OQPSK does, since for QPSK there will be transitions through zero. This means greater distortion by the HPA.

A scheme related to BPSK and OQPSK is **π/2-BPSK**. The real and imaginary axes are considered to rotate π/2 in a counter-clockwise direction with every transition. The two phasors of the constellation stay on the *I* axis. For transition to either possible value of the next bit, the next phasor is only π/2 away, so there are no transitions through zero. It is illustrated in Figure 12.19, where *T* is the symbol duration.

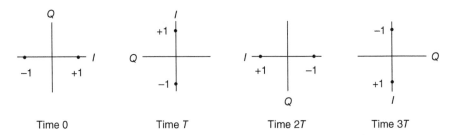

Time 0   Time T Time $2T$ Time $3T$

FIGURE 12.19 $\pi/2$-BPSK rotating constellation.

A similar scheme to $\pi/2$-BPSK but for QPSK is **$\pi/4$-QPSK** in which the axes rotate counterclockwise by $\pi/4$ with every transition.

Another PSK modulation scheme is **8-ary phase-shift keying (8PSK)**. Instead of only four phasors as in QPSK, there are eight, evenly spaced around the circle as illustrated in Figure 12.20. While this modulation scheme is spectrally efficient, it is not power-efficient because for *nearest neighbors* to be equally spaced the phasors have to be far out from the origin.

More complex than PSK is **8-ary amplitude-and-phase-shift keying (8APSK)**. As shown in Figure 12.21, the phasors are in three different rings, and the ratios of the ring diameters are parameters (ETSI EN 302 307-2).

16-ary quadrature-amplitude (16QAM) has 16 phasors arranged in a regularly spaced 4×4 grid as shown in Figure 12.22. This constellation has four phasors in "corners," which the HPA would most greatly distort. Thus, this constellation is not as good as the following 16-ary amplitude-and-phase-shift keying (16APSK) scheme when the HPA must be operated somewhat close to saturation. However, it is simpler to implement and demodulate.

More power-efficient than 16QAM is **16APSK** modulation, shown in Figure 12.23. The phasors are in two rings, where often the power of the outer ring is set so as to saturate the HPA. The modulation is also known as **(12,4)PSK**. The parameter γ is the ratio of the ring sizes in which the phasors lie. In DVB-S2, γ varies depending on the code rate (ETSI EN 302 307-1).

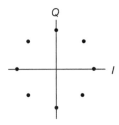

FIGURE 12.20   8PSK constellation.

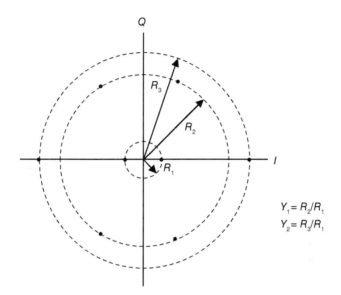

FIGURE 12.21   8APSK constellation (after [ETSI EN 301 545-2]).

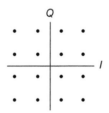

FIGURE 12.22   16QAM constellation.

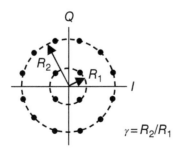

FIGURE 12.23   16APSK constellation.

The **32-ary amplitude-and-phase-shift keying (32APSK)** constellation is shown in Figure 12.24. The ratios of the ring sizes are parameters. In DVB-S2, γ_1 and γ_2 vary depending on the code rate (ETSI EN 302 307-1).

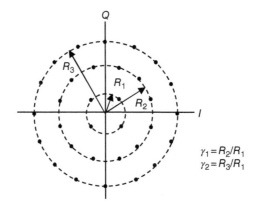

FIGURE 12.24 32APSK constellation (after [ETSI EN 302 307-1]).

DVB Satellite, 2nd Generation Extensions (DVB-S2X) has additional modulation schemes: **32APSK** with four rings of phasors, **64APSK** with four rings, **128APSK** with six rings, and **256APSK** with eight rings (ETSI EN 302 307-2). These are not discussed here.

12.8.2 Bit Mapper

Modulation schemes need a rule for how to assign bits from the coded bit stream to phasors of the constellation. For BPSK and OQPSK, no rule is needed, as only one bit at a time is used from the bit stream. The other schemes call for breaking up the bit stream into small sequences. The number of bits in a sequence equals $\log_2$ of the modulation order M. The commonly used mapping method, **Gray coding**, leads to a smaller BER than any other assignment does. Gray coding is based on the fact that the most likely kind of error that the receiver will make in deciding what phasor was transmitted is to mistake a corrupted phasor for the transmitted phasor's nearest neighbor. This is because smaller corruption is more likely than larger corruption. In Gray coding, the bit groups assigned to nearest-neighbor phasors differ in only one bit, so the most likely kind of decision error will correspond to only one bit error. Figure 12.25 gives an example of Gray coding for 8PSK. There is more than one way to Gray code.

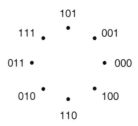

FIGURE 12.25 Example of Gray coding for 8PSK.

12.8.3 Differential Encoding of PSK Schemes

For PSK modulation schemes, the bit stream mapping into the phasors can be an absolute mapping. This means the mapping is directly from the bit stream to the phasor sequence. The other option is *differential encoding*. (This encoding has nothing to do with error correction.) The need for differential encoding exists when the receiver may not be able to recover the carrier phase absolutely but with some constant error. The sequence of bits that would, in an absolute mapping, determine a particular phasor, now determines the phase shift from the current phasor to the next. Differential encoding does not perform as well as absolute, requiring about 3 dB higher $C/(N + I)$ but allowing a simpler receiver (Benedetto et al., 1987).

Iridium Next (Section 19.2) uses differential encoding for OQPSK and BPSK.

12.8.4 Pulse-Shaping Filter

12.8.4.1 Introduction We now discuss how the discrete-time sequence of phasors from the modulation scheme is turned into a continuous-time baseband signal (Figure 12.16). The filter that does this is the **pulse-shaping filter**. (The name is somewhat of a misnomer in that the filter does not shape the pulse but actually produces the pulse.) The **pulse** is the impulse response $p(t)$ of the filter. We treat only real-valued pulses. The choice of pulse depends on ease of implementation, how much bandwidth is available, how much filtering the signal will see, and how far backed off the HPA will be.

We have to specialize the expression for the equivalent-baseband signal given in Figure 12.16 to the particular modulation scheme. For QPSK, BPSK, 8PSK, 16QAM, and 16APSK, the complex baseband signal is as in the figure, namely $\sum_k d_k p(t-kT)$, where d_k is the kth phasor. For OQPSK, where I and Q bits are used sequentially, it is more accurately written as $\sum_k \left[d_{kI} p\left(t - kT\right) + j d_{kQ} p\left(t - \left(k + \dfrac{1}{2}\right)T \right) \right]$, where d_{kI} and d_{kQ} are the kth I bit and Q bit, respectively.

We assume that the coded bits are independent of each other and that each is equally likely to be -1 or 1. Then, there exists the following simple relationship between the pulse transform $P(f)$ and the signal spectrum $S(f)$ (proof sketched in appendix in Section 12.A):

$$S(f) = E\left[\left|d_k\right|^2\right]\left|P(f)\right|^2 \text{ where } E\left[\left|d_k\right|^2\right] \text{ is the average power of the phasors}$$

We will lapse sometimes and refer to $|P(f)|^2$ as the **pulse spectrum**.

12.8.4.2 Rectangular Pulse The ideal rectangular pulse equals 1 during the symbol duration and equals 0 outside of it. This pulse is not realizable because its changes of value are instantaneous, so its transform has frequencies out to infinity. However, it is an easy pulse to understand so it is a good one to start with. We make the pulse now symmetric about zero time as shown in Figure 12.26a, in order to have its transform be real-valued and drawable. This makes the pulse non-casual, but we do not let this bother us, as the resulting explanation simplifications are worth it. $P(f)$ is the $\sin(x)/x$ function as shown in Figure 12.26b, and the spectrum is the square of this, as shown in Figure 12.27.

One way to limit the extent of this pulse's spectrum, not recommended, is simply to chop it off outside of some band. We multiply the spectrum by another abstraction, the theoretical **brickwall filter** shown in Figure 12.28.

Let us see what this chopping does to the pulse. Figure 12.29 shows the resultant pulse for when all but the mainlobe of the spectrum is chopped off. The pulse is no

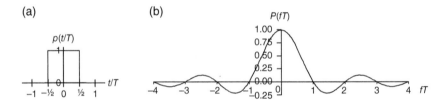

FIGURE 12.26   Rectangular pulse $p(t)$, symmetric about zero time, and its Fourier transform $P(f)$.

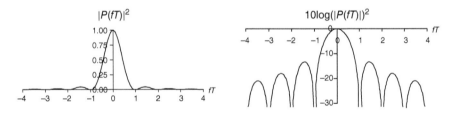

FIGURE 12.27 $|P(f)|^2$ for rectangular pulse, shown both on a linear scale and in dB.

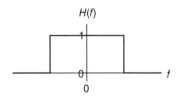

FIGURE 12.28   Transfer function $H(f)$ of theoretical brickwall filter, equivalent-baseband.

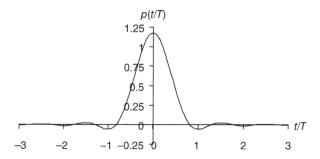

FIGURE 12.29   Bandlimited rectangular pulse $p(t)$, all spectrum sidelobes filtered out.

longer rectangular but is nearly triangular, which may be surprising since the main-lobe contains 90% of the signal power. The pulse is near zero outside the triangular part. The result of the bandlimiting on the signal is that the signal phasor no longer just sits at one of the constellation phasors and then instantaneously jumps to another phasor. Instead, most of the time it is simply near one or another of the ideal spots but it is always moving.

12.8.4.3   *Root Raised-Cosine Pulse*   The most commonly used family of pulses for satellite communications come from the **root raised-cosine (RRC) pulse-shaping filter**. The approach to design such a filter is to start with a good spectrum. The pulse spectrum has a *raised-cosine (RC)* shape. A case of the spectrum is shown in Figure 12.30 for one of the commonly used values of the **roll-off factor** α of the raised cosine, namely 0.35. For any α between 0 and 1, the shape is flat in the middle and is half of a cosine curve on either side, with the value 0.5 at $-1/(2T)$ and $+1/(2T)$. The factor α is defined so that $\alpha = 0$ gives the square spectrum and $\alpha = 1$ gives a purely RC shape, that is, no flat area.

A property of the Fourier transform is that generally the narrower a function is in the frequency domain, the wider its transform is in the time domain. Figure 12.31 shows the pulse for $\alpha = 0.35$ and 0.2. A small α like 0.2 gives a narrower spectrum but more signal-amplitude variation than the larger α does, leading to more nonlinear distortion by the satellite's HPA.

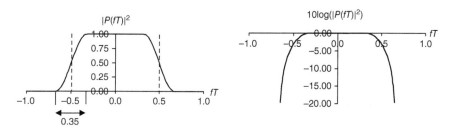

FIGURE 12.30   Pulse spectrum of raised-cosine shape with $\alpha = 0.35$.

(a)

(b)

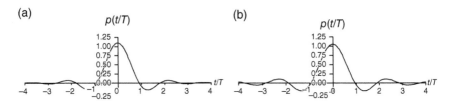

FIGURE 12.31   Pulse with RRC transform with (a) $\alpha = 0.35$, (b) $\alpha = 0.2$.

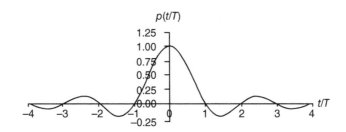

FIGURE 12.32   Pulse with RRC transform with $\alpha = 0.05$.

The DVB-S2X standard allows α to be as small as 0.05. The pulse for this value is shown in Figure 12.32. It leads to a signal with even more amplitude variation.

There have been only a few studies of these very small roll-off factors. The studies show that communications are only slightly degraded.

One study by a satellite owner put signals through a payload-channel emulator and state-of-the-art professional DVB-S2 transmitter and receiver. The study compared results for $\alpha = 0.05$ and 0.35. The emulator modeled a nonlinearized traveling-wave tube amplifier (TWTA) with the nonlinear performance of the DVB-S2 standard (ETSI EN 302 307-1). Results for a single carrier with QPSK modulation showed that the spectral regrowth out of the TWTA for $\alpha = 0.05$ was 1–2 dB higher, for all TWTA input backoffs. Results with one code rate for each DVB-S2 modulation scheme of order up through 32, on a linear channel, showed that $\alpha = 0.05$ required a $C/(N + I)$ greater by about 0.3 dB for the receiver to lock (Bonnaud et al., 2014).

Another study put signals through a Newtec user terminal and a geostationary (GEO) satellite. Results for $\alpha = 0.05$ and 0.20 were compared, for modulation schemes of 8PSK and 16APSK. The HPA backoff is not stated. The minimum required $C/(N + I)$ was higher by 0.2–0.3 dB for $\alpha = 0.05$ (Lee and Chang, 2017).

12.8.5 Carrier Modulation

In the next part of the transmitter, the carrier gets modulated as shown in Figure 12.33, where the carrier is assumed to be at intermediate frequency (IF). This is the method for both memoryless modulation and modulation with memory.

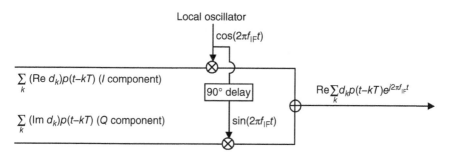

FIGURE 12.33 Modulating the carrier with the complex baseband signal.

12.9 MAXIMUM-LIKELIHOOD ESTIMATION

The signal then goes through the satellite payload and reaches the demodulation process in the receiving ground terminal (Figure 12.12). Before we describe this process for memoryless modulation in the next section, let us discuss the optimal method for two steps in the demodulation process.

The general ML method of making a decision can be used in various elements of the receiver: carrier recovery, symbol synchronization, demodulation of continuous-phase modulation (CPM) (Section 12.11), and convolutional decoding.

Maximum-likelihood estimation (MLE) is optimal for a parameter that is equally likely to take on any of its possible values. For memoryless modulation, it can be used to estimate the carrier phase and the *symbol epoch*, which is the time within a symbol duration at which the symbol is to be detected.

In MLE for carrier-phase recovery, the transmitted signal phase is taken to be that phase which maximizes the correlation between the transmitted signal and the received signal. If the transmitted signal includes a symbol sequence known to the receiver, these are *data-aided estimation*. For this purpose, at the beginning of a frame to be transmitted, a preamble is added which has a known symbol stream from memoryless modulation. If the transmitted signal does not contain a known symbol stream, the carrier-recovery loop makes a rough decision on it and uses it to estimate phase. This is *decision-directed estimation*. Not all carrier recovery is performed in the optimal fashion. However, in simulation it is (Section 16.3.6).

In MLE for symbol synchronization or timing recovery, the epoch is taken to be that timing which maximizes the correlation between the transmitted signal and the received signal (Proakis and Salehi, 2008). Not all symbol synchronization is performed in an optimal fashion. However, in simulation it is, and there are only a finite number of epochs to try, equal to the number of time samples per symbol.

The ML method for CPM demodulation and convolutional decoding, namely ML sequence estimation, is described in Section 12.12.

12.10   DEMODULATION FOR MEMORYLESS MODULATION

In this section, we discuss the processes marked as comprising demodulation in the bottom third of Figure 12.12, namely detection filtering through symbol decision.

12.10.1   Detection Filter

The **detection filter** is the filter in the demodulating receiver that cleans up the signal prior to sampling and symbol decision, basically by maximizing SNR. The detection filter removes spurious signals that are near-out-of-band or in nearby channels but does not remove inband ones. In this section, we look at what this filter should be and what the filtered signal looks like.

12.10.1.1   What to Use as Detection Filter   Figure 12.34 shows a simplified equivalent-baseband model of the large middle part of the end-to-end satellite communications system, from pulse-shaping filter $P(f)$ through to detection filter $Q(f)$ (compare to Figure 12.12). We ignore the HPA's nonlinearity but its filtering aspects (Section 16.3.7) are incorporated into the filter $H(f)$ along with the payload filters. The AWGN enters in two places, in the satellite electronics and in the ground electronics. Any noise-like interference is included in Gaussian noise.

The question of what the detection filter should be boils down to whether more noise is added on the uplink or the downlink, relative to the signal power. Usually, one or the other noise is dominant; it does not often happen that they are nearly equal. If only one link involves consumer equipment, usually that link has the dominant noise. Now, when the only signal impairment is one-location AWGN, the optimal detection filter is the **matched filter (MF)**, which maximizes the SNR into the symbol detector. Its transfer function $Q(f)$ is the complex conjugate of the total filtering that the signal has seen before the AWGN, times the inverse of the filtering after the AWGN (Proakis and Salehi, 2008). For us, neglecting the non-dominant noise source for now, $Q(f)$ would be the following for the two cases of the dominant noise location, where * denotes the complex conjugate:

$$Q(f) = \begin{cases} P^*(f)H^{-1}(f) \text{ when uplink noise is dominant} \\ P^*(f)H^*(f) \text{ when downlink noise is dominant} \end{cases}$$

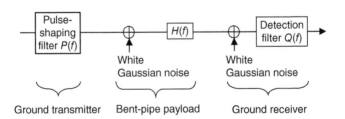

FIGURE 12.34   Simplified block diagram of middle part of end-to-end satellite communications system.

In practice, though, the filter $H(f)$ is ignored with the justification that it alters the signal only a little. When this filter is ignored, the detection filter is simply the MF for the pulse-shaping filter:

$$Q(f) = P^*(f), \text{in most applications}$$

This is not perfect but very good. Sometimes, though, perhaps to keep consumer terminals inexpensive or to keep using legacy terminals, the selected detection filter is less optimal:

$$Q(f) = \text{a filter with about same bandwidth as } P^*(f), \text{in some applications}$$

It is easy to see roughly why the filter matched to the pulse, namely $P^*(f)$, makes sense as the detection filter when the filters $H_1(f)$ and $H_2(f)$ can be neglected. Let us look at the equivalent-baseband transfer function of the signal pulse, for example, that shown in Figure 12.35, which is the RRC pulse with $\alpha = 0.35$. Now, the purpose of the detection filter is to maximize the SNR prior to the symbol decision, so the detection filter must keep as much of the signal power and cut out as much of the noise power as possible. Where the pulse transform is flat, the detection filter should be flat and at the highest level, where the transform is zero the detection filter should be zero, and where it is in between the detection filter should be in between.

12.10.1.2 What Detection Filter Output from Signal Looks Like We want to see what the signal part of the detection-filter output looks like. The examples given are only for MFs.

Since the pulse is the basic element of which the waveform for a symbol stream is constructed, we first look at the MF response $r(t)$ to the pulse $p(t)$, namely

$$r(t) \triangleq (p \circ q)(t) \text{ where } q(t) = p(-t)$$

Matched-filter responses to some pulses are illustrated in Figure 12.36. They are for four pulses, the first two for the rectangular pulse and the last two for RRC

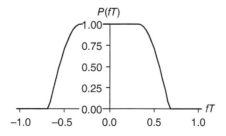

FIGURE 12.35 Transfer function of RRC pulse with $\alpha = 0.35$.

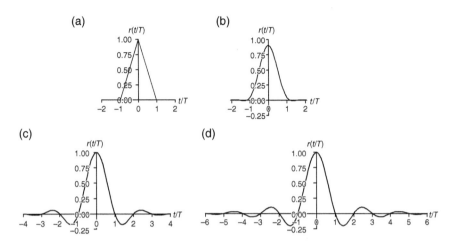

FIGURE 12.36   Matched filter output for pulse input: (a) rectangular pulse; (b) rectangular pulse with all but spectrum mainlobe filtered out; (c) RRC pulse with $\alpha = 0.35$; (d) RRC pulse with $\alpha = 0.2$.

pulses. The figure's second plot is actually for the variation on the rectangular pulse when only the mainlobe of the spectrum is kept.

The signal-only part $w(t)$ of the detection filter output is, for QPSK, BPSK, 8PSK, and 16QAM, in the equivalent-baseband representation, $\sum_k d_k r(t - kT)$, where d_k is the kth phasor. For OQPSK and minimum-shift keying (MSK–Section 12.11.1), where I and Q bits are used sequentially, $w(t)$ is $\sum_k d_{kI} r\left(t - kT\right) + j d_{kQ} r\left(t - \left(k + \dfrac{1}{2}\right)T\right)$, where d_{kI} and d_{kQ} are the kth I bit and Q bit, respectively. When $r(t)$ is zero at all non-zero multiples of T, then the value of $w(t)$ at any particular multiple of T has only one signal phasor contributing to it:

$$ w\left(iT\right) = \sum_k d_k r\left(it - kT\right) = d_i r\left(0\right) $$

for the first group of modulations. There is one clear time at which $w(t)$ gives us information about a particular phasor. Similarly for the second group. This property, which the rectangular pulse and the RRC pulses have, means that there is no ISI out of the detection filter. The bandlimited rectangular pulse does not have this desirable property, that is, the signal has ISI.

A pulse has no ISI out of its MF if and only if the pulse satisfies **Nyquist's (first) criterion** and the noise is AWGN (Proakis and Salehi, 2008; Benetto et al., 1987). The latter reference has a vivid drawing of what fulfilling the criterion means. $|P(f)|^2$ is chopped into segments R_s wide and they are all overlaid and added up, and the result is flat on a width R_s:

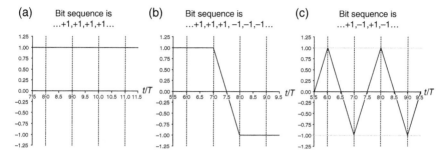

Optimal sampling times are marked by vertical dashed lines

FIGURE 12.37 Segment of matched filter output of BPSK signal with rectangular pulse, for various bit sequences.

$$\sum_k \left| P\left(f - \left(k + \frac{1}{2}\right) R_s \right) \right|^2 = \text{brickwall filter on } -R_s/2 \text{ to } +R_s/2$$

We look at segments of matched-filtered signal corresponding to different symbol streams, to see generally what the signal looks like and to notice the no-ISI property of appropriate pulses. Figure 12.37 shows a BPSK signal with rectangular pulse, for three different infinite-length bit sequences. The independent axis is time. The waveform is sampled once per T, that is, at rate R_s. In each plot in the figure, the dashed lines are at the optimal **sampling times**. These times are optimal because the absolute value of the signal level is maximum there, averaged long-term. Indeed, the absolute value of the samples is the same in all three plots, which is due to the lack of ISI.

Figure 12.38 shows a similar result for the RRC pulse with $\alpha = 0.35$. Once again, the absolute value of a sample at the optimal sampling time does not depend on the surrounding bits.

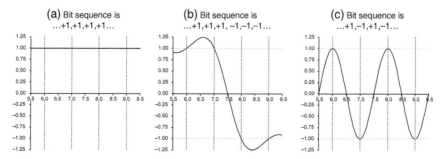

FIGURE 12.38 Segment of matched filter output of BPSK signal with RRC $\alpha = 0.35$ pulse, for various bit sequences.

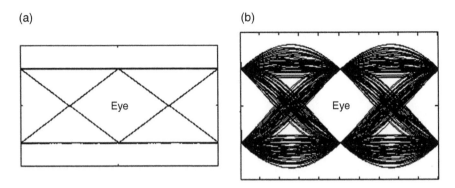

FIGURE 12.39 Example of BPSK eye patterns without noise for (a) rectangular pulse and (b) RRC $\alpha = 0.35$ pulse.

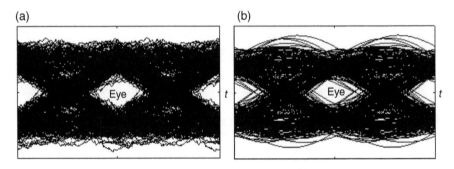

FIGURE 12.40 Example of BPSK eye patterns for E_b/N_0 of 9.6 dB and (a) rectangular pulse and (b) RRC $\alpha = 0.35$ pulse.

The **eye pattern** is a plot of the I or Q component of the detection-filter output in time, like the plots above but with many consecutive symbols' worth of the filter output overlaid. The optimal sampling time can be seen: it is the time at which the eye is open the widest. There the signal part of the detection-filter output has the greatest amplitude, so the SNR is greatest. Checking for an open eye is a quick way to verify that the transmitter and the receiver through the detection filter are probably working properly. The pattern is taken with a digital signal analyzer. Figure 12.39 gives examples of such patterns for BPSK without noise, for the rectangular pulse and the RRC $\alpha = 0.35$ pulse. Each pattern shows one eye in the middle surrounded by two half eyes. Figure 12.40 is similar but is for E_b/N_0 of 9.6 dB.

12.10.2 Carrier Recovery

The next step after cleaning up the signal with the detection filter is to generate good estimates of the received carrier's frequency and phase. **Carrier recovery** provides these. We have already introduced in Section 6.6.6.2 the basic PLL that reconstructs a carrier from a noisy input carrier. When the carrier is digitally modulated, there is no residual tone handy to lock to, so it must be created, then it can be fed to the PLL. This is discussed in some detail in Proakis and Salehi (2008).

The PLL will lock onto the phase of a random one of the signal phasors. If the modulation was PSK with differential encoding (Section 12.8.3), this is sufficient. The more common method is to have known symbol sequences in a frame so that the phase found by the PLL can be corrected to the unambiguous phase.

A feedback loop such as the PLL is designed to have a particular **order**, which determines what the loop can track with zero phase error and therefore what the demodulator can remove: a first-order loop can only correct a constant phase error, a second-order loop constant frequency error, and a third-order loop a constant frequency-rate error (Gardner, 2005). A second-order loop is sufficient for a GEO satellite since the frequency variation from a small orbital inclination is slow. For non-GEO satellites, there are two possibilities: a third-order loop or, since the ground station knows the orbit of the satellite, a continual VCO frequency correction and a second-order loop.

A second-order feedback loop has a native-baseband transfer function with (one-sided) bandwidth B_L. The **time constant** of the loop is about $1/B_L$. Roughly speaking, loop outputs separated in time by the time constant are independent. In the typical case of a high-gain second-order loop with damping factor $\zeta = 0.707$, the 3-dB-down point of the closed-loop response is at $B_L/3.33$ (Gardner, 2005). Approximately speaking, the loop tracks phase noise and spurious PM that have a variation rate less than $B_L/3.33$ and cannot track components that have a faster rate, as illustrated in Figure 12.41. Thermal noise within the loop bandwidth creates phase error. Any phase noise at a rate higher than roughly R_s will be averaged out by the detection filter, nullifying its effect. The detection filter is, roughly speaking,

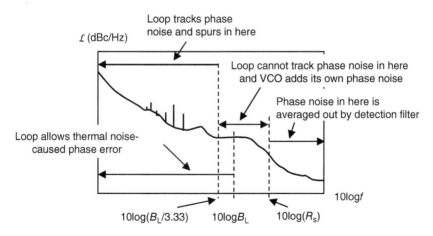

FIGURE 12.41 Approximately what happens to phase noise and phase spurs in various parts of spectrum for high-gain second-order PLL with $\zeta = 0.707$.

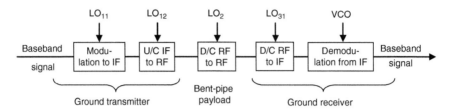

FIGURE 12.42   Phase noise sources in bent-pipe payload communications system.

FIGURE 12.43   QPSK phasors with phase noise on them.

an integration over the symbol duration T. For more accurate calculations or other ζ, refer to Gardner (2005).

Phase noise has several sources in the ground transmitter, the payload, and the ground receiver, as illustrated in Figure 12.42 for a bent-pipe payload. There are variations on the illustrated case; for example, the ground transmitter may modulate directly to IF, the payload may also have an upconversion, and the upconversion and the downconversion local oscillators may be derived from the same source, which reduces phase noise (Section 6.6.6.3).

Designing the carrier-recovery loop including choosing an appropriate value for B_L is a job for a ground station designer with knowledge of the phase noise characteristics of all system oscillators. For a second-order loop, choosing B_L is a tradeoff between tracking as much as possible of the phase noise (large B_L) and minimizing the thermal noise-caused phase error (small B_L).

Phase noise causes a zero-mean, Gaussian-distributed error on the signal's phase, as illustrated in Figure 12.43 for QPSK. While it is unlikely that phase noise itself will cause the wrong decision on what phasor was sent, in combination with thermal noise, signal distortion, and interference it will cause a wrong decision now and then.

12.10.3   Carrier Demodulation

The demodulation of the corrupted signal by means of the recovered carrier is illustrated in Figure 12.44. Demodulation consists of multiplying the detection-filtered signal by the recovered carrier's sine and cosine, separately, keeping the baseband parts, and forming a complex-baseband signal.

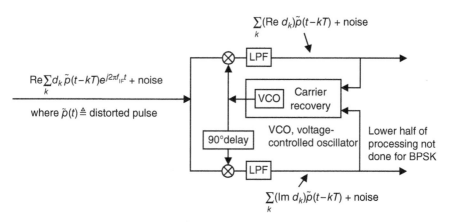

FIGURE 12.44   Demodulating the received signal.

12.10.4 Timing Recovery

The next step in the ground receiver is to generate a good estimate of the best time in a symbol duration at which to sample the complex-baseband signal. That sampling time, relative to some defined beginning of symbol, is the epoch. **Timing recovery,** also called **symbol synchronization,** recovers the epoch and the symbol rate R_s and provides them to the sampler. It is not necessary to track the derivative of symbol rate since this would have been removed by the carrier-recovery loop. (Doppler affects both carrier frequency and symbol rate by the same proportion.) Even with the carrier frequency tracked, there is still a small symbol rate variation left to track because in the transmitter the clock and oscillator are derived from different sources.

Timing recovery is conceptually similar to carrier recovery. In timing recovery, R_s corresponds to carrier frequency and *timing jitter* corresponds to carrier phase noise. The only sources of timing jitter are the transmitter clock and the timing recovery loop's voltage-controlled clock, when the payload is bent-pipe. For more information, see Proakis and Salehi (2008).

Symbol synchronization is often data-aided or decision-directed (Section 12.9). Commonly, there are known symbol sequences in a frame to allow data-aided synchronization.

12.10.5 Sampling

The sampler performs the sampling of the complex-baseband signal at the recovered epoch in every symbol duration. Out of the sampler, for QPSK, BPSK, 8PSK, 16QAM, and 16APSK, the equivalent-baseband stream of corrupted phasors is $\sum_k \tilde{d}_k \delta(t - kT)$. A **phasor diagram or scatter plot** of sampled

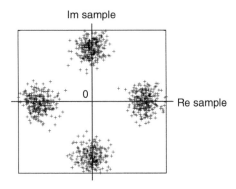

FIGURE 12.45   Example of QPSK scatter plot for E_b/N_0 of 9.6 dB and RRC pulse with $\alpha = 0.35$.

QPSK phasors in noise is shown in Figure 12.45. For OQPSK and MSK, where I and Q bits are used sequentially, the sampler output is

$$\sum_k \tilde{d}_{kI}\delta\left(t-kT\right) + j\tilde{d}_{kQ}\delta\left(t-\left(k+\frac{1}{2}\right)T\right).$$

12.10.6   Symbol Decision, SER, and MER

In some systems, symbol decision is performed on the samples and in others not. Most systems employ coding and directly input the samples of the noisy symbols into the decoder; such a decoder is **soft-decision**. The noisy symbols may be quantized. A **hard-decision** decoder does employ symbol decision, and so of course does an uncoded system. A hard-decision decoder does not perform as well as a hard-decision one. The three methods of data bit stream recovery are shown in Figure 12.46. This section applies to systems that do make symbol decisions.

The input to symbol decision is the stream of received, corrupted phasors, and the output is the best guess of the (undecoded) transmitted bit stream. The output bit stream is actually a sequence of groups of bits where each group corresponds to a detected phasor. So the operation is first to decide what the transmitted phasor probably was and second to undo the bit mapping.

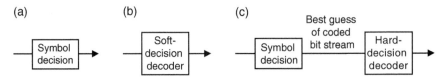

FIGURE 12.46   Data bit stream recovery: (a) uncoded data, (b) coded data with soft-decision decoding (most common), (c) coded data with hard-decision decoding.

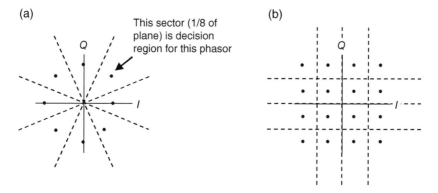

FIGURE 12.47   Decision regions for (a) 8PSK and (b) 16QAM.

The method of modulation-symbol decision is similar for all types of memoryless modulation we are treating. Let us start off by thinking about it for 8PSK and 16QAM. The complex plane is to be imagined, populated by the ideal phasors of the modulation scheme. The plane is divided into sectors that all contain a phasor. The edge between two sectors is equidistant from the two phasors in the sectors. The sectors are the **decision regions**. Now, whatever sector the received phasor falls into, the decision device decides that the ideal phasor in that sector was the one that was transmitted. This follows the rule that the nearest ideal phasor is selected. These regions are illustrated in Figure 12.47 for 8PSK and 16QAM. For QPSK, BPSK, and 16APSK, it is the same idea. For OQPSK, it is a little different because the I and Q bits do not transition at the same time, so their I and Q components can be thought of separately.

In digital communications systems, the most common quality measure of this recovered symbol stream is the **symbol error rate (SER)**. When the SER is low, it is almost always the case that when a symbol is detected wrong, the transmitted symbol was a nearest neighbor of the detected symbol (note that for 8PSK each symbol has two nearest neighbors, while for 16QAM some symbols have four, some have three, and the corner phasors have two). When the bits-to-symbol mapping was done with Gray coding, there is only one bit in error (Section 12.8.2):

$$\text{Uncoded BER} \approx \frac{R_{\text{s}}}{R_{\text{b}}}\text{SER} \quad \text{for low SER}$$

Plots of SER versus E_{b}/N_0 are given in Figure 12.48 for some memoryless modulation schemes plus MSK, where the only signal corruption is AWGN. QPSK, OQPSK, and MSK all have the same SER, represented by the second curve from the left.

A second performance measure used for digital television and digital radio is **modulation error ratio (MER)**. It is said to be a measure of SNR. Let $x(t)$ be the

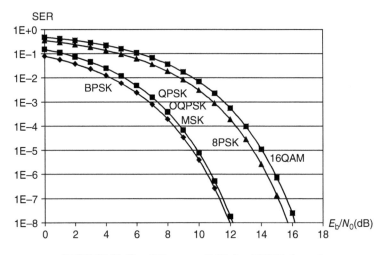

FIGURE 12.48 SER versus E_b/N_0 for AWGN channel.

transmitted, complex-baseband symbol stream and $y(t)$ be the recovered symbol stream, both of length N. Then MER is defined as follows:

$$\text{MER in dB} = 10 \log_{10} \left(\frac{P_{\text{reference}}}{P_{\text{error}}} \right)$$

$$\text{where } P_{\text{reference}} = \sum_{i=1}^{N} |x(i)|^2 \text{ and } P_{\text{error}} = \sum_{i=1}^{N} |y(i) - x(i)|^2$$

When a MER figure is quoted, it should be stated whether an equalizer was used (ETSI TR 101 290).

12.11 MODULATION WITH MEMORY

CPM is the most important subclass of modulations with memory. The phase is continuous. There is no amplitude variation, so it allows the payload's HPA to operate at or near saturation.

12.11.1 Modulation Schemes

The following discussion on CPM schemes comes from Anderson et al. (1986) and Proakis and Salehi (2008).

The memory in the modulation arises from the phase continuity. The phase φ at any time depends on all earlier phases:

$$\varphi(t) = \sum_{k=-\infty}^{n} 2\pi h d_k q(t - kT) \text{ for } nT \leq t \leq (n+1)T$$

The symbols d_k are real-valued and are selected from the alphabet ± 1, d$\pm 3, \ldots, \pm(M-1)$, where M is the order of the modulation scheme. The function $q(t)$ is the *phase-smoothing function* or the *phase-pulse*. If $q(t)$ is zero outside of the interval from 0 to T, then the modulation is *full-response* CPM, since the effect of the data symbol does not change past T. If $q(t)$ is non-zero past T, then the modulation is *partial-response* CPM. Beyond some time LT, where L is an integer, q equals 1/2. The number h is the *modulation index*. It tells by how much of the circle a modulation symbol can change the phase. It is a rational number usually less than 1.

The function $q(t)$ is a monotonically increasing function. It is usually thought of in terms of its non-negative derivative $g(t)$, the *frequency-pulse* or simply *pulse*, which gives the speed of the phase change:

$$q(t) = \int_0^t g(\tau)\,d\tau$$

MSK is a binary CPM with a constant rate of phase change over T, equal to one of two values. The rectangular phase pulse equals $1/(2T)$ from 0 to T and is zero outside, so MSK is a full-response CPM. The modulation index $h = 1/2$. Thus, during each transition, the phase goes at constant speed either ahead or behind by $\pi/2$. MSK is similar to OQPSK when the latter is considered to be a binary modulation at twice the symbol rate of the quaternary modulation.

Gaussian minimum-shift keying (GMSK) is not necessarily binary and is partial-response. The modulation index h is 1/2 (Neelamani and Iyer, 1996). The frequency pulse is as follows (Proakis and Salehi, 2008):

$$g(t) = \frac{Q\left[2\pi B\left(t - \dfrac{T}{2}\right)\right] - Q\left[2\pi B\left(t + \dfrac{T}{2}\right)\right]}{\sqrt{\ln 2}} \text{ where } Q(t) = \frac{1}{\sqrt{2\pi}}\int_x^\infty e^{-\frac{t^2}{2}}\,dt$$

This frequency pulse is symmetric about zero. It must be truncated beyond some $\pm LT/2$. To make it causal, it must be delayed by $LT/2$.

The GMSK frequency pulse represents a phase change that starts off slowly, speeds up, and brakes to a smooth stop. Depending on the parameter B, the pulse can be of duration about $2T$, for $BT = 1$, or $10T$, for $BT = 0.1$, for example. The European digital cellular standard GSM uses $BT = 0.3$ and truncates the phase pulse outside of a $3T$ duration (Proakis and Salehi, 2008).

CPM with the RC frequency pulse of duration LT is another choice, denoted LRC. The phase pulse is one cycle of cosine, raised up to be non-negative, and equals zero at 0 and LT (Proakis and Salehi, 2008).

The CPM specified in the DVB Return Channel Satellite, 2nd Generation (DVB-RCS2) standard is quaternary and partial-response with $L = 2$. Its frequency pulse can be any combination of the rectangular and the RC. The modulation index can take on various values less than 1/2 (ETSI EN 301 545-2).

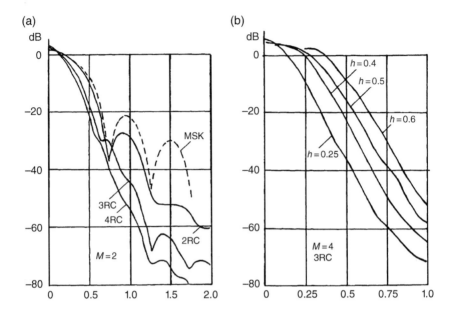

FIGURE 12.49 Power spectral density for CPM schemes: (a) binary with $h = 1/2$ and different frequency pulses; (b) quaternary with 3RC and different modulation indices. (©1981 IEEE. Reprinted, with permission, from Aulin et al. (1981)).

12.11.2 Modulation Spectrum

The spectrum shape of a CPM signal depends on all the modulation characteristics. The larger L is, the narrower the spectrum. The smoother the frequency pulse is, the narrower the spectrum. These two characteristics are illustrated in Figure 12.49a. The smaller h is, the narrower the spectrum, as illustrated in Figure 12.49b (Proakis and Salehi, 2008).

12.12 MAXIMUM-LIKELIHOOD SEQUENCE ESTIMATION

Maximum-likelihood sequence estimation (MLSE) is used for CPM demodulation and convolutional-code decoding. It is based on the assumption that all possible transmitted sequences are equiprobable. MLSE is optimal only for an AWGN channel but is widely used. The method decides on the most likely transmitted sequence given the received sequence. The most likely sequence is the one that is closest, in the sense of a distance metric, to the received one. On the surface, it would seem that the distance between the received sequence and all possible sequences would have to be computed.

The elegant **Viterbi algorithm** for MLSE drastically cuts down on the calculations. It treats the collection of all possible sequences as paths through a *trellis*.

States

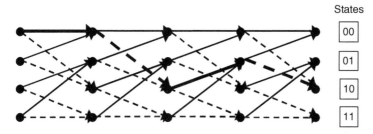

00

01

10

11

The solid lines indicate transitions where 0 is input
and the dashed lines where 1 is input

FIGURE 12.50   Simple trellis diagram (Wikipedia, 2020c).

A trellis path is an allowed sequence of *states*, one state at each *node* in the trellis. At each node, there are a number of possible states, where the state is the memory of the encoder or CPM modulator, respectively. From the state at one node, the path continues on to a state at the next node by means of the next bit(s) input to the encoder or the next symbol for the CPM modulator. A simple trellis diagram is shown in Figure 12.50. The nodes are the five vertical columns of states, at each node the same four states being possible. The solid lines indicate transitions where 0 was input and the dashed lines where 1 was input. An allowed path through the trellis is shown with a thick line, dashed in places.

The distance metric between the received sequence and a possible sequence is the sum of squared distances, one computed at each node along a trellis path. The algorithm remembers the metric up to each possible state at a node. It continues the trellis from these states to the states at the next node, in all possible ways. At this next node for each state, it keeps only the path to it that has the smallest metric and eliminates the other paths.

Minimizing the distance metric is equivalent to maximizing the correlation between the received sequence and all possible sequences (Anderson et al., 1986).

For practicality, in the Viterbi algorithm, the paths are *truncated*. This means that at each node, a decision is made for the node that is a fixed number of nodes back. This leads to a constant delay at the expense of decoding errors due to premature decisions.

12.13 DEMODULATION FOR MODULATION WITH MEMORY

For modulation with memory, a frame commonly contains interspersed, known symbol sequences to be used for carrier and timing recovery (example in Section 13.4.3.2). The European Telecommunications Standards Institute (ETSI) suggests an approach for receivers of DVB-RCS2 signals to synchronize on these sequences, based on correlation of the transmitted and received signals, as in Section 12.9 (ETSI TR 101 545-4).

After carrier and timing have been recovered achieved, the Viterbi algorithm of the previous section is applied to recover the symbol sequence. The trellis diagram for CPM has nodes spaced by the symbol duration T. The number of states is large: when the modulation index $h = n/p$ where n and p are integers without common factors, the number of states is $pM^{(L-1)}$ (Anderson et al., 1986).

A different technique for recovery of the symbol sequence can simultaneously recover the carrier phase. The technique is based on the *Laurent decomposition* of the CPM signal, which decomposes the signal as the sum of memoryless modulations. Only the largest contributions in the decomposition are retained. The phase recovery is decision-directed with a smaller delay in the decisions than is used for symbol sequence recovery, since the symbols are taken earlier and with less certainty for the phase recovery. The symbol sequence is then recovered by means of the Viterbi algorithm. Far fewer states than $pM^{(L-1)}$ are required per node (Colavolpe and Raheli, 1997). ETSI suggests this technique for CPM demodulation in receivers of DVB-RCS2 signals (ETSI TR 101 545-4).

12.14   BIT RECOVERY

This section applies to any type of modulation scheme because at this point the received signal has been reduced to a bit sequence. It may be helpful to refer to Figure 12.12 to see where we are in the satellite communications channel.

12.14.1   Descrambler and Deinterleaver

Descrambling and deinterleaving are simple to invert once the frame epoch is found.

12.14.2   Decoder and BER

The output of a decoder is a guess at the original data bits plus the baseband header bits, based on the recovered channel-bit stream. We are interested primarily in the guess of the data bits. The error rate in those bits gives the key performance measure of **data BER**. This is also called the **coded BER**.

A measure sometimes used to assess the strength of a code is to compare the data BER with the **channel-bit error rate (channel BER)**, which is the rate of errors in the bit stream before the decoder. It would also have been the BER if no coding had been used. It is also called the **uncoded BER**. The **coding gain** is the difference in dB between the $E_b/(N_0 + I_0)$ needed to achieve a given BER with coding and without coding, where the bits are data bits. When one sees a plot of BER versus some parameter, one must be clear which kind of BER it refers to. Usually, there is a direct relationship between the coded BER and the uncoded BER, that is, theoretical curves can be used in a simulation so it is not necessary to actually encode and decode.

See, for example, Proakis and Salehi (2008) for full descriptions of decoders.

Convolutional codes are decoded with the Viterbi algorithm (Section 12.12). First of all, if the code was punctured, then the missing bits are replaced by zeros (while present bits are either −1 or +1). Let us suppose that the convolutional encoder divided the input data bit stream into k parallel bit streams and output i bits for each k in. The input bit stream can then be thought of as a sequence of groups of k bits, and the coded bit stream can be thought of as a sequence of groups of i bits. While all input sequences are possible, not all output sequences are possible. The Viterbi algorithm can be implemented with either hard-decision or soft-decision input (Section 12.10.6).

The block decoders decide that the block sent was the one closest to the block received, by means of abstract algebra and not the Viterbi algorithm. The decoder outputs the block of bits that would have been input to the encoder to produce the decided-on transmitted block. There are also both hard-decision and soft-decision implementations of decoders (Proakis and Salehi, 2008).

When a convolutional code is concatenated with a block code, the decoding of the two codes is sequential, with deinterleaving in between.

Turbo decoding is different from the others in that decoding is iterative with the number of necessary iterations not known beforehand. Turbo decoding has soft-decision input. Typically, four iterations are enough if the achieved BER is 10^{-7}–10^{-6}, while eight to ten may be needed at BER of 10^{-5} (Proakis and Salehi, 2008).

LDPC decoding is also iterative. One simpler method uses hard decisions and the other more complex method uses soft decisions (Proakis and Salehi, 2008).

12.15 INTER-SYMBOL INTERFERENCE

ISI is the signal interfering into itself, namely each symbol is interfered into by nearby symbols. The main cause is filtering but imperfect synchronization also causes it. ISI from filtering can be mostly corrected by equalization in the ground receiver, a topic we do not discuss in this book. ISI from imperfect timing recovery cannot be corrected by equalization.

Some ISI sources will cause the I component of the baseband equivalent of an RF signal (Section 12.2.2) to interfere into itself and the Q component to interfere into itself. This is equivalent to the component of the RF signal that modulates cosine of the carrier frequency interfering into itself and the component of the RF signal that modulates sine interfering into itself. Other ISI sources will cause the I component to interfere into the Q and the Q into the I. This is called **I-Q crosstalk**. Both kinds can cause a larger-amplitude part of the signal to interfere into a smaller-amplitude part. Both kinds can occur at once.

We divide the ISI into three kinds: that caused by bandlimiting filters, which incidentally produce ISI, that produced by the pulse-shaping and detection filters, which is not incidental, and that caused by imperfect synchronization, which is again incidental.

12.15.1   From Bandlimiting Filters

Between the modulator and demodulator, the signal passes through many filters whose purpose is to bandlimit the signal. Every device has some bandwidth limitation, that is, only passes some range of frequencies, so it performs a filtering operation, no matter how benign. A filter smears the signal in time, creating linear distortion.

This section tells what filter characteristics produce ISI of the I signal component into itself and of the Q component into itself and what characteristics produce ISI of the I into Q and the Q into I.

As a way of introducing a concept about transfer functions, we start with a simpler, similar concept about real-valued functions. Some real-valued functions are **even functions** and some are **odd functions**, and the rest are a combination of even and odd parts. A function $x(f)$ is even when $x(f)$ is symmetric about zero, that is when $x(-f) = x(f)$. A function is odd when $x(-f) = -x(f)$. Every real function $x(f)$ can be separated into an even part $x_e(f)$ and an odd part $x_o(f)$ by the following method:

$$x(f) = x_e(f) + x_o(f) \qquad \text{where}$$

$$x_e(f) = \frac{1}{2}\left[x(f) + x(-f)\right] \quad \text{and} \quad x_o(f) = \frac{1}{2}\left[x(f) - x(-f)\right]$$

Figure 12.51 gives an example of this separation.

Corresponding terms for complex-valued functions are **conjugate symmetric** (or Hermitian symmetric (Proakis and Salehi, 2008)) and **conjugate antisymmetric**. A complex function $H(f)$ is conjugate symmetric when $H(-f) = H^*(f)$ and is conjugate antisymmetric when $H(-f) = -H^*(f)$. Every complex function $H(f)$ can be separated into a conjugate symmetric part $H_{cs}(f)$ and a conjugate antisymmetric part $H_{ca}(f)$ by the following method:

$$H(f) = H_{cs}(f) + H_{ca}(f) \qquad \text{where}$$

$$H_{cs}(f) = \frac{1}{2}\left[H(f) + H^*(-f)\right] \quad \text{and} \quad H_{ca}(f) = \frac{1}{2}\left[H(f) - H^*(-f)\right]$$

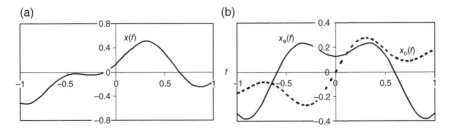

FIGURE 12.51   Example of (a) function, (b) separated into even and odd parts.

We now apply this separation to the filter transfer function, so we assume $H(f)$ is the baseband equivalent of an RF transfer function (Section 12.3.1). The importance of this separation is the following:

$$H(f) \text{ is conjugate symmetric} \quad \Leftrightarrow h(t) \text{ is real}$$
$$H(f) \text{ is conjugate antisymmetric} \Leftrightarrow h(t) \text{ is imaginary}$$

Thus, the conjugate symmetric part of $H(f)$ creates the real part $h_r(t)$ of the impulse response $h(t)$ and the conjugate antisymmetric part of $H(f)$ creates the imaginary part $h_i(t)$.

Now we divide a complex-baseband signal $u(t)$ into its real part $u_r(t)$ and imaginary part $u_i(t)$. From the formula for convolution, we have the following breakdown for the filtered signal:

$$v(t) = (h \circ u)(t) = \int_{-\infty}^{\infty} u(\tau) h(t-\tau) d\tau$$

$$= \int_{-\infty}^{\infty} \left[u_r(\tau) + ju_i(\tau) \right] \left[h_r(t-\tau) + jh_i(t-\tau) \right] d\tau$$

$$= \int_{-\infty}^{\infty} \left[u_r(\tau) + ju_i(\tau) \right] h_r(t-\tau) + \left[-u_i(\tau) + ju_r(\tau) \right] h_i(t-\tau) d\tau$$

Thus, a real-valued impulse response creates I-to-I (that is, real-to-real) and Q-to-Q (that is, imaginary-to-imaginary) ISI. An imaginary-valued impulse response creates I-Q crosstalk.

We want to see how to recognize a real or imaginary impulse response function from the filter's gain response $G(f)$ in dB and phase response $\varphi(f)$. The natural logarithm of $H(f)$ is $((\ln10)/20)G(f) + j\varphi(f)$. We divide $G(f)$ and $\varphi(f)$ into their even and odd parts, $G(f) = G_e(f) + G_o(f)$ and $\varphi(f) = \varphi_e(f) + \varphi_o(f)$. We now have to assume that $((\ln10)/20)G(f)$, which is about $0.115G(f)$, is small and that $\varphi(f)$, which is in radians, is also small in absolute value. $\varphi(f)$ is in the interval $(-\pi, \pi]$. By applying the Taylor series three times we obtain

$$H_{cs}(f) \approx 1 + \frac{\ln10}{20} G_e(f) + j\varphi_o(f) \quad \text{and} \quad H_{as}(f) \approx \frac{\ln10}{20} G_o(f) + j\varphi_e(f)$$

This means that for small gain (in dB, multiplied by 0.115) responses and small phase (in rads) responses,

- The even part of the gain response and the odd part of the phase response create I-to-I and Q-to-Q ISI
- The odd part of the gain response and the even part of the phase response create I-Q crosstalk.

FIGURE 12.52   Examples of filter gain response $G(f)$ and phase response $\varphi(f)$ that cause I–I and Q–Q distortion.

FIGURE 12.53   Examples of $G(f)$ and $\varphi(f)$ that cause I-Q crosstalk.

Figures 12.52 and 12.53 show examples of equivalent baseband transfer functions that exhibit these two kinds of distortion. The phase response $\varphi(f)$ in Figure 12.52 is a straight line, which corresponds simply to a delay, because the filter's group delay is related to phase by $\tau_g = -\mathrm{d}\varphi(f)/(2\pi\mathrm{d}f)$ (Couch II, 1990).

12.15.2   From Pulse-Shaping and Detection Filters

The pulse-shaping filter for memoryless modulations causes amplitude variations in the signal, more for smaller roll-off factor α of the RRC pulse. This can be seen in Figures 12.31 and 12.32. In an actual signal, the sequence of delayed pulses overlay each other, with various amplitudes and phases depending on the symbol constellation. The trailing end of a pulse for a large-amplitude phasor could almost cancel out the peak of the pulse for a smaller-amplitude phasor.

The amplitude variations in the signal are not harmful unless the payload's HPA is operating near saturation. The HPA will compress and rotate the signal more when the amplitude is large. User-terminal manufacturers have partially compensated for this in their proprietary symbol constellations.

If the payload's HPA is operating well backed off, the HPA does not distort the signal. In the ground receiver a MF, as the detection filter, would remove the ISI at the sampling points. This can be seen in Figure 12.36c,d, where the matched-filtered pulse has a large value at its center but is zero at all multiples of T away from the center.

If the detection filter is not the MF for the pulse, there will not be zeros at all multiples of T away from the detection-filtered pulse. Thus, there will be ISI, which could be removed by equalization.

12.15.3 From Imperfect Synchronization

The carrier-phase recovery and the symbol-epoch recovery are not perfect. The carrier-recovery loop has phase noise, some of which it tracks and removes and some of which it cannot track (Section 12.10.2). Similarly, the timing-recovery loop has timing jitter, some of which it tracks and some of which it cannot track. In addition, both kinds of loops are subject to thermal noise, which the loop tracks but should not.

Imperfect carrier-phase recovery causes I-Q crosstalk in the detection-filtered signal. Imperfect timing recovery causes I-to-I and Q-to-Q interference in the detection-filtered signal. The time constant of the variations is longer than a symbol duration because faster variations are averaged out by the detection filter, roughly speaking.

12.16 SNR, E_s/N_0, AND E_b/N_0

The related terms SNR, E_s/N_0, and E_b/N_0 can cause confusion. We give their general definition and then specifically their values out of the sampler.

12.16.1 General Definitions

The **SNR** is the ratio C/N of the signal power C to the noise power N at the output of a given filter. Sometimes "P" or "S" is used instead of "C." The filter can be actual or virtual. A virtual filter often used is the brickwall filter (Section 12.8.4.2). Alternatively, we may want to know what the ratio would be if a particular filter were placed at a particular point.

SNR can be measured with a power meter at filter output. Alternatively, it can be calculated as follows. Suppose the signal is $x(t)$ as a function of time, the impulse response of the filter of interest is $h(t)$, $y(t)$ is the virtually filtered signal, and the noise psd is N_0 over the filter's passband. Then, the calculated signal power and noise power are, respectively,

$$C = \int S_y(f)\,df = \int S_x(f)|H(f)|^2\,df \quad \text{and} \quad N = N_0 \int |H(f)|^2\,df$$

If the noise psd is not flat across the filter's passband, then the more general equation for noise power must be used:

$$N = \int S_n(f)|H(f)|^2\,df \quad \text{where } S_n(f) = \text{noise psd as a function of frequency}$$

Another way to compute SNR, a rougher way, is often used in signal-and-noise level budgets (Section 9.3). The estimated unfiltered signal power at a point in the signal path, the estimated N_0, and the noise bandwidth of an interesting filter are used.

Noise on an RF signal has two independent components, one on cosine and the other on sine of the carrier. It can equally be said to have a radial component and a circumferential component. Noise on the baseband equivalent of an RF signal is complex-valued. Phase noise is real-valued.

Electronic devices do not create noise with a flat psd. However, what matters is not the exact shape of the psd but the power of the noise out of a bandpass filter of interest. We must just be careful, when giving a number to the noise level into a filter, to use the average noise level over the filter passband.

Satellite communications does not provide an AWGN channel (Section 12.5). The way performance plots for an AWGN channel are translated to a satellite channel is that a hit to the plot's independent variable E_s/N_0, E_b/N_0, or SNR, is taken. That is, the E_s/N_0, E_b/N_0, or SNR which is actually required is larger by some number called "implementation loss" which takes into account the signal corruptions caused by things like filtering, the nonlinear HPA, spurious signals, and imperfect carrier recovery (Section 14.9).

E_s/N_0 and E_b/N_0 are, respectively, **symbol-energy-to-noise-psd** and **bit-energy-to-noise-psd**. They are more subtle terms than SNR because it appears that no filter is involved in the definitions. So what filter passband to take the average noise psd over? We must call to mind how E_s/N_0 is used, and that is primarily as the independent variable in SER or BER plots. Now, in the plots of theoretical SER or BER over an AWGN channel, E_s/N_0 is the SNR out of the MF at the optimal sampling time. It also equals $(C/R_s)/N_0$. However, on a channel with some other filters and a detection filter that is not the MF, these two quantities are not equal, as we will see in the following section. Also, as we have noted, the noise psd may not be constant. The truest definition of E_s/N_0 is the SNR out of the detection filter at the sampling time which maximizes the SNR, if the detection filter were placed at the point of interest. A more easily computable definition of E_s is C/R_s where C is unfiltered signal power at the point of interest, and N_0 is a weighting of the noise psd by the detection filter $P(f)$:

$$N_0 = \int S_n(f)|P(f)|^2 \, df \Big/ \int |P(f)|^2 \, df$$

Once E_s/N_0 is computed, then E_b/N_0 can be computed as $(E_s/N_0)(R_s/R_b)$.

12.16.2 Values at Sampler Output

We want to compare different cases of E_s/N_0 at the sampler output. Let us start with the ideal case, where there is no signal corruption besides AWGN and the detection filter is the MF of a pulse which satisfies the Nyquist criterion. For memoryless modulations, the equivalent baseband signal $x(t)$ into the MF is

$$x(t) = \sum_k d_k p(t - kT) + n(t)$$

We now filter this by the MF to get $y(t)$:

$$y(t) = \sum_k d_k \int p(\tau - kT) p(\tau + t) d\tau + \int n(\tau) p(\tau + t) d\tau$$

Sampled at the optimal sampling time zero for d_0 this is

$$y(0) = d_0 \int (p(\tau))^2 \, d\tau + \int n(\tau) p(\tau) \, d\tau$$

The following is proportional to its power:

$$E|y(0)|^2 = E\left(|d_0|^2\right)\left[\int (p(\tau))^2 \, d\tau\right]^2 + \int d\tau \int dt \, E\left[n(\tau) n^*(t)\right] p(\tau) p(t)$$

We note that $E|d_0|^2 = E|d_k|^2$ for any k.

The autocorrelation function of white Gaussian noise is the scaled delta function $N_0 \delta(t)$. So

$$E|y(0)|^2 = E\left(|d_0|^2\right)\left(\int (p(t))^2 \, dt\right)^2 + \int dt \, N_0 \left(p(t)\right)^2 \, dt$$

$$= E|d_0|^2 T^2 + N_0 T \quad \text{wher } p(t) \text{ has been scaled so that } \int (p(t))^2 \, dt = T$$

The E_s/N_0 out of the sampler is then $E|d_0|^2/(R_s N_0)$.

We now look at a non-ideal case, where the detection filter is matched to the same pulse as above but there has been non-negligible filtering $H_1(f)$, $H_2(f)$, and $H_3(f)$ (see Figure 12.12), so the filtered pulse is $\tilde{p}(t) \triangleq [p \circ h_1 \circ h_2 \circ h_3](t)$. Now the signal out of the detection filter $P(f)$ is

$$y(t) = \sum_k d_k \int \tilde{p}(\tau - kT) p(\tau + t) \, d\tau + \int n(\tau) p(\tau + t) \, d\tau$$

and, if $t = 0$ is still the optimal sampling time for d_0,

$$E|y(0)|^2 = E\left(|d_0|^2\right)\left[\sum_k \int \tilde{p}(\tau - kT) p(\tau) \, d\tau\right]^2 + \int d\tau \int dt \, E\left[n(\tau) n^*(t)\right] p(\tau) p(t)$$

$$= E\left(|d_0|^2\right)\left[\sum_k \int \tilde{p}(\tau - kT) p(\tau) \, d\tau\right]^2 + N_0 T$$

The E_s/N_0 out of the sampler in this case are then

$$E_s/N_0 = E\left(|d_0|^2\right)\left|R_s \sum_k \int \tilde{p}(\tau - kT) p(\tau) \, d\tau\right|^2 / (R_s N_0)$$

For comparison, E_s/N_0 computed right before the detection filter is something different:

$$E_s/N_0 = E\left(|d_0|^2\right)\left(R_s \int (\tilde{p}(t))^2 \, dt\right)^2 / (R_s N_0)$$

12.A SKETCH OF PROOF THAT PULSE TRANSFORM AND SIGNAL SPECTRUM ARE RELATED FOR MEMORYLESS MODULATION

We invoke probability theory to sketch the proof of an important property of the spectrum of the complex baseband signal $z(t)$, which we consider now to be a random or stochastic process (Papoulis, 1984).

For QPSK, 8PSK, 16QAM, and 16APSK, the baseband signal $z(t)$ can be written as $\sum_k d_k p(t-kT)$ where d_k is the kth phasor and $p(t)$ is the real-valued pulse function. We assume that the phasors d_k of the data sequence are chosen independently of each other and that each is equally likely to be any one from the phasor collection. Then, the autocorrelation function $R(\tau)$ is as follows:

$$R(\tau) = E\left[\sum_k d_k p(\tau + t - kT) \sum_i d_i^* p(t - iT)\right] = \sum_k E\left[d_k d_i^*\right] p(\tau + t - kT) \sum_i p(t - iT)$$

$$= \sum_k E\left[|d_k|^2\right] p(\tau + t - kT) p(t - kT) = E\left[|d_k|^2\right]\left[p(t) \circ p(-t)\right](\tau)$$

The only terms that matter in the product of the two sums above are those for which $k = i$; when $k \neq i$, $E\left(d_k d_i^*\right) = 0$ because d_i is equally likely to be any phasor or its negative.

For OQPSK and MSK, the baseband signal $z(t)$ can be written as follows:

$$z(t) = \sum_k d_{kI} p(t - kT) + j d_{kQ} p\left(t - \left(k + \frac{1}{2}\right)T\right)$$

where d_{kI} and d_{kQ} are the kth I bit and Q bit, respectively

We assume that the bits are all independent of each other and that each is equally likely to be -1 or 1. Then $R(\tau)$ for $z(t)$ is as follows:

$$R(\tau) = E\left\{\sum_k \left[d_{kI} p(t - kT) + j d_{kQ} p\left(t - \left(k + \frac{1}{2}\right)T\right)\right]\right.$$

$$\left.\sum_i \left[d_{iI} p(t - iT) + j d_{iQ} p\left(t - \left(i + \frac{1}{2}\right)T\right)\right]\right\}$$

$$= \sum_k \left(E d_{kI}^2\right) p(\tau + t - kT) p(t - kT)$$

$$+ \sum_k \left(E d_{kQ}^2\right) p\left(\tau + t - \left(k + \frac{1}{2}\right)T\right) p\left(t - \left(k + \frac{1}{2}\right)T\right)$$

$$= E\left[|d_k|^2\right]\left[p(t) \circ p(-t)\right](\tau)$$

where the d_k again are the possible phasors. So the result is the same as for the other modulation schemes.

We invoke properties of the Fourier transform, where $u(t)$ and $v(t)$ are any two time functions:

$$\left[\mathcal{F}(u \circ v)\right](f) = U(f)V(f) \quad \text{and} \quad \left[\mathcal{F}\left(v^*(-t)\right)\right](f) = V^*(f)$$

and we find the simple relationship between the signal spectrum and pulse transform for all of these modulation schemes:

$$S(f) = E\left[|d_k|^2\right]|P(f)|^2 \quad \text{where } E\left[|d_k|^2\right] \text{ is the average phasor power}$$

REFERENCES

Anderson JB, Aulin T, and Sundberg C-E (1986). *Digital Phase Modulation*. New York: Plenum Press.

Aulin T, Rydbeck N, and Sundberg C-E (1981). Continuous phase modulation--part II: partial response signaling. *IEEE Transactions on Communications*; 29 (3) (Mar.); 210–225.

Benedetto S, Biglieri E, and Castellani V (1987). *Digital Transmission Theory*. Englewood Cliffs, NJ: Prentice-Hall.

Biglieri E (2020). Private communication; Aug. 31.

Bonnaud A, Feltrin E, and Barbiero L (2014). DVB-S2 extension: end-to-end impact of sharper roll-off factor over satellite link. *International Conference on Advances in Satellite and Space Communications*; Feb. 23–27.

Colavolpe G and Raheli R (1997). Reduced-complexity detection and phase synchronization of CPM signals. *IEEE Transactions on Communicatons*; 45 (9) (Sep.); 1070–1079.

Couch II LW (1990). *Digital and Analog Communication Systems*, 3rd ed. New York: Macmillan Publishing Company.

Elbert BR (2004). *The Satellite Communication Applications Handbook*, 2nd ed. Boston, MA: Artech House.

ETSI TR 101 290, v1.3.1 (2014). Digital video broadcasting (DVB); measurement guidelines for DVB systems. July.

ETSI TR 101 545-4, v1.1.1 (2014). Digital Video Broadcasting (DVB); second generation DVB interactive satellite system (DVB-RCS2); part 4: guidelines for implementation and use of EN 301 545-2. Sophia-Antipolis Cedex (France): European Telecommunications Standards Institute.

Gardner FM (2005). *Phaselock Techniques*, 3rd ed. New Jersey: John Wiley & Sons, Inc.

Gilat Satellite Networks (2013). SkyEdge™ IP, IP router VSAT. Product information. On www.gilat.com/dynimages/t_brochures/files/skyedge-ip-290813-final.pdf. Accessed July 19, 2016.

Gilat Satellite Networks (2014). SkyEdge™II IP, high performance broadband router VSAT. Product information. On www.gilat.com/dynimages/t_brochures/files/SkyEdgeII-IP-280914-FINAL.pdf. Accessed July 19, 2016.

Howell A of Inmarsat (2010). Broadband global area networks. Viewgraph presentation. *Standards and the New Economy Conference* led by Cambridge Wireless; Mar. 25, 2010. Accessed Nov. 29, 2014.

Hughes Network Systems (2015). Hughes HT1300 Jupiter ™ system multi-band router. Product information. June. On www.hughes.com/resources/ht1300-jupiter-system-router-multi-band?locale=en. Accessed July 19, 2016.

Kienle F and Wehn N (2008). Macro interleaver design for bit interleaved coded modulation with low-density parity-check codes. *IEEE Vehicular Technology Conference*; May 11–14.

Lee JK and Chang DI (2017). Performance evaluation of DVB-S2X satellite transmission according to sharp roll off factors. *International Conference on Advanced Communications Technology*; Feb. 19–22.

Lin S and Costello, Jr DJ (1983). *Error Control Coding, Fundamentals and Applications*. Englewood Cliffs, NJ: Prentice-Hall.

Neelamani R and Iyer D (1996). Spectral performance of GMSK: effects of modulation index and quantization. *IETE Journal of Education*; 37 (4); 231–236.

Papoulis A (1984). *Probability, Random Variables, and Stochastic Processes*, 2nd, International student ed. Singapore: McGraw-Hill.

Proakis JG and Salehi M (2008). *Digital Communications*, 5th, International ed. New York: McGraw-Hill.

Unnikuttan A, Rathna M, Rekha PR, and Nandakumar R (2014). Design of convolutional interleaver. *International Journal of Innovative Research in Information Security*; 1 (5) (Nov.).

Wikipedia (2018). Satmode. Article. Dec. 19. Accessed 2019 May 1.

Wikipedia (2020a). BCH code. Article. Apr. 10. Accessed Apr. 11, 2020.

Wikipedia (2020b). Low-density parity-check code. Article. Mar. 26. Accessed Apr. 6, 2020.

Wikipedia (2020c). Convolutional coding. Article. Apr. 1. Accessed Apr. 27, 2020.

CHAPTER 13

SATELLITE COMMUNICATIONS STANDARDS

13.1 INTRODUCTION

This chapter addresses some aspects of a satellite communications system that go beyond digital communications theory. We introduce the Open Systems Interconnect (OSI) communications model, the basis for discussion in the rest of the chapter. Then, we give two examples of satellite systems with regenerative payload that use the first generation of satellite communications standards. Next, we summarize the second generation of Digital Video Broadcasting (DVB) standards for satellite communications. Finally, we summarize a simplified return-link standard that is in use. This chapter relies on the material in Chapter 12.

Another aspect of satellite communications, namely system connectivity among the satellite, user terminals, and ground stations, was discussed in Sections 1.2 and 10.1.2, and examples were given in Sections 10.3 and 10.4.

A note on terminology: a "ground terminal" is meant to be either a ground station or a user terminal.

Satellite Communications Payload and System, Second Edition. Teresa M. Braun and Walter R. Braun.
© 2021 John Wiley & Sons, Inc. Published 2021 by John Wiley & Sons, Inc.

13.2   BACKGROUND

13.2.1   OSI Model of Communications

Increasingly, satellites provide Internet connections. Even for those that do not but offer two-way communications, there must be a way for the user terminals to gain access to the return link, a topic we have not addressed yet. The **OSI** model of the communications functions of a telecommunications system offers a framework for understanding how these things work. The model divides communications into seven ascending **layers**, where each layer is conceptually built on top of the layers below. In this book, we care about only the lowest four layers (and, in all chapters before this one, only the first layer):

- **Layer 1** is the **physical layer**, with which we are most familiar, which has to do with modulation, coding, and frequencies. For time-domain multiplexed communications, timeslots and frames are also part of this layer.
- **Layer 2** is the *data link layer*. It provides node-to-node data transfer between two physically connected nodes (Wikipedia, 2020g). A familiar example is the Ethernet, which operates simultaneously on layers 2 and 1. The data link layer has two sublayers –
  a. **Medium access control (MAC) sublayer**
  b. **Logical link control (LLC) sublayer**, which encapsulates (Section 13.3.1) layer-3 packets.
  Transfer of data in layer 2 may occur by **circuit-switching** based on MAC addresses. A signal path dedicated to the message, called a **connection**, is established, used, then torn down.
- **Layer 3** is the *network layer*. The Internet is based on **packets**, which are fragments of a message along with network addresses of the source and destination. Transfer of data in layer 3 occurs by **packet-switching**, which may be connection-less, as with the **Internet Protocol (IP)**, or connection-oriented (Wikipedia, 2020g).
- **Layer 4** is the *transport layer*. Transfer of data in this layer occurs usually by one of two methods (Wikipedia, 2020d). One is the *Transmission Control Protocol (TCP)*, which delivers a *segment* of bytes from an application program over a reliable, error-free connection over which the segments arrive in order. The Internet's HTTP protocol uses it (Wikipedia, 2020f). The other method, the *User Datagram Protocol (UDP)*, delivers *datagrams*. It is for time-sensitive applications like voice over IP. It is unreliable and does not put datagrams back in their original order. It supports *multicast*, which is one ground terminal sending a message to a group of multiple ground terminals (Wikipedia, 2020c).

The air interface (Section 10.1.2) for each user link of a satellite system embodies layers 1 and 2.

The concept of the bottom three layers of the OSI model is shown in Figure 13.1 (The concept works the same when you include layers 4 through 7, too.) On the

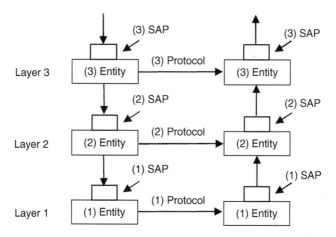

FIGURE 13.1   OSI model of communications between two directly connected computers, bottom three layers. (After Svobodova (1989)).

left-hand side is a stack of software entities for layers 1, 2, and 3 residing in the first computer. Similarly for the right-hand side of the figure and the second computer. The first computer has a message to send to the second computer. The message would have come down to layer 3 from higher layers. The horizontal lines mean that conceptually the first computer's entity for layer k communicates with the second computer's layer-k entity, using the protocol for layer k communications, for $k = 1, 2, 3$. In fact what happens is that the first computer's software for layer 3 asks the *service access point (SAP)* of its layer 2 for the service of converting the message into the layer-2 format (Tomás et al., 1987). This means adding a header and possibly reformatting. A similar thing happens between layers 2 and 1. The physical-layer message is transmitted to the physical-layer entity in the second computer. As the message goes up the stack in the second computer, the SAP of layer k removes the layer-k header and reformats the message for layer $k + 1$ and sends it upward, for $k = 1, 2$.

For a non-regenerative payload, the communications between two ground terminals are as illustrated in Figure 13.2. The satellite payload consists of two simple protocol stacks between the two ground terminals. The left-hand layer-1 entity of the payload understands the protocol of the signals sent from ground terminal 1 in terms of frequencies, beams, and timeslots. The payload converts the message into a signal that the downlink protocol understands and transmits it. Layer-2 functions are performed in the user terminal and the ground station while the network control center (NCC) controls the MAC function of layer 2.

For a regenerative payload, the communications between two ground terminals is as illustrated in Figure 13.3. The satellite payload is again between the two ground terminals. The payload consists of two stacks of level-1 and level-2 entities. The payload performs layer-2 functions, while the NCC controls the MAC function. We will see examples of this in Section 13.3.

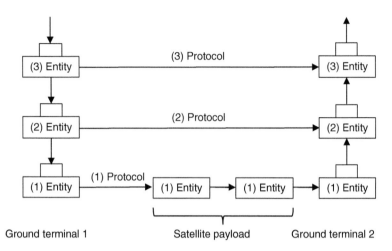

FIGURE 13.2   OSI model for one-hop communications between two ground terminals over non-regenerative payload.

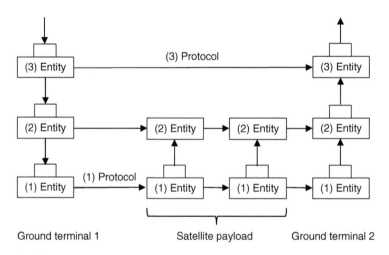

FIGURE 13.3   OSI model for one-hop communications between two ground terminals over regenerative payload.

13.2.2   Multiple Access Methods

In the OSI communications model, the means to access the satellite return-link resources belongs to layer 2, the data link layer. Various methods are used by satellite systems. A multiple access method provides (1) ways to allocate resources (MAC sublayer) and (2) ways to share the physical resources (LLC sublayer).

There are at least three ways to allocate the resources:

- Planning and scheduling.
- *Demand-assigned multiple access (DAMA)*, in which a user terminal sends a request to the NCC on a control or signaling channel, and the NCC returns a resource assignment to the user terminal. It is widely used in satellite communications, especially in very-small-aperture terminal (VSAT) systems (Wikipedia, 2020a). VSAT systems are described in Section 17.4.2.
- *Random access*. There is no assignment of resources to specific users.

There are various ways of sharing the physical resources that DAMA allocates. The ways are closely related to the ways of multiplexing described in Section 1.3:

- **Time-domain multiple access (TDMA)**, in which the user terminal transmits during the timeslots that have been assigned to it
- **Frequency-division multiple access (FDMA)**, in which the user terminal transmits on the frequency band that has been assigned to it
- **Multi-frequency TDMA (MF-TDMA)**, in which the user terminal transmits on the frequency band and during the timeslots that have been assigned to it
- **Code-division multiple access (CDMA)**, in which the user terminal applies the assigned spreading code to its data stream. It then transmits on the same frequency channel as the other user terminals.

The difference between TDMA and time-division multiplexing (TDM), for example, is that TDM is a physical technique in layer 1 of the OSI communications model.

In random access, the user terminal simply transmits data on a designated channel when it wants, so there can be conflicts. It is meant for short messages. One method is *slotted Aloha*, in which a user terminal starts transmitting at the beginning of any timeslot (Wikipedia, 2020e).

Examples of DAMA and random access are presented in Section 13.4.3.4.

13.2.3 Adaptive Coding and Modulation

Adaptive coding and modulation (ACM) is a technique for dynamically adapting user-link parameters to propagation and interference conditions. It dramatically improves both throughput (capacity) and link availability. In clear-air conditions, it uses higher-order modulations and higher coding rates, while in unfavorable conditions it uses the more robust lower-order modulations and lower coding rates. It is especially helpful at Ka/K-band because of the otherwise large rain margins necessary in a link with fixed parameters (Rinaldi and De Gaudenzi, 2004a). Many satellite systems implement it (Emiliani, 2020). An illustration of the benefit is given in Section 14.4.4.5.

Each possible combination of modulation and coding that the satellite system has at its disposal is called a **MODCOD**.

All ACM systems are proprietary. One reason for this is that they have different methods of transmitting the reverse channel required for ACM operation (Telesat, 2010).

It may be the NCC that commands the user terminal to a particular MODCOD for a link, based on a measure of the link's quality, or it may be the user terminal itself that decides on a MODCOD. The quality measure is $C/(N + I)$ in the receiver. Looking at the return link first, we see there are two ways the NCC can find out the user-uplink quality. One is that the ground station measures the end-to-end return link's $C/(N + I)$ (Section 14.7) and calculates the user-uplink $C/(N+I)$ from it. The second way is that the user terminal measures the forward link $C/(N+I)$ and sends the information to the ground station, which frequency-scales the rain attenuation from the user-downlink frequency to the user-uplink frequency (Section 14.4.4.1.6) and estimates the user-uplink quality. The quality of the forward link's user link can similarly be found out by the NCC in two ways.

ACM can be applied on just the uplink, just the downlink, or both.

An optimization and assessment study of ACM was done for the forward link. For a bent-pipe payload, whether processing or not, it was assumed that the processing necessary for ACM would take place in the ground station (or NCC), while for a regenerative payload it would take place in the payload. Communications would occur in a hybrid time-division multiplex, frequency-division multiplex way (TDM-FDM) (Section 1.3). The study case was for European coverage with 43 beams at 20 GHz, uniformly distributed traffic, and rain attenuation statistics as in central Italy. The downlink beam power was kept constant for all link conditions. It was found that the rain attenuation statistics had only a limited impact on the average system throughput. The reference case with fixed link parameters held a rain margin for 99.7% availability. The throughput for the ACM system was on average 2.5 times that of the fixed-parameter system. The worse the rain statistics or the higher the required link availability, the more benefit ACM can provide on the forward link (Rinaldi and De Gaudenzi, 2004a).

The same authors did a similar study for the return link. For return-link access, MF-TDMA and two CDMA schemes were treated. For a bent-pipe payload, it was assumed that the processing would take place in the ground station (or NCC), while a regenerative payload would perform it by itself. Link adaptation could be managed in more than one way. In a distributed way, the user terminal itself would decide on the MODCOD for the return link based on a link quality report provided by the ground station. In a centralized way, the MODCOD would be selected by the ground station and reported to the user terminal. The study case was the same as in the first study, but at 30 GHz. The throughput for the MF-TDMA system was found to be more than four times as great as the fixed-parameter system with uplink power control, with only 30% as much transmit power (one CDMA system needed only one third the transmit power of the MF-TDMA system) (Rinaldi and De Gaudenzi, 2004b).

13.3   APPLICATION EXAMPLES OF FIRST-GENERATION STANDARDS

13.3.1   How Communications Look to User Terminal

Good examples of the implementation of layers 1 through 3 for a regenerative pay-load come from Amazonas 2 and Spaceway 3. In both systems, a user terminal has an *IP entity* in it, which can receive and transmit IP packets. When a user terminal has IP packets to send, it performs *multi-protocol encapsulation (MPE)* on them, fragmenting the IP packets and adding a header to each fragment. This process creates a *transport stream (TS)* of *MPEG-2 transport packets* (ETSI TS 102 429-1 v1.1.1, 2006), as shown in Figure 13.4. MPEG is the Motion Picture Experts Group. The transport packets are at layer 2. The terminals have MAC addresses, and the payload routes the transport packets according to their destination MAC address. The receiving terminal receives the MPEG-2 packets and undoes the encapsulation, returning them to IP packets, and passes them to its IP entity (Garcia et al., 2006; Yun et al., 2010).

13.3.2   Amazonas 2

The AmerHis 2 processor onboard Amazonas 2 follows the European Telecommunications Standards Institute (ETSI) Regenerative Satellite Mesh (RSM) standard in version B (ETSI TS 102 602 v1.1.1, 2009a). The standard follows the DVB-S standard for the forward link and a specialization of the DVB Return Channel Satellite (DVB-RCS) standard for the return link (ETSI TS 102 429-1 v1.1.1, 2006) (see Section 13.4 on the second generation of these standards). The channel coding is concatenated convolutional and Reed Solomon and the signal modulation is quaternary PSK (QPSK) (Sections 12.6.2 and 12.8.1).

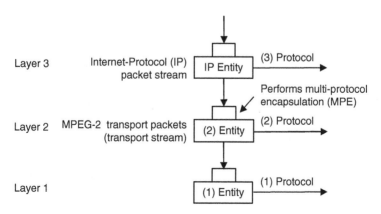

FIGURE 13.4   IP packet flow encapsulated into MPEG-2 transport packets, in Amazonas 2 or Spaceway 3 user terminal.

Amazonas 2 performs circuit-switching. Suppose a ground terminal generates or receives a sequence of IP packets to transmit to an IP destination. The terminal sends an alert to the payload, which passes it to the NCC. The NCC sets up the connection, a logical association of two terrestrial terminals identified by a pair of MAC addresses. The NCC commands the payload to set up appropriate channelization and routing to form a dedicated signal path through the payload. The terminal inserts the destination MAC address into its MPE transport packets as it encapsulates the IP packets of the sequence. The terminal then sends its signal over the path. The connection may be one-way (simplex) or two-way (duplex). The NCC dissolves the connection when the transmission is over (Yun et al., 2010; ETSI TS 102 602 v1.1.1, 2009a).

Amazonas 2 provides multicast as well, where one user terminal sends the same message to a group of multiple terminals (Garcia et al., 2006).

Thales Alenia has developed a third generation of AmerHis, which adds DVB-S2 options to the forward link (Yun et al., 2010).

The ground stations have the same kind of terminal as the users do.

13.3.3 Spaceway 3

Spaceway 3 follows an earlier version of the ETSI RSM standard, version A (Whitefield et al., 2006). RSM-A is not compliant with DVB (ETSI TS 102 188-1 v1.1.2, 2004). Spaceway 3 uses a different kind of MPE from Amazonas 2.

Let us first describe the packet-switching of the Spaceway 3 system. Sources for this section are Whitefield et al. (2006) and Fang (2011) except where indicated.

Spaceway 3 claims to perform packet-switching. Suppose a terrestrial terminal generates or receives a sequence of IP packets to transmit to an IP destination. The terminal makes a request to the payload for a bandwidth allocation (bandwidth on demand). The terminal is assigned a frequency and timeslot in its uplink beam. The terminal inserts the destination MAC address into its MPE transport packets as it encapsulates the IP packets of the sequence. The payload demodulates and decodes the transport packets and routes them to the queues for the destination beams and terminals. Each downlink beam has its own downlink queue. The downlink scheduler dynamically evaluates which downlink beams have bursts ready to transmit.

Spaceway 3 provides multicast as well (Whitefield et al., 2006).

Figure 13.5 shows Spaceway 3's multiple access methods, namely a combination of FDMA and TDMA on the uplink and TDMA on the downlink. The ground stations have the same kind of terminal as the users do.

13.4 SECOND-GENERATION DVB COMMUNICATIONS STANDARDS

The family of **DVB** standards for satellite communications is produced by the European Broadcasting Union (EBU), the Comité Européen de Normalisation Electrotechnique (CENELEC), and ETSI. DVB was originally designed for

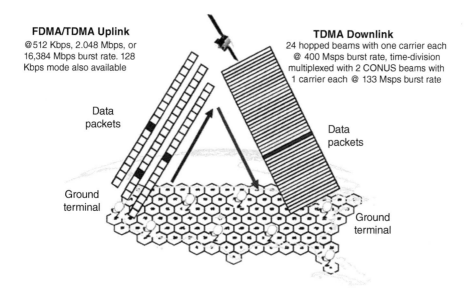

FDMA/TDMA Uplink
@512 Kbps, 2.048 Mbps, or
16,384 Mbps burst rate. 128
Kbps mode also available

TDMA Downlink
24 hopped beams with one carrier each
@ 400 Msps burst rate, time-division
multiplexed with 2 CONUS beams with
1 carrier each @ 133 Msps burst rate

Data
packets

Data
packets

Ground
terminal

Ground
terminal

FIGURE 13.5   Spaceway 3 uplink and downlink access techniques. (©2011 IEEE. Reprinted, with permission, from Fang (2011)).

terrestrial broadcast of digital television and is used by many broadcasters worldwide for this purpose (Wikipedia, 2018a). Today it is in wide use for sending both TV and data over satellite.

Many details of the digital communications aspects of the standards have already been pointed out in Chapter 12.

13.4.1   DVB-S2 Standard for Forward Link

DVB for Satellite, 2nd Generation (DVB-S2), is the second generation of the DVB standard for forward-link satellite communications (ETSI EN 302 307-1, v1.4.1, 2014). It has been called "the world's most successful satellite industry standard" (Hughes, 2020). In conjunction with a return-link channel, the standard is used for many interactive services. The main feature for this use case is that it allows the dynamic adaptation of the modulation and coding to the channel quality that each user is experiencing.

The standard defines the physical layer and the data link layer of the communications link, namely layers 1 and 2 of the OSI model. The standard specifies how one or more baseband digital signals are to be mapped into a RF signal suitable for the characteristics of the satellite channel.

In this section, we give an overview of the features of this standard, based on (ETSI EN 302 307-1, v1.4.1, 2014). We first describe the features which are the same for broadcast and interactive applications and then discuss the features specific to interactive services.

A block diagram of the major processes in the transmitter's physical layer is shown in Figure 13.6. These were described generally in Chapter 12. Here, we give some details specific to DVB-S2.

All the processes except pulse-shaping filtering and modulation are involved in the construction of the physical-layer frame. The **frame** is the unit of data transmission. To show more particularly what each process acts on and to point out the information included in headers that control the communications, we look a bit deeper and discuss the development of the frame, where the physical-layer frame is the final stage. The steps in the construction are shown in the descending rows of Figure 13.7. We discuss the pulse-shaping filter and modulator after the frame construction.

The original data stream can be a transport stream as defined by MPEG (MPEG-2 or MPEG-4) or a generic stream as defined in the DVB standards. The data stream is sliced into data fields.

The first row of the figure shows the baseband frame. The baseband header contains the root-raised cosine (RRC) roll-off factor, the type of data stream, whether the input stream is single or there are multiple, whether the modulation and coding

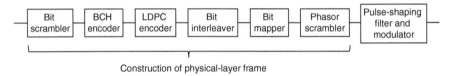

Bit scrambler
BCH encoder
LDPC encoder
Bit interleaver
Bit mapper
Phasor scrambler
Pulse-shaping filter and modulator
Construction of physical-layer frame

FIGURE 13.6 DVB-S2 transmitter physical-layer block diagram. (After ETSI EN 302 307-1, v1.4.1 (2014)).

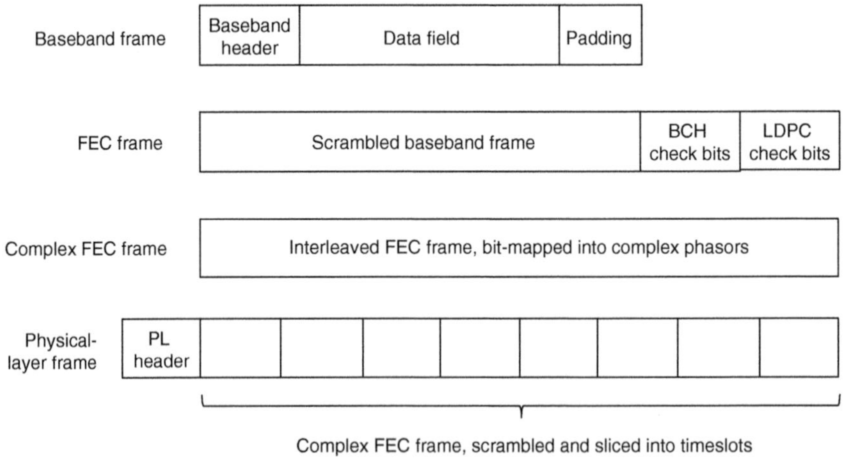

Baseband frame
Baseband header
Data field
Padding
FEC frame
Scrambled baseband frame
BCH check bits
LDPC check bits
Complex FEC frame
Interleaved FEC frame, bit-mapped into complex phasors
Physical-layer frame
PL header
Complex FEC frame, scrambled and sliced into timeslots

FIGURE 13.7 Steps in DVB-S2 physical-layer frame construction.

schemes will be constant or variable, and the length of the data field. The data field follows the header. Padding is required if the user data do not completely fill the available space.

The second row of the figure shows the forward error-correction (FEC) frame. Its first part is the baseband frame, whose bits have been scrambled. There are two FEC coders, an outer Bose–Chaudhuri–Hocquenghem (BCH) coder and an inner low-density parity-check (LDPC) coder. Both are systematic block codes. The output of a systematic coder consists of the input bits plus check bits. The input block for the BCH coder is the scrambled baseband frame. For the LDPC code, it includes additionally the check bits of the BCH code. Depending on the channel condition, varying code strengths are used. The BCH code is designed to correct up to 12 errors, except for very good channels, where the number of correctable errors is reduced to 8 or 10. The LDPC code rate can be selected from between 1/4 to 9/10 in ten steps. Its input block size varies such that the output block size is constant; for normal frames, it is 64,800 bits (there is also a short frame of 16,200 bits). The overall code rate varies between 0.247 and 0.898, that is, it is very close to the rate of the LDPC code alone.

The third row of the figure shows the complex-valued FEC frame. The FEC frame has been interleaved by a block interleaver, unless the modulation scheme will be QPSK. The interleaver has read the bits into a rectangular block row by row and read them out column by column. The number of columns equals the number of bits per symbol. The bit mapper then turns the bit stream into a stream of complex-valued phasors from the chosen modulation scheme.

The fourth row of the figure shows the physical-layer frame. The complex FEC frame has undergone complex-valued scrambling, via multiplication by the complex scrambling sequence. The elements of the scrambling sequence are of the form $e^{jk\pi/2}$, $k = 0, \ldots, 3$, so they rotate the phasors by integer multiples of $90°$. The scrambled FEC frame is divided into slots. In front of these slots, a physical-layer header is added in its own slot. The header contains a start-of-frame word, which is a unique 26-bit sequence that the receiver can use for synchronization, and some information about the signal format. This latter part is 64 bits long but contains only seven bits of information, with heavy protection against transmission errors. Five of those bits contain the MODCOD. The other two bits give the FEC frame length. With this information, the receiver will be in a position to demodulate and decode the rest of the frame. The physical-layer header is always transmitted with $\pi/2$-BPSK (binary phase-shift keying) modulation, which can be demodulated at low $C/(N + I)$ values.

At this point, we wish to clarify the meaning of the three types of bits in reference to Figure 13.7. The data bits make up the actual information to be sent, that fill the data field. The coded bits make up the entire FEC frame. The channel bits make up the entire physical-layer frame.

The final processes in the physical layer of the transmitter are the pulse-shaping filter and the modulator. The pulse shaping is RRC with a roll-off factor of 35, 25, or 20%. The possible modulation schemes are QPSK with absolute mapping, 8PSK with absolute mapping, 16APSK, and 32APSK. The carrier modulator puts the real and imaginary parts of the signal onto orthogonal carriers to finally generate the transmitted waveform. The symbol rate of the modulator is constant. For multi-plexed data streams, the modulation and coding formats remain constant over one frame, but they can change from frame to frame.

With the 11 LDPC code rates and four modulation schemes, a total of 44 modes would be possible. Of these, 28 modes are actually used as MODCODs. Based on an additive white Gaussian noise (AWGN) channel model (Section 12.5) the stand-ard defines an *ideal E_s/N_0*, at which each of these modes is used. It is the E_s/N_0 value at which quasi-error-free (QEF) transmission is achieved, which is defined as hav-ing less than one uncorrected error-event per transmission hour at the level of a 5 Mbps single-TV service decoder. This corresponds approximately to a packet-error ratio less than 10^{-7} for the MPEG-4 transport-stream packets of 188 bytes. Figure 13.8 shows the spectral efficiency of the MODCOD schemes as a function of their ideal E_s/N_0 values. The points for each of the four modulation schemes are joined by a line drawn for clarity. Together, these 28 combinations of coding and modulation allow reliable reception down to E_s/N_0 of −2.5 dB on an AWGN channel (ETSI EN 302 307-1, v1.4.1, 2014).

There are two cases where MODCODs are used. In broadcast applications, *vari-able coding and modulation (VCM)* may be applied to provide different levels of error protection to different service components (for example, standard-definition TV (SDTV) and high-definition TV (HDTV), audio, multimedia). This case will not be discussed further here. In ACM mode, the signal format is adapted in real time to varying channel conditions. This requires the transmission of signal quality informa-tion on the return link and, hence, is only applicable to interactive services.

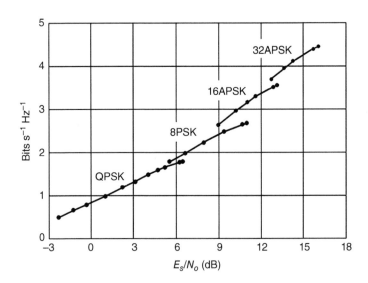

FIGURE 13.8   Spectral efficiency of DVB-S2 MODCODs versus ideal E_s/N_0 on AWGN channel.

When standard TV transponders with a bandwidth of 36 MHz are used for interactive services, the processing load on the receiver is manageable. The receiver can demodulate and decode the entire frame in real time and extract the IP packets that are addressed to it and discard the rest. However, high-throughput satellites (Chapter 18) have transponder bandwidths of 250 MHz. The decoding of the entire data stream requires too much processing capacity for consumer equipment. For this reason, a special operating mode called *time slicing* is defined for such terminals. In this mode, each frame contains data for a single receiver and it includes all information required for the processing of the frame, such as its length, coding rate, and modulation format. The physical-layer header is extended by another slot to transmit the required information. The terminal is still required to receive the physical-layer header correctly and completely and to perform coherent demodulation of the rest of the frame at the full speed of the transmitter. The time-consuming FEC decoding can be performed at the lower speed of the user data rate.

In ACM mode, the return-link standards DVB-RCS or DVB Return Channel Satellite, 2nd Generation (DVB-RCS2) may be used in conjunction with DVB-S2 or S2X. The user receiver in ACM mode has to provide signal quality information for the forward link to the NCC so that the NCC can select and command the MODCOD for optimum use of the bandwidth. To this end, the user terminal sends the $C/(N + I)$ available at its receiver, evaluated using known symbols. It is quantized with a resolution of 0.1 dB and should have a precision (mean error plus 3σ) better than 0.3 dB to take full advantage of the MODCOD resolution of 1 to 1.5 dB. As an option, the user terminal can also send a direct request for a particular MODCOD.

The DVB-S2 standard does not specify what return-channel protocol is to be used when needed. It can be DVB-RCS(2) or a proprietary standard.

Compared to the first-generation DVB-S2 standard, the second generation allows (1) transport packetization other than MPEG, (2) a much wider range of modulation and coding possibilities, and (3) ACM.

13.4.2 DVB-S2X Standard for Forward Link

In 2015, ETSI published extensions to the DVB-S2 standard under the acronym **DVB-S2X** (ETSI EN 302 307-2, v1.1.1, 2015). The standardization effort was started by Newtec and other VSAT manufacturers to take advantage of the progress in coding and modulation since the release of the DVB-S2 standard in 2005 (Willems, 2014). Newtec and Gilat both make VSATs that employ DVB-S2X with ACM, with modulation schemes all the way up through 256APSK on the forward link (ST Engineering, 2020; Globenewswire, 2020).

The following enhancements are included in the new standard (DVB Project Office, 2016):

- A finer gradation of code rates
- The additional modulation scheme $\pi/2$-BPSK, which allows operation at very low $C/(N + I)$, down to −10 dB

- Other additional modulation schemes 8APSK, 64APSK, 128APSK, and 256APSK
- Additional physical-layer scrambling options for critical co-channel interference situations
- Additional, smaller RRC roll-off options of 5 and 10%.

Figure 13.9 shows the spectral efficiency of the DVB-S2X MODCODs on the AWGN channel as a function of their ideal E_s/N_0 values (ETSI EN 302 307-2, v1.1.1, 2015). While the older standard requires a minimum E_s/N_0 of −3 dB, the new one can function down to −10 dB. This is attractive for transmission to mobile terminals with small antennas. At the high end the spectral efficiency keeps growing for E_s/N_0 beyond 15 dB up to 20 dB, with the new modulation formats. With the additional MODCODs the spectral efficiency comes very close to the Shannon bound (Section 12.6.2).

The new higher-order modulation schemes are primarily intended for professional services. They are mandatory for this user profile. They are optional for interactive services and satellite news-gathering. They are not applicable to broadcast services.

The finer-grained coding options necessitate better precision in the measured $C/(N + I)$ in the user terminal's receiver. The precision should be better than 0.2 dB.

13.4.3 DVB-RCS2 Standard for Return Link

13.4.3.1 *Introduction*   The **DVB-RCS2** standard defines coding, modulation, and multiple-access schemes for the return link for multiple users through a shared satellite channel. It extends the ACM feature to the return link and specifies the transmission of the forward-link quality information to the gateway.

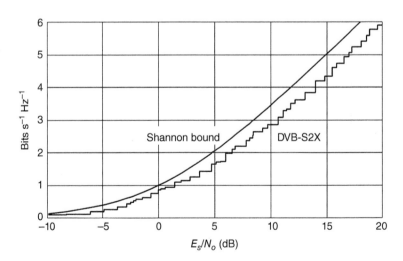

FIGURE 13.9   Spectral efficiency of DVB-S2X MODCODs versus ideal E_s/N_0 on an AWGN channel.

The standard is very powerful and flexible. It accounts for the requirements of several different application scenarios: consumer, multi-dwelling, corporate, industrial process *supervisory control and data acquisition (SCADA)*, backhaul, and institutional. It considers transparent and regenerative satellites (with and without switching), star and mesh network configurations, and more. The resulting standard is quite complex: the lower protocol-layer standard alone comprises 239 pages, and a companion document of lower-layer guidelines is 279 pages long. A paper coauthored by the chairman of the Technical Module RCS of the DVB consortium calls it "likely the largest standardization effort conducted for satellite communication systems" (Skinnemoen et al., 2013). A complex standard typically means an expensive implementation. This may be the reason why there seem to be no consumer Internet systems using this standard. Apparently, already the first version had a complexity issue: the only consumer system that seems to be using DVB-RCS, SES Broadband, uses only its resource allocation method, not its modulation format. Instead, it uses the Satmode modem specification (Section 13.5). The DVB-RCS standard, however, is in use by professional VSAT networks (NSSL Global Technologies, 2019).

The write-up here can only give a high-level description of a small portion of this epic work. We focus on the two lowest layers of the protocol. The description follows the lower-layer standard and guidelines (ETSI EN 301 545-2, v1.2.1, 2014).

The standard prescribes two modulation schemes for all terminals, linear modulation (memoryless modulation) and continuous-phase modulation (CPM).

13.4.3.2 *Memoryless Modulation and Coding* For memoryless modulation (Section 12.8), the formation of the transmitter physical layer is simple compared to DVB-S2, as shown in Figure 13.10.

A 16-state turbo code commonly called "turbo-phi" is used (Skinnemoen et al., 2013).

The modulation formats are $\pi/2$-BPSK, QPSK, 8PSK, and 16QAM. The pulse-shaping is RRC with a roll-off factor of 20%. 16QAM is a modulation format with large amplitude variation, so the satellite amplifier needs to operate in a backed-off mode. But since the return signal is comprised of multiple carriers, this is the case, anyway. In the case of BPSK, direct-sequence spectrum spreading (Section 1.3) may be applied with a spreading factor ranging from two to 16 (ETSI EN 302 307-2, v1.1.1, 2015).

To support carrier and timing recovery in the receiver, pilot blocks of known symbols are inserted in the beginning of a **burst** and throughout it, where a burst is transmission in a given timeslot and in a given frequency band.

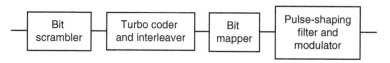

FIGURE 13.10 DVB-RCS2 physical-layer block diagram for memoryless modulation.

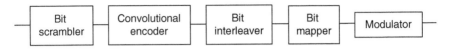

FIGURE 13.11   DVB-RCS2 physical-layer block diagram for modulation with memory.

13.4.3.3 *Modulation with Memory and Coding*   When the modulation is CPM (Section 12.11), the formation of the physical layer is different but still rather simple, as shown in Figure 13.11.

CPM was introduced for its potential for reduced cost of the terminals. CPM is a constant-amplitude modulation, so the SSPA of the user terminal can be operated in saturation; hence, a smaller and cheaper amplifier could be used to produce a particular output power (Skinnemoen et al., 2013).

An eight-state convolutional code is used, which can be switched between rates 5/7 and 15/17.

To support carrier and timing recovery in the receiver, short known sequences are inserted in the beginning of a burst and throughout it.

13.4.3.4 *Multiple Access*   The assignment of return-link resources to user terminals is by the DAMA method. The multiple access scheme is MF-TDMA. The NCC assigns to each active terminal a number of timeslot/carrier pairs, each defined by a number of parameters including modulation scheme, coding rate, carrier frequency, symbol rate, start time, and duration. This constitutes the transmission channel for a terminal. Note that the carrier bandwidth, symbol rate, and slot duration are carrier-specific, while the modulation scheme, code rate, and data-bit rate may change from burst to burst.

Besides DAMA, the NCC can also designate some timeslot/carrier pairs for random access, not just for control channels but also for user data. This results in faster response times since it avoids the request/response cycle required for DAMA and may be particularly beneficial to asymmetric traffic profiles, such as Web browsing.

TDMA requires time synchronization among all transmitters. The forward-link signal to the user terminals provides a reference, and each terminal computes the signal delay to the satellite from its own location and the satellite ephemeris data which are broadcast. The NCC sends correction messages to the terminals to fine-tune their timing.

13.5   SATMODE COMMUNICATIONS STANDARD

13.5.1   Satmode Standard

In about the year 2000, SES decided to set up a consortium to develop a low-cost, always-on, satellite return channel to support interactive television. It would replace the usual telephone back-channel (Satellite Today, 2003). The development was half funded by the European Space Agency (ESA). It started in 2003 and ended in 2006 (ESA, 2012a). The modem specification was standardized by CENELEC in the document EN 50478 (Wikipedia, 2018b)

As of 2012, no major TV broadcaster had made a mass deployment of Satmode for interactive TV. The ESA project then widened the scope of Satmode applications to include *triple-play services* and content delivery. Triple-play services bundle TV, telephone, and Internet over one broadband connection (ESA, 2012a).

Some of the characteristics of the OSI layer 1 of Satmode are as follows (source is Wikipedia (2018b) except where noted):

- Flexible turbo or turbo-like coding
- Programmable constituent codes with bypass possibilities
- Programmable interleavers
- Binary or quaternary CPM with programmable phase pulse, including GMSK
- Bit rate from 1 to 64 Kbps (ESA, 2012a)

Return-link access is random with the slotted Aloha protocol (Wikipedia, 2020e).

The hub is in the ground station. It usually implements one coding and modulation scheme and notifies the user terminals (Wikipedia, 2018b).

13.5.2 Implementation

SES Broadband provides broadband Internet services to Newtec's Sat3Play user terminals. Sat3Play implements part of the Satmode specification. We discuss Sat3Play, then SES Broadband.

Sat3Play is a multimedia platform for triple-play services. It provides a two-way Internet connection. The development was performed on an ESA project and took advantage of the results of the Satmode project. The project ended in about 2009. The primary market was felt to be home consumers, and the secondary market was professional applications (ESA, 2012b). Sat3Play merges some aspects of Satmode and some aspects of DVB-RCS. It became the first successful European satellite Internet access system for consumers (ESA, 2012b).

Modem characteristics are as follows (source is VSATplus (2020) except where noted):

- Maximum receive rate of UDP up to 20 Mbps (unicast or multicast)
- Maximum transmit rate of TCP up to 3.5 Mbps and same for UDP (Section 13.2.1)
- On receive, DVB-S2 ACM
- Receive physical layer
  - Modulation QPSK, 8PSK, 16APSK, 32APSK
  - Code rate from 1/4 to 9/10
  - RRC pulse roll-off 5, 10, 15, 20, 25, 35%
  - Symbol rate 3.6 to 63 Msps.

- On transmit, Satmode for coding and modulation and DVB-RCS for access (Wikipedia, 2020b)
- Transmit physical layer
  - Quaternary CPM with six different MODCODs, with adaptive return link
  - Channel bandwidth 128 KHz to 4 MHz.
- Ku-band and Ka-band.

Hub characteristics are as follows (VSATplus, 2020):

- Scalable by adding equipment for each additional carrier
- Transmit received by all terminals.

SES Broadband is a two-way satellite broadband Internet service provided across Europe by the Astra series of spacecraft. It offers *dual-play* (broadband Internet and telephone) as well as triple-play services to home consumers. It also offers Internet connections for a SCADA network (Section 13.4.3.1). Upload and download speeds can be symmetric if required. The user terminal transmits 500 mW of power. The system's central hub in an SES teleport has routers that connect to the Internet backbone. The hub uses DVB-S2 transmission formats. The SES Broadband maximum download speed has been increased to 10 Mbps and the upload speed to 256 Kbps. SES Broadband for Maritime offers the same service and is intended for smaller ships and operates mainly in the North and Baltic Seas and the northern Mediterranean (Wikipedia, 2020b).

REFERENCES

DVB Project Office (2016). DVB-S2X – S2 extensions. DVB fact sheet. On www.dvb.org/resources/public/factsheets/dvb-s2x_factsheet.pdf. Accessed Feb. 26, 2019.

Emiliani LD of SES (2020). Private communication, May 13.

ESA (2012a). Satmode. Project page. June 28. On artes.esa.int/projects/satmode. Accessed June 17, 2020.

ESA (2012b). Sat3Play. Project page. June 28. On artes.esa.int/projects/sat3play. Accessed June 19, 2020.

ETSI EN 301 545-2, v1.2.1 (2014). *Digital Video Broadcasting (DVB); second generation DVB interactive satellite system (DVB-RCS2); part 2: lower layers for satellite standard.* Sophia-Antipolis Cedex (France): European Telecommunications Standards Institute.

ETSI EN 302 307-1, v1.4.1 (2014). *Digital Video Broadcasting (DVB); second generation framing structure, channel coding and modulation systems for broadcasting, interactive services, news gathering and other broadband satellite applications; part 1: DVB-S2.* Sophia-Antipolis Cedex (France): European Telecommunications Standards Institute.

ETSI EN 302 307-2, v1.1.1 (2015). *Digital Video Broadcasting (DVB); second generation framing structure, channel coding and modulation systems for broadcasting, interactive services, news gathering and other broadband satellite applications; part 2: DVB-S2 extensions (DVB-S2X).* Sophia-Antipolis Cedex (France): European Telecommunications Standards Institute.

ETSI TS 102 188-1 v1.1.2 (2004). Satellite earth stations and systems (SES); regenerative satellite mesh—A (RSM-A) air interface; physical layer specification; part 1: general description.

ETSI TS 102 429-1 v1.1.1 (2006). Satellite earth stations and systems (SES); broadband satellite multimedia (BSM); regenerative satellite mesh—B (RSM-B); DVB-S/ DVB-RCS family for regenerative satellites; part 1: system overview.

ETSI TS 102 602 v1.1.1 (2009a). Satellite earth stations and systems (SES); broadband satellite multimedia; connection control protocol (C2P) for DVB-RCS; specifications.

Fang RJF (2011). Broadband IP transmission over Spaceway® satellite with on-board processing and switching. *IEEE Global Telecommunications Conference*; Dec. 5–9.

Garcia AY, Asenjo IM, and Piñar FJR (2006). IP multicast over new generation satellite networks. A case study: Amerhis. *International Workshop on Satellite and Space Communications*; Sep. 14–15.

Globenewswire (2020). Gilat announces availability of its flagship VSAT, achieving half a gigabit of concurrent speeds. On www.globenewswire.com/news-release/2020/06/18/20500006/en/Gilat-Announces-Availability-of-its-Flagship-VSAT-Achieving-Half-a-Gigabit-of-Concurrent-Speeds.pdf. Accessed Oct. 3, 2020.

Hughes (2020). HX systems: high-performance IP satellite broadband systems. Product information on VSAT systems. On www.hughes.com/technologies/broadband-satellite-systems/hx-systems. Accessed June 19, 2020.

NSSL Global Technologies (2019). SatLink VSATs. On www.sat.link/slproducts/satlink-vsats. Accessed May 1, 2019.

Rinaldi R and De Gaudenzi R (2004a). Capacity analysis and system optimization for the forward link of multi-beam satellite broadband systems exploiting adaptive coding and modulation. *International Journal of Satellite Communications and Networking*; 22 (3) (June); 401–423.

Rinaldi R and De Gaudenzi R (2004b). Capacity analysis and system optimization for the reverse link of multi-beam satellite broadband systems exploiting adaptive coding and modulation. *International Journal of Satellite Communications and Networking*; 22 (4) (June); 425–448.

Satellite Today (2003). Satmode raises the interactive stakes. Feb. 12. On www.satellitetoday.com/uncategorized/2003/02/12/satmode-raises-the-interactive-stakes. Accessed June 17, 2020.

Skinnemoen H, Rigal C, Yun A, Erup L, Alagha N, and Ginesi A (2013). DVB-RCS2 overview. *International Journal of Satellite Communications*; John Wiley; 31 (5).

ST Engineering (2020). Newtec Dialog, release 2.2. Product brochure. On www.idirect.net/wp-content/uploads/2020/03/ProductBrochure-NewtecDialog2-2.pdf. Accessed Oct. 3, 2020.

Svobodova L (1989). Implementing OSI systems. *IEEE Journal on Selected Areas in Communications*; 7 (7) (Sep.); 1115–1130.

Telesat (2010). Briefing on adaptive coding and modulation (ACM). White paper. On www.telesat.com/tools-resources/technical-briefings-white-papers. Accessed May 23, 2020.

Tomás JG, Pavón J, and Pereda O (1987). OSI service specification: SAP and CEP modelling. *ACM SIGCOMM Computer Communication Review*; 17 (1–2) (Jan.); 71–79.

VSATplus (2020). Newtec MDM2210 IP satellite modem. Specification. On www.vsatplus.com/collections/newtec/products/newtec-mdm2210-ip-satellite-modem?variant=6897123098678. Accessed June 19, 2020.

Whitefield D, Gopal R, and Arnold S (2006). Spaceway now and in the future: On-board IP packet switching satellite communication network. *IEEE Military Communications Conference*; Oct. 23–25.

Wikipedia (2018a). DVB-S2. Article. Nov. 29. Accessed Feb. 8, 2019.

Wikipedia (2018b). Satmode. Article. Dec. 19. Accessed May 1, 2019.

Wikipedia (2020a). Demand assigned multiple access. Article. Apr. 2. Accessed June 16, 2020.

Wikipedia (2020b). SES broadband. Article. Apr. 7. Accessed June 17, 2020.

Wikipedia (2020c). User datagram protocol. Article. Apr. 23. Accessed June 20, 2020.

Wikipedia (2020d). Transport layer. Article. May 7. Accessed June 20, 2020.

Wikipedia (2020e). ALOHAnet. Article. June 5. Accessed June 23, 2020.

Wikipedia (2020f). Transmission control protocol. Article. June 18. Accessed June 20, 2020.

Wikipedia (2020g). OSI model. Article. June 22. Accessed June 23, 2020.

Willems K of Newtec (2014). DVB-S2X demystified. White paper. Feb. 26. On www.newtec.eu/frontend/files/userfiles/files/Whitepaper%20DVB_S2X.pdf. Accessed Feb. 28, 2019.

Yun A, Casas O, de la Cuesta B, Moreno I, Solano A, Rodriguez JM, Salas C, Jimenez I, Rodriguez E, and Jalon A (2010). AmerHis next generation global IP services in the space. *Advanced Satellite Multimedia Systems Conference and Signal Processing for Space Communications Workshop*; Sep. 13–15.

CHAPTER 14

COMMUNICATIONS LINK

14.1 INTRODUCTION

This chapter discusses the communications links in a satellite communications system. A **link** consists of a signal path starting at the terminal of the transmitting antenna, over the propagation path through space and/or the atmosphere, to the receiving antenna, and terminating in the receiver before the demodulator. Into the receiver may also come interference. Recall that the forward link is actually two concatenated links, the uplink from a ground station and the downlink to a user terminal, and the return link is the concatenated uplink from the user terminal and the downlink to the ground station. A user-to-user link through the satellite is a concatenation of an uplink and a downlink.

On a satellite link, the equivalent isotropically radiated power (EIRP) in the direction of the receiving antenna and the gain and noise temperature of the receiving terminal will vary over time. Characteristics of the atmosphere will also change over time, as may also background radiation and interference as seen by the receiving antenna.

A note on terminology: a "ground terminal" is meant to be either a ground station or a user terminal.

Satellite Communications Payload and System, Second Edition. Teresa M. Braun and Walter R. Braun.
© 2021 John Wiley & Sons, Inc. Published 2021 by John Wiley & Sons, Inc.

14.2 PRIMARY INFORMATION SOURCES

The normative documents for analyzing link effects are those by the **International Telecommunication Union (ITU)**, a specialized body of the United Nations. Among other major responsibilities, the ITU assists in the development and coordination of worldwide telecommunications standards. Its **Radio Sector** is of special interest to us in this chapter, since it is the body that develops and publishes standards, called **recommendations**, on how to deal with the atmosphere, antenna noise, and interference. The documents are available online free of charge. Anyone about to perform an analysis of the communications link should start by searching for ITU-R documents on the topic. The Radio Sector documents are in series, and the series pertinent to this chapter are the following:

- P: radiowave propagation
- BO: broadcast satellite service (BSS)
- M: mobile satellite service (MSS)
- S: fixed satellite service (FSS)

Recall when reading ITU documents that the ITU uses the abbreviation "GSO" for geostationary (GEO). The ITU-R also publishes handbooks, also available online free of charge. The one most relevant to this chapter is the *Handbook on Radiometeorology*. It supplies background and additional information on radiowave propagation effects. It also serves as a guide to the ITU-R recommendations on propagation (ITU-R, 2014).

Another important propagation source is the book by Allnutt (2011). The author describes all effects in depth. He gives the current thinking on what causes the effect, what its characteristics are, and the reasons for and the assumptions of the analysis methods. He refers to the ITU standards.

A third good propagation source is the book by Ippolito (2017).

14.3 LINK AVAILABILITY

Link availability is the condition where the link provides $C/(N+I)$ into the demodulator that is sufficient for successful communications. When this is the case, the **link closes**, that is the **link is not out**. Some unusual atmospheric circumstances such as heavy rain or, in a dry location, unusually high humidity can cause the link to be unavailable. Another factor in availability is **sun outage**, where the sun is behind the satellite from the point of view of the ground terminal (Section 14.5). Temporarily increased noise or interference can also cause the link to go out. Aging of the payload and the bus's thermal control subsystem (Section 2.2.1.6) makes the uplink harder to close.

Various aspects of system design can help ensure that the link is only rarely unavailable:

- For the uplink
  - The payload can be designed to tolerate a wide range of flux-density values into the antenna. (This is usually the case.)
  - The transmitting ground terminal can have extra power available for use when necessary. (For example, Spaceway 3 performs uplink power control in a closed loop with the terminal—Section 10.4.3. This is a feature common in most satellite network systems today (Emiliani, 2020a).)
  - In a satellite system that supports adaptive coding and modulation (ACM— Section 13.2.3) on the return link, when weather and interference conditions on the user link are unfavorable the system can signal to a user terminal to make its return-link transmission more robust. (DVB-S2 and DVB-RCS2 communications standards for interactive communications employ this—Section 13.4.)
- For the downlink
  - The payload can have extra transmit power available. (For example, Spaceway 3 has a reserve pool in a hybrid closed- and open-loop system. In the closed-loop part of the system, a ground terminal can request an increase in downlink power for its beam. In the open-loop part, the payload preemptively increases downlink power to a beam, based on rain predictions coming from ground-based radar.)
  - In a satellite system that supports ACM on the forward link, when weather and interference conditions on the user link are unfavorable the system can signal to a user terminal that its forward-link signal parameters are being modified to be more robust. (DVB-S2 and DVB-RCS2 support this.)
- For either link
  - Interference among satellite systems is controlled (Sections 14.6.5 and 14.6.7).

For a system without ACM, extra transmit power is designed into the transmitter with the aim of accommodating most cases of rain attenuation provides **rain margin**. A more general term is **atmospheric margin**, which would aim at accommodating not just rain but the other atmospheric effects as well for almost all the time.

Customers of satellite services want high service reliability and availability and a low number of outage events. One measure of link availability is long-term average link availability. A large satellite-network provider sees requests for Ka/K-band of 99–99.7%, as well as requests for Ku-band TV broadcast of 99.9% in temperate regions and about 99.5% in tropical regions. Another measure of link availability is worst-month statistics, used by some satellite broadcasters but not in Europe (Emiliani, 2020a). Links that use ACM have higher availability than those that do not. The same satellite services provider counts outage events that last for at least a minute (Emiliani, 2020a).

The ITU provides a method for calculating long-term average atmospheric-attenuation statistics for GEO and non-GEO spacecraft. In the case where multiple satellites are visible at one time to the user terminal or ground station, an approximation can be made by assuming that the spacecraft with the highest elevation angle is used (ITU-R P.618, 2017).

If the customer of the satellite manufacturer leaves it up to the manufacturer to determine the various EIRP values needed across a coverage area to compensate for atmospheric effects, the customer may specify a particular rain model for use. The calculations to ensure an adequate payload design are complicated if the coverage area is large.

14.4 SIGNAL POWER ON LINK

14.4.1 Introduction

We wish to look at what determines the received signal power on the link. There are factors which are constant and factors which vary. Overviews of the two sets of factors are given in Tables 14.1 and 14.2, respectively, for the downlink (the uplink would be similar). The last column in the tables gives references for the factors except for those belonging to the ground terminal. Each end of a link must keep the

TABLE 14.1 Constant Losses in Downlink

Loss Type	Source	Ref Section
Average loss from payload antenna pointing error	Various	3.9 for horn antennas and single-beam reflector antennas, 11.2.2 for multi-beam antennas, and 11.13 for phased arrays
Average free-space loss	Inversely proportional to distance squared	14.4.2
Attenuation by atmospheric gas and clouds	Oxygen and water vapor, for carrier frequency above 10 GHz	14.4.4.2, 14.4.4.3
Loss from radome, reflector, and feed cover at BOL	—	N/A
Average loss from ground antenna pointing error	Installation issues and satellite movement	N/A
Average polarization-mismatch loss	Polarization ellipses different and/or in different orientations, of payload and ground antennas	14.4.5

TABLE 14.2   Varying Losses in Downlink

Variation Type	Source	Ref Section
Payload power-out instability (preamp in ALC mode)	Drift in preamp output power over life, variation of HPA's P_{out} with temperature, drift in HPA's P_{out} over life	8.10
Payload gain instability (preamp in fixed-gain mode)	Variation of HPA's P_{out} with temperature, drift in HPA's P_{out} over life	8.10
Jump in transmitted power	Switch over to redundant HPA (more switches and waveguide or coax post-HPA)	7.4.2.3
Payload antenna pointing error	Various	3.9 for horn antennas and single-beam reflector antennas, 11.2.2 for multi-beam antennas, 11.13 for phased arrays
Varying free-space loss	Non-GEO varying distance to ground terminal	14.4.2
Varying atmospheric attenuation	Ionospheric or tropospheric effects, depending on carrier frequency	14.4.3, 14.4.4
Varying ground antenna loss	Water or snow on feed, reflector, or radome; aging	Crane (2002), Crane and Dissanayake (1997)
Varying ground antenna mispointing	Installation issues and satellite movement	N/A
Varying polarization-mismatch loss	—	N/A
Jump in front-end loss	Switch over to redundant LNA or back to primary (different number of switches and more or less waveguide or coax)	N/A

variations it causes within bounds and contributes to accommodating the total variations. In the rest of this major section, we discuss the factors that have not been addressed elsewhere in the book.

14.4.2 Free-Space Loss

Free-space loss is conceptually the fraction of EIRP that does not arrive at the receiving antenna, as if the EIRP spread out equally in all directions from the transmit antenna. The surface area of a sphere centered about the transmit antenna and having radius R, the distance between the payload and the ground terminal, is $4\pi R^2$. What matters, actually, is the surface area in terms of the carrier wavelength, that is,

$4\pi(Rf/c)^2$ where f is the carrier frequency and c is the speed of light in vacuum. With some convenient scaling, the free space "gain" is defined as $[c/(4\pi Rf)]^2$. In dB, it would be 10 times the log of this.

It is evident that the free-space loss in dB is proportional to frequency in dBHz. For a very wide channel (on the order of 15% of center frequency), the resultant gain slope across the band will be significant. Fortunately, this gain slope is canceled by the gain slope of the transmit (or receive) antenna.

14.4.3 Atmosphere-Caused Attenuation for Carrier Frequencies from 1 to 10 GHz

In temperate regions, the only significant causes of radio-frequency (RF) signal impairment for carrier frequencies below 10 GHz occur in the ionosphere. The ionosphere reaches from about 60 to 1000 km altitude and has particles ionized by the sun (Wikipedia, 2020b). Very high-intensity rain can cause problems in X-band, and even in C-band when the required link availability is very high (Emiliani, 2020a). Section 14.4.4.5 illustrates examples of this. Rain is addressed in Section 14.4.4.1.

The ionospheric impairments are scintillation and the Faraday effect (ITU-R P.531, 2019).

The ionosphere is a region of ionized plasma that tends to be concentrated between 80 and 400 km altitude. The energy from the sun strips away electrons from some of the nitrogen and oxygen molecules, leaving them ionized. The most significant ionospheric effects on propagation occur at about 1 hour after sunset, when the electrons are combining back with the ionized molecules (Allnutt, 2011).

14.4.3.1 *Ionospheric Scintillation* **Ionospheric scintillation** is a relatively rapid and random fluctuation of the signal about a mean level. Scintillation is one of the most severe disruptions for frequencies below 3 GHz, and it can be observed occasionally up to 10 GHz. The effects can be particularly significant for non-GEO satellite systems operating at L- or S-band (ITU-R P.531, 2019).

Ionospheric scintillation is due to fluctuations of the refractive index along the propagation path, caused by inhomogeneities. The magnitude of ionospheric scintillation correlates well with sunspot activity, which has an 11-year cycle. It also correlates well with location and season where the sun is high in the sky at noon, because the sun creates a high level of ionization (Allnutt, 2011). Scintillation can be severe between about 20° north and 20° south of the geomagnetic equator and significant within about 30° of the poles. For equatorial ground terminals in years of solar maximum, scintillation occurs almost every evening (ITU-R P.531, 2019). In other years, it is strong within a month of the equinoxes (Davies and Smith, 2002). However, in quiet-sun years, scintillation can almost disappear. When the earth's magnetic field lines are parallel to the propagation path through the ionosphere, scintillation is worst (Allnutt, 2011).

A scintillation event can last from half an hour to several hours. The fading rate is about 0.1 to 1 Hz (ITU-R P.531, 2019), that is slow enough to be followed by an automatic gain-control unit (AGC) with enough range.

Amplitude scintillation data is usually presented as monthly statistics, with the worst month being at an equinox during the highest sunspot activity. For example, a 4-GHz link during a moderate sun-spot year will see a 2-dB attenuation for an average of 4 minutes a day in the worst month, which is 0.3% of that month. Scintillation can be severe at all elevation angles (Allnutt, 2011).

Phase scintillations follow a Gaussian distribution. When scintillation is weak or moderate, most observations in equatorial regions show that phase and amplitude scintillations are strongly correlated (ITU-R P.531, 2019). Phase scintillations are not a problem since the receiver's carrier-recovery loop can track it.

The ITU provides a rough calculation for ionospheric scintillation (ITU-R P.531, 2019).

The approximate size of scintillation under unfavorable atmospheric conditions but not the worst is given in Table 14.3.

14.4.3.2 Faraday Rotation Faraday rotation was mentioned in Section 3.3.3.4 in regard to antenna polarization selection. Here we explain it. The rotation is directly proportional to the integral of the product of the ionosphere's electron density and the component of the earth's magnetic field along the signal path (ITU-R P.531, 2019). A linearly polarized signal splits into two circularly polarized signals, which travel at slightly different speeds and along slightly different paths. When they recombine the linear polarization (LP) has been rotated (Davies and Smith, 2002). This is why LP is not often used at frequencies below C-band (Allnutt, 2011). Circular polarization (CP) is not affected. The rotation is large at L-band, small but significant at C-band, and insignificant at 10 GHz as can be seen from Table 14.3. Faraday rotation has relatively regular diurnal, seasonal, and 11-year solar-cycle variations and strongly depends on geographical location (ITU-R P.531, 2019).

TABLE 14.3 Estimated Ionospheric Effects near Equator, Elevation about 30° (ITU-R P.618, 2017)

Carrier Frequency (GHz)	Scintillation[a] (dB pk-to-pk)	Scintillation[a] (°rms)	Faraday Rotation[b] (°)
1	>20	>45	108
3	≈10	≈26	12
10	≈4	≈12	1.1

[a] Values observed near geomagnetic equator during early night-time hours at equinox under conditions of high sunspot number.

[b] Based on a high (but not highest) total electron content (TEC) value encountered at low latitudes in daytime with high solar activity.

14.4.4   Atmosphere-Caused Attenuation for Carrier Frequencies above 10 GHz

For frequencies between about 10 and 30 GHz, signal impairments occur only in the troposphere, which is where nearly all weather conditions take place. The worldwide average height of the troposphere is 13 km (Wikipedia, 2020c). The most important atmospheric effects on the signal power are from rain and, to a lesser extent, gas and clouds. Rain can be a factor at frequencies below 10 GHz (Section 14.4.3).

The top-level standard among the ITU Radio Section recommendations for this section is ITU-R P.618. It has formulas for computing most atmospheric effects, and for the more complicated ones it summarizes what is in the lower-level documents that it points to.

MATLAB has commands for calculating the attenuations due to the various atmospheric effects.

14.4.4.1 *Rain Attenuation*   Some antennas partially compensate for rain attenuation over their coverage area by providing higher gain toward rainy regions (Section 3.3.1).

14.4.4.1.1 **About Rain**   The material in this subsection comes from Allnutt (2011) except where noted.

Rain characteristics vary with location, but the physical processes that generate the different types of rain operate similarly from region to region, except for extraordinary events like hurricanes. The two most common rain structures are *stratiform rain* and *convective-cell rain*. Stratiform rain forms in a cloud with a large horizontal extent and its precipitation is relatively continuous and uniform in intensity. Convective rain forms in cells, which make small clusters. Cell separation is on the order of 5–6 km, and the cluster separation is roughly 11–12 km. Convective rain is more intense than stratiform rain and lasts for a shorter time. The two rain structures typically occur together.

Figures 14.1 and 14.2 are radar images of convective rain and stratiform rain, respectively. In these images, the maximum intensity of the convective rain is about ten times that of the stratiform rain.

The rain in the tropics is different from the rain in the temperate zones: the rain in temperate zones is mostly stratiform, while in the tropics it is mostly convective. A signal path may intersect two rain cells instead of just one or none (Mandeep and Allnutt, 2007). In addition, the rainfall rate is much higher.

The three most common types of rain events on satellite links are as follows:

- Stratiform rain, which usually comes from frozen particles melting as they fall below the 0 °C height. It falls from stratus clouds.
- Thunderstorms, which have a core with a high rainfall rate surrounded by less intense rainfall. A thunderstorm arises from a strong convective flow upward in its core, driven by solar heating. It comes from cumulonimbus clouds.

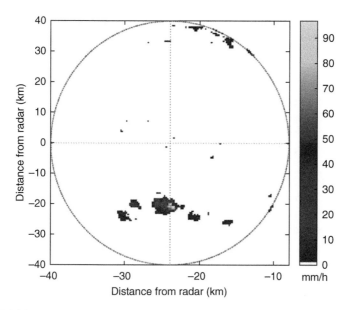

FIGURE 14.1   Radar image of convective rain. (Courtesy of Politecnico di Milano, from Prof. L. Luini).

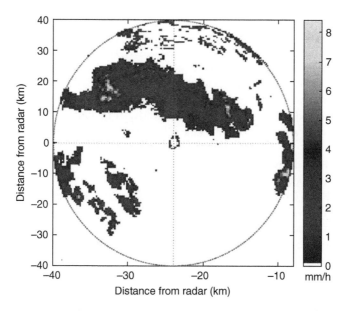

FIGURE 14.2   Radar image of stratiform rain. (Courtesy of Politecnico di Milano, from Prof. L. Luini).

- *Virga* rain, which condenses out at about the 0 °C height and begins to fall but evaporates before it reaches the ground. Forty-eight percent of the time it comes from altostratus clouds and 20% from cirrus (Wang et al., 2018).

The fraction of rain events that is virga is greater than 30% over arid regions including the deserts of Australia, the Arabian peninsula, Iran, and the western US, and about 10% over the Amazon region. Virga occurs much more frequently over land than ocean. Over the vast majority of the ocean, less than 20% of rain events are virga (Wang et al., 2018).

14.4.4.1.2   Rain Models Introduction   A **rain model** has two parts, a *rain rate model* and a *rain-attenuation model*, which in some models can be separately selected. The rain rate model provides long-term rain-rate probability distributions (rate versus **exceedance probability** or vice versa). The rain-attenuation model then factors in the carrier frequency, elevation angle to the satellite, and the polarization tilt angle with respect to the horizontal (Ippolito, 2017). The polarization tilt angle is the angle between the polarization ellipse's major axis and the local horizontal.

There are many different rain models but the two best-known ones are the ITU-R Rain Model and the Crane Global Model (Ippolito, 2017).

The ITU-R model is the normative model. It applies to frequencies up to 55 GHz and exceedances from 0.001 to 5% (ITU-R P.618, 2017). MATLAB has commands to calculate the model.

The Crane Global Model was the first published model (1980) that provided a self-contained rain-attenuation prediction procedure for the entire world. Its basic principles form the basis of many other models, including the present ITU-R models (Allnutt, 2011). It is used frequently, especially in the US (for example, Kymeta (2019)). Crane also has a Two-Component Model of rain rate and of rain attenuation, which he updated along with the Global Model in 1996 (Crane, 1996). The Crane models can also be found in Crane (2003b).

14.4.4.1.3   Rain Rate Models   We first describe the ITU-R rain rate models, then the Crane models.

The ITU-R has datasets and models applicable for estimating rainfall worldwide over various intervals of time.

On a monthly scale, the ITU-R provides worldwide datasets of long-term average rainfall for each month of the year (ITU-R P.837, 2017). From these, long-term average annual rainfall can be obtained.

It also provides a model for estimating the long-term average annual, one-minute-integrated rainfall rate $R_{0.01}$ exceeded only 0.01% of the time. This is a key input to the ITU-R rain attenuation model. The $R_{0.01}$ model takes input from worldwide datasets of long-term average rainfall for each month and long-term average temperature for each month measured 2 m off the ground, data which are provided in digital maps (ITU-R P.837, 2017). Temperature is measured off the ground so as to be insensitive to the ground's temperature. This recommendation also contains worldwide maps of $R_{0.01}$.

The worldwide datasets of rainfall are for rain that reaches the ground (Emiliani, 2020b).

It is possible in the tropics that the predicted rainfall rate for a given exceedance is greater on a zenith path than on a slant path. This may be due to the relatively greater incidence of convective rain cells than in temperate regions (Kumar et al., 2009). Another conjecture is that, during intense convective events, it is possible that a zenith path goes through the portion of the cell exhibiting the highest rainfall rate, while a path at a lower elevation could go through an area of the rain cell with lower rainfall intensity which would integrate to a lower total path attenuation (Emiliani, 2020a).

For those satellite service providers interested in long-term average statistics of the worst month of the year, the ITU-R provides two estimation models. One follows the procedure above for computing $R_{0.01}$ because along the way the monthly statistics are also determined (ITU-R P.837, 2017). The second procedure starts from long-term annual statistics (ITU-R P.841, 2019).

The seventh version of the key recommendation ITU-R P.837, published in 2017, came about because of the known weaknesses of the underlying climatic maps, as well as the increasing need to investigate the impact of monthly predictions on system design. The monthly rainfall-rate maps for land were derived from 50 years of data from the Global Precipitation Climatology Centre, which collects and analyzes rain-gauge data from more than 85,000 sites. The similar maps for the ocean were derived from 36 years of ERA Interim data of the European Centre for Medium-range Weather Forecast (ITU-R P.837, 2017), which comes from a large number of sources of various types. The seventh version of ITU-R P.837 also contains the important improvement that it no longer makes a distinction between stratiform and convective rain, which is not measurable and thus not verifiable, anyway. The one-minute integrated rainfall-rate data is improved and is provided as digital maps with finer resolution (ITU-R 3J/FAS/3-E, 2017).

Crucially, to ensure statistical stability, rainfall measurements must be collected over a sufficiently long period of time, typically more than 10 years (ITU-R P.837, 2017). Monthly rain amounts and monthly mean temperatures 2 m off the ground are required to be collected over at least 12 years (ITU-R 3J/FAS/3-E, 2017).

The Crane Global Model's rain rate model is embodied in a set of worldwide climate-zone maps and a list of zone properties (Crane, 1996).

The Crane Two-Component Model uses two rain rate distributions, one to describe the small (about 1 km) cells of intense rain and the much larger (roughly 30 km (Crane, 1993)) surrounding regions of lower intensity. All storms contain both types of rain but a given signal path may not intersect a high-intensity cell. These two types of rain areas correspond approximately to the convective rain cell and stratiform region, respectively, of other models (Crane, 2003a).

Crane also has a Local Model for rain rate, which is his initial Two-Component Model augmented by an adjustment for year-to-year variation (Crane, 2003a). At the time he published it, only in the US was there enough rain rate data to apply the model. The National Climate Data Center has been collecting measurement data at 105 US sites since 1965. The way to apply the model is to use the data for the nearest of those sites, following the principles set in Crane's books. The low-exceedance rain rates from the nearest site are good enough to use because low-exceedance rain levels are caused by high-intensity rain cells, which are similar enough at both locations.

14.4.4.1.4 Rain Attenuation Models   The derivation of the attenuation statistics from the rain statistics is the second part of a rain model.

The ITU-R attenuation model has since 1999 been based on the DAH rain attenuation model, named for its authors Dissanayake, Allnutt, and Haidara (Ippolito, 2017). The ITU-R changed to this model after a study comparing ten models (but not the Crane Global) against ITU-R propagation data showed that the DAH model provided the most consistent and accurate predictions (Feldhake, 1997). The DAH model was developed from a fit to all the rain attenuation data in the ITU-R propagation database, which at the time held altogether 120 years of rain attenuation observations from various places around the world (Crane and Dissanayake, 1997).

The ITU-R attenuation model takes as inputs $R_{0.01}$, the latitude, the altitude, the elevation angle to the satellite, and the frequency (ITU-R P.618, 2017).

The ITU-R attenuation model attempts to compensate for the fact that not all rain that causes attenuation on a satellite link reaches the ground and is counted in rainfall rate datasets and statistics. Virga rain is accounted for in the effective path-length model. The ITU-R model essentially assumes that over the path length the rainfall rate is constant. The path-length model is a topic of ITU-R discussion in the year of writing (2020) (Emiliani, 2020b).

The Crane Global Model attenuation model is based entirely on observations of rain rate, rain structure, and the vertical variation of atmospheric temperature, and not on attenuation measurements (Crane, 1996). It differs from the DAH model in being founded on the physics of rain attenuation (Crane and Dissanayake, 1997). Excel macros for both Global and Local attenuation models are to be found on the Internet at weather.ou.edu/~actsrain/crane/model.html.

14.4.4.1.5 Rain Variability   For a given exceedance in the range from 0.001 to 0.1%, the corresponding attenuation in dB varies from year to year from the long-term statistics by more than 20% rms, for the same signal path, frequency, and polarization (ITU-R P.618, 2017).

The ITU provides a method of predicting the year-to-year variation in annual statistics and worst-month statistics, compared to long-term statistics. It also gives a method of predicting the risk associated with a particular rain margin. The method for annual statistics is applicable for exceedance between 0.01 and 2% and frequency between 12 and 50 GHz (ITU-R P.678, 2015).

14.4.4.1.6 Frequency Scaling of Attenuation The ITU-R provides two ways to scale rain attenuation with frequency, for two different purposes (ITU-R P.618, 2017).

The first way is a simple relationship for scaling long-term rain attenuation at one frequency to another frequency on the same path, at any given exceedance probability. The scale factor depends not just on the two frequencies but also on the attenuation at the first frequency. Section 14.4.4.5 presents illustrations of this scaling.

The second way applies to a two-way link with uplink power control or ACM, that is, where the rain attenuation in one direction is known and the simultaneous attenuation in the other direction at the other frequency is needed. This method does not in fact provide a definite answer, since the relationship is not a deterministic one, but an answer with a chosen probability of error or risk.

14.4.4.1.7 Site Diversity Gain Intense rain cells that cause large attenuation often have horizontal dimensions of no more than a few kilometers. When the satellite can use an alternative ground station in case of heavy rain at the primary ground station (**site diversity**), the link availability can be improved considerably.

Diversity gain is the most convenient way of measuring the advantage provided by site diversity because it can serve as a line item in a link budget. Diversity gain is a function of the exceedance probability. It equals the rain attenuation in dB that can be accommodated by the dual site, minus the rain attenuation that can be accommodated by the primary site. The two sites are assumed to have the same rain statistics and the same-size ground-station antenna. If this condition is not satisfied, a different way of calculating the dual-site advantage is available (ITU-R P.618, 2017). A few months of measurements is often enough to obtain sufficiently good statistics for the diversity gain (Allnutt, 2011).

Depending on conditions and for all except low-elevation paths, the site separation that effectively provides the maximum diversity gain is 10–20 km, for frequencies up to 30 GHz (Allnutt, 2011). A more recent study has shown, using a three-dimensional atmospheric model (Section 14.4.4.6), that at 30 GHz in fact 30–40 km is desirable (Emiliani and Luini, 2016).

14.4.4.1.8 Differential Attenuation on Nearby Paths The ITU-R addresses the differential rain attenuation between propagation paths from a satellite to two nearby sites on the earth or between two nearby satellites and one ground site (ITU-R P.619, 2019). This is of interest when a nearby user terminal is interfering on the uplink, or a nearby satellite is interfering on a downlink, respectively. The ITU no longer says that differential rain attenuation is a minor issue. Instead, it offers a calculation for the joint statistics of rain attenuation on two paths from one satellite to two nearby ground sites (ITU-R P.1815, 2009).

14.4.4.2 Gas Attenuation In regards to attenuation, clear air may be thought of as consisting of two gases, dry air and water vapor (humidity). What matters in the dry-air component is the oxygen (ITU-R P.676, 2019). Gas attenuation at Ku-band is on the order of 1 dB in temperate regions.

The importance of gas attenuation increases with frequency. The attenuation from absorption by oxygen is relatively constant at any given location, elevation angle, and frequency, while the attenuation from water vapor varies, being typically highest during the season of maximum rainfall (ITU-R P.618, 2017).

The ITU provides three procedures for calculating gas attenuation on a given path (ITU-R P.676, 2019):

- A complex but accurate procedure for when atmospheric pressure, temperature, and water vapor density are known as a function of altitude
- A simplified estimation procedure based on the water-vapor density, the dry pressure, and the temperature at the surface of the earth
- Another simplified estimation procedure based on integrated water-vapor content along the path. This is the preferred procedure when local data are not available (Emiliani, 2020b).

The last two procedures can use data provided by the ITU. One document gives data on atmospheric pressure and water-vapor density as a function of altitude for the following reference atmospheres (ITU-R P.835, 2017):

- Mean global annual
- Low-latitude annual
- Mid-latitude summer
- Mid-latitude winter
- High-latitude summer
- High-latitude winter.

Another ITU document gives data on annual mean and monthly mean values of surface water-vapor density and of integrated water-vapor density along a zenith path. The data are given for various exceedance probabilities and as a function of latitude and longitude (ITU-R P.836, 2017).

Another ITU document gives data on the mean temperature at 2 m above the surface of the earth (ITU-R P.1510, 2017). Monthly mean and annual mean temperatures are given as a function of latitude and longitude.

Whenever the attenuation along a zenith path is given, it can be converted to another elevation angle greater than 5° by multiplying it by the cosecant of the elevation angle (ITU-R P.676, 2019).

14.4.4.3 Cloud and Fog Attenuation

Clouds and fog attenuate the signal due to the water vapor in them. The attenuation can be significant at frequencies well above 10 GHz or at low exceedance probabilities. At Ka-band clouds have a greater impact than gases (Emiliani, 2020a). Generally, the effect is worse at low latitudes. The ITU provides a method to calculate the attenuation for a given exceedance probability as a function of latitude and longitude, on an annual basis or for every month. One input to the method is the integrated cloud liquid water along the path (ITU-R P.840, 2019). The ITU provides a simplified method for airborne user terminals (ITU-R P.2041, 2013).

14.4.4.4 Tropospheric Scintillation There are clear-air effects that cause such a large variable signal impairment at low elevation angles that commercial satellite systems normally do not operate at these low angles. The usual minimum elevation angles are as follows (Allnutt, 2011):

- 5° for C-band
- 10° for Ku-band (11–14 GHz)
- 20° for Ka/K-band.

At elevation angles above the minimum, the only significant one of these effects is **tropospheric scintillation**. It is caused by small-scale fluctuations in the refractive index along the propagation path. The two phenomena that cause it are fluctuations in the lower troposphere and turbulent mixing near the edges of clouds of vapor-saturated air with dry air. Scintillation is worse in warm and humid locations. It has a diurnal and seasonal variation. Experienced scintillation intensity increases with the frequency and with decreasing ground-antenna diameter (Allnutt, 2011).

The ITU provides a method to calculate the fade depth due to scintillation at any given exceedance probability. Inputs are surface ambient temperature and surface relative humidity at the location, both averaged over a period of at least a month, as well as effective aperture size of the ground antenna. The model applies to frequencies from 7 GHz up to at least 20 GHz (ITU-R P.618, 2017).

14.4.4.5 Attenuation from Combined Effects The ITU-R provides a formula for the attenuation due to tropospheric effects (ITU-R P.618, 2017). The formula applies at all frequencies where the models of the individual effects apply (Emiliani, 2020e):

$$A_T(p) \doteq A_G + \sqrt{\left[A_R(p) + A_C\right]^2 + A_S^2}$$

where p is the exceedance probability of interest in the range 0.001–50%, A_T is the total attenuation in dB, A_G is the gas attenuation in dB, $A_R(p)$ is the rain attenuation in dB, A_C is the cloud attenuation in dB, and $A_S(p)$ is the tropospheric scintillation attenuation in dB. If there are good gas attenuation data for the location, then set A_G to $A_G(p)$; otherwise, for $p \geq 1\%$ use $A_G(50\%)$ and for $p < 1\%$ use $A_G(1\%)$. For A_C, for $p \geq 1\%$ use $A_C(50\%)$ and for $p < 1\%$ use $A_C(1\%)$ (ITU-R P.618, 2017). The model is available as Excel macros on the Internet site logiciel.cnes.fr/PROPA/en/logiciel.htm. The individual models on this site are not up to date but the sum formula is current (Emiliani, 2020b).

For a link operating at frequencies above about 18 GHz and especially if the elevation angle is low (less than 10°) or atmospheric margin is small, the total attenuation from the various atmospheric effects must be considered (ITU-R P.618, 2017).

This formula was developed empirically and not through physics. When the method was tested in 2002 with ITU-R rain predictions, it was found to match available measurement data with an rms error of about 35%, for p up to 1% and for all latitudes. The method was also tested with satellite link measurements, and the rms error was about 25% (Allnutt, 2011).

Figure 14.3 shows total link attenuation at uplink C-band for the very high link availability of 99.95% on two different parts of the earth, (a) for the Americas and (b) for Europe and Africa. For the Americas, the total attenuation is less than 1 dB in the temperate regions and less than 1.5 dB in the equatorial regions except for one small spot in Peru. For most of Europe and for central and southern Africa, the attenuation is less than 1 dB, but for the western-most parts of Morocco and the disputed territory of Southern Sahara, the attenuation is more than 2 dB. There is an area in the Pacific where the attenuation is greater than 2.5 dB and an area in the Atlantic where it is over 4.5 dB. These two maps illustrate the statement in Section 14.4.4 about rain attenuation being significant even at C-band in some areas when the link availability requirement is very high, since most or all of the attenuation at 6 or 6.5 GHz is due to rain.

Figure 14.4 is similar to the previous figure but for the Ku-band frequency of 12 GHz. What is most striking about these plots is how similar they are to the plots at C-band if you ignore the difference in the scales. This is because rain attenuation scales with frequency, as addressed in Section 14.4.4.1.6. The other thing to notice about these plots is that even at 12 GHz the total atmospheric attenuation can be a large number of dB. Now, for the Americas, the total attenuation is less than 6 dB, but for a large part of the South American equatorial region, the attenuation is 6–9 dB and even above 9 dB in spots. For most of Europe and for central and southern Africa, the attenuation is less than 4.8 dB, but for the western-most parts of Morocco and the disputed territory of Southern Sahara and parts of Algeria and Tunisia the attenuation exceeds 8 dB. These plots show the benefit of ACM (Section 13.2.3) at Ku-band for very high required availability, versus holding a link margin of several dB which goes unused most of the time. At higher frequencies, the required availability does not even need to be very high for ACM to provide a good benefit.

14.4.4.6 *Three-Dimensional Model of All Effects*   A three-dimensional model of fields of rain, clouds, and water vapor has been developed. The Atmospheric Simulator for Propagation Applications (ATM PROP) uses physically based approaches to synthesize high-resolution (1 km by 1 km horizontally and 100 m vertically) fields of all these effects simultaneously. It has been used to predict total

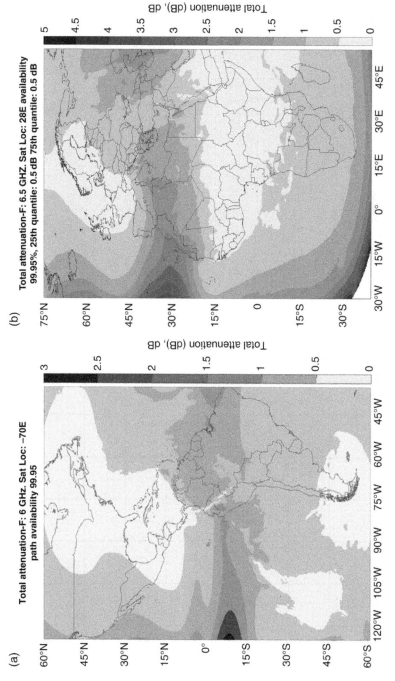

FIGURE 14.3 Total atmospheric attenuation for link availability of 99.95% at uplink C-band for (a) Americas and (b) Europe and Africa. (Used with permission of Luis Emiliani, from Emiliani (2020d)).

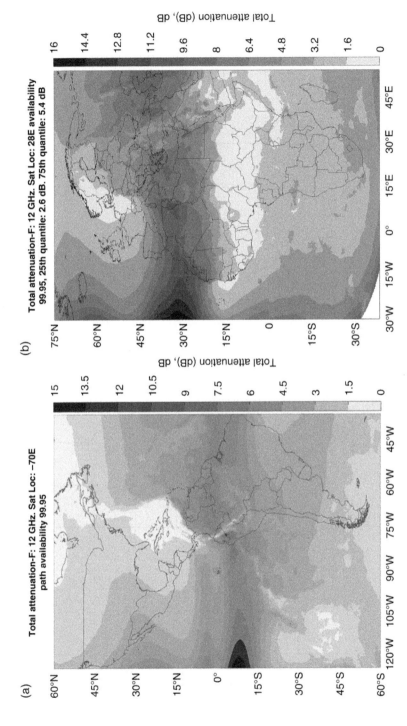

FIGURE 14.4 Total atmospheric attenuation for link availability of 99.95% at 12 GHz for (a) Americas and (b) Europe and Africa. (Used with permission of Luis Emiliani, from Emiliani (2020d)).

tropospheric attenuation. By 2017, it had seen initial testing against measured results from the Italsat campaign (Giannone et al., 1986). Comparisons with 7 years of measurement data for three Italian sites at 20, 40, and 50 GHz showed close agreement (Luini, 2017).

We have seen in Section 14.4.4.1.7 that the simulation has been used to study site diversity. It has also been put to good use in predicting the satellite resources required in an ACM system to guarantee a minimum bit rate for a given number of users, with some exceedance probability (Luini et al., 2011). It has also been used to show that in Africa the area with the highest attenuation shifts from the equatorial zone in June to the southern tropical zone by December. A possible application of the simulation would be the seasonal redistribution of satellite-transmitted power over a typical year. The sum of the powers to the areas would be lower, thus reducing payload mass and DC power needs (Luini et al., 2016).

14.4.5   Loss and Variation from Antenna Polarization Mismatch

For an antenna to be able to receive all the radiation incident on it from a transmitting antenna, the radiation and the receiving antenna must have exactly the same polarization and, for non-CP, the same alignment over the 180° range of possibilities. When this is not the case, there is **polarization mismatch loss**. For a GEO with a fixed ground terminal, the loss would be constant except during heavy rain, but in other situations it will most likely vary. The only time this loss is likely to be rather large is when one of the antennas is either a phased-array antenna or a reflector with feeds formed by a phased array and the beam is scanned (Section 11.7.2.2).

The relative amount of signal power received compared to the power available, due to polarization mismatch, is the following (Howard, 1975):

$$\rho = \frac{1}{2}\frac{\left(1\pm r_1 r_2\right)^2 +\left(r_1\pm r_2\right)^2 +\left(1-r_1^2\right)\left(1-r_2^2\right)\cos 2\theta}{\left(1+r_1^2\right)\left(1+r_2^2\right)}$$

where
   r_1 = axial ratio of elliptically polarized wave
   r_2 = axial ratio of elliptically polarized receiving antenna
   θ = angle between direction of maximum amplitude in incident wave and direction of maximum amplitude of elliptically polarized receive antenna
   + sign is to be used if and only if both the receiving and transmitting antennas have the same "hand" of polarization, otherwise − sign

Both r_1 and r_2 are between 0 and 1, equaling 0 for LP and 1 for CP (Section 3.3.3.4). When $r_1 = 1$, $\rho =\left(1+r_2\right)^2 /\left(2\left(1+r_2^2\right)\right)$. For polarization that is not linear, typically not r_1 and r_2 but $-20\log r_1$ and $-20\log r_2$ (in dB) are given. Normally, for example in a link budget, the polarization mismatch loss would be reported as $10\log\rho$. If the two polarizations are orthogonal (that is, $r_1 = r_2$, the "hands" are different, and

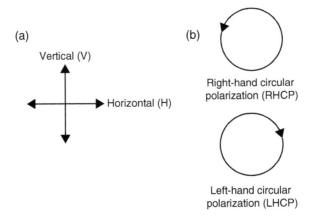

FIGURE 14.5   Pairs of orthogonal antenna polarizations: (a) linear, (b) circular.

$\cos2\theta = -1$), no radiation will be received. Examples of pairs of orthogonal polarizations are shown in Figure 14.5. If a CP antenna receives LP, it receives half the power and vice versa.

14.5 NOISE LEVEL ON LINK

The system noise temperature T_s on a given link, referred to as the receive antenna's terminal, has three components (Section 8.12.1):

- Antenna noise temperature T_a
- Ring around noise temperature T_r
- Thermal noise temperature $(F - 1)290$ where F is the noise figure (not in dB) of the part of the payload past the receive antenna terminal.

The second and third have been described in the above-referenced section. Here, we describe antenna noise.

The system noise temperature of the link will vary for several reasons. An overview of the variations is given in Table 14.4 for the uplink and in Table 14.5 for the downlink. Each end of a link must keep the changes it causes within a certain range, and the receiving end must be able to accommodate the changes outside its control.

The ITU-R characterizes antenna-noise temperature for both up- and downlinks (ITU-R P.372, 2019). The general calculation of antenna-noise temperature is the convolution of the antenna pattern with the *brightness temperature* of the background, where the brightness temperature is frequency-dependent. The ITU-R provides plots and maps and tells how to compute the brightness temperature. Following is general information from that document.

TABLE 14.4   Variations in Uplink System Noise Temperature

Variation Type	Source	Ref Section
Variation in payload antenna temperature	Brightness temperature of earth below varies; rain intensity varies	Current
Jump up in payload front-end noise figure	Switch to redundant receiver (more switches and perhaps more waveguide before LNA)	6.2.2
Jump in payload back-end noise figure (preamp in ALC mode)	Preamp ALC changes attenuation in reaction to change in input signal level	7.4.4.1

TABLE 14.5   Variations in Downlink System Noise Temperature

Variation Type	Source	Ref Section
Variation in ground antenna temperature	Sun and moon moving across sky; sun brightness temperature varies; rain intensity varies	Current
Jump up in-ground receiver front-end noise figure	Switch to redundant receiver (more switches and perhaps more waveguide before LNA)	N/A

For a payload receiving a signal with the earth as background, the brightness temperature is due to a combination of atmospheric radiation reflected from the earth and radiation emitted by the earth. A water surface has a brightness temperature from zenith of about 95 K at 1.4 GHz, 110 K at 10 GHz, and 175 K at 37.5 GHz, depending on the water surface temperature, salinity, and roughness. The land is brighter than water. The drier and the rougher the land is, the brighter it is. At 1.43 GHz, the temperature from zenith ranges from about 180 to 280 K.

For a payload receiving a signal from another satellite with space as the background, at frequencies above 1 GHz the brightness temperature is only a few degrees except in the direction of the sun, which has a very high temperature, and the moon, which varies from about 140 K at the new moon to 280 K at the full moon.

For a ground terminal receiving a signal from a payload, the sky is the background. In the range of frequencies from 1 to 30 GHz, the brightness temperature of the clear-sky atmosphere is maximum at about 23 GHz. There, for a zenith path, the temperature is about 28 K and for a path of 30° elevation, it is about 53 K. For frequencies above 20 GHz, heavy rain increases the sky temperature, up to a limit of an extra 290 K in a monsoon (Morgan and Gordon, 1989). For a GEO, the link will be unavailable once a day for several days around the time when the sun is nearly behind the spacecraft (sun outage) because of the extremely high brightness temperature of the sun (Maral and Bousquet, 2002).

14.6 INTERFERENCE ON LINK

14.6.1 Interference-Discrimination Domains

Interference can be divided into two categories based on source: self-inference and interference from other systems. Self-interference is a satellite system interfering with itself. For a GEO, the satellite system is the satellite, any collocated satellite(s) with the same owner and frequency bands, the ground station, and the user terminals. For a non-GEO constellation, it is the entire constellation, all the ground stations, and all the user terminals. Interference from other systems can come from other satellite systems and, in some frequency bands, from terrestrial communications systems. Interference can occur on the uplink and on the downlink.

When interference into a satellite system is a possibility, whether from self or not, the system design must create **discrimination** against the interference in at least one of three domains, as illustrated in Figure 14.6:

- Frequency, that is, the interfering signal is in a different frequency band from the signal of interest
- Antenna gain, that is, the antenna has lower gain toward the interfering signal than toward the signal of interest
- Polarization, that is, the interfering signal nominally has the orthogonal polarization to that of the signal of interest.

To describe the effect of interference on the signal, we define some notation (the same way it was defined in Chapter 12). C is signal power (elsewhere denoted by P or S), R_s is the rate of the modulation symbols, and E_s is energy per modulation symbol. We have $E_s = C/R_s$. N is the system noise power in a given bandwidth and N_0 is the RF one-sided power spectral density (psd) of the system noise. I is the interference power in the same bandwidth and I_0 is its psd. Interference is modeled as additive noise since the data on the interfering signal is independent

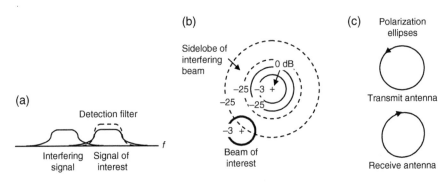

FIGURE 14.6 Three domains of discrimination against interfering signal: (a) frequency, (b) antenna gain, and (c) polarization.

TABLE 14.6   Interference Cases Treated

Interference Name	Frequency Band	Antenna Gain	Polarization
Adjacent-channel interference	Slightly overlapping	Same (worst case)	Same
Co-channel interference	Same	Different	Same (worst case)
Cross-polarized interference	Same (worst case)	Same	Different

of that of the signal of interest. In addition, often it is asynchronous and noncoherent in phase. A demodulator requires a minimum $C/(N + I)$ as measured at some point in the receiver. Equivalently, it requires a minimum $E_s/(N_0 + I_0)$. Sometimes the effect of interference is quantified as E_s/I_0 or N_0/I_0 or something similar.

When there is interference, increasing the power of the signal of interest would help but is usually not an option. When the signal of interest is weaker than the interfering signal, it is harder to bring E_s/I_0 into an acceptable range. For frequencies above about 10 GHz, attenuation from rain on the link can be significant, which can suppress the signal of interest relative to its interferers (Section 14.4.4.1.8).

Keeping total interference at least 10 dB below the noise level is a good starting value for ascertaining the tolerable level in any given situation. This corresponds to about 1 dB of increase from N to $N + I$ (FCC, 2017).

In the following three subsections, we look at three types of interference. Each is defined by having imperfect discrimination in one domain and no discrimination in a second. We assume the worst case in the third domain, which is worse than a normal situation but allows us to see why good discrimination is necessary. We assume that the interfering signal and intended signal have equal power into the transmit antenna terminal and that atmospheric attenuation is the same on the links of the two signals. The cases studied are given in Table 14.6.

14.6.2 Adjacent-Channel Interference

Adjacent-channel interference (ACI) is interference from an adjacent frequency band with the same polarization. We make the worst-case assumption that the antenna gain toward the interfering signal is the same as toward the signal of interest. Possible causes of significant ACI are (1) multiplexing of channel frequency bands that does not sharply enough filter out adjacent channel(s) and (2) spectrum spreading of the interfering signal by its high-power amplifier (HPA) (Section 9.5.2). ACI can arise in the transmitting ground terminal, the payload, and/or in the receiving ground terminal. Adjacent channels are normally separated by a guard band about 10% as wide as the channel-center spacing (Section 5.5.5.1). If the signals on the two channels have different spectral bandwidths but their power spectral densities are about the same, the wider signal is more likely to interfere with the narrower one than the other way around.

14.6.3 Co-Channel Interference

Co-channel interference is interference on the same frequency band as the signal of interest. It comes from an interfering signal that is either transmitted on a sidelobe and received on a main lobe or vice versa. We make the worst-case assumption that the polarization is the same. Figure 14.7 shows the self-interference scenarios which can lead to sidelobe interference, for the uplink and downlink. Figure 14.8

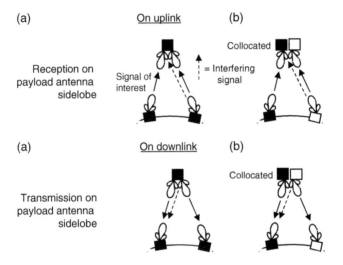

FIGURE 14.7 Scenarios of satellite system co-channel self-interference.

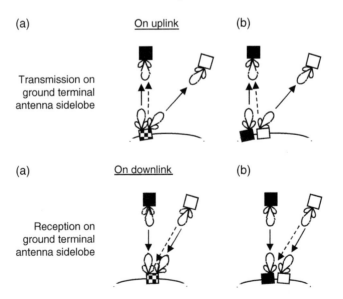

FIGURE 14.8 Scenarios of co-channel interference through ground terminal sidelobe.

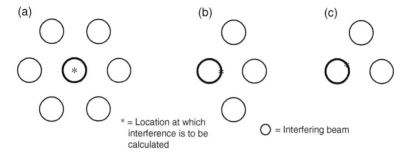

FIGURE 14.9 Significant co-channel interfering beams for various locations in a spot beam.

shows the scenarios for interference via the sidelobe of a ground terminal. Scenario Figure 14.8b involves two different ground terminals that are close enough to be in the same satellite beam. Only interference scenarios from a satellite system are shown in the figure; scenarios from a terrestrial system can exist but are not shown. The geometry of each uplink scenario is the same as that of a particular downlink scenario, shown below it in the figure.

Actually, a sidelobe need not be involved, since the low-gain outer edge of a main lobe could cause a similar situation. For a coverage area tiled with three-color spot beams (Section 11.2.1), in the spot of interest the interfering spot beam is probably transmitting/receiving on the side of its main lobe, while for a coverage area tiled with four-frequency-reuse spot beams, the interfering spot beam is probably transmitting/receiving in its first sidelobe (the latter depicted in Figure 14.6b) (Rao, 2003). Figure 14.9 shows probably the most significant interfering spot beams for various locations of interest relative to surrounding spot beams (the closer beams not shown are on different channel bands). When there are a few interfering spot beams, the total interference can be significant even though the interference from one beam is not (see Rao (2003) for example).

14.6.4 Cross-Polarized Interference

Interference from cross-polarized (cross-pol) signals is from the orthogonal polarization but with the same antenna gain as the signal of interest. We make the worst-case assumption that the frequency band is the same.

14.6.4.1 Self-Interference Cross-Pol
Cross-polarized self-interference can arise when two orthogonal polarizations are being used by one satellite or collocated satellites or one ground terminal. The transmit and receive antennas might not be polarized exactly the same.

Figure 14.10 shows the scenarios for cross-pol self-interference for uplink and downlink. The geometry of each uplink scenario is the same as that of a downlink scenario, shown one above the other. Scenarios (a) and (b) involve one satellite,

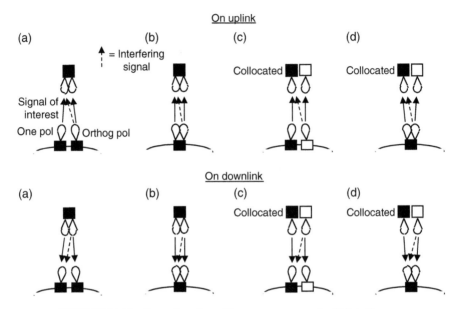

FIGURE 14.10 Scenarios of satellite system cross-pol self-interference.

while (c) and (d) involve collocated satellites. Scenarios (a) and (c) involve two ground terminals close enough to be in the same satellite beam, while (b) and (d) involve one ground terminal.

14.6.4.2 *Rain-Caused Cross-Pol*

A dual-polarized signal going through rain or ice particles may have its polarization corrupted, that is the signal may be *depolarized*. There would be a slight reduction of power in the intended polarization and a small amount of power that has crossed over to the orthogonal polarization. The amount of cross-polarization is given by the term *cross-polarization discrimination (XPD)*, which is the ratio of the power in the original polarization to the power in the orthogonal polarization in dB, in the depolarized signal.

Light rain or fog does not cause depolarization, as the particles are spherical. However, larger raindrops flatten as they fall, and they may be tilted to an angle off vertical (*canting angle*) by the wind. Tropospheric ice crystals are nonspherical as well. They lie in an extended, relatively thin sheet of cloud above stratiform rain (Allnutt, 2011).

A circularly polarized signal about to pass through the volume of rain or ice crystals can be conceptually decomposed into two orthogonal LPs. Since the medium has different characteristics along the two directions, the decomposed signal will experience differential attenuation and delay (phase shift) in the two components as it progresses through the medium, resulting in depolarization.

A linearly polarized signal will experience a tilt in its polarization as it goes through such a medium. This is also a depolarization.

The ITU presents a calculation for the long-term annual XPD caused by rain and ice particles together. It is valid for frequencies from 4 to 55 GHz and for elevation angle no greater than 60° (ITU-R P.618, 2017). The worst-month XPD can be calculated from the annual XPD by the same means as for rain (Allnutt, 2011).

Depolarization is correlated with rain attenuation but not by 100%, since some depolarization is caused by ice and not just rain. However, the ITU gives a probability distribution for XPD, not in dB but as an amplitude ratio, with a mean "very close to" the XPD for rain only and with a sigma of about 0.038 dB for rain attenuation between 3 and 8 dB (ITU-R P.618, 2017).

For a dual-polarized system, the sensitivity to rain-caused cross-pol reduces as the required E_s/N_0 goes down (Vasseur, 2000). We can conclude three things from this: (1) error-control coding reduces the sensitivity; (2) higher-order modulation increases the sensitivity; and (3) the ground terminals must have excellent polarization isolation.

14.6.5 Interference to GEO

The problems of GEOs and non-GEOs interfering with a GEO are treated in various ITU-R recommendation series:

- ITU-R BO series—for interference into a GEO providing BSS
- ITU-R M series—for interference into a MSS GEO
- ITU-R S series—interference into a GEO providing FSS.

The topic of satellites interfering with terrestrial communications is handled in other ITU-R series.

In regard to interference into a GEO system by another GEO system, GEO systems that strictly respect the frequency and coverage plans for FSS and BSS of each ITU region can avoid coordination. Otherwise, coordination between GEO systems is necessary (Emiliani, 2020a).

The requirement for coordination between non-GEO and GEO systems depends on the frequency band where the non-GEO system intends to operate (Emiliani, 2020a):

- In a specific set of frequency ranges including the C, Ku, and Ka/K-bands, non-GEO systems can operate without coordination with GEOs if the limits for operational power level and impact into other services are respected. In portions of the C, Ku, and Ka/K-bands, limits are provided in the form of an *equivalent power flux density* (*epfd*), which is related to *power flux density* (pfd), illustrated in Figure 14.11. However, epfd takes into account the emissions of all non-GEO satellites toward any ground terminal of interest and that terminal's antenna pattern (ITU-R, 2018).
- In other frequency bands, for example, 18.8–19.7 GHz and 28.6–29.5 GHz, non-GEO systems must coordinate with GEO systems and GEO must coordinate with non-GEO based on who is first.

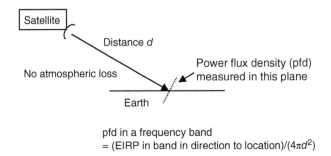

pfd in a frequency band
= (EIRP in band in direction to location)/$(4\pi d^2)$

FIGURE 14.11 Definition of power flux density reaching the earth.

The ITU-R has several procedures in place that regulate whether and how national administrations must coordinate among themselves and under which conditions coordination is not required. The ITU-R World Radiocommunication Seminars are a good source of information on what is happening. These are available free of charge on the Web site www.itu.int/en/ITU-R/seminars/Pages/default.aspx (Emiliani, 2020c).

Each non-GEO system operator files documents with the Federal Communications Commission (FCC). In its Attachment A, Technical Information to Supplement Schedule S (or some similar name), if the system is obliged not to interfere with GEOs the operator demonstrates compliance with ITU epfd limits and FCC rules.

14.6.6 Non-GEO Constellation Self-Interference

The constellation of satellites in a low earth orbit (LEO) and/or medium earth orbit (MEO) system has a complex self-interference problem. For one thing, the constellation is always shifting, so the self-interference is always changing. To get a good picture of it requires a simulation over time. Even at just one instant, though, it is complicated to figure out. The situation depends on the altitude of the satellites, how many orbit planes, how many satellites per orbit plane, the traffic distribution and amount, the multi-beam antennas the satellites have, and what kind of antennas the users have, that is, omnidirectional or with a shaped beam (Loreti and Luglio, 2002).

Not many papers have addressed this difficult problem. The best available paper is the source material for this section (Loreti and Luglio, 2002). It makes the following assumptions:

- Feeder links have so much margin that they can be ignored
- User terminal communicates with satellite closest to zenith
- User terminals are uniformly distributed in the beam.

On the forward link (down), the interferers are all the satellites in the system whose signals can be received by the user terminal of interest. There may be interfering signals on the same satellite as the user terminal of interest intends to receive from, for example, if the beam transmits on two polarizations or another beam interferes. Such interfering signals and the signal of interest propagate on the same path so see the same propagation impairments. There may be interfering signals from other satellites radiating from the sidelobes of their beams. In this case, they go on different paths from the signal of interest.

On the return link (up), all the user terminals whose signals are received by the satellite serving the user terminal of interest, are potential interferers. They all have different propagation paths.

The paper points out and discusses many system-level techniques to minimize interference.

The analysis performed in the paper is for a 48-satellite constellation of LEOs at 1414 km altitude, in eight orbital planes inclined at 52°. Both code-division multiplexing (CDM) and frequency-division multiplexing (FDM) (Section 1.3) are treated for the user links. Globalstar-2 uses CDM on user links on both forward and return. DVB-S2 uses FDM on the user forward link and DVB-RCS2 uses it on the user return link. With FDM only a subset of the users acts as interferers at one time. The analysis results are different for CDM and FDM, as expected.

14.6.7   Non-GEO Constellations' Mutual Interference

The topic of non-GEO constellations' mutual interference is even more complicated than that of non-GEO system's self-interference since two complicated systems are involved.

The ITU requires that non-GEO systems coordinate with other non-GEO systems, with exceptions as noted in the ITU Radio Regulations (Emiliani, 2020a).

The FCC intends that non-GEO operators should mutually coordinate to avoid interference. An operator of communications satellites over the US must apply for a license if US-based or for market access if not. One way to coordinate is *look-aside* (or *avoidance-angle mitigation* or *satellite diversity*), where at least one of the mutually interfering systems uses a different satellite to communicate with the user terminal, different from the nominal satellite. In general, a user terminal would communicate with the satellite closest to zenith (ITU-R S.1431, 2000). If coordination fails, the FCC requires the burdensome *band-splitting*, where interfering systems equally divide the common frequency band (FCC, 2017). Some countries require completion of a successful coordination for market access to be granted (Emiliani, 2020b).

Look-aside can be used on both the uplink and the downlink, but it is different on the two. On the uplink, a system can only establish a link from a user terminal to a satellite if no satellite of another system is within a prescribed number of degrees from the first satellite as seen from the user terminal. On the downlink, a system can only establish a link from a satellite to a user terminal if no satellite of another system is within a prescribed number of degrees from the user terminal as seen by the first satellite. In this case, in general, the satellites of the second system would have a lower altitude than that of the first satellite (Fortes and Sampaio-Neto, 2003).

In the case of two satellite systems, having both systems practice look-aside simultaneously does not enhance the situation (ITU-R S.1431, 2000).

The ITU suggests that for a given permissible link outage, 10% of the outage time be allocated to interference and 90% to atmospheric attenuation. The ITU also provides methods to calculate the probability distribution of allowable interference (ITU-R S.1323, 2002).

Each non-GEO system operator files documents with the FCC. In Attachment A, Technical Information to Supplement Schedule S (or some similar name), the operator explains how it will be able to coordinate with other non-GEO systems.

Published studies on the topic of non-GEO systems interfering with each other are rarer than those on interference by non-GEO systems into GEO systems. We summarize some results of two of them, the first for Ka/K- and V-band systems and the second for Ku-band systems. This is the order in which they were published.

The first study was thought at the time to be the first publicly available probabilistic risk assessment (Tonkin and de Vries, 2018). The analysis was for Ka/K-band or V-band systems that the authors considered reasonably likely to come into operation. (The IEEE calls V-band the range of frequencies between 40 and 75 GHz. The V-band range used by these satellites is popularly called "Q-band," which ranges from 33 to 50 GHz, but neither the IEEE nor the ITU recognizes a Q-band (Wikipedia, 2019).) The systems studied are summarized in Table 14.7. For the systems where user links used more than one frequency band, each satellite used all of them.

The greatest interferer at Ka/K-band was OneWeb, while at V-band it was OneWeb and Starlink. OneWeb interfered so much because its orbit is higher than the others, so more OneWeb satellites were visible to a user terminal and a OneWeb satellite could see more user terminals. Additionally, OneWeb was supposed to have a large number of satellites: a site might see more than 100 satellites above the minimum elevation angle (Tonkin and de Vries, 2018).

TABLE 14.7 Non-GEO Constellations Analyzed for Mutual Interference at Ka/K- and V-Bands (Tonkin and de Vries, 2018)

Constellation	Orbit	No. of Active Satellites	Details	User-Link Frequencies
O3b in 2018	MEO	16	8062 km altitude, equatorial	Ka/K-band
OneWeb	MEO	2560	About 8500 km altitude, about 45° inclination	Ka/K- and V-bands
Starlink	LEO	4425	3200 satellites at about 1150 km altitude, about 53° inclination. Also, 1125 at about 1300 altitude, 70–81° inclination	Ka/K- and V-bands
Telesat	LEO	117	45 satellites at 1248 km altitude, 37.4° inclination. Also, 72 at 1000 km altitude, 99.5° inclination	Ka/K- and V-bands

If a large constellation triggers band-splitting by interfering with a small constellation, the large constellation suffers a greater throughput decrease than the small constellation.

The study conclusions were that mutual interference among these four systems may not be a significant problem. It would decrease throughput by less than two percentage points for all constellations at all sites. The reference throughput is based on poor atmospheric transmission but no interference (Tonkin and de Vries, 2018). *Throughput degradation* appears to be a plausible metric for mutual interference among non-GEO systems. It can be calculated as the change in $C/(I+N)$ from a reference $C/(I+N)$ or as the decrease in C/N to $C/(I+N)$ (FCC, 2017).

For ACM systems, throughput degradation is a weighted average using all the modulation and coding combinations (MODCODs) (Section 13.2.3) of the system (Emiliani, 2020b).

The second study was for the more complicated situation of the Ku-band systems, but just for the downlink (Braun et al., 2019). It took into account all recently proposed Ku-band systems, of which there were eight. The systems were very different from each other in terms of numbers of satellites, ranging from 2 to 4425, and there

TABLE 14.8   Non-GEO Constellations Analyzed in Tonkin and de Vries (2018) for Mutual Interference at Ku-Band

Constellation	Orbit	No. of Active Satellites	Details	Reference
Kepler	LEO	140 minus spares	575 km altitude, polar, sun-synchronous	Wikipedia (2020a)
OneWeb	MEO	2560	About 8500 km altitude, about 45° inclination	Tonkin and de Vries (2018)
OneWeb	LEO	720	About 1200 km altitude, 86.4° inclination	del Portillo et al. (2018)
Starlink	LEO	4425	3200 satellites at about 1150 km altitude, about 53° inclination. Also, 1125 at about 1300 altitude, 70–81° inclination	Tonkin and de Vries (2018)
Theia	LEO	112	800 km average, 98.6° inclination, sun-synchronous	Fargnoli (2016)
Karousel	GSO	12	63.4° inclination	Norin (2016)
Space Norway	HEO	2	63.4° inclination, 16-h orbit, 43,500 km by 8100 km altitude orbit	Henry (2019)
New Spectrum Satellite	HEO	15	63.4° inclination, 8-h orbit, apogee 26,172 km altitude, eccentricity 0.605	Brosius (2017)

were different orbits including elliptical. The systems are summarized in Table 14.8. This study was also the first of its kind. The authors used the software that the authors of the first study had developed. The worst case of user terminal placement was assumed, namely that the user terminal of the system of interest and the user terminals of the interfering systems were collocated. ACM was also assumed.

A Starlink user terminal at Miami, where all the constellations would provide coverage, had a median throughput degradation of 60% caused by the other systems, where almost all of the reduction was due to interference by Kepler. Look-aside did not help. If the only interferer to the user terminal was OneWeb, the median degradation was about 35% even with look-aside.

Upon finding that the originally proposed parameters of systems caused heavy interference, the authors proposed alternative, tuned parameters. The constellations' transmit gain, transmit power, and receive gain were modified to cause less interference. The largest modifications were as follows:

- Kepler transmit antenna gain increased by 26 dB, transmit power decreased by 45 dB, and receive antenna gain increased by 14 dB
- OneWeb LEO transmit antenna gain increased by 25 dB, transmit power decreased by 14 dB, and receive antenna gain increased by 14 dB
- Starlink transmit antenna gain increased by 13 dB
- OneWeb MEO transmits antenna gain increased by 6 dB, transmit power decreased by 9 dB, and receive antenna gain decreased by 6 dB.

In the tuned case, Starlink had a very low median throughput-degradation even without look-aside. On the other hand, Starlink itself was a strong interferer for the other systems. Kepler had moderate throughput degradation without look-aside, varying with the ground site. The degradation increased with look-aside since Kepler does not have many satellites. The beam sizes of these four systems were decreased in the proposal, so entire system redesigns would be necessary.

The second study's conclusions were that the Ku-band inter-constellation downlink interactions are too complicated to manage with the simple techniques of look-aside and band-splitting. Look-aside can be beneficial to large constellations but is detrimental to small ones. More sophisticated solutions and potentially stricter regulation will be needed (Braun et al., 2019).

14.7   END-TO-END $C/(N_0 + I_0)$

We derive the well-known formula for the composite C/N_0 of an uplink, a bent-pipe payload, and a downlink.

The quality of the (impaired) signal on a link can be represented to first order by $C/(N_0 + I_0)$. We follow the usual convention of referring this fraction to the receiver's antenna terminal. Whether the link quality is good enough is told by comparing the link's $C/(N_0 + I_0)$ with the $C/(N_0 + I_0)$ required by the receiver. For simplicity of notation in the rest of this section, we fold I_0 into N_0. We treat the repeater as an electronic element described by its gain and noise figure, where by repeater we mean the payload minus its antennas. All we need for the ground receiver's RF front end is its noise figure. Any noise added to the signal by the ground transmitter is negligible.

Recall from Section 14.5 that a link's system noise temperature T_s is given by

$$T_s = T_a + T_r + (F-1)290$$

Recall also that

$$N_0 = \kappa T \text{ where } \kappa \text{ is Boltzmann's constant}$$

Figure 14.12 shows the calculation of uplink C/N_0 from the system noise temperature T_1 and the noise figure F_1 of the repeater, which at this point can be either bent-pipe or processing.

Figure 14.13 shows the similar calculation of downlink C/N_0 from the system noise temperature T_2 and the noise figure F_2 of the receiver on the ground.

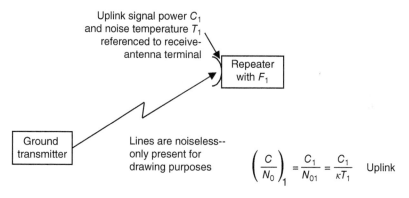

FIGURE 14.12   Uplink C/N_0 derived from repeater noise figure.

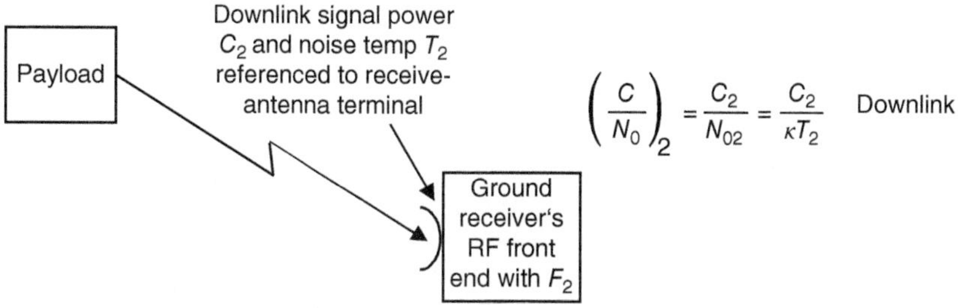

FIGURE 14.13   Downlink C/N_0 derived from ground-receiver noise figure.

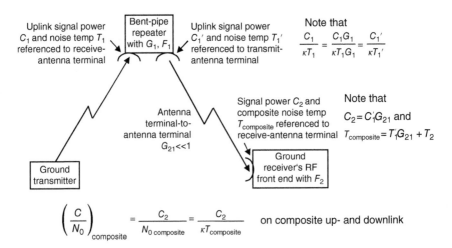

FIGURE 14.14   Composite C/N_0 for up- and downlink combined, for bent-pipe payload.

The number important to end-to-end performance for a bent-pipe payload is the composite C/N_0 of the signal that goes through both links. Figure 14.14 shows this calculation. Downlink propagation is equivalent to a big loss without noise figure.

We need to show that the composite C/N_0 represented in the figure conforms to the well-known formula for the composition of the uplink and downlink C/N_0's:

$$\left(\frac{C}{N_0}\right)^{-1}_{\text{composite}} = \left(\frac{C}{N_0}\right)^{-1}_{1} + \left(\frac{C}{N_0}\right)^{-1}_{2}$$

From the figure, we have

$$\left(\frac{T}{C}\right)_{\text{composite}} = \frac{T_{\text{composite}}}{C_2} = \frac{T_1' G_{21} + T_2}{C_2} = \frac{T_1'}{C_2 / G_{21}} + \frac{T_2}{C_2} = \frac{T_1'}{C_1'} + \frac{T_2}{C_2} = \frac{T_1}{C_1} + \frac{T_2}{C_2}$$

So it is correct.

14.8   LINK BUDGET

We now put together what we have discussed so far in this chapter into a **link budget**, which is a tallying of the transmitted power, power losses, non-ideal hardware effect, and noise and interference contributions to a bottom-line $E_s/(N_0 + I_0)$ and then comparison of this with the $E_s/(N_0 + I_0)$ necessary for adequate performance.

Term		Unit	Value
GEO transmit			
TWTA Pout		dBW	16.8
Post-TWTA loss		dB	−1.7
Redundancy switch		dB	−0.3
Antenna gain		dB	55.3
Pointing loss		dB	−0.3
	EIRP	dBW	69.8
Propagation			
Free space loss		dB	−210.3
Gas attenuation		dB	−0.5
Rain atten (0.01%)		dB	−5.5
Polarization mismatch		dB	−0.5
Pointing loss		dB	−0.3
Antenna gain		dB	69.0
	C	dBW	−78.3
Noise			
System noise temperature, inverse		dB(1/K)	−28.9
Boltzmann's constant, inverse		dB(K-Hz/W)	228.6
	C/N_0	dBHz	121.4
Interference			
C/I before discrims applied		dB	0.0
Frequency discrimination		dB	0.0
Antenna gain discrimination		dB	10.0
Polarization discrimination		dB	10.0
Symbol rate		dBHz	99.8
	C/I_0	dBHz	119.8
Combined			
	$C/(N_0+I_0)$	dBHz	117.5
Symbol rate, inverse		dB(1/Hz)	−99.8
	$E_s/(N_0+I_0)$	dB	17.7
Margin			
Implementation loss		dB	−1.5
Risk margin		dB	−2.0
Required $E_s/(N_0+I_0)$, inverse		dB	−13.0
	Excess margin	dB	1.2

FIGURE 14.15 Sample downlink budget.

The link budget we present in Figure 14.15 is a simplified one for the downlink. In most cases, interference is well modeled as an additional noise-type term, which is why N_0 and I_0 are usually summed as they are here. The correct sign is put on entries so that they can all be summed, that is, if an item is called a "loss" it is still shown negative. (Other people's budgets may not follow this convention.) The term "C/I before discrims applied" is the ratio of power input to the antenna terminal for C relative to that for I, before any discrimination is considered. The choice to have no discrimination in the frequency domain and 10 dB in each of the antenna-pattern and polarization domains is random and just for illustration. The implementation-loss line-item is discussed in the following section.

TABLE 14.9 Summary of Signal Corruptions by Hardware in Bent-Pipe-Payload Satellite Communications System

Signal Distortion Type	Occurs in Ground Transmitter	Occurs in Payload	Occurs in Ground Receiver	Source	Specifics	Ref Section
Linear distortion	Maybe	Yes	Yes	Filters	Input and output multiplexing; detection filter not matched to the pulse-shaping filter	12.15.1, 12.15.2
Clock jitter	Yes	No	Yes	Clocks	Symbol timing and timing recovery	12.15.3
Phase noise and spurious phase modulation	Yes	Yes	Yes	Oscillators	Modulation; frequency conversion; carrier recovery; demodulation	6.6.6; 12.10.2
Long-term frequency instability	Perhaps in consumer equipment	Yes	Perhaps in consumer equipment	Oscillators	Aging	8.6
Frequency converter-caused spurious signals	Yes	Yes	Yes	Frequency converters	LO harmonics, signal harmonics, cross-products of signal and LO, leakage of the input signal and LO	8.7
Inband and near out-of-band intermodulation products	Maybe	Yes	No	HPA	Bandpass nonlinearity	6.3.1, 9.5.1
Nonlinear distortion of the signal	Maybe	Yes	No	HPA	Bandpass nonlinearity	7.2.1
HPA-caused spurious signals	Maybe	Yes	No	HPA and its power supply	Various	8.9

Ordinarily, there would also be explanatory or informational entries that are not part of the summation but are used in the calculation of the entries that are summed. Also, one may want to make maximum and minimum signal-level cases. Uncertainties or variations could also be included (Section 9.3.4).

For a bent-pipe payload, it is common to make a link budget for the uplink and a budget for the downlink and form a composite excess margin.

For a link that employs DVB with ACM, the budget philosophy would be different. If the user has a guaranteed minimum bit rate with a given outage probability, the budget might be constructed so as to show closure for that bit rate, the lowest-order of modulation, the lowest code rate, and rain attenuation with exceedance equal to the outage probability (Luini et al., 2011). A second way to construct the budget is to make two, for the most and least robust MODCODs (Telesat, 2010). A third, more thorough way is to make two kinds of analysis. One is a link budget with the most efficient MODCOD and either a required throughput value or a target of long-term annual link availability. The second analysis is a chart in which all options are analyzed, showing availability and throughput associated with each option (Emiliani, 2020b).

14.9 IMPLEMENTATION LOSS ITEM IN LINK BUDGET

A link budget contains a line item for **implementation loss**. This is a catch-all item to account for the less-than-ideal behavior of the system hardware which is not well characterized. It includes payload imperfections, ground-terminal imperfections, and payload-ground interaction imperfections that are not captured elsewhere in the link budget.

Table 14.9 summarizes the sources of signal corruption in the system hardware. Some of the sources may have their own line items in the link budget, but those that do not are accounted for in the implementation loss item. The table is for a bent-pipe payload but it is obviously alterable for a processing payload.

REFERENCES

Allnutt JA (2011). *Satellite-to-Ground Radiowave Propagation*, 2nd ed. London: The Institution of Engineering and Technology.

Braun C, Voicu AM, Simić L, and Mähönen P (2019). Should we worry about interference in emerging dense NGSO satellite constellations? *IEEE International Symposium on Dynamic Spectrum Access Networks (DySPAN)*; Nov. 11–14.

Brosius JW, acting CTO of New Spectrum Satelllite (2017). Letter of intent to FCC, technical narrative. July 26.

Crane RK (1993). Estimating risk for earth-satellite attenuation prediction. *IEEE Proceedings*; 81 (June): 905–913.

Crane RK (1996). *Electromagnetic Wave Propagation through Rain*. New York: John Wiley & Sons.

Crane RK and Dissanayake AW (1997). ACTS propagation experiment: attenuation distribution observations and prediction model comparisons. *IEEE Proceedings*; 85 (June): 879–892.

Crane RK (2002). Analysis of the effects of water on the ACTS propagation terminal antenna. *IEEE Transactions on Antennas and Propagation*; 50 (July): 954–965.

Crane RK (2003a). A local model for the prediction of rain rate statistics for rain-attenuation models. *IEEE Transactions on Antennas and Propagation*; 51 (Sep.): 2260–2273.

Crane RK (2003b). *Propagation Handbook for Wireless Communication System Design.* Boca Raton, FL: CRC Press.

Davies K and Smith EK (2002). Ionospheric effects on satellite land mobile systems. *IEEE Antennas and Propagation Magazine*; 44 (Dec.): 24–31.

del Portillo I, Cameron BG, and Crawley EF (2018). A technical comparison of three low earth orbit satellite constellation systems to provide global broadband. *International Astronautical Congress*; Oct. 1–5.

Dissanayake A, Allnutt J, and Haidara F (1997). A prediction model that combines rain attenuation and other propagation impairments along Earth-satellite path. *IEEE Transactions on Antennas and Propagation*; 45: 1546–1558.

Emiliani LD of SES (2020a). Private communication. May 13.

Emiliani LD of SES (2020b). Private communication. May 29.

Emiliani LD of SES (2020c). Private communication. June 2.

Emiliani LD of SES (2020d). Computer program. Oct. 18.

Emiliani LD of SES (2020e). Private communication. Oct. 21.

Emiliani LD and Luini L (2016). A combined rain and cloud attenuation field simulator and its application to gateway diversity analysis at Ka, Q, and V bands. *International Communications Satellite Systems Conference*; Oct. 18–20.

Fargnoli JD, CTO of Theia Holdings A (2016). Application to FCC, technical narrative. On licensing.fcc.gov/myibfs/download.do?attachment_key=1158366. Accessed Mar. 14, 2010.

FCC (Federal Communications Commission) Technological Advisory Council (2017). A risk assessment framework for NGSO-NGSO interference. Satellite Communication Plan Working Group. Dec. 6.

Feldhake G (1997). Estimating the attenuation due to combined atmospheric effects on modern earth-space paths. *IEEE Antennas and Propagation Magazine*; 39: 26–34.

Fortes JMP and Sampaio-Neto R (2003). Impact of avoidance angle mitigation techniques on the interference produced by non-GSO systems in a multiple non-GSO interference environment. *International Journal of Satellite Communications and Networking*; 21: 575–593.

Giannone B, Matricciani E, Paraboni A, and Saggese E (1986). Exploitation of the 20-40-50 GHz bands: propagation experiments with Italsat. *Communications Satellite Systems Conference*; Mar. 17–20.

Henry C (2019). Space Norway in final procurement for two highly elliptical orbit satellites. *SpaceNews;* Apr. 10.

Howard W. Sams & Co, Inc. (1975). *Reference Data for Radio Engineers.* Indianapolis: Howard W. Sams & Co.

ITU-R (International Telecommunication Union, Radiocommunication Sector) (2014). *Handbook on Meteorology*, 2013 ed. ITU-R Study Group 3, Working Party 3J.

ITU-R (2017). Document 3J/FAS/3-E. Concerning the rainfall rate model given in Annex 1 to Recommendation ITU-R P.837-7. Working Party 3J fascicle. Apr. 3.

ITU-R (2018). Equivalent power flux density limits (epfd). Viewgraph presentation. *ITU World Radiocommunication Seminar;* Dec. 3–7.

ITU-R. Recommendation P.372-14 (2019). Radio noise. Aug.

ITU-R. Recommendation P.531-14 (2019). Ionospheric propagation data and prediction methods required for the design of satellite services and systems. Aug.

ITU-R. Recommendation P.618-13 (2017). Propagation data and prediction methods required for the design of Earth-space telecommunication systems. Dec.

ITU-R. Recommendation P.619-4 (2019). Propagation data required for the evaluation of interference between stations in space and those on the surface of the earth. Aug.

ITU-R. Recommendation P.676-12 (2019). Attenuation by atmospheric gases. Aug.

ITU-R. Recommendation P.678-3 (2015). Characterization of the variability of propagation phenomena and estimation of the risk associated with propagation margin. July.

ITU-R. Recommendation P.835-6 (2017). Reference standard atmospheres. Dec.

ITU-R. Recommendation P.836-6 (2017). Water vapor: surface density and total columnar content. Dec.

ITU-R. Recommendation P.837-7 (2017). Characteristics of precipitation for propagation modelling. June.

ITU-R. Recommendation P.840-8 (2019). Attenuation due to clouds and fog. Aug.

ITU-R. Recommendation P.841-6 (2019). Conversion of annual statistics to worst-month statistics. Aug.

ITU-R. Recommendation P.1510-1 (2017). Mean surface temperature. June.

ITU-R. Recommendation P.1815-1 (2009). Differential rain attenuation. Oct.

ITU-R. Recommendation P.2041 (2013). Prediction of path attenuation on links between an airborne platform and space and between an airborne platform and the surface of the earth. Sep.

ITU-R. Recommendation S.1323-2 (2002). Maximum permissible levels of inerference in a satellite network (GSO/FSS; non-GSO/FSS; non-GSO/MSS feeder links) in the fixed-satellite service caused by other codirectional FSS networks below 30 GHz.

ITU-R. Recommendation S.1431 (2000). Methods to enhance sharing between non-GSO FSS systems (except MSS feeder links) in the frequency bands between 10-30 GHz.

Ippolito LJ (2017). *Satellite Communications Systems Engineering: Atmospheric Effects, Satellite Link Design, and System Performance*, 2nd ed. UK: John Wiley & Sons Ltd.

Kumar LS, Lee YH, and Ong JT (2009). Slant-path rain attenuation at different elevation angles for tropical region. *International Conference on Information, Communications and Signal Processing*; Dec. 8–10.

Kymeta Corp (2019). Link budget calculations for a satellite link with an electronically steerable antenna terminal. 793-00004-000-REV01. White paper. June 1. On www. kymetacorp.com/wp-content/uploads/2019/06/Link-Budget-Calculations-2.pdf. Accessed May 27, 2020.

Loreti P and Luglio M (2002). Interference evaluations and simulations for multisatellite multibeam systems. *International Journal of Satellite Communications*; 20: 261–282.

Luini L, Emiliani L, and Capsoni C (2011). Planning of advanced satcom systems using ACM techniques: the impact of rain fade. *European Conference on Antennas and Propagation*; Apr. 11–15.

Luini L, Emiliani L, and Capsoni C (2016). Worst-month tropospheric attenuation prediction: application of a new approach. *European Conference on Antennas and Propagation*; Apr. 10–15.

Luini L (2017). A comprehensive methodology to assess tropospheric fade affecting earth-space communication systems. *IEEE Transactions on Antennas and Propagation*; 65 (July): 3654–3663.

Mandeep JS and Allnutt JE (2007). Rain attenuation predictions at Ku-band in South East Asia countries. *Progress in Electromagnetics Research*; 76: 65–74.

Maral G and Bousquet M (2002). *Satellite Communications Systems*, 4th ed. Chichester, UK: John Wiley & Sons Ltd.

Morgan WL and Gordon GD (1989). *Communications Satellite Handbook*. New York: John Wiley & Sons.

Norin JL (2016) on behalf of Karousel LLC. Application to FCC to launch and operate. Nov. 15. On assets.fiercemarkets.net/public/007-Telecom/karousel.pdf. Accessed Mar. 14, 2020.

Rao SK (2003). Parametric design and analysis of multiple-beam reflector antennas for satellite communications. *IEEE Antennas and Propagation Magazine*; 45 (Aug.): 26–34.

Telesat (2010). Briefing on adaptive coding and modulation (ACM). White paper. On www.telesat.com/tools-resources/technical-briefings-white-papers. Accessed May 23, 2020.

Tonkin S and de Vries JP (2018). NewSpace spectrum sharing: assessing interference risk and mitigations for new satellite constellations. *TPRC46, Research Conference on Communications, Information and Internet Policy*; Sep. 21–23.

Vasseur H (2000). Degradation of availability performance in dual-polarized satellite communications systems. *IEEE Transactions on Communications*; 48: 465–472.

Wang Y, You Y, and Kulie M (2018). Global virga precipitation distribution derived from three spaceborne radars and its contribution to the false radiometer precipitation detection. American Geophysical Union's *Geophysical Research Letters*. May 1.

Wikipedia (2019). Q band. Article. Sep. 9. Accessed Mar. 6, 2020.

Wikipedia (2020a). Kepler communications. Article. Feb. 19. Accessed Mar. 14, 2020.

Wikipedia (2020c). Ionosphere. Article. Oct. 7. Accessed Oct. 21, 2020.

Wikipedia (2020b). Troposphere. Article. Oct. 15. Accessed Oct. 21, 2020.

CHAPTER 15

PROBABILISTIC TREATMENT OF DOWNLINK MARGIN FOR MULTI-BEAM PAYLOAD

15.1 INTRODUCTION

This chapter presents a link-level analysis of a multi-beam, bent-pipe, geostationary-orbit (GEO) payload that shows lower equivalent isotropically radiated powers (EIRPs) to be necessary than a traditional, overly conservative analysis would dictate. The analysis treats variations via probability theory. (Here, the word "variation" is used instead of "uncertainty" as in Chapter 9, even though the definition of variation is the same as the definition of uncertainty, because "variation" is the term used in a payload specification.) Even relatively more EIRP can be saved if the downlinks are subject to rain loss and the traditional EIRP requirements are replaced with a link-availability requirement. Recall that the downlink is available if the link closes, that is, if the signal in the earth-station receiver meets a minimum $C/(N + I)$ requirement.

In a traditional analysis of this kind, worst-case value is piled upon worst-case value and held against C and used to build up I_{self} (= self-interference—Section 14.6). This characterizes an extreme situation that occurs with miniscule probability. A probabilistic characterization more fully and accurately portrays the on-orbit payload performance. If the links are subject to rain loss, traditionally a rain margin is separated out from a payload-variation margin and applied one-to-one to increase

Satellite Communications Payload and System, Second Edition. Teresa M. Braun and Walter R. Braun.
© 2021 John Wiley & Sons, Inc. Published 2021 by John Wiley & Sons, Inc.

the specified EIRPs. In reality, though, most of the time the payload's performance is fairly close to nominal, so for the rare times when the rain attenuation is uncommonly large, the chance is high that the payload is performing near nominal.

In this chapter, we deal with the following determinants of the power that reaches the earth station:

- Varying performance of repeater, that is, payload minus antennas
- Varying performance of payload antennas (Sections 3.9, 3.11, 11.2.2, and 11.13)
- Atmosphere, for carrier frequencies above about 10 GHz (Section 14.4.4).

The I_{self} that matters most in C/I_{self} variation is for the interferer that creates the smallest C/I_{self} for the particular C of interest, at the coverage-area location under consideration. C/I_{self} includes the discrimination that C has against the self-interference, but the C/I_{self} variation does not. At any randomly chosen location, there is usually just one significant interferer, if any. There may well be isolated points in the coverage area at which more than one interferer is significant.

15.2   MULTI-BEAM-DOWNLINK PAYLOAD SPECIFICATIONS

There are different kinds of multi-beam-downlink payloads and ways to specify them. The basic specification is either EIRP or link availability. For carrier frequencies below 10 GHz, it has to be EIRP since there is no significant atmospheric loss (for exceptions see Section 14.4.3). For frequencies above 10 GHz, it can be either EIRP or link availability. Whichever it is, other specifications follow along. First, we discuss the kinds of specifications and then the set of locations on which they must hold.

If the basic specifications are a set of minimum EIRPs, then these EIRPs compensate rain attenuation, if any. Other specifications are maximum C variation, minimum C/I_{self}, and maximum C/I_{self} variation. (With the probabilistic approach given in this chapter, it should be easy to meet the variation specifications.)

C is the EIRP toward the location of interest on the ground. All other powers are referred to the antenna output, too. The reason is that the goal is to find a complete set of EIRPs that meets requirements, that the payload can provide with its high-power amplifiers (HPAs) and transmit antennas.

If the basic specification is a minimum link availability, then a particular statistical rain model is also specified. Also specified is the minimum threshold of atmosphere-attenuated $C/(I_{self} + I_{other} + N)$ for the earth-station receiver to perform acceptably, where I_{other} is other interference besides self-interference. The threshold equals the value necessary to the receiver if the only signal impairment were additive white Gaussian noise, augmented to make up for the less-than-ideal implementations of the earth stations and the payload and for their interactions. N in the no-rain case and I_{other} and are specified, both constant and both referred to the same point as C, namely at the payload-antenna output. Atmosphere-attenuated $C/(I_{self} + I_{other} + N)$ must exceed the threshold for a percentage of the time at least equal to the specified minimum availability, long-term.

Link availability instead of EIRP is recommended as the basic specification for a multi-beam-downlink payload with carrier frequencies 10 GHz or higher. Availability is fundamentally what counts. Making a satellite create and transmit more power than necessary carries a cost penalty.

In either case, there is also a set of maximum EIRP specifications, to ensure that the power flux density reaching the earth (Section 14.6.5) meets regulations, within and in some cases outside of the coverage area.

Now, on what set of locations do the specifications apply? Multiple beams may be all of the same size (as seen from the payload) in a hexagonal grid or of different sizes in an irregular pattern. For carrier frequencies below 10 GHz, the specifications can apply to "every square inch" of the coverage area because it is easy to see what the possible worst locations are. They are the same in every beam relative to its center. It is comparatively easy to check those few locations and make adjustments in payload design. However, for carrier frequencies above 10 GHz, the specifications can only apply to a (large number of) discrete locations. The beam pattern is necessarily irregular at the least in the sense of varying power levels in the beams to compensate atmospheric attenuation. It is usually not possible to look at an irregular beam pattern and the irregular rain map and see where the tricky locations are. It is humanly impossible to check specifications on an infinite number of locations. All one can do is check a large number of locations.

The type of lifetime requirement for the payload specifications matters (Section 9.3.4.5). Let us define δ as in Section 9.3.4.9, namely as the expected change in average performance from beginning of life (BOL) to middle of life, in dB. For the average-over-life requirement, the variation is averaged over life and the nominal value is δ relative to BOL average. For the at-end of life (EOL) requirement, both the variation and the nominal are for EOL, so the nominal is 2δ relative to BOL average.

15.3   ANALYSIS METHOD

We perform the analysis in three stages. The first is for the repeater, to obtain its contribution to the nominal EIRP δ relative to BOL average and the variation about this. The second is to obtain the similar information about the antenna and combine it with the repeater numbers. The third is to probabilistically combine the payload performance statistics with rain statistics.

The analysis of the payload-caused variation in C and C/I_{self} is based on the probabilistic treatment of uncertainty in payload budgets of Section 9.3.4. The analysis is made at a point in time at which units are being built but the traveling-wave tube amplifiers (TWTAs) are complete, but with small changes, the analysis could be adapted to any other time in the payload build (Section 9.3.1). We also find the

TABLE 15.1   Independent Components of Repeater-Caused C Variation, as Analyzed

Independent Components of Repeater-Caused Variation
Constant but unknown offsets
Diurnal variations
Lifetime drifts unrelated to temperature

nominal value of C and C/I_{self} relative to their BOL average values (Section 9.3.4.9). For one interfering signal I_{self}, we also analyze the C/I_{self} variation, which we need in the later analysis of the $C/(I_{self} + I_{other} + N)$ variation.

Payload results are primarily shown for the average-over-life requirement but adaptations for the at-EOL requirement are stated. We will find that the repeater has a nonzero contribution to δ but the payload antenna contributes zero.

The orientation of the GEO satellite in space relative to the earth and the sun at various times of year is illustrated in Figures 2.9, 2.10, and 2.11. The variation of C has a term which is constant but unknown over life, a diurnal swing (different at BOL and EOL), a seasonal variation, and a drift of the year's average value over the years of life. We make the pessimistic assumption that every day has the EOL, worst-day diurnal spread. The seasonal variation is minor, so we make the slightly pessimistic simplification of only treating the worst day of the year. The constant but unknown offset and the variations on the remaining time scales are now independent and are listed in Table 15.1.

15.4   ANALYSIS ASSUMPTIONS

The payload is assumed to be on a GEO spacecraft. The repeater is assumed to be bent-pipe and has the transponder block diagram presented in Figure 1.3 and the layout in the spacecraft shown in Figure 2.5. The antennas are reflectors. The channel amplifiers (CAMPs) are operated in automatic-level-control (ALC) mode. The C variation due to the repeater is the following sum:

$$C \text{ variation due to repeater in dB} = \text{Variation in TWTA } P_{out}$$
$$+ P_{out} \text{ measurement error}$$
$$+ \text{ Variation in repeater loss past TWTA,}$$
$$\text{all in dB}$$

That is, the variation comes only from the CAMP (since it drives the TWTA), the TWTA, the output multiplexer (OMUX), and the rest of the post-TWTA hardware (often just waveguide) including the antenna. The units on the signal path before the CAMP, including the receive antenna, do not matter because the CAMP compensates input signal-level variation since it is in ALC mode. The variation in I_{self} is

TABLE 15.2   Rise of Nominal Unit Temperatures Over Life (Example)

Unit	Nominal Temperature at EOL Relative to BOL Average (°C)
CAMP	+35
TWTA	+35
OMUX on east panel	+27
OMUX on west panel	+27

TABLE 15.3   Temperature Variation of Single Units at EOL (Example)

Unit	EOL Diurnal Temperature Swing (°C)
CAMP	0
TWTA	0
OMUX on east panel	±27
OMUX on west panel	±27

similar to that for C but from different payload equipment. We assume that the drive levels into the TWTAs stay very nearly the same over life.

The CAMPs and TWTAs are on the north and south panels and the OMUXes are on the east and west panels. There is temperature variation of the spacecraft mounting plates to which the repeater's unit baseplates are attached. The north and south panels have heat pipe systems which are connected, so the temperatures of units on the two panels are always nearly the same. The east and west panels have their own heat pipe systems, separate from each other and the heat pipes of the north and south. There will be more temperature difference at any one time among the OMUXes than among the CAMPs or among the TWTAs. The nominal temperatures of these units are shown in Table 15.2.

For the C variation due to temperature change, we are interested in the temperature variations of the individual units in the signal path, while for the C/I_{self} variation we are interested in the variation of the temperature delta between two units of the same kind. Only one unit in any signal path, the OMUX, has a diurnal temperature swing. Some examples of realistic numbers for single units are given in Table 15.3. Between two OMUXes, if they are on the same panel the difference is zero (that is, they are completely positively correlated) and if they are on different panels the limits are twice as great as for single units (that is, they are completely negatively correlated).

15.5   REPEATER-CAUSED VARIATION OF C AND C/I_{self} AND NOMINAL VALUE

The analysis flow for the repeater is as follows. Sections 15.5.1 through 15.5.5 break down the C variation and the C/I_{self} variation due to the repeater units, each into three independent components. Section 15.5.6 summarizes the units' contributions to the

TABLE 15.4 C Variation Due to CAMP's Settability Resolution (Example)

	Constant but Unknown Offset
	In general form of a random number uniformly distributed over $-Q_1$ to $+Q_1$ where Q_1 is
TWTA at saturation	0 dB
TWTA at small-signal	0.25 dB

variations, representing them by random numbers. Section 15.5.7 combines the units' contributions into the composite variations and computes the standard deviations. Also derived is the parameter δ. A numerical example is carried throughout. In the following tables and equations, "negligible" means less than 0.05 dB.

15.5.1   Variation Contributions from CAMP

The CAMP preamplifies the signal to the level desired for input to the HPA. Recall that in our example all CAMPs are operated in ALC mode.

15.5.1.1   CAMP Settability Resolution   The CAMP's output power is settable to steps with a resolution of 0.5 dB, so without particular information on the settability error, we assume it is uniformly distributed between −0.25 and +0.25 dB. Settability resolution affects only the signals with unsaturated TWTAs, where it leads to an error in the TWTA's P_{out}. Half the values of P_{out} variation due to settability resolution are beneficial to C relative to I_{self} and half are detrimental.

Table 15.4 shows examples of the variation in P_{out} due to one CAMP's settability resolution. For two signals, it is likely that either the settabilities are the same, since the CAMPs may come from the same build, or independent. Independence is worse so we assume that. Then the variation in the ratio of two signals' P_{out} due to their CAMPs' settability resolutions is the sum of the two individual, independent random numbers.

15.5.1.2   CAMP Temperature Compensation   The CAMP attempts to compensate its ALC-mode output power for the TWTA's P_{out} variation with temperature but it does not compensate exactly. Over the operating temperature range from −5 to 60 °C, CAMP output power rises linearly by 0.5 dB. This temperature compensation is taken into account in the section below on TWTA variation over temperature.

15.5.2   Nominal-Value Contribution from TWTA

For most TWTAs in a flight set with a constant drive level, P_{out} goes down a little as the TWTA's temperature rises, due to the TWTA's loss increasing. A TWTA will experience no diurnal variation (see Table 15.3), so there is also no diurnal variation between a pair of TWTAs. There is in fact a seasonal variation that we have disposed of by looking at the worst day of the year. All the TWTAs will always have nearly the same temperature because of the heat pipes linking the north and south panels.

Over the satellite's life, the TWTAs' temperatures will rise due to the decreasing efficiency of the satellite's thermal subsystem, the effect of which is to gradually change the average daily P_{out}.

The CAMP's temperature-compensation circuit is intended to counteract the TWTA's temperature sensitivity for ALC mode. As discussed in the section above on the CAMP, the compensation causes the TWTA's input drive to go up by 0.5 dB over the CAMP's operating temperature range of 65 °C. For the few TWTAs whose P_{out} goes up with temperature, the CAMP's compensation will be in the wrong direction.

In our example, each TWTA's EOL P_{out} relative to BOL P_{out} can be computed from the formula below:

$$\text{EOL } P_{out} \text{ rel to BOL } P_{out}$$
$$= \left(\frac{35}{85}\right)\left[P_{sat\Delta} + gs(\text{OBO})(P_{ss\Delta} - P_{sat\Delta})\right] + \left(\frac{35}{65}\right)0.5\,gs(\text{OBO})$$

where $gs(\text{OBO})$ is TWTA gain slope in dB/dB at the operating output backoff (OBO) and $P_{sat\Delta}$, $P_{ss\Delta}$ are the TWTA's measured P_{out} variation at saturation and small signal, respectively, over 0 to 85 °C. When the TWTA is saturated the function $gs(\text{OBO}) = 0$, and for small signal it equals 1. The formula can yield positive and negative values, depending on the parameter values. On the flight set of TWTAs, measured $P_{sat\Delta}$ ranged from −0.08 to −0.04 dB/dB, and measured $P_{ss\Delta}$ from −1.0 to −0.4 dB/dB. The 35° in the formula is the EOL temperature drift from BOL temperature, from Table 15.2. The 85° is the TWTA's operating temperature range. This formula does not go into the variation but affects the nominal value of P_{out}. If the lifetime requirement is average-over-life half of this difference should be used, while for at-EOL the whole difference is used.

15.5.3 Measurement Uncertainty in P_{out}

We assume in our example that the size of P_{out} measurement error is to be taken into account (Section 9.3.4.4). The actual error is unknowable. Its probability distribution is approximately Gaussian with zero mean. Half the values of measurement error are beneficial to C and half are detrimental.

Table 15.5 shows an example of the uncertainty in a P_{out} measurement. The uncertainties of two measurements are assumed to be independent (worse than positively correlated), so the variation in the difference of two measurement uncertainties is the sum of the two individual, independent random numbers.

TABLE 15.5 C Variation Due to P_{out} Measurement Uncertainty (Example)

Constant but Unknown Offset
In general form of Gaussian-distributed number with mean 0 and std dev Q_2 where Q_2 is 0.125 dB

TABLE 15.6 C/I_{self} Variation Due to OMUXes' Delta Diurnal Insertion Loss Variation (Example)

	Delta's Diurnal Swing
	In general form of sinewave with limits $\pm H$ where H is
OMUXes on same panel	0 dB
One OMUX on east panel, other on west	0.07 dB

TABLE 15.7 C Variation Due to OMUX's Manufacturing Tolerance (Example)

Constant but Unknown Offset
In general form of Gaussian-distributed number with std dev Q_3 where Q_3 is 0.1 dB

15.5.4 Variation Contribution from OMUX

The OMUX's insertion-loss change with temperature is a known function of temperature: only 0.1 dB over the operating temperature range of 80 °C. The EOL diurnal swing has the panel-illumination pdf given in the book Appendix Section A.3.7. The swing's limits are only ±0.03 dB, which is negligible. The situation between two OMUXes is shown in Table 15.6. If they are on different panels the delta's swing limits are ±0.07 dB, with the pdf given in the book Appendix Section A.3.8.

At the point in time that we are assuming, manufacturing tolerance in the OMUX is also a contributor to uncertainty. An example of this uncertainty is given in Table 15.7. The uncertainties of two OMUXes are assumed to be independent (worse than positively correlated), so the uncertainty in the difference of two tolerances is the sum of the two individual, independent random numbers.

15.5.5 Variation Contribution from Other Post-TWTA Hardware Besides OMUX

The post-TWTA hardware besides the OMUX consists of waveguide, possibly a few payload-integration components, and possibly another filter. At the point in time that we are assuming, manufacturing tolerance is the main contributor to uncertainty, and a small diurnal swing is comparatively negligible. An example is given in Table 15.8.

15.5.6 Summary of Repeater Units' Variation and Nominal-Value Contribution

For our example, Table 15.9 summarizes the constituents of the three components of the C variation. The three components as analyzed are constant-but-unknown offset, diurnal variation, and lifetime drifts unrelated to temperature.

TABLE 15.8 C Variation Due to Manufacturing Tolerance of Other Post-TWTA Repeater Hardware Besides OMUX (Example)

Constant but Unknown Offset
In general form of Gaussian-distributed number with std dev Q_4 where Q_4 is 0.1 dB

TABLE 15.9   Summary of Constituents of Repeater-Caused C Variation (Example)

	Constituents of Repeater-Caused C Variation		
Unit	Constant but Unknown Offset	Diurnal Swing	Lifetime Drift Unrelated to Temperature
CAMP	Random number uniformly distributed over $-Q_1$ to $+Q_1$ (*settability*)	Negligible	Negligible (*aging and radiation*)
TWTA	N/A	0	Negligible (*aging and radiation*)
Power measurement	Random number Gaussian-distributed with std dev Q_2 (*msmt uncertainty*)	N/A	N/A
OMUX	Random number Gaussian-distributed with std dev Q_3 (*manuf tolerance*)	Negligible	N/A
Other post-TWTA hardware	Random number Gaussian-distributed with std dev Q_4 (*manuf tolerance*)	Negligible	N/A

The constant-but-unknown offset has four constituents (from CAMP settability, power-measurement uncertainty, and manufacturing tolerances in OMUX and other post-TWTA hardware) that are independent of each other. The diurnal variation and lifetime drift unrelated to temperature are negligible.

For our example, each constituent of the C variation is independent of the corresponding constituent of the I_{self} variation except for the diurnal swing of OMUX insertion loss.

The only nonzero contributor to δ in the nominal value calculation is the TWTA's P_{out} sensitivity to the rise in average TWTA temperature.

15.5.7   Repeater-Caused Variation and Nominal Value of C and C/I_{self}

We now form the composite repeater-caused variation of C and C/I_{self}. The three variation components as defined are mutually independent. We characterize the composite variation by its standard deviation. In what follows, we assume there is only one significant interferer, but to treat more, the standard deviations for all would be rss'ed together.

We also characterize the nominal value about which the variation is centered in terms of δ.

15.5.7.1 Repeater-Caused C Variation and Nominal Value

The first component of the C variation due to the repeater, the unknown constant offset, is characterized as follows:

$$\text{Constant but unknown component of } C \text{ variation in dB}$$
$$\approx \text{Random number with std dev } \sqrt{\frac{1}{3}Q_1^2 + \sum_{i=2}^{4}Q_i^2}$$

The second component, diurnal swing, and the third, lifetime drift, are negligible. Therefore, the composite C variation due to the repeater is characterized as follows:

$$\text{Variation in dB} \approx \text{Random number with std dev } \sigma_r = \sqrt{\frac{1}{3}Q_1^2 + \sum_{i=2}^{4}Q_i^2}$$

The parameter δ equals half the drift over life from the TWTA's P_{out} sensitivity to the rise in average TWTA temperature.

Example

Suppose that the signal of interest has a well backed-off TWTA and an OMUX on the east panel, and that the strongest interferer has a saturated TWTA and an OMUX on the west panel. Then the parameter values for calculating the C and I_{self} variations are given in Table 15.10. Table 15.11 shows the terms in the sums whose square roots are the standard deviations. The standard deviations of the resultant variations are $\sigma_{r1} = 0.24$ and $\sigma_{r2} = 0.19$, respectively, where the subscript 1 means for C and 2 means for I_{self}. If the C variation is Gaussian, then 95.4% of the time C will lie within $\pm 2\sigma_{r1}$ of its nominal value (book Appendix Section A.3.5), and similarly for I_{self}. Suppose that the parameter values for calculating the C and I_{self} EOL average values relative to BOL averages are given in Table 15.12. Then the values of δ_1 and δ_2 are -0.07 and -0.02, respectively.

TABLE 15.10 Parameter Values for Calculating Repeater-Caused C and I_{self} Variations (Example)

Variation Constituent Source	Parameter	For C Value (dB)	For I_{self} Value (dB)
CAMP settability	Q_1	0.25	0
Power-measurement uncertainty	Q_2	0.125	0.125
OMUX manufacturing tolerance	Q_3	0.10	0.10
Other post-TWTA hardware manufacturing tolerance	Q_4	0.10	0.10

TABLE 15.11 Repeater-Caused C and I_{self} Variation Terms (Example)

Variation Constituent Source	Term	For C Value (dB)	For I_{self} Value (dB)
CAMP settability	$\frac{1}{3}Q_1^2$	0.021	0
Power-measurement uncertainty	Q_2^2	0.016	0.016
OMUX manufacturing tolerance	Q_3^2	0.010	0.010
Other post-TWTA hardware manufacturing tolerance	Q_4^2	0.010	0.010
Sum	σ_r^2	0.057	0.036

TABLE 15.12 Parameter Values for Calculating Repeater-Caused C and I_{self} EOL Values Relative to BOL (Example)

Parameter	For C Value (dB/dB)	For I_{self} Value (dB/dB)
$P_{\text{sat}\Delta}$	−0.06	−0.08
$P_{\text{ss}\Delta}$	−1.0	−0.7

15.5.7.2 Repeater-Caused C/I_{self} Variation and Nominal Value

The composite C/I_{self} variation due to the repeater is approximately given by

$$\frac{C}{I_{\text{self}}} \text{ variation in dB} \approx \text{Random number with std dev } \sqrt{\sigma_{r1}^2 + \sigma_{r2}^2 + \frac{1}{2}H^2}$$

where σ_{r1} and σ_{r2} equal σ_r for the signal of interest and the interferer, respectively, and $\pm H$ are the limits of the diurnal swing between the two OMUXes. When the OMUXes are on the same panel $H^2/2 = 0$, and when they are on different ones it is 0.002. Compared to $\sigma_{r1}^2 + \sigma_{r2}^2$ both values are negligible. Thus, the C, I_{self}, and C/I_{self} variations have no significant temperature-sensitive components, at least in our example. Composite C/I_{self} variation due to the repeater is then approximately given by

$$\frac{C}{I_{\text{self}}} \text{ variation in dB} \approx \text{Random number with std dev } \sqrt{\sigma_{r1}^2 + \sigma_{r2}^2}$$

The parameter δ in the nominal value calculation equals the difference in δ_s:

$$\frac{C}{I_{\text{self}}} \text{ average value at middle of life rel to BOL in dB} = \delta_1 - \delta_2$$

Example

Suppose that the signal of interest and the strongest interferer have the characteristics from the previous example. The C/I_{self} variation has standard deviation of 0.07 dB. If the variation is Gaussian, then 97.7% of the time C/I_{self} will be within ±0.14 dB of nominal. In regard to the nominal value, we have $\delta_1 - \delta_2 = -0.05$ dB.

15.6 COMBINING ANTENNA-CAUSED VARIATION AND NOMINAL VALUE INTO REPEATER-CAUSED VARIATION

In this section, we discuss the payload antenna-caused variation and fold it into the repeater-caused variation of C and C/I_{self} and similarly for the nominal values. The complete list of payload sources of C and C/I_{self} variations is given in Table 15.13. They are approximately independent.

The antenna makes no contribution to δ in the nominal value calculation because the antenna's performance does not vary from year to year. So the composite payload δ is simply the value from the repeater, given in Section 15.5.7.1. The average performance of the payload over any year of life is based on antenna gain that has been averaged over antenna-pointing error.

15.6.1 Variation Contribution from Antenna-Gain Inaccuracy

For the point in the build at which we are doing analysis, antenna-gain inaccuracy comes from manufacturing tolerance, computer-model error, and thermal distortion. On orbit, it will be thermal distortion only.

The C variation contribution from antenna-gain inaccuracy has a Gaussian distribution:

> C variation in dB from antenna gain inaccuracy ≈ Gaussian random number with std dev σ_g

Thermal distortion is not the largest contributor to overall antenna-gain loss (Section 3.11.5), so we simplify and use the worst-case value.

In regard to C/I_{self} variation, we need to think about how C and I_{self} are correlated. There seems to be no reason why they would be negatively correlated, so we make the worst-case assumption that they are independent.

TABLE 15.13   Independent Components of C Variation Due to Payload, as Analyzed

Sources of Independent Components of Payload-Caused Variation
Repeater
Antenna-gain inaccuracy
Antenna-pointing error

15.6.2   Variation Contribution from Antenna-Pointing Error

Pointing error for a reflector antenna was discussed in Section 3.9 and, specifically for a multi-beam antenna, in Section 11.2.2. We make the simplifying approximations here that the two orthogonal components of pointing error are independent and have the same Gaussian pdf.

The nominal value of antenna gain toward a given location is a function of the two-dimensional pointing error off boresight. The nominal gain is found by averaging over the pointing error. Antenna-gain variation is not generally Gaussian distributed.

The antenna-pointing error causes the nominal gain to be less than the no-error gain.

The variation in antenna gain is relative to the nominal gain.

The mean value and variation of antenna gain at a given off-pointing direction can be estimated with a Matlab program. The antenna pattern of a spot beam is only a function of off-boresight angle θ. The given off-pointing direction points from the center of the circle toward the large dot in Figure 15.1. The two orthogonal pointing-error axes to average over are shown in bold, in two images. In the left image, one axis is a half circle that goes through boresight and the other axis is along the half circle where the plane, shown in profile, intersects the hemisphere. In the right image, 90° around the hemisphere, the first axes is shown as the vertical line and the second as the smaller half circle. Gauss–Hermite integration (book Appendix A.4) is used to average gain as a function of Gaussian pointing error along each axis. Average antenna gain in dB for the given off-pointing direction would then be the average of the gains along the two axes. At the same time as average gain is found, the average squared value of gain along each axis would also be found. From this and the mean, the standard deviation σ_p can be calculated.

For C/I_{self}, there are two cases, namely when the two beams have uncorrelated pointing errors and when the two beams have basically the same pointing error at any given time. They have the same pointing error if they are on the same reflector. Also, they have basically the same pointing error if they are on different antennas that do not track.

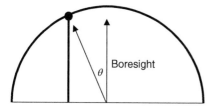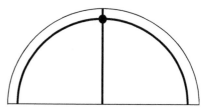

FIGURE 15.1   Geometry for estimating nominal gain and gain variance from two-dimensional antenna pointing error.

The uncorrelated case is easy. The standard deviation is the rss of the two separate ones:

$$\frac{C}{I_{\text{self}}} \text{ variation in dB} \approx \text{Random number with std dev } \sqrt{\sigma_{\text{p1}}^2 + \sigma_{\text{p2}}^2}$$

where the subscript 1 is for the signal of interest and 2 is for the interferer.

The case when the two beams off-point together is more complicated. We must keep in mind that if the interfering beam is a spot beam, at the location of interest the interferer is either on the steep side of the mainlobe of its antenna pattern or on one of the sidelobes (Section 14.6.3). Even if it is on a sidelobe, the pattern roll-off is probably sharper than that of the signal of interest, which is on the mainlobe of its antenna pattern.

Thus, if the interfering beam is a spot beam, the C/I_{self} variation in dB is approximately equal to σ_{p2}. If the interfering beam is not a spot beam, the C/I_{self} variation in dB is approximately equal to σ_{p1}. Therefore, a rough approximation and/or worst case is that, over the two-dimensional pointing error, the gain variation's standard deviation is the larger of σ_{p1} and σ_{p2}.

15.6.3 Payload-Caused C Variation

The composite variation in C from all of the payload is as follows:

$$C \text{ variation in dB} \approx \text{Random number with std dev } \sigma = \sqrt{\sigma_{\text{r}}^2 + \sigma_{\text{g}}^2 + \sigma_{\text{p}}^2}$$

15.6.4 Payload-Caused C/I_self Variation

We have shown that with minor worst-case assumptions, the component of C variation due to the repeater is uncorrelated with the corresponding component of I_{self} variation, and similarly for the components due to the antenna. Therefore, the composite variation in C/I_{self} from all of the payload is as follows:

$$\frac{C}{I_{\text{self}}} \text{ variation in dB} \approx \text{Random number with std dev } \sqrt{\sigma_1^2 + \sigma_2^2}$$

where the subscript k is 1 for C and 2 for I_{self}.

Let us generalize this to the case where there is more than one significant interferer, so

$$I_{\text{self}} = \sum_{l=1,n} I_l$$

The Taylor series for a linear approximation to a function of multiple variables is as follows:

$$g\left(\eta, \mu_1, \ldots, \mu_n\right) \approx g\left(\eta_{nom}, \mu_{1nom}, \ldots, \mu_{nnom}\right)$$
$$+\left(\eta - \eta_{nom}\right)\frac{\partial g}{\partial \eta}\bigg|_{nom} + \sum_l \left(\mu_l - \mu_{lnom}\right)\frac{\partial g}{\partial \mu_l}\bigg|_{nom}$$

where

$$\eta = 10\log C \qquad \mu_l = 10\log I_l$$
$$g\left(\eta, \mu_1, \ldots, \mu_n\right) = 10\log\frac{C}{I_{self}} \qquad \text{where } \log \triangleq \log_{10}$$

Then

$$\frac{C}{I_{self}} \text{ variation in dB} \triangleq 10\log\frac{C}{\sum_l I_l} - 10\log\left(\frac{C}{\sum_l I_l}\right)_{nom}$$
$$\approx \left(10\log C - 10\log C_{nom}\right) - \sum_l \alpha_l\left(10\log I_l - 10\log I_{lnom}\right)$$
$$= \left(C \text{ variation in dB}\right) - \sum_l \alpha_l\left(I_l \text{ variation in dB}\right)$$

where $\alpha_l = I_{lnom}/I_{self\ nom}$. The composite variation in I_{self} can be characterized as follows:

$$I_{self} \text{ variation in dB} \approx \text{Random number with std dev } \sigma_2 = \sqrt{\sum_l \alpha_l^2 \sigma_{l2}^2}$$

Thus, the composite variation of C/I_{self} in dB can be written in the same form as for one interferer.

15.6.5 Payload-Caused Variation and Nominal Value of $C/(I + N)$

Now we extend the analysis of the previous section on C and C/I_{self} to $C/(I_{self} + I_{other} + N)$, where N includes I_{other} (Sections 15.1 and 15.2). N and I_{other} are given and are referred to the payload-antenna output, as C and I_{self} are. N is calculated from space loss and the earth receiver's G/T_s.

Making a calculation similar to the one in the section above, we find that the composite variation of $C/(I_{self} + N)$ in dB can be characterized in the following way, where $\beta = I_{self\ nom}/(I_{self\ nom} + N)$:

$$\frac{C}{\left(I_{self} + N\right)} \text{ variation in dB} \approx \text{Random number with std dev } \sqrt{\sigma_1^2 + \beta^2 \sigma_2^2}$$

where subscript 1 means for C and 2 means for I_{self}. The composite nominal value of $C/(I_{self} + N)$ in dB is the following:

$$\left[\frac{C}{(I_{self} + N)}\right]_{nom} \text{ in dB} \approx \left(C_{nom} \text{ in dBW}\right) - \beta\left(I_{self\,nom} \text{ in dBW}\right)$$

Example

We consider only the repeater-caused component of variations and use the numbers from Section 15.5.7.2. We further assume that $N = I_{self\,nom}$, which means $\beta = 0.5$. Then the standard deviation is 0.38 dB.

15.7 COMBINING ATMOSPHERE-CAUSED VARIATION INTO PAYLOAD-CAUSED VARIATION

This section combines a statistical model of atmospheric attenuation (Section 14.4.4) with the probabilistic model of payload performance. The usual EIRP requirements are assumed to have been replaced by a long-term link-availability requirement or equivalently by a maximum long-term link-outage probability q_{spec} in %.

Two aspects of the atmosphere are most likely to cause significant link attenuation at frequencies above 10 GHz: the dominant one is rain and the other is atmospheric gas, especially water vapor.

Link margin is the ratio of atmosphere-attenuated $C/(I_{self} + I_{other} + N)$ to the threshold value r_{th} required by the receiver for good performance. As in the previous section, we fold I_{other} into N. The atmosphere attenuates C and I_{self} by the same amount $a > 1$, but N (including the part from I_{other}) is not affected. N includes the additional antenna noise that occurs from the heavy rain corresponding to the atmospheric attenuation at spec value (Section 14.5). So the link margin is as follows:

$$m_{link} = m_{link}\left(C, I_{self}, a\right) = \frac{\left(\dfrac{C}{a}\right)}{\left(\dfrac{I_{self}}{a + N}\right)} \frac{1}{r_{th}}$$

The link margin is greater than or equal to 1 (nonnegative, if in dB) if and only if the link is available. The link margin is less than 1 (negative, if in dB) if and only if the link is suffering an outage. When the link margin is greater than 1 (positive, if in dB), there is some room for C, I_{self}, and a to vary. The calculated link margin is a little lower than the actual number when the atmospheric attenuation is lower than the spec value.

The atmospheric margin m (not in dB) is something different from link margin. It is the atmospheric attenuation at which the link margin equals 1 (zero, if in dB):

$$m = m\left(C, I_{\text{self}}\right) = \frac{r_{\text{th}}^{-1} - \left(\dfrac{C}{I_{\text{self}}}\right)^{-1}}{\left(\dfrac{C}{N}\right)^{-1}}$$

It depends on the instantaneous values of C and I_{self}. Define $M = 10\log m$, the atmospheric margin in dB.

The long-term probability of link outage is the link outage averaged over the probability distributions of C, I_{self}, and atmospheric attenuation a. We need the nominal values and standard deviations of C and I_{self}, all in dBW. The computational technique is that we set C and I_{self} to values in their ranges, then conditioned on these values we compute the link-outage probability, then we average over the C and I_{self} probability distributions. To make the problem tractable, we assume that the payload's performance variation in dB is Gaussian distributed. It is unknowable how good this approximation is because there are so many contributors to the variation, each with not-so-well-known probability distributions, but it is most likely a fairly good approximation. Partial precedence for this approach is the computation of antenna-pointing error (Section 3.9). We use Gauss–Hermite integration (book Appendix A.4) to perform the two-dimensional averaging. We write "I" for "I_{self}." The steps for computing the link-outage probability at the location are as follows:

- Decide how many points to use for each one-dimensional Gauss–Hermite integration (try 4 and 6). Obtain the corresponding set of factors η_i and the set of weights w_i.
- Compute C_{nom}. Compute σ_1 characterizing the payload variation (Section 15.6.3).
- Compute the discrimination factor (Section 14.6) that C has against each of its possible significant interferers. Identify the significant ones and compute $I_{l\,\text{nom}}$, I_{nom}, and α_l according to Section 15.6.4. Compute σ_2.
- Compute and store the constants for evaluating the inverse atmospheric-attenuation function (Section 15.9.2).
- Compute all $C_i = C_{\text{nom}} 10^{\wedge}(\eta_i \sigma_1 / 10)$ and all $J_i = I_{\text{nom}} 10^{\wedge}(\eta_i \sigma_2 / 10)$. For each index pair (i,l), compute the atmospheric margin $M_{il} = M(C_i, J_l)$. For atmospheric attenuation $A = M_{il}$, evaluate the inverse atmospheric-attenuation function to yield the link-outage probability q_{il} (see Section 15.9.2). Perform the two-dimensional weighted averaging to obtain the long-term average link-outage probability q_0:

$$q_0 = \sum_i w_i \sum_l w_l q_{il}$$

15.8 OPTIMIZING MULTI-BEAM-DOWNLINK PAYLOAD SPECIFIED ON LINK AVAILABILITY

We wish to optimize a multi-beam-downlink payload that has been specified by a minimum long-term link availability. A large set of locations is given on which the specification must hold (Section 15.2). There are so many different constraints on the payload design that the optimization must be done iteratively. We sketch out how to iteratively determine a set of C_{nom}, one for each location, that meets the availability requirement while the signal levels vary in the manner discussed in the preceding sections. Once the final set of C_{nom} is found, the corresponding set of C_{BOL} is calculated from the difference between C_{nom} and C_{BOL} at every location.

The payload does not change configuration between iterations but only at the beginning of a set of iterations. "Configuration" means the CAMPs, HPAs, and antennas. The idea is that a provisional payload configuration is set, a few iterations are done here until a problem is seen with the configuration, the payload is reconfigured, and so on. "I_{self}" is written simply "I" here.

A. To do once:
   • Decide how many points to use for the Gauss–Hermite integrations. Obtain the corresponding set of factors η_i and the set of weights w_i.
   • Set the iteration adjustment factor ξ_0 for upward adjustments of C_{nom}, referred to in the iteration details presented in Section 15.9.3. It is a positive number up to 1. A value of 0.5 may be good.
B. Steps to do once for every location, for the given configuration:
   • Obtain an initial C_{nom}, for example, the one that exactly meets link availability without consideration of self-interference. Compute and store σ_1 (Section 15.6.3).
   • Identify and store the small number of indices of the potentially significant interfering beams.
   • Compute and store the constants for evaluating the approximate forward and the inverse atmospheric-attenuation functions (Section 15.9.2).
   • Initialize the "previous" iteration-quality measure to zero.
C. Steps to do at every iteration of finding a whole new set of C_{nom}:
   • For every location,
      ○ Check its small set of potentially significant interfering beams. Compute the discrimination factor that C has against each of them. Identify the significant ones and compute the α_i. Compute I_{nom} and σ_2.
      ○ Compute and store $C_{nom\ new}$ (Section 15.9.3).
      ○ If $C_{nom\ new}$ causes the HPA to exceed its capacity, reconfigure the payload and go to step B.
      ○ Otherwise, replace C_{nom} with $C_{nom\ new}$.
   • Compute the iteration-quality measure of this iteration, which is the number of locations that meet requirements. Check to see that it is better than

the value from the previous iteration. If not, adjust the factor ξ_0 and start this iteration again. Set the "previous" value of the measure to the current value.

- Make a plot of all the C_{nom}.

D. After a few iterations, if C_{nom} has been consistently rising for all locations in some subset, then the discrimination of each against the others is not sufficient. Increase the discriminations in at least one of the discrimination dimensions (frequency band, antenna pattern, polarization) and go to step B. Also check all the C_{nom} for possibly exceeding the maximum-allowed flux density on the earth.

E. After the iterations have converged, steps to do for every location:

- Compute $P_{sat\Delta}$ and $P_{ss\Delta}$ and from them compute C_{nom} relative to the C_{BOL} without pointing error. Compute the increase required to C to compensate the average loss from antenna-pointing error. Compute C_{BOL} from C_{nom} and both adjustments.

If convergence seems impossible, consider reducing the C variations by tightening up unit specifications.

15.9 APPENDIX. ITERATION DETAILS FOR OPTIMIZING MULTI-BEAM PAYLOAD SPECIFIED ON LINK AVAILABILITY

15.9.1 Approximate Rain-Attenuation Function and Its Inverse

The rain attenuation R for a location is a function of the exceedance probability p (Section 14.4.4.1.2), that is, $R = R(p)$. R is in dB and p is in %. Since both $R(p)$ and its inverse function $p(R)$ will be evaluated repeatedly, we wish to replace them with efficient approximations.

For the International Telecommunication Union (ITU) rain-attenuation model and therefore presumably for other models, $\log R$ and $\log p$ are nearly linearly related over a not-so-small range. The range of interest is for p near the specified long-term link-outage probability q_{spec}. With the actual rain-attenuation model, we compute and store $R_{spec} = R(q_{spec})$. Then for p near q_{spec} we have the following good approximation:

$$\log R \approx B'' + D' \log p \quad \text{where } D' = \left[\frac{d \log R(p)}{d \log P} \right]_{p=q_{spec}}$$

$$\text{and } B'' = \log B' \quad \text{where } B' \triangleq R_{spec} q_{spec}^{-D'}$$

The forward relationship is the following:

$$R = R(p) \approx B' p^{D'}$$

The inverse relationship is the following:

$$p = p(R) \approx B R^{D} \quad \text{where } D \triangleq 1/D' \text{ and } B \triangleq R_{spec}^{-D} q_{spec}$$

We compute and store D' and B' for use in the forward-function approximation and D and B for use in the inverse function.

If the rain model changes expression at $p = q_{spec}$ then the tangent line may not be defined, that is, be different on the lower side and the upper side of q_{spec}. Then we could define two tangent lines and know that they apply in different regions. Or it may work to just define one tangent line with the slope equal to the average of the slopes on either side.

15.9.2 Atmospheric Attenuation Function and Its Inverse

The atmospheric attenuation A for a location is a function of the long-term link-outage probability q (Section 14.3). We assume that the other atmospheric attenuations besides the rain attenuation R are constant. We obtain and store them. Then A and R are related one-to-one, that is, $A = A(R)$ and $R = R'(A)$. The forward atmospheric-attenuation function is $A = A(R(q))$ and the inverse function is $q = p(R'(A))$, where $R(p)$ and $p(R)$ are given in the preceding section.

15.9.3 Details of Iteration

Compute all $C_i = C_{nom}10^{\wedge}(\eta_i\sigma_1/10)$ and all $J_i = I_{nom}10^{\wedge}(\eta_i\sigma_2/10)$.

For each index pair (i,l), compute the atmospheric margin $M_{il} = M(C_i,J_l)$. For atmospheric attenuation $A = M_{il}$, evaluate the inverse atmospheric-attenuation function to yield the link-outage probability q_{il} (Section 15.9.2). Perform the two-dimensional weighted averaging to obtain the long-term average link-outage probability q_0 for the current C_{nom}:

$$q_0 = \sum_i w_i \sum_l w_l q_{il}$$

Use the forward atmospheric-attenuation function (Section 15.9.2) to compute the total atmospheric attenuation a_0 (not in dB) corresponding to q_0.

Now, we are trying to achieve a one-point-equivalent atmospheric margin equal to the attenuation at the specified outage probability, a_{spec}, but in this iteration, we reached a_0. All we do during an iteration is calculate the new C_{nom} for the next iteration, so the adjustment we need, we can only attribute to the new C_{nom}. C_{nom} needs to be multiplied by approximately the factor a_{spec}/a_0. If we have too much atmospheric margin, we can decrease C_{nom} by the full amount without making a negative impact on other locations. On the other hand, if we do not have enough atmospheric margin, then we have to increase C_{nom}, but since this may have a negative impact on other locations, we only increase it part way. In symbols: if $a_0 > a_{spec}$, set the iteration adjustment factor ξ to 1; otherwise set it to ξ_0. Then compute and return $C_{nom new}$ for the next iteration:

$$C_{nom\,new} = C_{nom}\left(\frac{a_{spec}}{a_0}\right)^{\xi}$$

CHAPTER 16

MODEL OF END-TO-END COMMUNICATIONS SYSTEM

16.1 INTRODUCTION

This chapter discusses how to model the end-to-end communications system at layer 1 of the Open Systems Interconnect (OSI) communications model, the physical layer (Section 13.2.1). System performance is assumed to be evaluated in terms of bit error rate (BER) (Section 12.16), symbol error rate (SER), or modulation error ratio (MER) (Section 12.10.6). Modeling may be done to help develop the payload specifications, to assess the impact of a payload parameter not meeting its requirement, to figure out what could be causing an on-orbit anomaly, or to try out a system upgrade on a testbed before committing to implementation.

There are three general ways to perform analyses: by hand mathematically or with Microsoft's Excel, by software **simulation** on the computer, and by hardware **emulation**. It is possible and even sometimes advisable to combine techniques. For example, hand calculations may show the levels of interfering signals compared to the level of the signal of interest or how interfering signals can be modeled in a simplified way; hardware can be characterized and the measurements input to a simulation; or performance of the system in the nominal operating condition may be found by emulation, and this nominal performance used to calibrate a simulation, which is then used to assess system performance sensitivity to a parameter variation.

Satellite Communications Payload and System, Second Edition. Teresa M. Braun and Walter R. Braun.
© 2021 John Wiley & Sons, Inc. Published 2021 by John Wiley & Sons, Inc.

16.2   CONSIDERATIONS FOR BOTH SOFTWARE SIMULATION AND HARDWARE EMULATION

16.2.1   System Model

An analysis model of the end-to-end satellite communications system with a bent-pipe payload consists of three main parts, the ground transmitter, the satellite payload, and the ground receiver. (Atmospheric effects are handled separately, usually.) With a regenerative payload, there would be four, as the payload would be split into a receiving and demodulating model and a modulating and transmitting model. With a regenerative payload, the noise added in the payload on the downlink is negligible, so the only significant noise is in the receiver, as for the uplink. For the rest of this section, we consider only the bent-pipe payload. We do not treat framing.

Figure 16.1 gives block diagrams of typical analysis models of the ground transmitter, payload, and ground receiver. There is both uplink and downlink noise, coming from the payload and ground receiver, respectively. There is interference coming in on the links. The main nonlinear element in the system is the payload high-power amplifier (HPA) in this typical case. The individual elements in the payload have been discussed in earlier chapters, and the elements on the ground in Chapter 12. In an emulation, all of the elements in the three models would be present. In a simulation, they would be, too, but the representations would be more abstract (see Section 16.3.2).

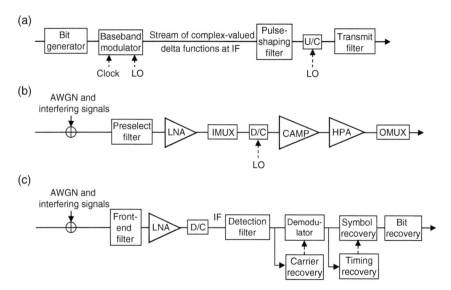

FIGURE 16.1   Typical system model for bent-pipe payload: (a) ground transmitter, (b) payload, (c) ground receiver.

Additional system elements must sometimes be modeled:

- Error-control coder and decoder
- Interfering signals entering low-noise amplifiers (LNAs)
- Interfering signals and spurious signals entering elsewhere in system
- Payload-unit performance variation with temperature and/or over life
- Antenna noise
- Antenna pointing error
- Doppler and Doppler rate on link(s)
- Nonlinear HPA in ground transmitter
- Limiter in payload front end
- Limiter in ground-receiver front end
- Atmospheric effects on link(s).

16.2.2 Know the Whole Communications System

It is crucial, even if the analysis focus is on the payload, to model every element of the end-to-end communications system as accurately as possible. If not, the system performance results can be so misleading as to show nothing about the actual system but something about a quite different system, and a false conclusion is drawn. The reason this can happen is that one aspect of the system can magnify the performance effect of another aspect or diminish it (Braun and McKenzie, 1984). This is discussed further in Sections 16.3.12 and 16.3.13.

Assuming that the payload characteristics are known in detail, the rest of what has to be known (aside from the possible additional system elements listed above) includes the following:

- Uplink and downlink levels of signal of interest and the ground-receiver noise figure
- Modulation and demodulation schemes and detection filter. Large ground stations normally have excellent performance so can often be modeled as ideal, but this must be ascertained. Consumer equipment is typically not near-ideal. Some ground receivers may have legacy equipment, which is older and possibly optimized for an earlier transmitter or not up to current performance standards.
- All other ground filters
- Carrier and timing recovery schemes
- Phase noise characteristics of ground local oscillators (LOs) and clock jitter characteristics of ground clocks
- Interference scenario
- What the coding, decoding, and interleaving schemes are, if any
- What the ground receiver's antenna noise is
- What the losses are in the ground transmitter's and ground receiver's antenna gain from pointing error, if significant
- Nonlinear behavior of ground transmitter's HPA and ground receiver's front end, if significant.

All system elements are to be in nominal operation unless specifically a worst case is being studied. Thus, element specification data is not sufficient; measured data is required. The reason is again the magnifying or diminishing effect of one parameter on another.

16.2.3   What Results Modeling Can Provide

Neither simulation nor emulation, and certainly not hand analysis, can provide performance to match that of the actual analog hardware system being modeled. Simulation can never match hardware because hardware has second- and third-order behavior, including interactions, that is untested, unknown, and thus unmodeled. (The author remembers that in about 1994 we had gotten our first results on a new-technology testbed and the results did not match the customer's simulation. The customer was shocked and thought the hardware results were wrong and wanted us to work through the Thanksgiving weekend. This was the beginning of years of discussion about simulation accuracy.) Emulation – using one set of hardware and test equipment and operating under one set of environmental conditions and having a particular warm-up time – can never match another set of hardware and test equipment and conditions, even if all is nominally identical. Measurement error alone will often obscure the "real results" by tenths of a dB (Chie, 2011).

So if simulation and emulation cannot match performance of the modeled system, what can they do? They are at their most useful in predicting sensitivity, namely, what the change in performance is when one system parameter is varied (basically a partial derivative). This is why the model must match the real system as closely as possible. Simulation and emulation can also possibly say that one version of unit x is better than another version of unit x. Emulation can also show that new hardware can work in at least some conditions.

Once the model is as realistic as possible, one still has to calibrate the model to get it operating with the same performance as the real system, for example, the same BER. Frequently the only way to model the otherwise unmodeled aspects is to inject more noise in an appropriate place. This is discussed further in Section 16.3.13.

16.2.4   Generating Symbol Stream Plus Noise

Generating the bit stream by a **maximal-length pseudo-noise (PN) sequence** (Jeruchim et al., 2000) is more efficient than by a random-number generator. A PN sequence is a periodic bit sequence whose autocorrelation function looks much like that of a randomly generated bit sequence. It is generated with a feedback shift register. Suppose the feedback shift register has q memory locations. After each shift, it computes a linear, base-two combination of some of the q bits in memory and outputs that bit. The PN sequence generated has maximal length if all possible runs of 2^q bits are represented in the repeated sequence. However, the run of q zeros cannot be present, since then the whole sequence would be only zeros. So the PN sequence must be altered to add one more zero to the run of $q - 1$ zeros.

The symbol stream is then generated from the bit stream, each symbol from m consecutive bits where $m = \log_2 M$ and M is the size of the modulation scheme's alphabet.

We need to determine how large q must be. Now, the satellite communications system has memory. Without regard to the HPA, phase error, timing jitter, spurious signals, and so on, the system from pulse-shaping filter through demodulator can be represented by one filter. Say its memory is n symbols long, which corresponds to mn bits. Nominally, all bit sequences of length mn must be present in the modified PN sequence, so we need q to be at least a few times mn. One way to find out what the symbol-memory length n is, is to send an impulse down the system without noise and look at the response in the time domain. The number n should be set to the length of the main content of the response. This method would work for a channel that is not highly nonlinear.

If the channel is highly nonlinear, a trial-and-error approach can be used, in which some value of m is tried and the SER is obtained, a larger value of m is tried and that SER is obtained, and if the answers are nearly the same the first m was big enough.

If the modeling is either by emulation or Monte Carlo simulation (Section 16.3.4), then the modified PN sequence is used over and over to generate the data. In the former case, the hardware itself adds noise, while in the latter case "random" noise must be generated and added in. We investigate now how long the total noisy symbol sequence has to be in order for the SER obtained by emulation or Monte Carlo simulation to be a good approximation. A precondition for the following rule is that the symbol errors seen in the simulation must be independent of each other (Jeruchim et al., 2000). Suppose it is desired that the actual value of SER lies within a factor of $1/\rho$ to ρ about the simulated SER with 95% probability, where $\rho > 1$. Then 10 symbol errors are enough for ρ of 2 and 100 enough for ρ of 5/4 (Jeruchim et al., 2000).

However, because of the system memory, successive symbols are in fact correlated, so their errors are correlated. Now, for an uncoded link carrying data, not voice, the target SER may be about 10^{-5}, while for a coded link it may be 10^{-2} before decoding and a doubly coded link 10^{-1} before any decoding. For SER of 10^{-5} and 10^{-2}, anyway, the errors are uncommon enough that the independent-error assumption is not far from correct. For the doubly coded case, it may be necessary to specifically model the outer (Section 12.6.2) encoder and decoder.

16.2.5 Modeling Interferers

Other signals may come into the end-to-end system and degrade recovery of the signal of interest. These other signals may be self-interference, interference from other communications systems, other signals going through the HPA with the signal of interest, spurious signals from a frequency converter, and so on. For simplicity just in the rest of this chapter we call them all "interference."

Approximating interferers as additive noise, when appropriate, simplifies a system model significantly and cuts down on the computer run-time, with little or no loss in accuracy.

One important thing to keep in mind when deciding how to model an interferer is what filtering the interferer will see, compared to the signal of interest. The main filter is the detection filter, but sometimes other filters can be significant such as the output multiplexer (OMUX) or a post-HPA notch filter. Whatever portion of the interferer's signal spectrum is within the noise bandwidth of the filters, averaged over that noise bandwidth, can be treated as an I_0 level to be added to the N_0 level.

This approximation is sufficient for modulated-signal interferers that come into the signal of interest after the payload's HPA or for weak interferers, but large interferers that come in before the HPA must be explicitly modeled through the HPA.

The Central Limit Theorem helps us deal with more than one interferer. This theorem is one of the most useful in satellite communications analysis. It says that the sum of a number n of random variables that are mutually independent and identically distributed becomes closer and closer to Gaussian as n becomes larger. In fact, the number of random variables does not need to be large nor do they need to be identically distributed to add up to something that is close to Gaussian (book Appendix Section A.3.9). Therefore, the Gaussian approximation for a sum of a few independent random variables is a good one. The caveat is that the pdf approximation may only be good within $\pm 2\sigma$ from the mean. When we do not know the individual pdfs accurately, which is often the case, there is even more reason to accept the approximation.

One application of this is that when the signal of interest is amplified in the HPA with a few other signals, those other signals can often be modeled as additive noise.

16.3 ADDITIONAL CONSIDERATIONS FOR SIMULATION

Among the reasons for performing a simulation with focus on the payload are the following: (1) to help determine the payload specifications; (2) to assess the BER sensitivity to a range of values for a payload parameter to support a request for specification relief; and (3) to assess the BER impact of a payload parameter already known to be out of spec.

Simulation is often done with a software package such as the MathWorks' product Matlab. For simple simulation, PTC's product Mathcad is sufficient. Both of these packages operate alternately in the time and frequency domains as is convenient. Two new tools explained in the open literature may be useful. One is a demonstration tool developed by the Aerospace Corp., a hardware-accelerated simulation tool (HAST) consisting of field-programmable gate arrays (FPGAs) to perform computation-intensive operations, coupled to Matlab (Lin et al., 2005, 2006). Another, a commercial product, is Keysight's Advanced Design System Ptolemy simulator, a time-domain simulator capable of "co-simulating" with Matlab (examples of usage in Braunschvig et al. (2006), Gels et al. (2007)).

Communications system simulation is completely described in (Jeruchim et al., 2000).

16.3.1   Pitfalls of Simulation

First some words of serious caution about simulation. Communications system simulations sometimes drive important or mid-level decisions, for example, that a particular hardware design is satisfactory, that a particular hardware design is better than another, or that a particular parameter being out of spec in the built hardware doesn't matter in end-to-end system performance. However, relying on simulation results is a risky proposition: the author has been provided more garbage as supposedly final results than valid answers. This happens less often with hardware tests. There are several reasons for this:

- At two companies the author has worked at, people newly out of graduate school were given the job of simulating. At best they may have studied communications theory so they know the basic concepts but not the fine points. Additionally, almost every new graduate does not understand the importance of reaching the right answer in payload calculations (this also applied to the author early on). In school, getting the theory right is more important than getting the answer numerically correct. Specifically for simulation, the new people do not know yet that they must check every step of their modeling and every result every way they can think of (see below). In contrast, with hardware testing, no one would think of just giving the test equipment manuals to a new employee who would then, unsupervised and before learning the discipline of validating elements and partial integrations, provide his first measurement results to an engineer to base a decision on.
- The most common mistake of people new to communications simulation is to use filters that are either twice as wide or half as wide as they should be. This is because of the existence of various definitions of bandwidth (radio-frequency (RF) versus baseband, one-sided versus two-sided – Sections 12.2 and 12.3). There are many other possible mistakes involving noise power, signal-to-noise ratio (SNR) versus E_s/N_0, aliasing, sampling rate, and sample-run length.
- Communications system simulation packages are deceptively easy to use. Ease of use is a selling point for the packages. This is dangerous. Some people without a communications background, who do not even know the basics let alone the fine points, and without the other crucial knowledge of discrete-time-and-frequency signal processing, think they can get valid results because they are able to put together something that runs.
- A communications simulation can be put together and run so fast that even a knowledgeable and experienced user can easily make mistakes. It is easy to forget to check that every element from an earlier, similar simulation is still appropriate to the current simulation. It is tempting not to take the time to think through the details of the best way to do a new simulation but just to plunge right in. In contrast, hardware testing takes time so there is more opportunity to think.
- Customers of simulation may not have the background to judge whether or not results are valid.

Any customer can be almost sure that simulation results are valid if he takes great care in accepting them. No matter how much experience or education the simulator has, the customer must require the simulator to do the following at each results presentation:

1. Show the overall simulation model in some detail. In an appendix, show every modeling assumption made of every element, including plots as appropriate that are fully labeled.
2. Describe confirmation of the simulation model including every element and partial system integrations. Confirming an element model means confirming that the chosen value of every model parameter yields the expected characteristic.
3. Show exhaustive confirmation of results including comparison against theoretical results for simple but similar cases, journal article results for one-parameter variation cases, and if available hardware results and earlier proven simulation results.
4. If the simulator is rather new to it, he must show that he has talked over his model, his model confirmation, and his results with an experienced person.

16.3.2 System Model Specialized to Simulation

Figure 16.2 shows a typical simulation model of a satellite communications system with a bent-pipe payload. The simulation works on the complex baseband equivalent of RF and intermediate frequency (IF) (Sections 12.2 and 12.3). It uses discrete time and frequency. Some operations are carried out in the time domain, such as data-bit stream generation, modulation, HPA action, and from the demodulator through to symbol decision. Filtering is done in either the time or frequency domain. Noise is added in either domain or is in some cases handled analytically (Section 16.3.4). In

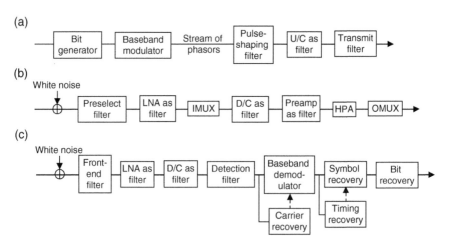

FIGURE 16.2 Typical simulation model of system with bent-pipe payload: (a) ground transmitter, (b) payload, (c) ground receiver.

this particular model, residual phase noise and residual timing jitter, where residual is what is left after tracking in the ground receiver, are represented as a part of the white noise added to the ground-receiver front end. This is only appropriate if they are believed to be insignificant. Spurious signals that arise from the payload are also represented as a part of that white noise. However, in some circumstances, for example, for a coded system, this is not an appropriate way to model (Chie, 2011).

Other options would be to more explicitly model residual phase error, residual timing jitter (Section 16.3.9), or spurious signals (Section 12.10), especially if the payload's contribution is under study. If there are a few spurious signals that matter, it may be possible to model them as additive noise, as for interferers (Section 16.2.5). Additional system elements that must sometimes be modeled are listed above in Section 16.2.1.

16.3.3 Basic Signal-Processing Considerations

When the simulation alternates between the time and frequency domains, the signal must be defined on discrete, not continuous, points in both domains. The following applies to the signal when it is first considered as defined on continuous time and continuous frequency:

$$\text{Signal has period } T_{\text{per}} \Leftrightarrow \text{Fourier transform is delta functions at}$$
$$\text{multiples of } 1/T_{\text{per}} = \Delta f$$

$$\text{Signal is delta functions at multiples of } \Delta t$$
$$\Leftrightarrow \text{Fourier transform has period } 1/\Delta t = F_{\text{per}}$$

Then considered as defined on discrete time and discrete frequency, the delta functions are replaced by complex numbers. The relationships among the discrete times and the discrete frequencies are illustrated in Figure 16.3. The signal actually does not exist at the intermediate times and frequencies. If interpolation must be done, in the time domain, say, it must be done in a way that does not increase the signal's frequency extent.

When the simulation alternates between the time and frequency domains, the signal is constructed so as to be periodic in both domains. Usually, this means

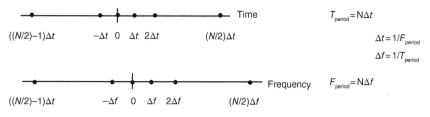

Lines are for ease of visualization only

FIGURE 16.3 Relationships among discrete times and discrete frequencies in simulation.

that it goes to zero at the edges. When constructing a signal in the time domain, one must be careful that the signal's extent in the frequency domain would be contained in the available extent, otherwise aliasing occurs (Section 10.2.1). In aliasing, the parts of the frequency domain signal that lie outside the available extent overlap the parts inside the available extent. A similar consideration also applies to constructing a signal in the frequency domain and thinking about its extent in the time domain.

16.3.4   Simulating Additive Noise

The uplink C/N is set by the payload. Since one of the repeater requirements is on noise figure referred to the receive-antenna terminal (Section 8.12), the equivalent noise level there can be easily calculated. The signal level there can be obtained from the signal level into the payload and the antenna gain. There are two cases, the channel amplifier (CAMP) in automatic level control (ALC) mode and the CAMP in fixed-gain mode (FGM). In the first, the repeater's noise figure depends on the ALC gain, which depends on the input signal level. In the second case, the gain is fixed, so the noise figure is independent of input signal level.

In modeling, additive noise can be used to represent not just thermal noise but also antenna noise and some signal distortions. There are two main ways to model additive noise, besides ignoring it.

One is the **Monte Carlo** technique, where samples of noise are explicitly generated by a random-number generator and added to the signal samples (Jeruchim et al., 2000).

The difficulty in modeling the additive noise for a system with a bent-pipe payload is the existence of the nonlinear element, the HPA, in the middle of the system. If it were not there, that is, the system were linear, the thermal noise and the other noises could be evaluated at the places they come into the system, added up, and modeled as one additive noise in the ground-receiver front end. Monte Carlo would not be necessary (see just below). However, the HPA is a difficulty. It changes the characteristics of the noise (partially suppressing the radial component when the HPA is near saturation), and the presence of the noise in the HPA changes the way the signal reacts to the HPA. If uplink noise must be modeled, Monte Carlo is the best way to do it.

Thus, to model noise, the first step is to assess the uplink and downlink noise. It is not often the case that the uplink C/N and the downlink C/N are about equal. If one is much bigger than the other, it can be assumed ideal. Otherwise, both uplink and downlink noise must be treated.

The good thing about downlink noise is that it can be modeled **semi-analytically** – the noise need not be explicitly generated. (This also applies to the uplink noise in a regenerative payload.) For every recovered symbol phasor out of the ground-receiver's sampler, the SER and/or the BER is calculated, for example, for quaternary PSK (QPSK) by the complementary error function (erfc). The running sum of the SER/BERs is kept and at the end divided by the total number of symbols/bits generated. (Each calculated SER/BER is actually a *conditional* SER/BER, conditioned on the set of nearby symbols. The number of symbols in the set is the system memory length, as in Section 16.2.4 above.) In the case that both uplink and downlink noise must be modeled, a two-dimensional Monte Carlo simulation, which would be prohibitive, can thus be avoided.

Let us summarize the way to model both uplink and downlink noise when that is necessary. The signal is generated at a sample rate of usually 8 or 16 samples per symbol. The uplink noise is modeled via a Monte Carlo simulation. A stream of correlated noise samples is generated at the same sample rate and added to the signal samples. The noisy signal samples are put through all the pre-nonlinearity filters. The memoryless baseband nonlinearity performs its action on each noisy sample. The nonlinearity output consists of a signal component, noise, and intermodulation products (IMPs). The post-nonlinearity samples pass through the filter at the payload output, the ground receiver front-end filter, the baseband demodulator, the detection filter, and the symbol recovery. The downlink noise is treated by the semi-analytic method within the Monte Carlo simulation.

16.3.5   Filtering in General

Digital filtering was discussed in Section 10.2.3.

16.3.6   How to Incorporate a Measured Filter into Simulation

Incorporating a measured distortion filter into a simulation takes a few steps. You may want to combine the distortion filter into the other filters the signal sees, but in our description we assume you want to keep it separate. The steps:

- In either case, you must experiment first to see how many symbols long you feel the causal impulse response needs to be. Make it an even number *distpulslen*.
- Set *nfreqprs* (number of frequency samples per symbol rate R_s) to an even number at least as great as *distpulslen* (greater buys you nothing except perhaps convenience). Set *deltaf* to R_s divided by *nfreqprs*.
- Read points from the file containing the transfer function file of the distortion filter, using the frequency sample delta *deltaf*. Obtain points to as far out from the center frequency as necessary, the same number on either side of the center frequency. If the detection filter's transfer function has limited extent, as a root raised-cosine (RRC) pulse-shaping filter does, you only need to read points out to those edge frequencies. Interpolate points if necessary (Section 16.3.3). Obtain an odd number of points n of the transfer function.
- An input to the routine is the number *nsampsym* of time samples per symbol duration that you are using in the rest of the simulation. Then, the fast Fourier transform (FFT) size *FFTsize* will be *nsampsym* times *nfreqprs*.

- Fill out the transfer function with the same number of zeros to either side so you have *FFTsize* +1 frequency points. Delete the last point as it is a repeat of the first point. Swap the first and second halves of the transfer function. Take the FFT to obtain the non-causal impulse response.
- Delete the middle points of the impulse response, that is, shorten it, so it has length *distpulslen* number of symbol durations. Swap its first and second halves so that it becomes causal. It will cause a delay of *distdelay* equal to half *distpulslen* number of symbols in the received signal stream compared to the transmitted signal stream.

16.3.7 HPA Model

If the HPA includes a linearizer, the HPA model is of the combined linearizer and amplifier, that is, the two are not modeled separately.

A memoryless nonlinearity is completely described by P_{out} versus P_{in} and phase shift versus P_{in} curves (Section 7.2.1). The HPA model is this simple if the HPA's small-signal frequency response $H_{ss}(f)$ and the gain-only saturated-signal frequency response $|H_{sat}(f)|$ are nearly flat over the spectrum band of the signal of interest. If there is a large interferer to be dealt with, the responses would have to be practically the same over its spectrum band as over the spectrum band of the signal of interest.

If the conditions are not satisfied, the recommended HPA model is the **three-box model**, consisting of a filter $H_1(f)$ followed by a memoryless nonlinearity and another filter $H_2(f)$, as illustrated in Figure 16.4. $H_{ss}(f)$ is the frequency response of the HPA where the input continuous wave (CW) has a fixed power level at 15-dB input backoff (IBO) or weaker across the band. The gain-only frequency response $|H_{sat}(f)|$ is measured with a CW power that saturates the HPA at midband. $H_1(f)$ and $H_2(f)$ are scaled so that both have unity gain and zero phase at midband (Silva et al., 2005).

There were much study and experimentation for several years on the best way to represent the HPA, and many models were proposed. The three-box model matches the HPA's response at small signal and at large signal but not especially in between. However, the model has big advantages: it is easily characterized (indeed, the necessary measurements are commonly required of the HPA manufacturer by the payload manufacturer), quick-executing, and independent (for good or ill) of the exact

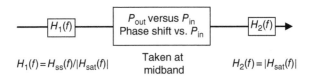

FIGURE 16.4   Three-box model for mid- to wide-band HPA (after Silva et al. (2005)).

signals put through the HPA. The model is accurate enough for all simulations except possibly when the signals are very wideband (on the order of 15% of the center frequency) or when the simulation's focus is on detailed HPA performance. The only other practical contender is the family of polyspectral methods. They are more accurate than the three-box model and quick-executing, but they are applicable only to the signal(s) with which a special time-domain HPA characterization is made (Silva et al., 2005).

Large CW or modulated-signal interferers that go through the HPA with the signal of interest must be explicitly simulated, as they create significant IMPs.

16.3.8 Carrier and Timing Recovery

In simulation, carrier-phase recovery and symbol-epoch recovery are performed with maximum-likelihood estimation (Section 12.9). This is the optimal approach for an additive white Gaussian noise (AWGN) channel, where "channel" has the communications-theory meaning. Unless the payload HPA is backed off to the linear region, the satellite communications channel is not AWGN, but this approach is still used.

16.3.9 Imperfect Synchronization

Residual phase error after carrier recovery and residual timing error after symbol synchronization have Gaussian pdfs so are modeled with Gauss–Hermite integration (book Appendix A.4).

For each abscissa of Gauss–Hermite integration, a conditional SER is obtained by a simulation run. The conditional SERs are weighted and added to obtain an estimate of the unconditional SER averaged over the residual error. Similarly, the unconditional MER or BER can be obtained. For residual phase error, the rotation and averaging can be performed after symbols have been sampled. For residual timing error, resampling must be done.

Since residual phase error and residual timing error are independent, variation of both parameters can be performed by multi-dimensional conditioning, weighting, and summing.

16.3.10 How to Generate Random Numbers of Any Probability Density

Sometimes, it may be desired to generate random numbers from a pdf other than uniform or Gaussian. The first step is to integrate the pdf to obtain the (cumulative) distribution function (cdf). Next is to generate a random number between 0 and 1 from the uniform density function. Last step is to input this number to the inverse function of the cdf. Without an analytical inverse of the cdf, one can find the inverse at each use by several iterations of Newton's method.

16.3.11   Coding, Decoding, and Interleaving

Some codes require, for best performance, that the errors in the decoder input be mutually independent. There are at least two situations where independence may not exist. One is when the decoder input includes burst errors, for example, from fading or jamming. The other situation is when the decoder input is actually the output of another decoder and this other decoder puts out error bursts, as a Viterbi decoder of a convolutional code does.

All decoders require that the rate of input errors not be too large, and what is too large depends on the code and the decoding method.

Upon finding out that either requirement may be lacking, a system designer would insert an appropriate interleaver and matching deinterleaver into the system (Sections 12.6.3 and 12.14.1).

If errors into the decoder can be assumed independent, coding/decoding need not be explicitly simulated but uncoded BER-to-coded BER curves can be used.

16.3.12   Some Instructive Simulation Results

We present here four sets of simulation results that were obtained with DVB-S2 conditions.

The first set of results, presented in Figures 16.5, 16.6, and 16.7, illustrate the point made in Section 16.2.2, that performance sensitivity to a **distortion parameter** depends on the reference conditions. By distortion parameter, we mean any one of the satellite-channel imperfections. The three figures show **SNR degradation**, that is, how much the SNR has to increase to keep the same symbol error rate with an increase in signal distortion. Each figure shows the SNR degradation for two different reference cases, one with no distortions besides the linearized TWTA (LTWTA), labeled the "no-distortion case," and one with a few reasonable distortions, labeled the "nominal" distortion case. Figure 16.5 is for pre-LTWTA gain tilt, which in the payload comes from waveguide, coax, the CAMP, and the TWTA (Section 8.2.2). Figure 16.6 is for pre-LTWTA parabolic phase, which in the

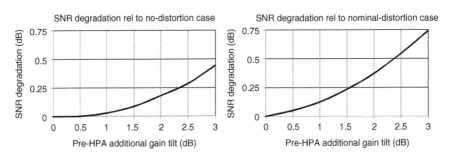

FIGURE 16.5   SNR degradation for additional pre-HPA gain tilt, for two different reference payload models.

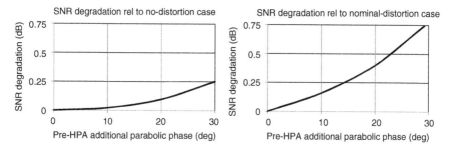

FIGURE 16.6   SNR degradation for additional pre-HPA parabolic phase, for two different reference payload models.

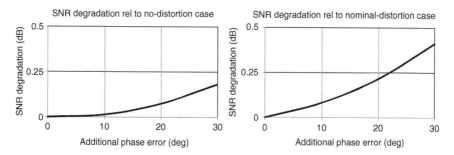

FIGURE 16.7   SNR degradation for additional carrier phase error, for two different reference payload models.

payload comes from coax, the CAMP, and the TWTA (Section 8.3.2). Figure 16.7 is for carrier phase error, which in the payload comes from frequency conversion. For all three parameters, the sensitivity to one distortion parameter's increase is much higher in the nominal-distortion reference case than in the no-distortion case. The exact numerical results, at least, would be different for different signals.

The simulations were run with the following general conditions:

- One signal plus downlink noise
- QPSK modulation
- Root-raised-cosine pulse-shaping filter with roll-off factor $\alpha = 0.2$
- Three-box model of typical LTWTA of Figure 7.13
- LTWTA input backoff = 1.2 dB, according to Newtec's MODCOD calculator for DVB-S2 with 3/4 code rate
- Matched filter of pulse
- SNR = 4.42 dB in no-distortion reference case, according to Newtec's MODCOD calculator for DVB-S2 with 3/4 code rate (Newtec, 2015). SNR increased by 0.22 dB for nominal-distortion reference case so it has same SER as no-distortion case.

TABLE 16.1   Distortion Conditions for Simulation Runs

Distortion Parameter	Value in No-Distortion Case	Value in Nominal Case
Pre-LTWTA gain tilt	0	0.5 dB
Post-LTWTA gain tilt	0	0.5 dB
Pre-LTWTA parabolic phase	0	20°
Post-LTWTA parabolic phase	0	0°
Pre-LTWTA gain and phase ripple	0	1 dB and 7° phase
Post-LTWTA gain and phase ripple	0	0
Phase error	0	3°

The values of the distortion parameters for the case of no distortion besides the LTWTA and the case of nominal distortions are shown in Table 16.1. The non-zero values are reasonable.

A second result found from these simulation runs was that the SNR degradation from increased distortion in one parameter depends in only a minor way on whether the additional distortion was pre- or post-LTWTA.

A third result was that the SNR degradation from the combined introduction of the LTWTA and the linear distortions was about equal to the sum of the separate SNR degradations of introducing the LTWTA and of introducing the linear distortions.

A fourth set of results was intended as a check on how well simulation calibration works in terms of then giving good sensitivity results. The first part of the idea was that the nominal-distortion reference case of the first set of results above would represent hardware that has not been well characterized. SNR would be known, and the resultant SER would be measured. The SNR was 4.64 dB as above. The second part of the idea was to calibrate a no-distortion (except for LTWTA) simulation by decreasing its SNR below 4.64 dB until the SER was the same. Of course, the SNR had to decrease by 0.22 dB down to 4.42 dB. The hope was that the sensitivity of these two cases to distortion would be the same or at least similar. They are not, as illustrated in Figures 16.5, 16.6, and 16.7.

16.3.13   Calibrating a Simulation

We have just seen in the fourth set of simulation results above that it is not possible to calibrate a no-distortion simulation by decreasing SNR, to make its distortion sensitivity comparable to that of hardware with distortions. The sensitivity can be much higher in the hardware. We reiterate the point made in Section 16.2.2: as much of the end-to-end communications system must be known as possible and the simulation must match that. If some hardware is not well characterized, it is better to put some reasonable distortion values into the simulation as the reference case and then calibrate the simulation by decreasing SNR by a small amount.

16.4 ADDITIONAL CONSIDERATIONS FOR EMULATION

16.4.1 Emulating Uplink

In emulating the system with hardware, care must be taken to obtain each link's proper C/N_0.

One simple way, sometimes sufficient, to emulate the uplink is shown in Figure 16.8. In the figure, the block marked "attenuators" between the ground-transmitter emulator and the bent-pipe payload emulator, could just as well be any cascade of passive elements including coaxial cable or waveguide. The cascade has to have the appropriate gain to reduce the signal power to the desired value C_1 into the payload emulator. If the total gain of the passive elements is sufficiently small, then the noise temperature into the payload emulator is nearly as low as 290 K. This is fine as long as the T_{10} to be emulated is 290 K. T_{10} is supposed to be the payload's antenna-noise temperature T_{a1}.

When the noise temperature T_{10} into the payload emulator must be larger than 290 K, then a noise source has to be added to the hardware setup, as shown in Figure 16.9. The noise source must provide a noise temperature much higher than 290 K. The required value of the attenuation after the ground transmitter emulator is easily found. The attenuator "gain" between the noise source and the hybrid (that is, 3-dB) coupler is as follows:

$$G_{12} = \frac{T_{10} + 290 - (1 - G_{11})G_{13}\,290}{(T_{\text{source}} - 290)G_{13}}$$

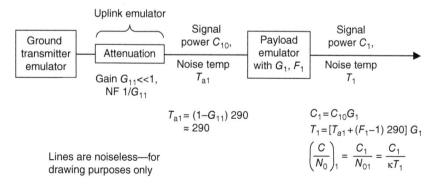

FIGURE 16.8 Uplink C/N_0 emulation for when noise temperature into payload needs to be 290 K, for bent-pipe payload.

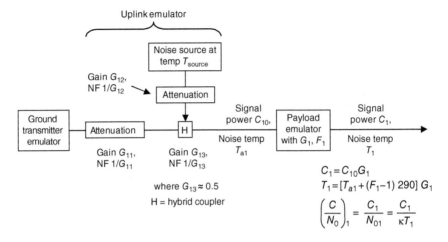

FIGURE 16.9 Uplink C/N_0 emulation for when noise temperature into payload needs to be greater than 290 K, for bent-pipe payload.

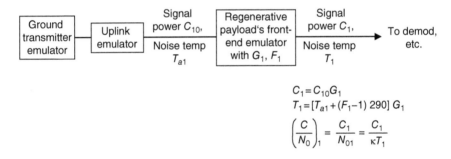

FIGURE 16.10 Uplink C/N_0 emulation for regenerative payload.

For a regenerative payload, the uplink emulation is similar except that the payload emulator is only of the payload's front end, as shown in Figure 16.10.

16.4.2 Emulating Downlink

Figure 16.11 shows the downlink emulator for any kind of payload. The issue of antenna noise for the link emulator is similar to that on the uplink.

Figure 16.12 shows the composite up- and downlink emulation for a bent-pipe payload and confirms that the end-to-end C/N_0 comes out to its proper value.

16.4.3 Matching Gain Tilt and Parabolic Phase

In emulators of the ground transmitter, the payload, and the ground receiver's front end, for convenience the lengths of waveguide and coax are usually not the same as in the actual hardware. Over a frequency band, both waveguide and coax

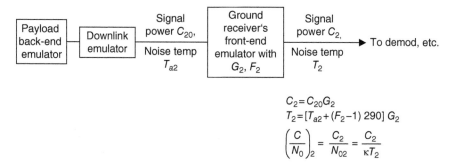

$$C_2 = C_{20} G_2$$
$$T_2 = [T_{a2} + (F_2 - 1)\ 290]\ G_2$$
$$\left(\frac{C}{N_0}\right)_2 = \frac{C_2}{N_{02}} = \frac{C_2}{\kappa T_2}$$

FIGURE 16.11   Downlink C/N_0 emulation for both bent-pipe and regenerative payloads.

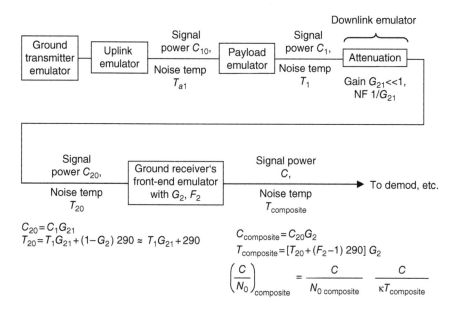

$$C_{20} = C_1 G_{21}$$
$$T_{20} = T_1 G_{21} + (1 - G_2)\ 290 \approx T_1 G_{21} + 290$$

$$C_{composite} = C_{20} G_2$$
$$T_{composite} = [T_{20} + (F_2 - 1)\ 290]\ G_2$$
$$\left(\frac{C}{N_0}\right)_{composite} = \frac{C}{N_{0\ composite}} = \frac{C}{\kappa T_{composite}}$$

FIGURE 16.12   System C/N_0 emulation for bent-pipe payload.

introduce gain tilt and perhaps parabolic gain distortion, and waveguide introduces parabolic phase distortion (Sections 8.2.2 and 8.3.2, respectively). These distortions degrade the communications performance (Section 12.15.1). The lengths of the waveguide or coax used to emulate the links can potentially make up for some of the difference. In any case, the effects must be assessed. If necessary, a gain and/or phase equalizer can be added to a link emulator to better mimic the actual system.

In the actual system, the gain tilt from one antenna exactly compensates the gain tilt from space loss. The gain tilt from the other antenna needs to be emulated.

REFERENCES

Braun WR and McKenzie TM (1984). CLASS: a comprehensive satellite link simulation package. *IEEE Journal on Selected Areas in Communications*; 2 (Jan. 1); 129–137.

Braunschvig E, Casini E, and Angeletti P (2006). Co-channel signal power measurement methodology in a communication and payload joint simulator. *AIAA International Communications Satellite Systems Conference*; June 11–14.

Chie CM, retired from Boeing Satellite Center (2011). Private communicaton, Oct. 30.

Gels B, Andrews M, and Hendry D (2007). Simulation of the effects of Q-band amplifier nonlinearities on non-constant envelope SATCOM waveforms. *Proceedings, IEEE Military Communications Conference*; Oct. 29–31.

Jeruchim MC, Balaban P, and Shanmugan KS (2000). *Simulation of Communications Systems: Modeling, Methodology, and Techniques*, 2nd ed. New York: Kluwer Academic/ Plenum Publishers.

Lin VS, Speelman RJ, Daniels CI, Grayver E, and Dafesh PA (2005). Hardware accelerated simulation tool (HAST). *IEEE Aerospace Conference*; Mar. 5–12.

Lin VS, Arredondo A, and Hsu J (2006). Efficient modeling and simulation of nonlinear amplifiers. *IEEE Aerospace Conference*; Mar. 4–11.

Newtec (2015). Universal Modcod calculator v.2.20.Beta.v3.3. On www.newtec.eu. Accessed Sep. 3, 2015.

Silva CP, Clark CJ, Moulthrop AA, and Muha MS (2005). Survey of characterization techniques for nonlinear communication components and systems. *IEEE Aerospace Conference*; Mar. 5–12.

PART III

SATELLITE COMMUNICATIONS
SYSTEMS

CHAPTER 17

FIXED AND BROADCAST SATELLITE SERVICES

17.1 INTRODUCTION

In 2016, 91% of global satellite services revenue came from satellite systems providing fixed satellite service (FSS) or broadcast satellite service (BSS). By comparison, the revenue from mobile satellite and earth-observation satellite services was small, accounting for 3 and 1.6%, respectively. Of the total services revenue, 82% came from communications services to consumers and 14% from communications services to businesses (Bryce, 2017).

Some satellites exclusively broadcast television in the BSS; this may be more common in the US than elsewhere. Others offer exclusively FSS, for nominally fixed users. FSS includes not only broadcasting television but also interactive services. Finally, some satellites provide both BSS and FSS; this is common in Europe (Emiliani, 2020a). The primary difference between BSS and FSS for television broadcast is in the frequency bands.

For FSS data users, very small-aperture terminals (VSATs) are the primary type of user terminal and have a different relationship to their grounds station from that of consumer-grade user terminals.

In this chapter we discuss more or less traditional satellites that offer BSS and FSS and provide examples. The high-throughput satellites, which offer FSS, are addressed in the next chapter.

Satellite Communications Payload and System, Second Edition. Teresa M. Braun and Walter R. Braun.
© 2021 John Wiley & Sons, Inc. Published 2021 by John Wiley & Sons, Inc.

17.2   SATELLITE TELEVISION

Satellite television accounted in 2019 for 75% of satellite services revenue. The amount of revenue from television has been slowly falling, though, in the last few years, from $98 billion in 2016 to $93 billion in 2019 as more people stream video over the Internet (Bryce, 2017; SIA, 2020).

Satellite TV is the only revenue source for BSS. Satellite systems that provide FSS have TV as their major revenue source.

Sometimes, the distinction is made that TV broadcast in the BSS is a *direct broadcast satellite (DBS)*, and TV broadcast in the FSS is **direct to home (DTH)**. In Europe, both are commonly known as DTH (Emiliani, 2020a).

The architecture of a satellite TV system is as shown in Figure 17.1. The programming to be broadcast can arrive at the satellite ground station on fiber from playout centers, other satellites providing backhaul, cloud storage, and the Internet (Emiliani, 2020a). A playout center aggregates content and programs. A backhaul satellite may receive a program such as a football game from a satellite news gathering truck, which it retransmits to the broadcast-satellite ground station to be broadcast. The sources provide the *contribution* to TV. The ground station is sometimes called the *broadcast facility* or the *uplink facility*. It is here that the various shows are conditioned and synchronized, advertisements are inserted, and the signals are made ready for transmission to the broadcast satellite. The broadcast by the satellite is called the *distribution* of TV. The signals may go directly to people's homes or they may go to the *headends* of cable TV companies, which redistribute the signals to consumers over cable (Dulac and Godwin, 2006). Both types of distribution can go over either BSS or FSS (Emiliani, 2020c). The antenna of the home consumer satellite-TV receiver in the US and Europe ranges from 18 inches (45 cm),

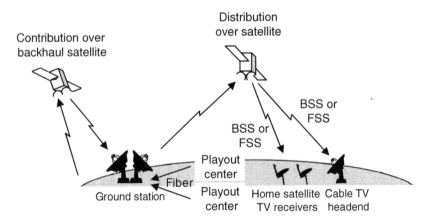

FIGURE 17.1   Television-distribution satellite-system architecture.

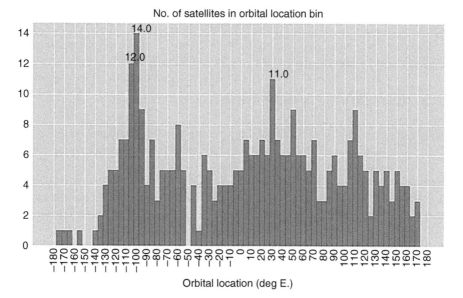

FIGURE 17.2 Number of commercial GEO communications satellites by longitude. (Used with permission of Luis Emiliani, from Emiliani (2020b)).

round, for reception from a single satellite, to 24 inches by 36 inches (60 cm by 90 cm) for reception from three satellites.

It is interesting to look at the distribution of commercial geostationary-orbit (GEO) communications satellites with longitude. This is shown, as of April 4, 2020, in Figure 17.2, collected in 10° bins. There are three bins with 10 or more satellites. The popularity of a longitude has to do with the markets that can be reached. For example, around −100° E and −60° E, there are quite a few satellites delivering video services to cable headends and DTH consumers. There are similar hot spots over Europe, for example, the set 16° E and 19.2° E, which support DTH. That is a contributor to the number of satellites around 20° E and 25° E being close to eight. Around 30° E, there are quite a few satellites responsible for video and data services toward Europe and the Middle East, with satellites from Arabsat, Eutelsat, Intelsat, and SES in those slots (Emiliani, 2020b).

Over the US, television is usually carried on 36 MHz-wide satellite channels with centers spaced 40 MHz apart, although some is carried on 24 MHz-wide channels. When the digital video and sound have been compressed according to the Moving Picture Experts Group-4 (MPEG-4) standard and processed according to the DVB-S2 standard (Section 13.4.1), in 36 MHz about 26 standard-definition (SD) TV channels can be carried or about 9 high-definition (HD) TV channels. When the compression method is high-efficiency video coding (HEVC), in 36 MHz about 15 HD channels or three ultra-high-definition (UHD) channels can be carried

(Eutelsat, 2020a). HEVC is a successor to MPEG-4 and is an International Telecommunication Union (ITU) standard (Wikipedia, 2020b, ITU-T H.265, 2019).

Over Europe, most DTH channels are carried on 33 MHz-wide channels in the BSS and on 26 MHz-wide channels in the FSS (Emiliani, 2020a).

17.3 REGULATIONS IN GENERAL

The Radiocommunication Sector of the International Telecommunication Union, ITU-R, has **allocated** some frequency bands to particular satellite communications services including the FSS, the BSS, and the mobile satellite service (MSS) (Chapter 20). The same bands, as well as others, are also allocated to particular terrestrial services, such as fixed service, broadcast service, and mobile service. There are other allocated uses of frequency bands, such as astronomy and earth observation. The ITU Radio Regulations contain the allocation tables (ITU, 2016a).

Most frequency bands are allocated to one or more services on a **primary basis** and to others on a **secondary basis**. A secondary service may not cause harmful interference into primary services.

The allocations vary by ITU-R **Region**, of which there are three as follows:

- **Region 1** includes Europe, Africa, the western part of the Middle East, and the ex-Soviet Union
- **Region 2** includes the Americas and Greenland
- **Region 3** includes the eastern part of the Middle East, the rest of Asia, and Australia.

Each national spectrum administration has the right to authorize the use of a frequency band by a specific entity. An authorized use is an **assignment** of the band to the entity (ITU-R, 2012). The administrations coordinate among themselves to limit the risks of interference (Eutelsat, 2019b).

Article 1 of the Radio Regulations volume 1 defines a **FSS** as "a radiocommunication service between earth stations at given positions, when one or more satellites are used; the given position may be a specified fixed point or any fixed point within specified areas; in some cases, this service includes satellite-to-satellite links, which may also be operated in the inter-satellite service; the FSS may also include feeder links for other space radiocommunication services."

The same article defines a **BSS** as "a radiocommunication service in which signals transmitted or retransmitted by space stations are intended for direct reception by the general public. In the broadcasting-satellite service, the term 'direct reception' shall encompass both individual reception and community reception."

17.4 FIXED SATELLITE SERVICE

Satellites providing FSS may be GEO or non-GEO (Hayden, 2003).

17.4.1 Services

As of 2019, the three largest FSS operators together had 54% by revenue of the global FSS market. The operators were SES with 21% share, Intelsat with 19%, and Eutelsat with 14% (Eutelsat, 2019b). Each provided a different mix of the same services.

SES provides services in both the FSS and BSS. In a 2020 investor presentation, SES divided its business into video and networks. Three-fourths of the video revenue came from broadcast and one fourth from other video services. The video business brought in a little over 60% of the company revenue. The network's revenue came about 40% from government, less than 35% from fixed data, and more than 25% from mobility, which is connecting aeronautical and maritime customers (SES, 2020).

Intelsat, whose creation was instigated by US President John F. Kennedy in 1964, filed for chapter 11 bankruptcy in May 2020 but was expected to emerge from bankruptcy (Wikipedia, 2020c). In its report to the US Securities and Exchange Commission for the year 2019, Intelsat divided its business into media, network services, and government. The media sector provided satellite capacity for TV broadcasters, which includes cable. Intelsat believed itself to be in 2019 the leading provider of satellite service capacity for the delivery of cable television programming to cable headends in North American network services. This sector brought in 43% of company revenue in 2019. The network services sector provided communications for fixed and wireless telecommunications providers and VSAT (see below) networks and accounted for 37% of revenue. The company was the largest provider of such services in 2019. The government sector brought in 18% of revenue and was the largest provider of these commercial services (Intelsat, 2020a).

In a 2020 investor presentation, Eutelsat divided its core business into two elements: (1) broadcast and (2) fixed data and professional video. In 2019, DTH was the main way that TV was received in Eutelsat's combined coverage area of Europe, Russia, North Africa, and the Middle East. The same was true of just the western European coverage area, where one-third of homes received TV by satellite. Sixty percent of Eutelsat's revenue came from broadcast. Eutelsat's data services were used by corporate VSAT networks including those of the oil and gas industry, mining, and banking. They were also used for mobile telephone backhaul and Internet backbone connection (trunking). The professional video services were used for TV transmission to ground stations of broadcast satellites (Eutelsat, 2019b, 2020a).

17.4.2 VSAT

A **VSAT** is an example of a FSS user terminal, a type mostly used by businesses on land and ships on the sea. VSAT networks carry video, voice, and data over GEOs. Applications include retail sales networks, lottery networks, banking automated teller machine (ATM) networks, internet protocol (IP)-based traffic, cellular backhaul, and emergency backup communications (Berlocher, 2010; Comsys, 2016).

TABLE 17.1   European Standards for VSAT Antennas

Frequency Band (GHz)	Maximum Aperture Diameter (m)	Polarization	Source
4 and 6	7.3	LP or CP	ETSI (1998)
11/12/14	3.8	LP	ETSI (1997)
30	1.8	LP or CP	ETSI (2016)

TABLE 17.2   Antenna Sizes for Newtec VSAT Dialog MDM2000 Series (ST Engineering, 2020)

Frequency Band	Aperture Diameter
C-band	1.8–2.4 m
Ku-band	75 cm–1.2 m
Ka/K-bands	75 cm–1.2 m

VSATs most often operate in C-band, lower Ku-band (i.e., 14/11 GHz), or Ka/K-bands. The antenna diameters may vary due to standards of regional or national regulatory bodies, satellite spacing, and coordination agreements reached with operators of adjacent satellites (ITU-R S.2278, 2013). Table 17.1 gives maximum antenna diameters and polarizations for Europe. Aperture diameters are often smaller, however. Examples of actual diameters of VSAT apertures are given in Table 17.2 for low-cost, high-throughput terminals of one major manufacturer. Ku- and Ka/K-band frequencies are generally preferred, except in tropical regions, where C-band is often preferred because it experiences little rain loss (Elbert, 2004).

Sometimes the term VSAT is taken to include consumer terminals, but we do not follow that usage, as the terminal price and monthly service price are higher than for a consumer-grade terminal and monthly service.

Maximum and guaranteed minimum data rates differ considerably. Until recently, maximum forward-link data rates appeared to range between about 2 and 20 Mbps, while minimum guaranteed data rates ranged between 64 Kbps and 20 Mbps, depending on the rate plan. Return-link data rates are lower than forward-link. In 2020, though, both Gilat and Newtec announced VSATs with a maximum forward-link data rate of 400 Mbps (Globenewswire, 2020; ST Engineering, 2020). (ST Engineering now owns both Newtec and iDirect.)

Some information on VSATs has already been provided in the book: modulation schemes of particular manufacturer's VSATs in Section 12.7, multiple access methods in Section 13.2.2, and communications standards in Sections 13.4 and 13.5.

The retail giant Walmart was one of the early adopters of the technology. In 1987, Walmart completed what was at the time the largest VSAT network in the US. The network linked all the operating units and headquarters with voice, data, and one-way video (Wailgum, 2007). The network was built by M/A-Com, which was later acquired by Hughes (Berlocher, 2010).

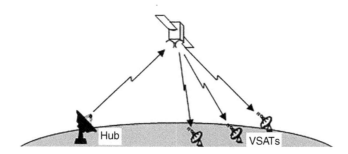

FIGURE 17.3   VSATs and hub communicating in outbound direction.

By the end of 2016, 4.6 million enterprises and star broadband-data VSATs had been shipped and were operating at more than 1.5 million sites. Hughes had a 44% share of the market in terms of terminals shipped (Comsys, 2016). VSAT networks are typically proprietary and do not interoperate with others (Elbert, 2004).

The ground station for the system is called the **hub**. For a large enterprise, the hub is located at the corporate headquarters or the data center. The forward link is said to be **outbound** and the return link **inbound**. Figure 17.3 shows outbound communications in a VSAT network. The hub controls the network communications. If all communications go through the hub, the network has a star topology. Some VSAT networks also provide a mesh topology, in which VSATs can communicate in a single hop over satellite without going through a hub (Wikipedia, 2020a).

Instead of one private hub for each satellite, an enterprise can use a "virtual hub" service, which isolates that enterprise's communications from those on the other virtual hubs in the physical hub. Some ground stations (teleports) offer this service, many of which have the advantage of being collocated with a gateway to terrestrial communications including the Internet (Berlocher, 2011).

Hughes offers IP broadband VSAT systems in star and/or mesh topologies. The hub is compliant with DVB-S2 with adaptive coding and modulation (ACM) (Section 13.4.1) for outbound traffic, which is received by all VSATs. The VSATs use frequency-division multiple access/time-division multiple access (FDMA/TDMA) channels to communicate to the hub or each other. The hub can multicast, and it can be connected to an Internet gateway (Hughes, 2020a).

17.4.3   Regulations Specific to FSS

The frequency bands wider than 100 MHz where FSS has a primary allocation are shown in Table 17.3, for frequencies between 1 and 31 GHz. These frequencies are found in C-, Ku-, K-, and Ka-bands. Not included are the frequencies in X-band, which in many countries has primary allocation to government use, especially military, which is outside the scope of this book (Wikipedia, 2019). The allocations contain many notes specializing the allocations, which are not given here.

TABLE 17.3 Frequency Bands Wider than 100 MHz Where FSS Is Primary (ITU, 2016a)

Band Name	Region 1 Frequency Band (GHz)	Region 1 Direction	Region 2 Frequency Band (GHz)	Region 2 Direction	Region 3 Frequency Band (GHz)	Region 3 Direction
S-band	—	—	2.500–2.655	Down	—	—
C-band	3.400–4.200	Down	3.400–4.200	Down	3.400–4.200	Down
C-band	4.500–4.800	Down	4.500–4.800	Down	4.500–4.800	Down
C-band	5.091–5.250	Up	5.091–5.250	Up	5.091–5.250	Up
C-band	5.725–6.700	Up	5.850–6.700	Up	5.850–6.700	Up
C-band	6.700–7.075	Both	6.700–7.075	Both	6.700–7.075	Both
Ku-band	10.70–11.70	Both	10.70–12.20	Down	10.70–11.70	Down
Ku-band	12.50–12.75	Down	—	—	12.2–12.75	Down
Ku-band	12.75–13.25	Up	12.7–13.25	Up	—	—
Ku-band	13.40–13.65	Down	—	—	—	—
Ku-band	13.75–14.80	Up	13.75–14.80	Up	13.75–14.80	Up
Ku-band	15.43–15.63	Up	15.43–15.63	Up	15.43–15.63	Up
Ku-band	17.3–18.4	Up	17.3–18.4	Up	17.3–18.4	Up
Ku-, K-bands	17.3–21.2	Down	17.7–21.2	Down	17.7–21.2	Down
K-band	24.65–25.25	Up	24.75–25.25	Up	24.65–25.25	Up
Ka-band	27.5–31.0	Up	27.0–31.0	Up	27.0–31.0	Up

TABLE 17.4   FSS Frequency Bands that May Be Used for High-Density Applications (ITU, 2016a)

Band Name	Region 1 Frequency Band (GHz)	Direction	Region 2 Frequency Band (GHz)	Direction	Region 3 Frequency Band (GHz)	Direction
Ku-band	17.3–17.7	Down	18.3–19.3	Down	—	—
K-band	19.7–20.2	Down	19.7–20.2	Down	19.7–20.2	Down
K-band	27.5–27.82	Up	28.35–29.10	Up	—	—
Ka-band	28.45–28.94	Up	—	—	28.45–29.10	Up
Ka-band	29.46–30.00	Up	29.25–30.00	Up	29.46–30.00	Up

The FSS frequency bands are shared on a primary basis with terrestrial services and other satellite services, so are subject to interference. They were not initially planned for television (Evans, 1999).

Certain FSS applications are known as *high-density fixed satellite service* (HDFSS). High-throughput satellite systems represent the objectives of HDFSS (Emiliani, 2020a). HDFSS systems are characterized by a large number of small, low-cost user terminals with small antennas, which can be rapidly deployed over a large geographical extent. HDFSS is also characterized by high frequency reuse. The ITU has designated the frequency bands shown in Table 17.4 for HDFSS. They are in the Ku-, K-, and Ka-bands. Other bands at higher frequencies are not shown. In particular, the bands were chosen to include 18.8–19.3 and 28.6–29.1 GHz because GEO and non-GEO systems have equal rights there, so these bands represent the best opportunity for non-GEO HDFSS systems. HDFSS was thought to offer great potential to developing countries to establish their telecommunications infrastructure rapidly. The satellite systems could provide broadband communications with gateways to the Internet and terrestrial telephone (Hayden, 2003).

At the 2019 World Radiocommunication Conference, the ITU decided to allocate new FSS bands to GEO satellites for broadband communications with mobile users. These users are collectively called *earth stations in motion* (ESIM). The new bands are 27.5–29.5 GHz for the uplink and 17.7–19.7 GHz for the downlink. The broadband communications include Internet connectivity. A typical data rate of such an existing terminal was said to be around 100 Mbps, much higher than those provided historically by satellites in the MSS, which use the L- and S-bands. Maritime ESIM need Internet connectivity for cruise ship passengers and general broadband connectivity for those that operate the ship. Aeronautical ESIM are on aircraft. Terrestrial ESIM are on all types of land vehicles (ITU, 2019). In the uplink band, ESIM may not cause unacceptable interference to rules-following terrestrial services. In the downlink band, ESIM may not claim protection from rules-following non-GEO FSS systems, BSS feeder links, and terrestrial services. The operation of

ESIM within a territory must be authorized by that territory's spectrum administration (ITU-R, 2019).

The ITU requires a satellite in the FSS to keep its longitude within ±0.1° of nominal. If the orbit is slightly inclined, it is the longitude where the orbit plane crosses the equator that must be kept to this limit (ITU-R S.484, 1992).

17.4.4   Examples of Satellites Providing FSS

We present examples of more or less traditional FSS satellites. They have only FSS payloads onboard. They all have US coverage because that meant the operators had to file applications with the Federal Communications Commission (FCC) and thus provide some technical information that is effectively unavailable for satellites without US coverage. To have a feel for the orbital locations of these satellites in terms of US coverage, it is useful to know that the contiguous US (CONUS) extends from about 65 to 125° W longitude.

17.4.4.1 *Intelsat 34/Hispasat 55W-2*   Intelsat 34 is a geostationary FSS satellite launched in 2015 and located at 55.5° W longitude (Gunter, 2017a). It has both C-band and Ku-band payloads (Intelsat, 2020b), which is usual for a FSS satellite. The Ku-band capacity on the satellite is known as Hispasat 55W-2 (WikiZero, 2020).

The spacecraft has two Ku-band reflectors deployed off one panel. One reflector is circular and the other is oblong. A C-band reflector is deployed off the opposite panel. Each reflector has a subreflector fixed to the earth deck and a feed fixed to the panel (Gunter, 2017a; Spaceflight101, 2015). The spacecraft keeps to its authorized position within an error of ±0.05° north–south and east–west (Shambayati, 2014).

The C-band coverage is of the Americas minus Alaska, plus western Europe and the northwest corner of Africa. The Ku-band Brazil beam covers all of Brazil. The Ku-band North Atlantic (NAOR) beam covers the southern half of North America, Central America, a thin northern slice of South America, the Caribbean, the north Atlantic, western Europe, and a bit of northwestern Africa (Intelsat, 2020b).

The services offered are video, audio, and data. The C-band beam and the Ku-band Brazil beam provide media services to Latin America, and the NAOR beam provides broadband communications to aeronautical and maritime companies (Gunter, 2017a; Spaceflight101, 2015).

The rest of what follows on Intelsat 34 comes from Shambayati (2014) except where noted.

The C-band payload uses the Intelsat teleport in Riverside, California, about 100 km east of Los Angeles. The Ku-band payload uses the Intelsat teleport in the state of Maryland, about 110 km northwest of Washington, DC (Intelsat, 2020b). However, the signals transmitted on the NAOR beam are uplinked from a teleport outside the US.

The frequencies and polarizations are shown in Table 17.5. The Brazil and NAOR beams transmit on non-overlapping Ku-band frequency bands. The C-band

TABLE 17.5   Intelsat 34 Frequency Bands and Polarizations (Shambayati, 2014)

	Uplink		Downlink	
	Frequencies (GHz)	Polarization	Frequencies (GHz)	Polarization
C-band	5.925–6.425	Dual LP	3.70–4.20	Dual LP
Ku-band Brazil beam	14.0–14.5	Dual LP	11.70–12.2	Dual LP
Ku-band NAOR beam	14.0–14.25	Horizontal	11.45–11.70	Vertical

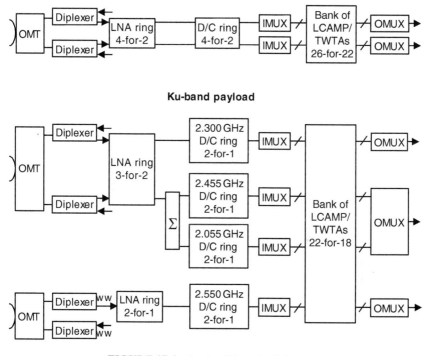

FIGURE 17.4   Intelsat 34 payload diagram.

payload has channels with bandwidths of 36, 41, and 72 MHz. The Ku-band payload has channels with bandwidths of 36 and 72 MHz.

Figure 17.4, or something much like it, represents the payload block diagram of Intelsat 34. It is a simple payload. Please refer to earlier chapters for abbreviations and meanings. All three antennas have orthomode transducers (OMTs) to separate horizontal and vertical polarizations and diplexers to separate transmit and receive signals. The rightmost arrows, out of the output multiplexers

(OMUXes), connect to the arrows going left into the diplexers. The top part of the Ku-band section is for the Brazil beam, which uses one antenna and both linear polarizations on receive and transmit. The bottom part of the Ku-band section is for the NAOR beam, which uses the other Ku-band antenna and only one polarization each for receive and transmit. The linearizer-channel amplifiers (LCAMPs) can be operated in either fixed-gain mode (FGM) or automatic level control (ALC) mode.

The satellite system employs uplink power control. At C-band, a beacon goes out on a global horn antenna. At Ku-band two beacons go out on a Ku-band global horn antenna. The beacons are linearly polarized. A user terminal adjusts its transmit power based on the received power of a beacon.

17.4.4.2 Eutelsat 65 West A

Eutelsat 65 West A is a FSS geostationary satellite launched in 2016 and located at 65° W longitude. It has C-, Ku-, and Ka/K-band payloads.

The spacecraft has two reflectors deployed off the east panel, two deployed off the west panel, and one on the earth deck. The earth-deck antenna has a subreflector and so does at least one side-panel antenna (Gunter, 2017b).

The services that the satellite is meant to provide are summarized as follows (Eutelsat, 2020b):

- At C-band, cross-continental video contribution and distribution
- At Ku-band, corporate connectivity in Central America, the Caribbean, and the Andean region as well as SD and HD TV to Brazil
- At K-band, broadband access for Latin America.

We now describe the beams (SatBeams, 2020a) and some of the services they carry. The C-band downlink beam covers Latin America, most of the US coastal areas, and western Europe. SD and HD video are uplinked from a United Teleports teleport in southeastern Florida and from at least one teleport in Europe to the satellite for broadcast (Lyngsat, 2020; Hennriques, 2017; Eutelsat, 2020c). At Ku-band there are two beams: the Americas beam covering Central America and the Andes region, and the Brazil beam covering Brazil. Both beams are uplink and downlink (Zúñiga, 2018). For the Americas beam, one teleport is in southeastern Florida that uplinks pay-TV that is broadcast (McNeil, 2016; Eutelsat, 2019a). For the Brazil beam, one teleport is in the Brazilian state of São Paolo to support professional video services (Eutelsat, 2015). As regards K-band, up to 24 spot beams cover the most populous urban and suburban areas of Brazil as well as coastal areas of other Latin American countries. Hughes has leased the entire capacity of the Brazilian spot beams to provide high-speed Internet service to consumers and businesses (Hughes, 2020b). It is unknown to the author if the Ka/K-band beams are downlink only.

The frequencies and polarizations are shown in Table 17.6. The author could find no information on the Ka/K-band frequencies or polarizations.

TABLE 17.6   Eutelsat 65 West A Frequency Bands and Polarizations (McNeil, 2016; Hennriques, 2017)

	Uplink		Downlink	
	Frequencies (GHz)	Polarization	Frequencies (GHz)	Polarization
C-band	6.725–7.025	Dual LP	4.500–4.800	Dual LP
Ku-band	12.75–13.25	Dual LP	10.70–10.95 and 11.20–11.45	Dual LP
Ka/K-band	Unknown	Unknown	Unknown	Unknown

The author could find little information about the payloads. The C-band payload has ten 54-MHz channels. On both uplink and downlink, the channels on opposite polarizations overlap fully. The Ku-band payload has twenty-four 36-MHz channels. Twelve channels are switchable between the South America and Brazil beams (McNeil, 2016). The Ku-band uplink data have modulation schemes quaternary phase-shift-keying (QPSK) and 8-ary phase-shift keying (8PSK) (FCC, 2016). On both uplink and downlink, the channels on the opposite polarizations overlap fully in most cases (Frequencyplansatellites, 2020). The author could find no information on the Ka/K-band payload.

17.4.4.3 SES-4   SES-4 is a geostationary FSS satellite launched in 2012 and located at 22° W longitude. The satellite has both C-band and Ku-band payloads. The payload is unusually large for a FSS satellite, with the equivalent of 52 C-band transponders and 72 Ku-band transponders, measured in terms of 36-MHz bandwidth. At the time of launch, it became the largest satellite in the SES fleet of 50 (de Selding, 2012).

The spacecraft has six offset-fed, Gregorian reflectors. Two each are deployed off the east and west panels, where one of the two is somewhat larger than the other. Of the two on the earth-deck tower, one is also somewhat larger than the other, and both are somewhat smaller than the pair on each side panel (Aliamus, 2020). The two largest ones must be for C-band and the four others for Ku-band. There are also two C-band horns on the tower, one for receive and one for transmit of a global beam (GB) (Aliamus, 2020). The spacecraft keeps to its authorized position within an error of ±0.05° north–south and east–west (SES World Skies, 2011).

All beams are for both transmit and receive. There are three C-band beams and four Ku-band beams, and all except the GB are contoured (SES World Skies, 2011):

C-band beams:

- West hemisphere (WH), covering eastern North America, Central America, and South America
- East hemisphere (EH), covering Europe, the Middle East, and Africa
- GB.

Ku-band beams:

- Europe/Middle East beam (EU), covering Europe, the Middle East, northern Africa, and part of Russia
- North America (NA) beam
- Southern cone beam (SC), covering Latin America
- West Africa beam (WA), covering western and central Africa.

The channel bandwidths, channel spacings, and uplink-and-downlink beam connections are given in Table 17.7 for all the beams (SES World Skies, 2011). The C-band uplink GB only connects to itself for the downlink. The other two C-band uplink beams, WH and EH, connect to themselves, each other, EU, and WA for the downlink. The Ku-band uplink EU and WA beams connect to themselves and all other beams except the SC beam and the GB. The Ku-band uplink SC beam connects only to itself and the NA beam. The table shows "N/A" for the spacing of global-beam channel centers because those channels are widely spaced. Most of the channels in a polarization completely overlap with channels in the opposite polarization.

The frequencies and polarizations of the links are given in Table 17.8. In C-band, there is only one uplink band and one downlink band. In Ku-band, there is only one uplink band but four downlink bands. All beams have dual polarization.

TABLE 17.7 SES-4 Channel Bandwidths, Channel Spacings, and Beam Connections (SES World Skies, 2011)

Uplink Band	Uplink Beam	Channel Bandwidth (MHz)	Channel Spacing (MHz)	Connects to Which Downlink Beams
C-band	WH	54, 72	60, 79	WH, EH, EU, WA
	EH	54, 72	60, 79	WH, EH, EU, WA
	GB	36	N/A	GB
Ku-band	EU	36, 54, 62	40, 60, 70	WH, EH, EU, WA, NA
	NA	36, 54, 62	40, 60, 70	EU, WA, SC, NA
	SC	54, 62	60, 70	SC, NA
	WA	36, 54, 62	40, 60, 70	WH, EH, EU, WA, NA

TABLE 17.8 SES-4 Frequency Bands and Polarizations (SES World Skies, 2011)

	Uplink		Downlink	
	Frequencies (GHz)	Polarization	Frequencies (GHz)	Polarization
C-band	3.625–4.200	Dual CP	5.850–6.425	Dual CP
Ku-band	13.75–14.50	Dual LP	10.95–11.20	Dual LP
			11.45–11.70	
			11.70–12.20	
			12.50–12.75	

TABLE 17.9   SES-4 Transponder Count (SES World Skies, 2011)

Uplink Band	Channel Bandwidth (MHz)	No. of Transponders on Transmit
C-band	36	12
	54	16
	72	8
Ku-band	36	6
	54	38
	62	6

Information on the number of transponders of each channel bandwidth is given in Table 17.9. There are a total of 36 at C-band and 50 at Ku-band, on transmit. The C-band traveling-wave tube amplifiers (TWTAs) are in a 42:36 redundancy ring and the Ku-band TWTAs in a 58:50 redundancy ring (SES World Skies, 2011).

Modulation and coding schemes of the system are QPSK with a code rate from 0.5 to 0.75 and 8PSK with a code rate of 0.816 (SES World Skies, 2011).

17.5   BROADCAST SATELLITE SERVICE

17.5.1   Regulations Specific to BSS

The frequency bands wider than 100 MHz where BSS is primary are shown in Table 17.10, for frequencies between 1 and 31 GHz. These are found in S-, Ku-, and K-bands. The allocations contain many notes specializing the allocations, which are not given here.

TABLE 17.10   Frequency Bands Wider than 100 MHz Where BSS Is Primary (ITU, 2016a)

Band Name	Region 1	Region 2	Region 3
	Frequency Band (GHz)	Frequency Band (GHz)	Frequency Band (GHz)
S-band	2.520–2.670	2.520–2.670	2.520–2.670
Ku-band	11.7–12.5 (27 MHz per channel[a])	12.2–12.7 (24 MHz per channel[a])	11.7–12.2 (27 MHz per channel[a])
Ku-band	—	—	12.5–12.75
Ku-band uplink	14.5–14.8 (to be used outside of Europe[b])		14.5–14.8
Ku-band uplink	17.3–18.1[b]	17.3–17.8	17.3–18.1[b]
K-band	21.4–22.0	—	21.4–22.0

[a] ITU (2016b).
[b] ITU (2016c).

In ITU Region 2, BSS has a primary allocation on the band 17.3–17.8 GHz, in the US on only 17.3–17.7 GHz (FCC, 2020). Before these allocations came into effect in 2007, in Region 2 GEO broadcast satellites used this band only for feeder uplinks (it was paired with 12 GHz for downlinks). It was allocated on a primary basis to FSS and this was the only use allowed. This usage still exists. When the band is used for broadcast instead it is called *reverse band* BSS. It is paired in Region 2 with the primary FSS band 24.75–25.25 GHz for the feeder uplinks, and the combination is referred to as 17/24 GHz (Cornell, 2020). The satellite DirecTV 16 uses this band combination.

If a satellite carrier wants to use a US Copyright Act section 119 license, it must provide local service in all 210 local TV markets (US Copyright Office, 2019). The Nielsen Company, which generates TV show ratings, divided the US into 210 *designated market areas* (DMA*s*), which are mutually exclusive and exhaustive markets. They cover the entire CONUS, Hawaii, and part of Alaska (Nielsen, 2020). DMAs are widely used in the sale of advertising (Crawford, 2015). A section 119 license is required for broadcasting programming to households, vehicles, and campers that are underserved by network stations. Both DirecTV and Dish use such a license (Collins, 2019).

In terms of potential interference from other primary communications-satellite services sharing BSS bands, it is a mixed bag. At S-band, in Region 1 BSS is the only primary satellite service. At lower Ku-band, in all Regions, there is some bandwidth in which only BSS is primary, with the largest bandwidth being in Region 1. In the upper Ku-band, Region 2 is the only Region that allows BSS, and BSS shares 100 MHz of that with FSS. In K-band, only Regions 1 and 3 allocate any bandwidth to satellite services, and BSS is alone. All bands are shared with primary terrestrial services.

17.5.2 Examples of Satellites Providing BSS

We present two examples of satellites that offer BSS, the first only BSS and the second both BSS and FSS. They both have US coverage because that meant the operators had to file applications with the FCC and thus provide some technical information that is effectively unavailable for satellites without US coverage.

17.5.2.1 EchoStar 16
Until May 2019, EchoStar owned the BSS satellites that Dish Network used in order to be the second-largest satellite provider of broadcast television in the US. EchoStar sold these nine satellites to Dish because of the decreasing revenue from satellite TV (Nyirady, 2019). EchoStar 16 is one of these satellites. It was launched in 2012 and is located at 61.5° W (SatBeams, 2020b). It is a good example of a BSS satellite because it uses only the traditional Ku-band DBS frequencies and it has spot beams.

The following information comes from Minea (2011) except where otherwise noted.

FIGURE 17.5   EchoStar 16 image. (Used with permission of Maxar Technologies).

The BSS geostationary spacecraft is shown in Figure 17.5. It has two antennas deployed off the east panel and two deployed off the west panel. There is one reflector clearly larger than the others. The spacecraft stays in its orbital location within ±0.05° north–south and east–west. It is one of several satellites that EchoStar owns at this position. It was designed and built so that it could operate at other EchoStar orbital slots instead, if needed.

The coverage area is CONUS and Puerto Rico. The satellite is too far east to cover Alaska and Hawaii.

The satellite broadcasts HD TV, with national stations to CONUS and Puerto Rico and local stations to some of the 210 US DMAs (Dish as a whole covers all of the 210).

The frequency bands and polarizations for communications are shown in Table 17.11. As usual for a BSS satellite, the uplink band is a FSS band (Section 17.5.1). The payload channels are 26 MHz wide with 29.16-MHz

TABLE 17.11   EchoStar 16 Frequency Bands and Polarizations (Minea, 2011)

Uplink		Downlink	
Frequencies (GHz)	Polarization	Frequencies (GHz)	Polarization
17.3–17.8	Dual LP	12.2–12.7	Dual LP

spacing. The uplink and downlink bands are divided into 32 channels, 16 on each polarization. Channels on opposite polarizations are offset by half a channel separation.

There is one downlink beam for CONUS and Puerto Rico, the CONUS+ beam. There are also 71 downlink spot beams, of which 67 are for DMAs in CONUS and the other four are for coverage in Puerto Rico, Bermuda, Mexico, and parts of the Caribbean. The payload can operate in any of three modes:

1. CONUS+ beam and all spot beams simultaneously, where the lower 16 channels are on the CONUS beam and the upper 16 are on the spot beams
2. CONUS+ beam only, on all 32 channels with normal transmit power
3. CONUS+ beam only, on 16 channels with twice the transmit power of mode 2.

There are altogether six ground stations around CONUS and six corresponding uplink spot beams. For the CONUS+ beam, the two ground stations in Wyoming and Arizona can each provide all 32 channels, 16 on each polarization. For the spot beams, in mode 1, all six ground stations provide channels for the downlink spot beams, each providing channels for a different subset of the downlink spot beams. Four ground stations can each provide 32 channels, 16 on each uplink polarization; another can provide only the upper 16 channels, eight on each uplink polarization; and the last one can provide nine of the lower channels and eight of the upper channels on one uplink polarization. Each spot beam carries between one and six channels.

For the CONUS+ beam, the payload performs a single frequency shift by 5.100 GHz from the uplink frequency band to the downlink frequency band. For the downlink spot beams, the same is true for the upper 16 channels. The frequency shifts for the lower 16 channels do not seem to follow any simple rules.

The payload has 96 TWTAs: fifty-five 151-W TWTAs, thirty-six 90-W TWTAs, and five 35-W TWTAs. In payload operation mode 2, single TWTAs are used to transmit to CONUS+, while in mode 3, two TWTAs combined are used for each channel.

One antenna is for the CONUS+ beam, possibly the one with the single feed for the subreflector, the lower one on the right-hand side of Figure 17.5. The other three antennas are for spot beams. It may be that those antennas have subreflectors that are fed by phased arrays. Alternatively, it may be that what appear to be subreflectors of these antennas are actually phased arrays that form the feeds to the reflectors.

The modulation schemes used are QPSK and 8PSK, with QPSK being combined with rate-5/6 inner turbo encoding and 8PSK being combined with rate-2/3 inner turbo encoding

17.5.2.2   *DirecTV 16/AT&T T-16*   In 2020, DirecTV was the largest provider of broadcast television in the US. However, in May 2020, news articles were saying that stockholders were pressuring AT&T, the owner of DirecTV, to sell DirecTV.

Stockholders wanted AT&T to reduce its tremendous debt. As mentioned earlier in this chapter, satellite TV is losing customers. Dish Network chairman Charles Ergen was reported to have seemingly accepted that DirecTV and Dish have to merge to survive (Munson, 2020; Barnes, 2020).

DirecTV 16, also known as AT&T T-16, is a good second example of a satellite offering BSS because it uses the reverse band for its uplinks (Section 17.5.1). In 1997 DirecTV filed a petition to the FCC to allocate spectrum for the 17/24-GHz BSS, and DirecTV was the first to seek authorization for it from the FCC (Pontual, 2014). The satellite is also a good example because it is part of a trend to combine a FSS Ka/K-band payload with a BSS payload on one satellite so that the satellite can potentially serve in multiple orbital locations as a supplement to or a replacement of another satellite.

The geostationary satellite DirecTV 16 was launched in 2019 and is positioned at 100.85° W (Gunter, 2019; Dulac, 2019). It keeps to its orbital position within ±0.025° east–west and ±0.05° north–south. It is meant to incorporate redundancy into DirecTV's HD TV broadcast ability. It is designed to operate at any of DirecTV's orbital locations of nominally 101° W, 110° W, and 119° W. These positions are nominal because when a satellite operator has multiple satellites at one nominal longitude, the operator asks permission of the spectrum administrator to put them at slightly different longitudes near the nominal so they do not collide. The seven satellites nominally at these longitudes in 2018 had payloads in the same three bands as DirecTV 16: 12/17 GHz, 17/24 GHz, and Ka/K-band. The frequency-band pair 12/17 GHz is the usual DBS pair for Region 2. At the 100.85° W orbital slot, the satellite does not operate the 12/17-GHz payload except for telemetry, tracking, and command (TT&C). The Ka/K-band payload is in the FSS and broadcasts DTH (Regan, 2018).

The frequency bands and polarizations of the various payloads are given in Table 17.12. All beams have dual circular polarization. The channels that are broadcast at 12 GHz are all 24 MHz wide and separated by 29.16 MHz, and channels on opposite polarizations are offset by half. The channels that are broadcast at 17 and

TABLE 17.12 DirecTV 16 Frequency Bands and Polarizations (Regan, 2018)

	Uplink		Downlink	
	Frequencies (GHz)	Polarization	Frequencies (GHz)	Polarization
Ku-band (12/17 GHz)	17.3–17.8	Dual CP	12.2–12.7	Dual CP
Ku/K-band (17/24 GHz)	24.75–25.15	Dual CP	17.3–17.7	Dual CP
Ka/K-band	28.35–28.6, 29.25–29.29, and 29.5–30.0	Dual CP	18.3–18.59 and 19.7–20.2	Dual CP

20 GHz are all 36 MHz wide and separated by 40 MHz, and channels on opposite polarizations overlap completely (Regan, 2018).

All three payloads broadcast national channels. The same programming is broadcast on one beam to CONUS and Alaska and on two separate spot beams to Hawaii and Puerto Rico.

The 12/17-GHz payload is capable of supporting 32 channels, 16 on each polarization. All channels would carry national HD TV. The programming could be uplinked from ground stations in Los Angeles and a city in Colorado. Any uplink channel can be received from either ground station.

The 17/24-GHz payload broadcasts up to 18 channels, nine on each polarization. The programming is uplinked from Washington state and New Hampshire.

The Ka/K-band payload is capable of broadcasting 38 channels, 19 on each polarization. It currently broadcasts on only 28. Whether it will ever broadcast more depends on how it is integrated with DirecTV's other Ka/K-band satellites. The programming is uplinked from Los Angeles and a city in Colorado. Any uplink channel can be uplinked from either ground station.

The consumer terminals have reflectors effectively 65 cm wide and have fixed pointing.

REFERENCES

Aliamus M of SSL (2020). Private communication. Oct. 19.

Barnes J (2020). AT&T may be selling DirecTV soon, sources say. May 23. *Cord Cutters News*. On www.cordcuttersnews.com/att-may-be-selling-DirecTV-soon-sources-say/. Accessed July 18, 2020.

Berlocher G (2010). Advances keep VSATs relevant in changing market. *Via Satellite*; Sep. 1.

Berlocher G (2011). VSAT hubs: "virtual" benefits becoming apparent. *Via Satellite*; Oct. 1.

Bryce Space and Technology (2017). 2017 State of the satellite industry report. *Satellite Industry Association*; June. On www.nasa.gov/sites/default/files/atoms/files/sia_ssir_2017.pdf. Accessed July 21, 2020.

Collins D, ranking minority member of US House of Representatives Committee on the Judiciary (2019). Views concerning section 119 compulsory license, sent to director of US Copyright Office. May 28. On www.copyright.gov/laws/hearings/views-concerning-section-119-compulsory-license.pdf. Accessed July 28, 2020.

Comsys (2016). The Comsys VSAT report: VSAT statistics. On www.comsys.co.uk/wvr_stat.htm. Accessed June 27, 2020.

Cornell University (2020). 47 CPR § 25.264 - Requirements to facilitate reverse-band operation in the 17.3-17.8 GHz band of 17/24 GHz BSS and DVB service space stations. Law School, Legal Information Institute. On www.law.cornell.edu/cfr/text/47/25.264. Accessed July 21, 2020.

Crawford GS (2015). The economics of television and online video markets. Working paper no 197. University of Zurich, Dept of Economics. On www.econ.uzh.ch/static/wp/econwp197.pdf. Accessed July 28, 2020.

de Selding PB (2012). Long-delayed SES-4 launched successfully. *SpaceNews;* Feb. 15.

Dulac S (2019). DirecTV application to FCC requesting special temporary authority for additional 30 days to drift T-16 to permanent location. FCC file no SAT-STA-20190826-00081.

Dulac SP and Godwin JP (2006). Satellite direct-to-home. *Proceedings of the IEEE*; 94 (1) (Jan.); 158–172.

Elbert BR (2004). *The Satellite Communications Applications Handbook*, 2nd ed. Boston, MA: Artech House.

Emiliani LD (2020a). Private communication. Sep. 28.

Emiliani LD (2020b). Computer program with input of Union of Concerned Scientists satellites database of 2020 Apr 1. Oct. 6.

Emiliani LD (2020c). Private communication. Oct. 9.

European Telecommunications Standards Institute (ETSI) (2016). EN 301 459 v2.1.1. Satellite earth stations and systems (SES); harmonised standard for satellite interactive terminals (SIT) and satellite user terminals (SUT) transmitting towards satellites in geostationary orbit, operating in the 29,5 GHz to 30,0 GHz frequency bands covering the essential requirements of article 3.2 of the Directive 2014/3/EU. May.

ETSI (1997). Technical basis for regulation (TBR) 28. Satellite earth stations and systems (SES); very small aperture terminal (VSAT); transmit-only, transmit/receive or receive-only satellite earth stations operating in the 11/12/14 GHz frequency bands. Dec.

ETSI (1998). TBR 43. Satellite earth stations and systems (SES); very small aperture terminal (VSAT) transmit-only, transmit-and-receive, receive-only satellite earth stations operating in the 4 GHz and 6 GHz frequency bands. May.

Eutelsat (2015). Speedcast Serviços Multimedia selects Eutelsat 5 West A for professional video services. Aug. 25. On news.eutelsat.com/pressreleases. Accessed July 15, 2020.

Eutelsat (2019a). Eutelsat 65 West A selected by Ultra DTH for new pay-TV platform across the Caribbean and the Andean region. July 31. On news.eutelsat.com/pressreleases. Accessed July 15, 2020.

Eutelsat (2019b). Universal registration document 2018-2019. Presentation of Eutelsat Communications group activities, main markets and competition. Report. On www.eutelsat.com/en/investors/financial-information.html?#investor-presentation. Accessed July 7, 2020.

Eutelsat (2020a). Eutelsat Communications investor presentation. Presentation package. July On www.eutelsat.com/en/investors/financial-information.html?#investor-presentation. Accessed July 7, 2020.

Eutelsat (2020b). Satellite Eutelsat 65 West A, multi-mission satellite for Latin America. Datasheet. On www.eutelsat.com/en/satellites/eutelsat-65-west.html. Accessed July 7, 2020.

Eutelsat (2020c). Eutelsat digital platform, Brazil, Eutelsat 65 West A. On www.eutelsatamericas.com/files/PDF/brochures/BCAST_EDP_E65WA_2P_EN.pdf. Accessed July 15, 2020.

Evans BG, editor, (1999). *Satellite Communications Systems*, 3rd ed. London: The Institute of Electrical Engineers.

Federal Communications Commission (FCC) of US (2016). Radio station authorization. Aug. 5. FCC file no SES-LIC-20160513-00427.

FCC (2020). FCC online table of frequency allocations. June 18. On transition.fcc.gov/oet/spectrum/table/fcctable.pdf. Accessed July 21, 2020.

Frequencyplansatellites (2020). Eutelsat 65 West A provisional frequency plan. On frequencyplansatellites.altervista.org/Eutelsat/Eutelsat_65_West_A.pdf. Accessed July 15, 2020.

Globenewswire (2020). Gilat announces availability of its flagship VSAT, achieving half a gigabit of concurrent speeds. On www.globenewswire.com/news-release/2020/06/18/20500006/en/Gilat-Announces-Availability-of-its-Flagship-VSAT-Achieving-Half-a-Gigabit-of-Concurrent-Speeds.pdf. Accessed Oct. 3, 2020.

Gunter DK (2017a). Intelsat 34 (Hispasat 55W-2). On space.skyrocket.de/doc_sdat/intelsat-34.htm. Dec. 11. Accessed May 6, 2020.

Gunter DK (2017b). Eutelsat 65 West A. On space.skyrocket.de/doc_sdat/eutelsat-65-west-a.htm. Dec. 11. Accessed July 14, 2020.

Gunter DK (2019). AT&T T-16 (DirecTV 16). June 27. On space.skyrocket.de/doc_sdat/DirecTV-16.htm. Dec. 11. Accessed July 20, 2020.

Hayden T (2003). Draft U.S. proposal on WRC-03 agenda item 1.5. Submitted to WRC-2003 Advisory Committee IWG-4. Document IWG-4/016(12.12.01). On transition.fcc.gov/ib/wrc-03/files/docs/advisory_comm/mtg6/wac087.pdf. Accessed June 9, 2020.

Hennriques H for United Teleports (2017). Technical appendix. Part of application to FCC for earth station license for C-band Eutelsat 65 West A. Feb 27. FCC file no SES-STA-20170228-00209.

Hughes (2020a). HX systems: high-performance IP satellite broadband systems. Product information. On www.hughes.com/technologies/broadband-satellite-systems/hx-systems. Accessed June 19, 2020.

Hughes (2020b). Hughes 65W. Technology description. On www.hughes.com/technologies/hughes-high-throughput-satelite-constellation/eutelsat-65-west-a. Accessed July 16, 2020.

Intelsat (2020a). Form 10-K provided to US Securities and Exchange Commission, annual report for 2019. Feb 20. On investors.intelsat.com/financial-information/sec-filings. Accessed July 8, 2020.

Intelsat (2020b). Intelsat 34 at 304.5° E. Coverage map. On www.intelsat.com/fleetmaps/?s=IS-34. Accessed July 13, 2020.

International Telecommunication Union (ITU) (2016a). *Radio Regulations*, vol. 1 Articles.

ITU (2016b). *Radio Regulations*, vol. 2. Appendices. Appendix 30, rev.WRC-15.

ITU (2016c). *Radio Regulations*, vol. 2. Appendices. Appendix 30A, rev.WRC-15.

ITU (2019). Satellite issues: earth stations in motion (ESIM). Dec. Media Centre. On www.itu.int/en/mediacentre/backgrounders/Pages/Earth-stations-in-motion-satellite-issues.aspx. Accessed July 24, 2020.

ITU Radiocommunication Sector (ITU-R) (2012). International Telecommunication Union. Presentation. On www.itu.int/en/ITU-R/workshops/regional/RRS-13-Americas/Documents/Tutorial. Accessed July 10, 2020.

ITU-R (2019). World Radiocommunication Conference 2019 (WRC-19), Final acts. ITU Publications. On www.itu.int/pub/R-ACT-WRC.14-2019. Accessed July 24, 2020.

ITU-R. Recommendation S.484-3 (1992). Station-keeping in longitude of geostationary satellites in the fixed-satellite service.

ITU-R. Report S.2278 (2013). Use of very small aperture terminals (VSATs). Oct.

ITU Telecommunication Standardization Sector. Recommendation ITU-T H.265 (2019). Series H: Audiovisual and multimedia systems; infrastructure of audiovisual services - coding of moving video; high efficiency video coding. Nov.

Lyngsat (2020). Eutelsat 65 West A at 65.0°W. TV channels listing. On www.lyngsat.com/Eutelsat-65-West-A.html. Accessed July 15, 2020.

McNeil SD for United Teleports (2016). Narrative statement and Technical appendix. Parts of application to FCC for earth station license for Ku-band Eutelsat 65 West A. FCC file no SES-LIC-20160513-00427.

Minea A (2011). Narrative and Schedule S tech report. Parts of application to FCC to launch and operate EchoStar 16. Sep. 2. FCC file no SAT-LOA-20110902-00172.

Munson B (2020). AT&T under pressure again to sell DirecTV; report. May 22. *FierceVideo*. On www.fiercevideo.com/video/at-t-under-pressure-again-to-sell-DirecTV-report. Accessed July 18, 2020.

Nielsen Corp (2020). DMA® regions. On www.nielsen.com/us/en/intl-campaigns/dma-maps/. Accessed Aug. 12, 2020.

Nyirady, A. (2019). Dish Network acquires EchoStar's broadcast satellite service business. *Via Satellite*; May 20.

Pontual R (2014). Narrative. Part of DirecTV application to FCC for milestone extension. June 24. FCC file no SAT-MOD-20140624-00075.

Regan B (2018). Narrative and Schedule S tech report. Parts of DirecTV application to FCC to launch and operate T16. Sep. 13. FCC file no SAT-RPL-20180913-00071.

SatBeams (2020a). Eutelsat 65 West A. On www.satbeams.com/satellites?norad=41382. Accessed July 15, 2020.

SatBeams (2020b). EchoStar 16. On www.satbeams.com/satellites?norad=39008. Accessed July 21, 2020.

Satellite Industry Association (SIA) (2020). Summary of 2020 State of the satellite industry report. On sia.org/category/press-releases/. Accessed July 21, 2020.

SES (2020). Investor presentation. May. On www.ses.com/investors/presentations. Accessed July 7, 2020.

SES World Skies (2011). Schedule S technical report and Technical appendix to FCC application for US market access for SES-4. July 8. FCC file no SAT-PPL-20110620-00112.

Shambayati R of Intelsat (2014). Engineering statement. Jan. 10. Part of application to FCC for Intelsat 34. FCC file no SAT-LOA-20140114-00005.

Spaceflight101 (2015). Intelsat 34. News article. On spaceflight101.com/spacecraft/intelsat-34/. Accessed July 13, 2020.

ST Engineering (2020). Newtec Dialog, release 2.2. Product brochure. On www.idirect.net/wp-content/uploads/2020/03/ProductBrochure-NewtecDialog2-2.pdf. Accessed Oct. 3, 2020.

US Copyright Office (2019). Satellite television community protection and promotion act of 2019. On www.copyright.gov/licensing/stcppa.html. Accessed July 28, 2020.

Wailgum T (2007). 45 years of Wal-Mart history: a technology time line. Oct. 17. *CIO from IDG Communications*. On www.cio.com/article/2437873/45-years-of-wal-mart-history--a-technology-time-line.html. Accessed June 27, 2020.

Wikipedia (2019). X-band satellite communication. Article. Sep. 18. Accessed July 10, 2020.

Wikipedia (2020a). Very-small-aperture terminal. Article. Jan. 19. Accessed Apr. 20, 2020.

Wikipedia (2020b). High efficiency video coding. Article. June 19. Accessed July 13, 2020.

Wikipedia (2020c). Intelsat. Article. July 7. Accessed July 8, 2020.

WikiZero (2020). Intelsat 34. Article. On www.wikizero.com/pt/Hispasat_55W-2. Accessed July 28, 2020.

Zúñiga AP for Globecomm (2018). Technical annex. Sep. 11. Part of Globecomm application to FCC for Eutelsat 65W A ground station Ku-band operation. File no SES-MFS-20180911-02588.

CHAPTER 18

HIGH-THROUGHPUT SATELLITES

18.1 INTRODUCTION

This chapter addresses high-throughput satellites (HTS) in geosynchronous orbit. Satellite constellations in lower orbits also achieve a high system-wide throughput since they need many satellites to cover the service area. Such constellations will be treated in Chapter 19, so they will not be included in this chapter.

There is no exact definition for the term HTS. It is generally applied to satellites with a throughput that is many times larger than the throughput of a traditional fixed satellite service (FSS) spacecraft using the same amount of bandwidth.

The traditional satellite for this comparison is a satellite with a few beams covering the service area. HTS use many narrow beams to cover the service area. The same carrier frequency is reused in many beams which results in a high compound data rate per Hertz of assigned bandwidth. The narrow beams require satellite antennas with a high gain, which in turn reduces the transmit power (per beam) required to produce the same equivalent isotropically radiated power (EIRP) as a single beam covering the service area. Effectively, the total transmit power of all beams remains roughly constant when the number of beams servicing a given area is increased despite the fact that the throughput grows linearly with the number of beams.

The really high throughputs are of interest for essentially one customer segment: the consumer residential market for Internet access. The satellites provide this service wherever the population density is too low for terrestrial services (typically digital subscriber line (DSL) or cable) to be available or the terrestrial service

Satellite Communications Payload and System, Second Edition. Teresa M. Braun and Walter R. Braun.
© 2021 John Wiley & Sons, Inc. Published 2021 by John Wiley & Sons, Inc.

providers offer only relatively low data rates. In 2017, 78% of US households and 84% of Canadian households had Internet access. In northern and central Europe, the values were between 86% (Belgium) and 97% (Denmark). In Central and South America, the values were around 50% (World Bank, 2019). In the second and third quarter of 2018, the average download speed for US Internet users was 96 Mbps. In Canada, the average speed was 76 Mbps. Similar data rates were measured in northern and central Europe. On the other hand, the value for Mexico was 20 Mbps and for many countries in Central America less than 8 Mbps (Ookla, 2018). Satellite Internet rates were more typically 25 Mbps at that time. In other words, there is a sizable market for such services in less developed countries and in rural areas everywhere.

The satellites serving the residential Internet market typically operate in the Ka/K-band. Narrow beams require large antennas (many wavelengths diameter), which can only be realized at high carrier frequencies, which favors this frequency choice. It is a highly competitive business, so profit margins are low (Newtec, 2017). Other services in the FSS business, such as enterprise communication (very small-aperture terminal (VSAT)), mobility (mostly commercial aircraft and ships), and cellphone backhaul (connecting cellphone base stations to the rest of the provider's infrastructure) offer higher margins but typically require higher data rates and higher service quality.

Satellite operators focusing more on the higher-margin services have traditionally operated in the Ku-band, where high availability is easier to achieve. They often offer integration with their wide-beam services so that gaps in coverage can be avoided. Some systems offer mesh connectivity, while all systems focusing on residential Internet have a star architecture with gateway stations providing the connection to the terrestrial network (Section 1.1). The operators of such systems are moving into the Ka/K-band, too, with SES-17, Inmarsat's GX-5, and Eutelsat's Konnect VHTS satellites.

On the other hand, some Ka/K-band satellites whose primary market is residential Internet offer also mobility services. In the rest of this chapter, we will use the terms "residential Internet HTS" and "VSAT services HTS" to address the two types, but it should be kept in mind that these names do not fully describe the target markets of the two systems. (See Section 17.4.2 for a description of VSATs.)

Table 18.1 shows selected HTS systems for both frequency bands. The "type" column indicates whether the primary service is residential Internet (R) or VSAT services (V). Not all operators provide the same information about system capacity: some show the composite bandwidth, other show total throughput. In both cases, it is the sum of the forward and return link. Bandwidth is the more objective measure since most systems adapt the data rate to the signal quality, hence the quoted throughput may only apply under ideal conditions. Many operators of VSAT services may not be able to quote a data rate because they lease out at least part of the bandwidth to network operators. It is obvious from this table that the system capacity has been growing fast over the last two decades.

TABLE 18.1 Selected HTS Systems

Satellite	Launch	Band	Type	No. of Beams	Bandwidth (GHz)	Throughput (Gbps)	References
Anik F2	2004	Ka/K	R	45	3.8		Bertenyi and Tinley (2000)
Thaicom 4 (IPStar)	2005	Ka/K	R	84		45	Sawekpun (2003)
Spaceway-3	2007	Ka/K	R	38		10	Hughes (2020a)
Eutelsat 9A Ka-Sat	2010	Ka/K	R	82		90	Guan et al. (2019)
ViaSat-1	2011	Ka/K	R	72		140	ViaSat Inc. (2018)
EchoStar XVII	2012	Ka/K	R	60		100	Rehbehn (2014)
HYLAS 2	2012	Ka/K	R	25	11.72		Avanti (2020)
Inmarsat 5 Global Xpress F1–F4	2013	Ka/K	V	95	2.88		Koulikova and Roberti (2012)
Sky Muster	2015/16	Ka/K	R	110		92	Wikipedia (2020)
Jupiter-2/EchoStar XIX	2016	Ka/K	R	138		200	Hughes (2020b)
Eutelsat 172B	2017	Ku	V	36		1.8	Spaceflight 101 (2021)
SES-15	2017	Ku	V	45	10		Sabbagh et al. (2017)
SES-12	2017	Ku	V	68	14		Sabbagh et al. (2017)
Intelsat 33e Epic	2016	Ku	V	63	9.2		Spaceflight 101 (2016)
ViaSat-2	2017	Ka/K	R	?		260	ViaSat Inc. (2019a)
Y hsat Al Yah 3	2018	Ka/K	R	53			Orbital ATK (2015)
Telstar 19 Vantage	2018	Ka/K	R	50	54 (est.)		Godles (2016)
SES-14	2018	Ku	V	44	12		Sabbagh et al. (2017)
Kacific-1	2019	Ka/K	R	56		60	Kacific (2019)
Inmarsat GX5	2019	Ka/K	V	72			Satbeams (2020a)
Eutelsat Konnect	2020	Ka/K	V	65		75	Nyirady (2020)
SES-17	2021	Ka/K	V	~200	80^{a}?		Krebs (2020)
ViaSat 3	2021	Ka/K	R			1000	ViaSat Inc. (2020)
Jupiter-3/EchoStar XXIV	2021	Ka/K	R			500	Satbeams (2020b)
Eutelsat Konnect VHTS	2021	Ka/K	V			500	Eutelsat (2021)

[a] The total bandwidth was estimated as follows: 16 gateway stations, 2.5 GHz bandwidth, 2 polarizations. So, it is clearly an upper bound.

18.2 FREQUENCY AND BANDWIDTH

Bandwidth is an essential resource for HTS systems. The user terminals for all applications rely on a "blanket license" which permits their use within the covered territory without a separate licensing procedure for each terminal. For this, a slice of frequency spectrum is required where other licensed systems are not interfered with and where interference to the user terminals is manageable. This means that the transmit frequencies of the user terminals should not be shared with other systems of the same or higher priority in the same area. The receive frequencies are not quite as critical since it may be possible to use a different channel for the affected user.

HTS systems operate in the FSS spectrum (Section 17.4.3). Table 18.2 shows the assignment in the Ka/K and Ku frequency bands which are not shared with other services of the same or higher priority, for the countries organized in CEPT (European Conference of Postal and Telecommunications), USA, and Canada. These bands are particularly suitable for user links. As can be seen from the table, the bandwidths are quite similar between Ka/K- and Ku-bands. On the uplink, at least 500 MHz is available. Since all systems use four-color schemes of frequency reuse (two frequencies and two polarizations), the available uplink bandwidth per beam is at the minimum 250 MHz. Many satellites use more bandwidth than shown in this table, as will be discussed later.

The gateway links need more bandwidth than the user links since one gateway station should serve many beams. Ideally, it does not take away bandwidth from the user beams. The gateway antennas have fewer interference issues than user terminals due to the larger antenna size. Also, since the gateways are not blanket-licensed the operator can place them where other co-primary systems are not causing problems and selectively reduce interference through shielding. This substantially increases the available bandwidth. For some systems, it is possible to place the gateways outside the service area, as is done for most gateways of ViaSat-1 (Barnett, 2008). This allows the gateways to reuse the user frequencies. ViaSat-2 places the gateways in the service area but uses a separate set of antennas on the satellite with

TABLE 18.2   Frequency Bands Best Suited for User Links (CEPT 2017; Federal Communications Commission 2018; Canada Government 2018)

	Ka/K-band		Ku-band	
	Earth to Space (GHz)	Space to Earth (GHz)	Earth to Space (GHz)	Space to Earth (GHz)
Europe (CEPT)	29.50–29.90	19.70–20.20	12.50–12.75 13.75–14.50	
USA	28.35–28.60 29.50–30.00	18.60–19.30 19.70–20.20	13.75–15.63	11.70–12.20
Canada	29.50–30.00	19.70–20.20	13.75–14.50	11.70–12.20

TABLE 18.3 Frequencies of Typical Ka/K HTS Systems

Satellite	Gateway		User		References
	Uplink	Downlink	Uplink	Downlink	
	Frequencies (GHz)	Frequencies (GHz)	Frequencies (GHz)	Frequencies (GHz)	
Anik–F2	28.35–28.6	18.3–18.8	29.5–30.0	19.7–20.2	Bertenyi and
	29.25–29.5				Tinley (2000)
IPStar	27–27.55	18.3–18.7	14.0–14.375	12.2–12.75	Sawekpun (2003)
	29.5–30.05	19.7–20.1			NBTC (2020)
	28.35–28.6	20.0–20.2			acma (2020)
Ka-Sat	28.83–29.50	18.4–19.7	29.5–30.0	19.7–20.2	Badalov (2012)
ViaSat–1	28.1–29.1	18.3–19.3	28.35–29.1	18.3–19.3	Barnett (2008)
	29.5–30.0	19.7–20.2	29.5–30.0		
ViaSat–2	27.5–29.1	17.7–19.3	28.35–29.1	19.7–20.2	Janka (2015)
	29.5–30.0	19.7–20.2	29.5–30.0		Janka (2017)
EchoStar XIX	27.85–28.35	18.3–18.8	29.25–30.0	18.3–18.8	Baruch (2011)
	28.35–28.6	18.8–19.3		18.8–19.3	
	28.6–29.1	19.7–20.2		19.7–20.2	
	29.25–30.0				

a diameter of 5 m versus the 2 m for the user beams (Janka, 2015). The much narrower gateway beams can then reuse the colors of all adjacent user beams without interfering with them if the gateway station is not too close to the edge of the user beam. An additional approach which is being described in the literature but has not been implemented yet in any system is to place the gateway beams in the Q/V-band; that is in the bands 37.5–43.5 and 47.2–51.4 GHz (Mignolo et al., 2011). Clearly, there is considerable bandwidth available in these bands, but the atmospheric losses will require site redundancy for the gateway stations.

Frequency assignments for some typical HTS systems are shown in Table 18.3. The following comments are in order:

- ViaSat-1 and ViaSat-2 as well as EchoStar XIX use spectrum in the 18.8–19.3 GHz band for both gateway and user terminal downlinks in the United States (ViaSat-2 for gateway links only). This band is allocated in the United States for non-geostationary satellite orbits (NGSO) FSS operations on a primary basis, with no secondary allocation for geosynchronous orbit (GSO) FSS operations. At the time when the satellites were designed there were very few NGSO satellites. Now, several NGSO systems are under development or already in operation (see Chapter 19) and there will be interference issues. Despite the clear priorities as stated above, the NGSO operators promise in their FCC applications to coordinate the use of this band with the GSO operators.

- Similarly, the same three satellites operate in the 28.6–29.1 GHz band. In this band, NGSO FSS have priority over GEO FSS, but again, the NGSO operators will protect the GEO links.
- ViaSat-1 uses the entire user frequency range also for the gateways. This is possible because the gateway stations are outside the service area. This frees more bandwidth for the user beams.
- ViaSat-2 also uses the same frequencies for the user and the gateway beams. This system uses larger antennas for the gateway beams, as explained above.
- IPStar uses Ka/K-band frequencies for the gateways and Ku-band frequencies for the user links.

18.3 RESIDENTIAL INTERNET HTS

18.3.1 Typical System Architecture

As mentioned before, all HTS systems in this category use a star architecture; that is all user terminals communicate with the gateways and not with each other. This is well suited for the main service these systems provide. It is less suited for VSAT applications for which a single-hop communication between the headquarters and the branch offices of a company may be required. A typical example is the Ka-Sat infrastructure shown in Figure 18.1. There are 10 gateway stations (SG in the figure) connected by a high-speed MPLS (multiprotocol label-switching) network and connected to the Internet infrastructure at PoPs (points of presence) spread over the service area. Typically, there are some backup gateway stations in the network to provide service when one station is off-line for maintenance or repair or experiences extreme atmospheric conditions. In the Ka-Sat system, 8 out of 10 stations are active at a time (Astrium, 2012).

Each gateway station transmits and receives the signal to and from a number of beams. In the Ka-Sat case, there are 82 beams in total as shown in Figure 18.2. Each beam has a diameter of approximately 250 km (Fenech et al., 2016).

The gateways use large antennas, in the range of 4–13 m (Table 18.4) and high transmit power so that the gateway-to-satellite link has a high signal-to-noise-plus-interference ratio (SNIR). The overall SNIR is then practically equal to the downlink SNIR on the forward link. In addition, the gateways compensate for uplink fades by adjusting the transmit power (uplink power control), provided the coordination limits set by the international telecommunication union (ITU) are not reached. In many cases, they also compensate for downlink fades in the same way. This requires, however, that the satellite can handle the additional output power. On the return link, the overall SNIR is determined by the user uplink since the gateway receiver has a high G/T. As an example, Table 18.5 shows simplified forward and return link budgets for ViaSat-2. The gateway antenna gain is 26 dB higher, and the G/T is 21 dB higher than the user terminals'. Since the gateway can compensate the uplink fades via the uplink power control, the forward link margin of 7.2 dB is entirely available for downlink fades.

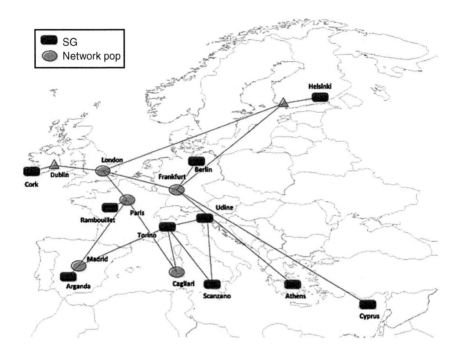

FIGURE 18.1 Terrestrial infrastructure of KA-Sat. (© 2016, John Wiley and Sons. Reprinted, with permission, from Fenech et al. (2016)).

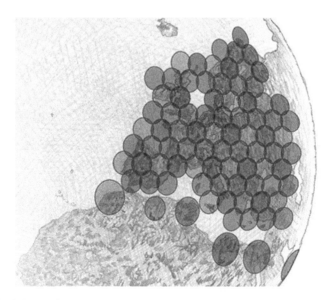

FIGURE 18.2 Ka-Sat beam pattern (Drawing courtesy of Eutelsat).

TABLE 18.4   Gateway Antenna Size of Ka/K-Band Systems

System	GW Antenna Size (m)	References
Anik-F2	8.1	Telesat Canada (2004)
Ka-Sat	9.0	Fenech et al. (2016)
ViaSat-1	7.3	Barnett (2008)
ViaSat-2	4.1–9.2	ViaSat Inc. (2016)
EchoStar XIX	5.6, 8.1, 13.2	Comsearch (2014)

TABLE 18.5   Sample Forward and Return Link Budgets, Simplified (Based on Janka (2017))

Forward Link Budget (Clear Sky)			Return Link Budget (Clear Sky)		
General	Units		**General**	Units	
Modulation		8-PSK	Modulation		QPSK
Carrier bandwidth	MHz	500	Carrier bandwidth	MHz	3.125
Uplink			**Uplink**		
Uplink frequency	GHz	27.8	Uplink frequency	GHz	27.8
Gateway antenna diameter	m	7.3	User terminal antenna diameter	m	0.67
Gateway EIRP per carrier	dBW	74.7	User terminal EIRP per carrier	dBW	48.4
Uplink losses	dB	214.7	Uplink losses	dB	214.0
Satellite G/T (peak)	dB/K	16.1	Satellite G/T	dB/K	19.5
C/l – intra-system	dB	25.5	C/l – intra-system	dB	14.3
Downlink			**Downlink**		
Downlink frequency	GHz	18	Downlink frequency	GHz	18
EIRP per carrier toward user terminal	dBW	67.9	EIRP per carrier toward gateway	dBW	45.9
Downlink losses	dB	210.2	Downlink losses	dB	210.95
User terminal antenna diameter	m	0.67	Gateway antenna diameter	m	7.3
User terminal G/T	dB/K	16.7	Gateway G/T	dB/K	38.0
System noise temp (LNA + Sky)	K	224	System noise temp (LNA + Sky)	K	250
C/l – intra-system	dB	18.7	C/l – intra-system	dB	13.3
End-to-End			**End-to-End**		
C/N – thermal uplink	dB	18.5	C/N – thermal uplink	dB	18.5
C/l Up – adjacent satellites	dB	19.2	C/l up – adjacent satellites	dB	15.6
C/N – thermal downlink	dB	16.8	C/N – thermal downlink	dB	37.6
C/l down – adjacent satellites	dB	17.1	C/l down – adjacent satellites	dB	37.1
C/(N+I) – total actual	dB	10.8	C/(N+I)–total actual	dB	9.0
C/N – required	dB	3.6	C/N – required	dB	4.9
Excess margin	dB	7.2	Excess margin	dB	4.1

18.3.2 Payload Architecture

All current residential Internet systems use a bent-pipe architecture for the satellite payload. The forward uplink consists of a small number of carriers, each of which is routed to a different beam for the downlink. On the return link, there is a larger number of carriers in each beam, each one time-shared by a number of active users, all in the same transponder band. These signals are amplified together and sent as one multi-carrier signal to the gateway.

We use the payload of the WildBlue-1 satellite to illustrate payload design aspects. The description is based on (Hudson, 2006). This satellite is at the low end of HTS throughput: the composite bandwidth for all beams and both directions is 6 GHz. It is stationed at 111° west and is collocated with Anik-F2. The two satellites use the same frequency bands but different circular polarizations. WildBlue-1 has a total of 35 user beams covering the 48 contiguous US states. There are six gateway stations, each exclusively serving a number of user beams.

The frequency plan is shown in Figure 18.3. The gateway stations transmit in two bands of 250 MHz each. The forward downlink, as well as the uplink and downlink for the return path, have contiguous 500-MHz bands. Each user beam has a bandwidth of one or two times 62.5 MHz, depending on the expected traffic load.

The payload diagram for the forward link is shown in Figure 18.4. The forward transponder receives uplink signals from the six gateway beams. The low-noise amplifiers (LNAs) are paired for redundancy. The down-conversion translates the uplink frequency to the downlink frequency in one step. Because of the gap between the sub-bands two different local oscillator (LO) frequencies are required. Frequency-selective input multiplexers separate the appropriate channels for connection to the correct user downlink beams. Each downlink-beam signal is amplified in a low-level channel amplifier with gain control and then a high power traveling-wave tube amplifier (TWTA) before being transmitted by the downlink antenna. In cases where a shared TWTA is used, the two signal paths are separated at the TWTA output using an output multiplexer.

The return-link payload, shown in Figure 18.5, receives its signals from the 35 uplink user beams. After amplification, the signals going to the same gateway are multiplexed together, amplified, and transmitted. While the figures show redundancy

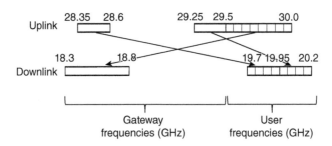

FIGURE 18.3 WildBlue-1 frequency plan.

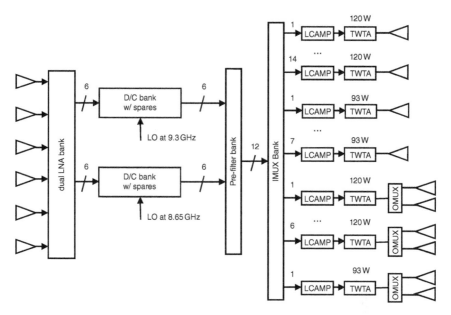

FIGURE 18.4   Wildblue-1 forward-link payload. (adapted from Hudson (2006)).

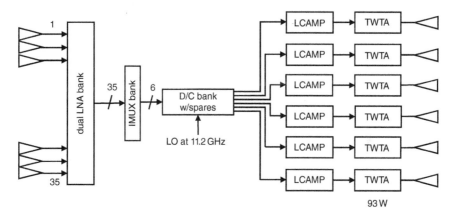

FIGURE 18.5   WildBlue-1 return link payload. (adapted from Hudson (2006)).

for the LNAs and the downconverters, they do not show redundancy for the power amplifiers. However, it can safely be assumed that they are also grouped in redundancy rings.

This payload design is typical for residential Internet HTS in the following respects:

- Large number of identical user beams forming a regular pattern over the service area (see Figure 18.6).

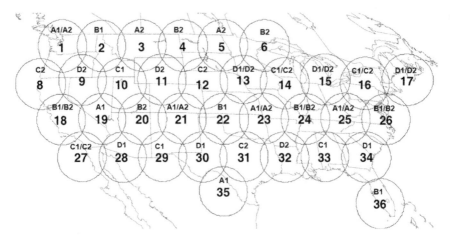

FIGURE 18.6   WildBlue-1 beam pattern and user frequency plan. (adapted from Hudson (2006)).

TABLE 18.6   Parameters of Residential Internet HTS Systems

System	Number of Gateways	Number of User Beams	Gateway BW (MHz)	User Beam BW (MHz)	References
Anik-F2	6	45	336, 392, 448	56, 112	Bertenyi and Tinley (2000)
WildBlue-1	6	35	500	62.5, 125	Hudson (2006)
Ka-Sat	10	82	2000	250	Gidney et al. (2009)
ViaSat-1	23	72	1000	500	Wikipedia (2018), Barnett (2008)
ViaSat-2[a]	40	180	2000	500	Janka (2015), Janka (2017)
EchoStar XIX	22	138	250	125	Hughes (2018), Doiron (2014)

[a] The publicly available information about ViaSat-2 is sketchy, so some of the parameters are estimates. There are 40 gateway stations in the USA, but there may be more elsewhere. The individual user beams are not shown in the FCC application, only the service area. The satellite serves the same users as ViaSat-1, so the user signal format is most likely the same, at least over land. The satellite also covers a large area of the North Atlantic, where a different signal format might be used. The number of user beams is estimated based on a ViaSat statement that the satellite's capacity is 2.5 times as much as ViaSat-1's (de Selding, 2013).

- More or less regular frequency reuse plan on user beams: the different bandwidth requirements for different regions result in some deviations from a strictly regular pattern. The designations A1, A2, B1, ... in Figure 18.6 indicate the frequency subbands used in the beams, where the subbands are, from low to high, A1, A2, B1, B2, ..., D1, D2.

- Separate frequency bands for gateway and user links.
- Bent-pipe design.
- Star architecture: all communication is between a gateway and a user.
- One gateway beam serving many user beams.

More recently launched satellites have more beams and more bandwidth per beam, but the above concepts remain mostly unchanged. Table 18.6 provides parameters for a variety of residential Internet HTS systems.

18.3.3 Antennas

The multi-beam antennas are covered in Chapter 11, so this section only addresses issues specific to HTS and certain details related to various satellites.

All operational HTS use solid parabolic dish antennas (with the exception of ViaSat-2, see below) and a single feed per beam. The antenna diameters of some Ka/K-band satellites are given in Table 18.7. Accommodation and manufacturing issues limit the size of solid reflectors to 3.5 m (Corbel et al., 2014) which is considerably larger than the largest solid dish in the table.

A four-color scheme with a separate dish for each color is the norm. While early HTS, such as Anik-F2, used separate antennas for receive and transmit, all later models use the same antenna for both directions. In most cases, gateway and user beams use the same antennas. An exception is the ViaSat-2 satellite: it uses unfurlable mesh antennas with a diameter of 5 m for the gateway beams for the reasons given in Section 18.2.

TABLE 18.7   Antenna Diameters, Beamwidth, and Edge Gain for HTS

Satellite	Antenna Diameter	Beamwidth	Edge Gain	References
Anik-F2	1.5 m	**0.85°**	**−4.5 dB**	Bertenyi and Tinley (2000)
WildBlue-1	1.45 m	**0.84°**	**−4.0 dB**	Hudson (2006)
Ka-Sat	**2.6 m**	0.43°	−3.5 dB	Fenech et al. (2011)
ViaSat-1	**2.6 m**	0.24°	**−1.0 dB**	est. from Barnett (2008)[a]
ViaSat-2[b]	**5.0 m** (mesh reflector)	0.24°	−4 dB	de Selding (2018)
	2.0 m (solid reflector)	0.6°	−4 dB	est. from Janka (2015)
EchoStar XIX	2.5 m	**0.47°**	−4 dB	est. from Baruch (2011)

Bold values come from references, the others are estimated.

[a] The estimation of the satellite antenna sizes was done as follows: The FCC filings include antenna footprints in "GXT" files, which can be opened with the "GIMS" tool from ITU. For two points on the −4 dB contour on a straight line through the beam center, the coordinates were digitized. The beamwidth was then computed and the corresponding antenna size determined.

[b] The FCC filings for ViaSat-2 and EchoStar XIX do not provide the peak and edge gain of the beams. So, it is not possible to calculate the beamwidth from the antenna size.

18.3.4 Air Interface

Three air-interface specifications for residential Internet HTS systems can be found in the technical literature. There are two standards: (1) Digital Video Broadcasting (DVB)-S paired with Digital Video Broadcasting–Return Channel Satellite (DVB-RCS), standardized by the European Telecommunications Standards Institute (ETSI), which is described in more detail in Section 13.4; and (2) internet protocol over satellite (IPoS) specified by Hughes Network Systems (ITU Working Party 4B, 2006) and standardized by the US Telecommunications Industry Association (TIA), the International Telecommunication Union (ITU), and ETSI. Both standards have evolved into second generations: DVB-S2/DVB-RCS2 and IPoS-B. For DVB-S2, there is even a newer extension, DVB-S2X. The third specification is proprietary and was produced by ViaSat Inc. It is sometimes referred to as S-DOCSIS or DOCSIS-S, for example (OFCOM, 2011). It is used in the SurfBeam system which is in service over ViaSat, Eutelsat (Wikipedia, 2019a), SES Americom (ViaSat Inc., 2019b), and possibly other satellites. It has also been updated to a second version, called Surfbeam 2 (ViaSat Inc., 2010).

The IPoS standard is based on the open systems interface (OSI) standard of the International Standards Organization (ISO) and separates the protocol stack into satellite-dependent and satellite-independent layers, as shown in Figure 18.7. SI-SAP is the satellite-independent service access point. DOCSIS-S has a similar structure, following the Data-over-Cable System Interface Specification (DOCSIS)

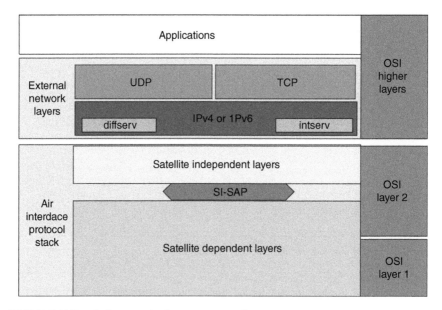

FIGURE 18.7 IPoS protocol reference model. (Source: Recommendation ITU-R S.1709-1 (01/2007)).

standard in the higher layers and implementing proprietary technology on the lower layers. We focus our description on the physical layer and the data-link layer of the OSI standard (Section 13.2.1).

18.3.4.1 Forward Link
The IPoS-B standard uses DVB-S2 for the forward link (ITU Working Party 4B, 2006), so no further discussion is required here.

DOCSIS-S uses the same access method as DVB-S2, that is time-division multiplexing (TDM) (Section 1.3) with a single carrier per satellite beam. For ViaSat-1, the forward links consist of a 500-MHz-wide carrier for each beam. Since the bandwidth allocated to the gateway stations is 1 GHz[1], each gateway can support four user beams using both circular polarizations (Barnett, 2008). The modulation formats and error correction coding are similar but not identical to DVB-S2: The data rate is 416.67 Msps with 4, 8, and 16 phase modulation formats and forward error correction (FEC) code rates varying between 0.247 and 0.6642. Adaptive coding and modulation (ACM) is implemented (Barnett, 2008). The fact that only phase modulation is used suggests that the satellite HPA is driven into the nonlinear region.

18.3.4.2 Return Link
The return link access of IPoS-B is based on TDMA (Section 13.2.2). The user terminal selects a return link frequency from the channels available in its beam and communicates its choice to the hub. The hub then assigns time slots on this frequency when the terminal requests it. There is no spontaneous transmission for user data. All transmissions are based on demand-assignment multiple access (DAMA) (Section 13.2.2).

IPoS-B uses constant-envelope offset quaternary phase-shift-keying (OQPSK) modulation with root raised-cosine pulse shaping and a roll-off factor of 0.45. OQPSK has only small envelope variations, particularly with a roll-off factor as high as 0.45. The signal is hard-limited in the modulator to remove even the remaining envelope variation. The power amplifier in the remote terminal can be operated at saturation, which may reduce the cost. Two coding schemes are specified: an inner turbo code with puncturing to achieve the rates 1/3, 1/2, 2/3, and 4/5 combined with a Bose–Chaudhuri–Hocquenghem (BCH) outer code with 39 bits of redundancy or a low-density parity-check (LDPC) code with code rates 1/2, 2/3, and 4/5. A 16-bit cyclic redundancy check (CRC) code for error detection is added in both cases. ACM (Section 13.2.3) is supported.

The return link of ViaSat-1 uses MF-TDMA (Section 13.2.2) with a variety of bandwidths and data rates. There are from 20 to 640 carriers in the 500 MHz channel (Barnett, 2008). ACM is supported. The modulation is PSK with 2, 4, or 8 phases, and the FEC code rate varies between 0.375 and 0.833 (Barnett, 2008).

---

[1] As pointed out in Section 18.2, the gateways use some frequencies where NGSO satellites have priority. Whenever a NGSO satellite signal may be interfered with, the gateway link bandwidth is reduced to 500 MHz.

Again, only phase modulation is used, which reduces the linearity constraint for the user terminal amplifier.

For the return link, the satellite HPA is always required to operate in linear mode since the signal to the gateway is a frequency-division multiplex of a number of carriers.

18.4   VSAT SERVICES HTS

The HTS systems to be discussed here offer services other than residential Internet access:

- Enterprise services
- In-flight services to commercial airplanes
- Maritime communications (e.g. cruise ships and commercial fleets)
- Government and defense communication.

These services require higher flexibility in coverage, better quality of service, and, often, mesh connectivity rather than a star topology. The throughput is considerably smaller than for the most powerful residential Internet HTS. Many have multiple payloads, for example global or regional beams at C- or Ku-band and broadcast channels in addition to the HTS payload. While the residential Internet systems generally use proprietary air-interface protocols and user equipment, at least some of the systems to be discussed here let the customers select their own ground equipment. This allows customers to use their installed base, and they can transfer the service to other open systems (Hudson, 2018).

A typical family of VSAT services HTS is the Intelsat Epic^{NG} family. It originally consisted of six satellites. However, Intelsat 29E, which provided service for the Americas and the north Atlantic Ocean, failed on April 8, 2019. At the time of writing in 2020, it is not clear yet whether and how it will be replaced. Despite this fact, the description of these HTS will be mostly based on the Intelsat 29E because the FCC filing (Hindin, 2013) provides a better description than can be found for other satellites.

Besides the Ku-band HTS payload, Intelsat 29e had a C-band payload with 14 transponders with a total bandwidth of 864 MHz (24 transponder equivalents, one transponder equivalent being 36 MHz wide). It covered South America with a single beam (Wikipedia, 2019b). It also had a Ka/K-band transponder providing global coverage with a single beam of 500-MHz bandwidth which could connect large earth stations with antennas in the 7–9 m range (Hindin, 2013).

The Ku-band HTS payload had 45 user beams and 6 gateway beams. The beam centers are shown in Figure 18.8. Large circles combine a user beam and a gateway beam. (The user and gateway beams near Washington DC were not exactly aligned; all others were.) The beam width can be estimated from the antenna gain given in Hindin (2013) as 1.5°, which is considerably larger than those shown in Table 18.7

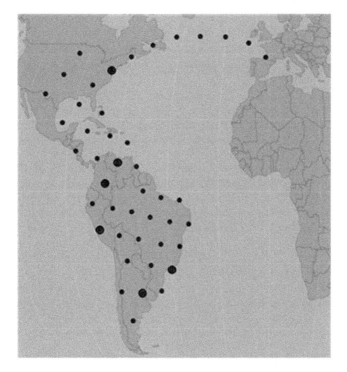

FIGURE 18.8   Intelsat 29E spot beam centers.

for Ka/K-band HTS. The payload of the Intelsat Epic[NG] satellites includes a programmable digital channelizer (Section 10.1.7). This device can link C-band and Ku-band uplink channels from different beams to a single C-band or Ku-band downlink channel. The uplink channel may be in the same beam as the downlink channel. According to Henry (2016), this connectivity is available in any bandwidth increment. According to Hindin (2013), the channel bandwidths can be 36, 62.5, 125, 187.5, 250, 300, 375, or 500 MHz. The channels through the satellite are transparent and linear, which allows the customer to use any kind of equipment if he is operating a closed system. If he needs access to other networks (e.g. the Internet) through the gateway stations, his equipment has to use a prescribed protocol. Most likely, this is DVB-Satellite 2nd Generation (DVB-S2), but it is nowhere stated. While earlier Epic[NG] satellites have fixed power per beam, the last satellite in the Intelsat Epic[NG] series, Intelsat Horizons 3e, includes multiport amplifiers (Section 11.12.3) which allow the operator to direct the transmit power from poorly loaded beams to fully loaded beams (Bleakley, 2018).

As pointed out in the introduction, some operators offer the higher-value services now on satellites in the Ka/K-band. SES-17 is a typical example of this trend. It is described in more detail in Section 10.3.

REFERENCES

ACMA (2020). Australian communications and media authority, register of radiocommunications licences. On https://web.acma.gov.au/rrl/register_search.main_page. Accessed July 15, 2020.

Astrium (2012). Use of Ka-band for satellite communications systems and services, the Astrium experience. On https://www.itu.int/dms_pub/itu-r/md/12/iturka.band/c/R12-ITURKA.BAND-C-0003!!PDF-E.pdf. Accessed Apr. 11, 2019.

Avanti (2020). Hylas 2. On https://www.avantiplc.com/technology/satellites/hylas-2. Accessed July 13, 2020.

Badalov K (2012). KA-SAT services in Europe. On https://www.itu.int/dms_pub/itu-r/md/12/iturka.band/c/R12-ITURKA.BAND-C-0004!!PDF-E.pdf. Accessed July 15, 2020.

Barnett RJ for ViaSat (2008). Narrative-tech. annex of ViaSat 1 license application, Federal Communications Commission. FCC file number SAT-AMD-20080623-00131.

Baruch SD for EchoStar (2011). Part of application to FCC for satellite operation. On https://fcc.report/IBFS/SAT-LOI-20110809-00154/910919.

Bertenyi E and Tinley R (2000). The triple-band Anik F2 spacecraft. *51st International Astronautical Congress*. Rio de Janeiro, Brazil.

Bleakley T (2018). A connected pacific: horizons 3e is final piece in Intelsat Global EpicNG network. On https://www.intelsat.com/resources/blog/a-connected-pacific-horizons-3e-is-final-piece-in-global-intelsat-epicng-network. Accessed Apr. 6, 2019.

Canada Government (2018). Canadian table of frequency allocations. On https://www.ic.gc.ca/eic/site/smt-gst.nsf/eng/sf10759.html. Accessed Sep. 24, 2018.

CEPT (2017). The European table of frequency allocations. On https://efis.cept.org/sitecontent.jsp?sitecontent=ecatable. Accessed Sep. 24 2018.

Comsearch (2014). Exhibit E of EchoStar XIX license application, Federal Communications Commission. FCC file number SAT-MOD-20141210-00126. Accessed March 2, 2021.

Corbel E, Charrat B, Dervin M, Garnier B, Baudoin C, Combelles L, Merour J-M (2014). 2016 – 2020 high-throughput satellite systems on the right track. *20th Ka and Broadband Communications Conference*.

Doiron S for EchoStar (2014). Narrative, part of FCC application for Jupiter 97 W. FCC file number SAT-MOD-20141210-00126.

Eutelsat (2021). Konnect VHTS. On https://www.eutelsat.com/en/satellites/future-launches.html?#konnect-vhts. Accessed Jan. 14, 2021.

Federal Communications Commission (2018). FCC online table of frequency allocations. On https://transition.fcc.gov/oet/spectrum/table/fcctable.pdf. Accessed Sep. 24, 2018.

Fenech H, Tomatis A, Serrano D, Lance E, and Kalama M (2011). Antenna requirements as seen by an operator. *5th European Conference on Antennas and Propagation*.

Fenech H, Tomatis A, Amos S, Soumpholphakdy V, and Serrano Merino JL (2016). Eutelsat HTS systems. *International Journal of Satellite Communications and Network*; 34 (4); 503–521.

Gidney P, Jones D, Paullier T, and Fenech H (2009). Performance optimization of multibeam broadband payloads. *14th Ka and Broadband Conference*.

Godles JA for Telstar (2016). FCC filing for Telstar 19 Vantage. FCC file number SAT-PPL-20160225-00020.

Guan Y, Geng F, and Saleh JH (2019). Review of high throughput satellites: market disruptions, affordability-throughput map, and the cost per bit/second decision tree. *IEEE Aerospace and Electronic Systems Magazine*; 34 (5); 2019.

Henry C (2016). Intelsat starts multi-tiered Ku-band system with first HTS satellite, *Via Satellite Magazine, 8 Feb. 2016*. On https://www.satellitetoday.com/innovation/2016/02/08/intelsat-starts-multi-tiered-ku-band-system-with-first-epicng-satellite. Accessed May 17, 2019.

Hindin JD (2013). for Intelsat. FCC Application for Intelsat 29E fixed satellite service, FCC file number SAT-LOA-20130722-00097.

Hudson C (2018). Ku-band vs. Ka-band - separating fact from fiction, Intelsat General. On https://www.intelsatgeneral.com/blog/ku-band-vs-ka-band-separating-fact-from-fiction. Accessed May 5, 2019.

Hudson E for WildBlue (2006). Technical exhibit. Part of application to FCC for satellite operation. FCC file number SES-MFS-20060811-01347.

Hughes (2018). EchoStar XIX, Hughes high-throughput satellite constellation. On https://www.hughes.com/technologies/hughes-high-throughput-satellite-constellation/echostar-xix. Accessed Aug. 31, 2018.

Hughes (2020a). Spaceway 3, Hughes high-throughput satellite constellation. On https://www.hughes.com/technologies/hughes-high-throughput-satellite-constellation/spaceway-3. Accessed July 13, 2020.

Hughes (2020b). High-throughput satellite constellation. On https://www.hughes.com/products-and-technologies/high-throughput-satellite-fleet/jupiter-2. Accessed July 13, 2020.

ITU Working Party 4B (2006). Draft revision of recommendation ITU-R S.1709. On https://www.itu.int/rec/R-REC-S.1709/en. Accessed July 26, 2020.

Janka JP for ViaSat (2015). Narrative. Part of application to FCC for satellite operation of ViaSat-2. FCC file number: SAT-AMD-20150105-00002.

Janka JP for ViaSat (2017). Attachments of ViaSat-2 FCC Filing. FCC file number: SAT-MOD-20160527-00053.

Kacific (2019). Technology – latest Ka band with 56 beams supporting 5G speeds. On https://kacific.com/technology. Accessed July 13, 2020.

Koulikova Y and Roberti L (2012). Global mobile broadband. On https://www.itu.int/md/meetingdoc.asp?lang=en&parent=R12-ITURKA.BAND-C&source=Yulia%20Koulikova%2C%20Laura%20Roberti. Accessed July 13, 2020.

Krebs GD (2020). SES-17, Gunther's space page. On https://space.skyrocket.de/doc_sdat/ses-17.htm. Accessed July 13, 2020.

Mignolo D, Re E, Ginesi A, Almanac AB, Angeletti P, and Harveson M (2011). Approaching terabit/s satellite capacity: a system analysis. *Proceedings of Ka Broadband Conference*.

NBTC (2020). Frequency licence database of Thailand, office of the national broadcasting and telecommunications commission – 2020. On https://www.nbtc.go.th/Home.aspx?lang=en-us. Accessed July 15, 2020.

Newtec (2017). Getting the most out of high throughput satellites. On https://www.newtec.eu/article/article/getting-the-most-out-of-high-throughput-satellites-hts. Accessed Mar. 3, 2021.

Nyirady A (2020). Eutelsat KONNECT satellite enters into service. On https://www.satellitetoday.com/broadband/2020/11/16/eutelsat-konnect-satellite-enters-into-service. Accessed Jan. 14, 2021.

OFCOM (2011). Understanding satellite broadband quality of experience – final report. On https://www.scribd.com/document/357094239/Understanding-Satellite-Broadband-Quality-of-Experience-Final-Report. Accessed Apr. 30, 2019.

Ookla (2018). Fixed broadband speed test data. On https://www.speedtest.net/reports. Accessed May 20, 2019.

Orbital ATK (2015). Al Yah 3. On https://www.sky-brokers.com/uploads/f9/a6/f9a6fd6ca58 6fd88fd4b88c58cf02052/Datasheet-YahSat-Al-Yah-3-satellite-built-by-OSC.pdf. Accessed July 13, 2020.

Rehbehn D. (2014). High throughput satellites + the APAC region. On http://www.satmagazine.com/story.php?number=1777232412. Accessed Jan. 14, 2021.

Sabbagh KM, De Hauwer C, Collar S Halliwell M, McCarthy P (2017). SES investor day 2017. On https://www.ses.com/sites/default/files/2017-06/170627_IR%20DAY%20 2017_FINAL%20Web.pdf. Accessed July 13, 2020.

Satbeams (2020a). Intelsat GX5. On https://www.satbeams.com/satellites?norad=448017. Accessed Oct. 31, 2020.

Satbeams (2020b). Echostar 24. On https://www.satbeams.com/satellites?id=2727. Accessed July 13, 2020.

Sawekpun T (2003). The iPSTAR broadband satellite project, AIAA 21st. *International Communications Satellite Systems, Conference and Exhibit.*

de Selding PB (2013). ViaSat-2's 'first of its kind' design will enable broad geographic reach. On http://spacenews.com/35369viasat-2s-first-of-its-kind-design-will-enable-broad-geographic-reach. Accessed Mar. 21, 2018.

de Selding PB (2018). ViaSat's Mark Dankberg: cause of defect on two ViaSat-2 antennas remains a mystery, Space Intel Reports. On https://www.spaceintelreport.com/viasats-mark-dankberg-cause-of-defect-on-two-viasat-2-antennas-remains-a-mystery. Accessed Aug. 14, 2020.

Spaceflight 101 (2016). Intelsat 33e. On https://spaceflight101.com/ariane-5-va232/intelsat-33e. Accessed July 13, 2020.

Spaceflight 101 (2021). Eutelsat 172B. On https://spaceflight101.com/ariane-5-va237/eutelsat-172b. Accessed Jan. 14, 2021.

ViaSat Inc. (2010). SurfBeam® 2 high-performance, high-capacity broadband satellite system. On https://www.viasat.com/files/assets/surfbeam2_Overview_018_web.pdf. Accessed Apr. 30, 2019.

ViaSat Inc. (2016). Fixed earth station license application. FCC file number SES-LIC-20160610-00519.

ViaSat Inc. (2018). Fact sheet ViaSat-1. On https://www.viasat.com/files/assets/news/web/ Fact%20Sheet%20High-Cap_Sat_VS-1.pdf.

ViaSat Inc. (2019a). DOCSIS-based broadband satellite system for SES AMERICOM. On https://www.viasat.com/news/docsis-based-broadband-satellite-system-ses-americom. Accessed Apr. 30, 2019.

ViaSat Inc. (2019b). ViaSat-2 at a glance. On https://www.viasat.com/sites/default/files/media/ documents/770853_vs-2_2019_infographic_009_lores.pdf. Accessed July 13, 2020.

ViaSat Inc. (2020). Viasat-3 platform will take our service around the world. On https:// www.viasat.com/news/going-global. Accessed July 13, 2020.

Wikipedia (2018). ViaSat-1. On https://en.wikipedia.org/wiki/ViaSat-1. Accessed July 5, 2018.

Wikipedia (2019a). Tooway 4. On https://en.wikipedia.org/wiki/Tooway. Accessed Apr. 30, 2019.

Wikipedia (2019b). Intelsat 29e. On https://en.wikipedia.org/wiki/Intelsat_29e. Accessed June 7, 2019.

Wikipedia (2020). Sky Muster. On https://en.wikipedia.org/wiki/Sky_Muster. Accessed July 13, 2020.

World Bank (2019). Internet access (% households), TCdata360. On https://tcdata360. worldbank.org/indicators/inet.acc?country=BRA&indicator=31&viz=line_chart&years= 2005,2016#table-link. Accessed May 20. 2019.

CHAPTER 19

NON-GEOSTATIONARY SATELLITE SYSTEMS

19.1 INTRODUCTION

This chapter focuses on communication satellite constellations in medium earth orbits (MEO) and low earth orbits (LEO). We exclude systems based on CubeSats, that is very small satellites made up of multiple 10 cm by 10 cm by 10 cm cubes and serving niche applications.

We can distinguish two different categories of non-geostationary orbit (NGSO) satellite systems, as they are called by International Telecommunication Union (ITU): The early systems were designed to provide voice and/or low-rate data services. The user terminals are typically small, such as hand-held devices. Iridium (Section 19.2) and Globalstar (Section 19.3) are two systems in this category that became operational. Orbcomm is a third system in this category, but it is not discussed here because it operates at frequencies below 1 GHz, which are not covered in the rest of the book. The first operational system was Iridium, which started service in 1998. All three systems went through bankruptcy around the year 2000, but have survived, have been updated over time, and are still operational. There seem to be no new projects underway which address this market, however.

The frequencies these systems use are in the Mobile Satellite Service (MSS) band. At L-band there are 50 MHz with the same assignment for MSS up- and

Satellite Communications Payload and System, Second Edition. Teresa M. Braun and Walter R. Braun.
© 2021 John Wiley & Sons, Inc. Published 2021 by John Wiley & Sons, Inc.

downlinks in all three ITU regions, so the satellites can provide service globally. In S-band there are another 16.5 MHz for MSS downlinks, again in all ITU regions.

The second, more recently introduced category of NGSO constellations provides broadband services, so these systems are in competition with high-throughput satellites. Several projects have been started or announced, with mixed results. Right now, there exists only one operational system, O3b, which is presented in Section 19.4. However, there is a very large interest in such systems and at least eight organizations have plans for such services and some have already launched satellites. An early project was OneWeb, which survived bankruptcy in 2020 after launching 74 satellites. It will be presented in Section 19.5. Another early project was LeoSat, which folded in 2019, before its first launch. SpaceX is the current frontrunner and its Starlink system will be presented in Section 19.6. Amazon (Project Kuiper), Boeing, SES, and Telesat Canada have projects in early stages. Only the last one is presented, in Section 19.7.

These newer systems use frequencies in the fixed satellite service (FSS) band for the user links, currently in the Ku- and Ka/K-bands. The bandwidths available in these bands are over 1 GHz. Some systems may add higher frequencies in the future, where even more bandwidth is available. The frequencies in the Ku- and Ka/K-bands are mostly harmonized across the three ITU regions, but not completely.

Since practically all such systems want to provide service on US territory, the Federal Communications Commission (FCC) is a good source of information. It is too early to tell which ones of these projects will be successful, not as much technically as economically. The focus in this chapter, as in the rest of the book, will be on systems that succeeded to become operational. Starlink and OneWeb have shown that they are technically viable, so they are being included. Telesat LEO has not reached that point yet, but it is not too far off and it has the backing of the Canadian government, so we are optimistic that it will be implemented.

There are several reasons for the interest in LEO and MEO orbits:

- The geostationary arc is practically booked out, at least over the continents and at the desirable frequencies.
- Since LEO and MEO systems require a large number of satellites to provide continuous service such systems offer automatically a very high total throughput and spectral efficiency.
- These constellations can bypass all terrestrial communication infrastructure if they have crosslinks among the satellites. This can improve the privacy of communication.
- Many of the communication services benefit from lower signal delays provided by satellites in lower orbits. LEO and MEO satellite systems have not just lower signal delays than geostationary orbit (GEO) satellites: constellations with crosslinks have also lower delays than terrestrial fiber-optic systems since the speed of light in vacuum is approximately 50% higher than in a fiber-optic cable (3×10^8 vs. 2×10^8 m/s).

- All LEO and MEO constellations (operational and planned) use non-equatorial orbits, at least for some of their satellites. They typically provide service up to higher latitudes than GEO satellites. Most constellations provide global coverage.

There are hurdles and challenges. A major hurdle is the up-front investment required to enable the service. Several companies stumbled over this: Iridium, Globalstar, and OneWeb went through bankruptcy proceedings and Teledesic and LeoSat, folded. Not all of the companies that applied for spectrum will turn their venture into a viable business. The satellite fleet needs constant renewal: satellites at an altitude of 300 km have a life expectancy of five years. A third issue is the coexistence with GEO satellites and among NGSO constellations. This topic is covered in Section 14.6.

In what follows a number of NGSO systems will be described.

19.2   IRIDIUM

19.2.1   Overview

Iridium is a LEO constellation at an altitude of 780 km which provides telephony and low-rate data services. It started service in 1998 and became fully operational in 2002. A second generation of satellites, called Iridium Next, was launched between January 2017 and January 2019.

Iridium provides full-earth coverage. The satellites are in 6 almost polar orbital planes with 11 satellites each, for a total of 66[1] operational satellites. There are nine in-orbit spares of the second generation (Iridium, 2019a).

The voice and data communication between two Iridium user terminals flows completely within the system: each satellite has crosslinks to its four neighbors and forwards the traffic towards its destination. Iridium has four commercial gateway stations (Alaska, Arizona, Chile, Norway) (Wikipedia, 2019) and one in Hawaii for US Government traffic only (Satcom Direct, 2019). Communication with terrestrial stations (phones, Internet, etc.) flows through these gateways.

19.2.2   Spacecraft

An artist's rendering of the Iridium Next spacecraft is shown in Figure 19.1. Major elements are labeled: (A): main mission antenna (L-band), (B): feeder link antennas (Ka/K-band), (C):cross-link antennas (K-band), (D): secondary rotation joint for solar array wing (see Section 2.1.4), (E): hosted payload (Aireon, a global air traffic surveillance system). The satellites were built by Thales Alenia Space. Their design lifetime is 15 years.

---

[1] The original design called for seven orbital planes with 11 satellites each, or a total of 77 operational satellites. This was the basis for the name of the system since the element Iridium has the order number 77.

There are four cross-link antennas, one pointing forward and one back along the orbital arc and two pointing sideways left and right. The first two are fixed while the second two are movable. There are two feeder-link antennas for redundancy.

The orbit is almost polar: the inclination is 86.4°. The main axis of the satellite body is in the direction of flight, and the solar panels extend east and west.

19.2.3 Antennas

The main mission antenna is a phased array consisting of 120 transmit/receive modules on a flat panel, shown in Figure 19.1. It produces 48 fixed beams in a cellular pattern, as shown in Figure 19.2. Only 32 of these beams can be used at the same time. The field of view is ±60° (Lafond et al., 2014). The north-south extent of this pattern is about 33° of latitude, corresponding to approximately 3650 km or 1/11 of the earth's circumference, so the 11 satellites per orbit do not leave any gap. As the satellites move toward the poles, some beams are switched off to prevent interference with other satellites of the constellation. As discussed in Section 11.7 it is a direct radiating array with a close spacing of the radiating elements (Lafond et al., 2014). According to the above reference, the element spacing in phased arrays is typically in the range of 0.55–0.8 wavelengths, depending on the beamwidth. Judging from pictures of the Iridium satellites the spacing must be approximately 0.58 wavelengths. Since the satellite orbit period is 100 minutes, hand-off to a new satellite happens at least every 7 minutes at the equator. Hand-offs between beams are required about every 50 seconds. While hand-offs to other beams are not noticeable to the user, satellite hand-offs lead to a quarter-second interruption (Wikipedia, 2019).

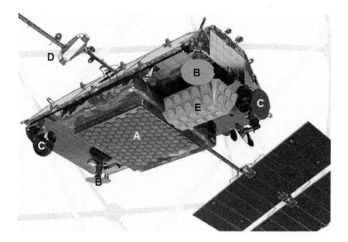

FIGURE 19.1 Iridium Next satellite (Used with permission of Iridium Communications Inc.).

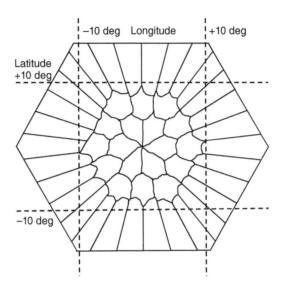

FIGURE 19.2   Iridium main mission antenna pattern. (Adapted from Buntschuh (2013), used with permission of Iridium Communications Inc.).

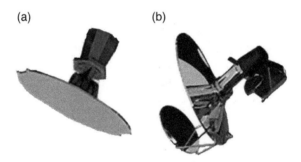

FIGURE 19.3   Iridium Next antennas (a) crosslink, (b) feeder link. (Extracted from Figure 19.1).

The K-band antennas for the crosslinks and the Ka/K-band antennas for the feeder links (FL) are dishes of approximately 40 cm diameter, as shown in Figure 19.3. The feeder link antennas each have a subreflector.

19.2.4   Radio Interfaces

There are three types of radio links on the satellite: the links from and to the users, the links from, and to the gateway stations and the intersatellite links. They will be described separately below.

We use the term "radio interface" instead of "air interface" in this chapter in order to be able to include satellite-to-satellite protocols.

19.2.4.1   *User Links*   Iridium Next seems to have two signal formats: one for communication with legacy user terminals and one for newer equipment. Since each satellite uses many carriers some may be used with old equipment and others with new equipment. We will focus on the new signal format.

The user links are in L-band, at 1616.0–1626.5 MHz, with time division duplex (Section 11.12.2) between the up- and downlinks. There are nominally 252 carriers in this band with a spacing of 41.667 KHz. The bandwidth of each carrier varies between 35 and 36 KHz. Frequency bands can be combined, up to 288 KHz, that is the equivalent of 8 carriers. The polarization is right-hand circular polarization (RHCP) for both links (Buntschuh, 2013).

Four user channels share a carrier in time-division multiple access (TDMA) (Section 13.2.2). The frame length is 90 ms, divided into an uplink-subframe and a downlink-subframe. Each user channel is assigned one uplink and one downlink timeslot of 8.27 ms. At the start of each frame is a 20.3 ms downlink simplex channel for pager messages and call alerts (Wikipedia, 2019). The frame structure is shown in Figure 19.4.

There are five carriers that are used for downlink communication only. One is for ring alert for phone calls and the other four are messaging carriers for paging and acquisition (Buntschuh, 2013). The phones monitor the ring alert channel when they are not engaged in a phone call and the ring alerts in the simplex time slots are only for phones already in a call. Only one of the four paging channels is used at any time. The ring alert channel is transmitted at a higher power level than the downlink traffic channels (Zehl and Schneider, 2016) because the phone must receive it without extended antenna and possibly while stored in a pocket or other disadvantaged location. The pager channel has even more power because pagers do not have big antennas, have to work indoors, and need a high probability of success because there is no feedback when the message transmission fails.

The modulation for the traffic channels is differentially encoded quaternary phase-shift-keying (QPSK) (Section 12.8.3). For acquisition and synchronization differentially encoded binary phase-shift-keying (BPSK) is used on the uplink (Wikipedia, 2019). Forward error correction is provided, with code rate of 4/5 for the lowest symbol rate (30 Ksps) and 2/3 at higher symbol rates (60 and 240 Ksps are listed in (Buntschuh, 2013)) The link budgets are laid out for a target bit error rate of 10^{-7} for up- and downlink (Buntschuh, 2013).

FIGURE 19.4   Iridium Next user link frame structure (Buntschuh, 2013). (Used with permission of Iridium Communications, Inc.).

TDMA requires time synchronization among the user terminals on the uplink. From the geometry of the antenna footprint, it can be estimated that without closed-loop timing control the time difference of signal arrival at the satellite could be as high as 7 ms. Since the guard time between bursts is only 0.233 ms the satellite sends timing error information to the user terminal (Zehl and Schneider, 2016).

The satellite motion induces a Doppler frequency shift which must be corrected. The user terminals have to handle up to ±45 KHz of frequency shift on the downlink and pre-correct the uplink frequency by the same amount (Buntschuh, 2013). Most likely, the frequency offset on the forward link is measured by the user terminal and the return link frequency is shifted by the same amount in the other direction. This should work well since both signal transmissions use the same carrier frequency. Also, the satellite communicates the uplink frequency error to the user terminal (Zehl and Schneider, 2016).

19.2.4.2 Gateway (Feeder) Links The FL operate in 19.4–19.6 GHz for the downlink and 29.1–29.3 GHz for the uplink. There are 13 channels each on the uplink and downlink spaced at 15 MHz with a bandwidth of 14 MHz. The assignment of uplink and downlink transponders is independent. The uplink polarization is RHCP and the downlink polarization is left-hand circular polarization (LHCP) (Buntschuh, 2013). The symbol rate is 11.7 Msps. The modulation is QPSK with rate 2/5 or rate 4/5 or 8PSK with rate 2/3 coding. For the uplink only, there is also 16APSK with rate 2/3 coding (Buntschuh, 2013). Each Ka/K-band feeder uplink/downlink transponder pair on a given satellite can support all of the traffic routed through that satellite (Buntschuh, 2013).

On this link, timing is no issue, but the frequency shift has to be corrected by the feeder station. On the uplink, the frequency shift is in the range ±750 KHz and on the downlink ±470 KHz (Buntschuh, 2013).

19.2.4.3 Intersatellite Links For the 4 intersatellite links there are altogether 8 transmit and receive channels in the 22.18–22.38 GHz band. The center frequencies of these transponders are spaced at 25 MHz and each transponder has a bandwidth of 21.6 MHz. Each transmit and receive intersatellite link can be assigned independently of the other links. Horizontal polarization is used for both transmit and receive. The symbol rate is 18 Msps, the modulation is 8PSK and the code rate is 2/3 (Buntschuh, 2013). A half-duplex protocol is used; that is a single frequency channel carries traffic in both directions like time-division duplex but the duration of transmission in one direction is not fixed. The transmitting side informs the other side when it relinquishes the channel. Such protocols are common for voice radio.

Timing is again no issue on this link. Frequency shift is not a major issue, either, because the two satellites in the same orbital plane have exactly the same speed and the satellites in the adjacent planes have a small speed relative to each other if they move in the same direction. Where they travel in opposite directions, between orbital planes 1 and 6, there is no crosslink (Buntschuh, 2013).

19.2.5 Payload Architecture

Since connections between users are routed through the space segment, a processing payload is required. All data streams are demodulated, decoded, and routed to the appropriate cross- or downlink by this processor which was produced by Seakr, Inc. Figure 19.5 shows its architecture. The four modem cards on the right perform the A/D and D/A and filtering, modem/codec functions, and routing. They process the receive and transmit signals of all links. There are three space-grade XILINX Virtex-5QV FPGAs on each card. According to Thales Alenia Space (2018), each one is dedicated to one of the three link types (user, gateway, and crosslink). The four modem cards can process a total RF bandwidth in excess of 600 MHz.

The three cards on the left coordinate the tasks on the modem/codec cards and perform higher-level L-band traffic routing. They are based on PowerPC (Medusa) microprocessors. The SpaceWire, a fault-tolerant, point-to-point interconnect protocol, provides connections from each processor card to each processor and modem/codec card. With this arrangement, tasks can be flexibly distributed based on the traffic load and any component failures can be bypassed (Murray et al., 2012). Figure 19.6 shows the processor hardware. The total processing capacity is approximately 1 TFlop. The modem is completely reconfigurable from the ground so that new modulation and coding formats can be uploaded in the future.

19.2.6 Services

Iridium Next offers voice service for handheld terminals and various data services for bigger terminals. There are currently no handheld voice terminals available that use the enhanced capabilities of the new satellite constellation and there is no

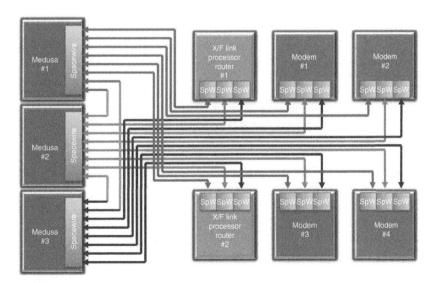

FIGURE 19.5 Iridium Next processor architecture. (© 2012 IEEE. Reprinted, with permission, from Murray et al. (2012)).

FIGURE 19.6 Iridium Next processor hardware (Seakr Inc., 2016). (Used with permission of SEAKR Engineering, Inc.).

indication from Iridium that this will change any time soon. The service offered now compresses the voice signal to 2.4 Kbps using the Advanced Multi-Band Excitation (AMBE) vocoder developed by Digital Voice System Inc. This voice encoder (vocoder) is tailored to the Iridium communication channel. It provides good quality audio performance with a nominal mean opinion score (MOS) of 3.5 under typical operating and channel conditions (Meza, 2006). This is surprisingly high, considering that Global System for Mobile Communications (GSM) codecs need at least 5.9 Kbps to achieve the same MOS, as shown in Figure 19.7[2].

The handheld terminals also offer SMS messaging and low-rate data services up to 13 Kbps (IEC Telecom, 2021).

The call setup procedures, mobility management, etc. are based on GSM protocols. The phones have a SIM, just like terrestrial mobile phones. A home location register (HLR; in the subscriber's home network) keeps track of the visited network of the subscriber and the visited location register (VLR) points to the actual location of the subscriber so that signaling and call data can be forwarded.

Iridium offers a short-burst data service which is optimum for messages of 270 characters for mobile-terminated messages and 340 characters for mobile-originated messages (Iridium, 2019b). (According to Iridium (2003) message size for mobile-originated is between 1 and 1960 bytes, and message size for mobile-terminated is between 1 and 1890 bytes. The author did not find an explanation of what "optimum" means in this context.)

The new service offering provided by the Iridium Next satellites is Iridium Certus (Iridium, 2018). It offers two service classes, Iridium Certus Midband,

---

[2] Acronyms used in the figure: AMR, advanced multi-rate; NB, narrowband audio (300–3400 Hz); WB, wideband audio (50–7000 Hz); HR, half rate; FR, full rate; EFR, enhanced full-rate.

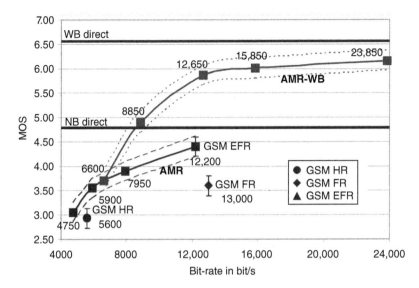

FIGURE 19.7 MOS of various GSM codecs. (© 2010 IEEE. Reprinted, with permission, from Rämö (2010)).

with data rates up to 134 Kbps, requiring smaller, low-gain antennas (diameter 20 cm) (Iridium, 2014), and Iridium Certus Broadband, with data rates of 176 Kbps–1.4 Mbps and an antenna diameter of 36 cm (Thales, 2020). It is not clear when the highest data rate will actually become available. As of spring 2020, the highest available rate is 704 Kbps (Iridium, 2020). The service can be used in parallel for a number of voice connections and for data applications, such as Internet access.

19.3 GLOBALSTAR

19.3.1 Overview

Globalstar is another LEO constellation for the provision of telephony and low-rate data communication services. The original constellation of 48 satellites was completed in February 2000. The satellite altitude was 1410 km, which means that they passed through the South Atlantic anomaly (Section 2.2.3). The radiation received by the satellites most likely affected the hardware and resulted in early failures of some forward-link transmitters. Globalstar contracted with Thales Alenia Space to produce a second generation of satellites. Between 2010 and 2013, 24 satellites of this design were launched into the same orbit. As of 2019, the 24 second-generation satellites are in operation. There are still eight first-generation satellites in the constellation (Union of Concerned Scientists, 2020). They may be used as in-orbit spares. No additional satellites have been ordered.

Coverage is not global, for three reasons: (1) the orbit inclination is 52°, which does not cover the poles, (2) the satellites have a bent-pipe architecture, so they must be in contact with one of the 24 feeder stations to provide service, and (3) with the new satellite generation the constellation has been thinned out. With the original constellation, the temperate regions were served by at least two satellites at any time, and all points on earth within ±70° of latitude were served by at least one satellite (Schiff and Chockalingam, 2000). For the new constellation, Globalstar claims that the service is available 75% of the time everywhere between latitudes 70° north and 55° south (Navarra, 2008).

The coverage area depends on the service and the user device. A voice coverage plot can be found on the Globalstar homepage. It distinguishes between primary coverage areas with good signal quality and fringe areas with weaker signals (Globalstar, 2014). The Americas, Europe, and Australia are all in the primary coverage area, but Africa and South Asia are not covered at all.

The 24 satellites are in 8 orbital planes with three satellites per plane (Globalstar, 2017). The orbits are circular with an inclination of 52°. The orbital period is 114.1 minutes (N2YO.ORG, 2019).

19.3.2   Spacecraft

The satellites, shown in Figure 2.7, are small and weigh only 700 kg. The earth deck of the spacecraft is rectangular and the four side panels slant away from it to attach to the smaller anti-earth deck. The purpose of this shape, instead of a rectangular box, is to allow six spacecraft to fit into one launch-rocket fairing. Four satellites are on a lower deck and two more are on an upper deck. The launch vehicle was a Soyuz-2 with a Fregat upper stage.

19.3.3   Antennas

The antennas are clearly visible in Figure 2.7. The two pyramids of phased arrays are part of the S-band transmit antenna for the user link. They form 16 beams in the configuration shown in Figure 19.8a. The sides of the taller pyramid, slanted at an angle of 48° off nadir, form the outer ring of nine beams and the horn in the center forms the center beam. The shorter pyramid, with panels slanted at 27° off nadir, forms the inner ring of six beams (Croq et al., 2009). The X and Y designations refer to the mapping of the user links onto the gateway links and need not be discussed here. The transmit power of the beams increases from the center of the pattern to its edge to compensate for the increased range. The antenna is strictly passive. In Globalstar 1 the power amplifier was integrated into the antenna; hence, it was exposed to radiation, which may have caused the early failures of some of them.

The flat phased array at the bottom is the receive antenna for the L-band user link. It also forms 16 beams, but the pattern is different as shown in Figure 19.8b. Again, the X and Y designations indicate the mapping onto the feeder link. The antenna includes 52 radiating elements spaced at 0.6 wavelengths and the same

(a)   (b)

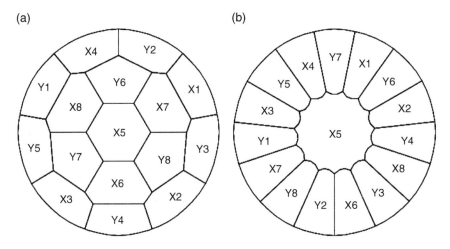

FIGURE 19.8 User link antenna patterns: (a) transmit, (b) receive (Navarra, 2008).

number of filters and LNAs on the top of the antenna structure. The beam-forming section is on the bottom of the antenna structure (Croq et al., 2009).

The field of view of both user link antennas is 108° (Croq et al., 2009), which provides coverage of any point on the visible earth with an elevation angle of 10° or higher.

The two small dishes in the bottom right are the antennas for the C-band gateway link. They provide earth coverage (Navarra, 2008).

19.3.4 Radio Interfaces

Globalstar uses direct-sequence code-division multiplexing (CDM) for multiplexing the forward link signals. A general description of this scheme is given in Section 1.3. The particular scheme used by Globalstar is the one that was defined in the terrestrial mobile radio standards IS-95 and CDMA2000, which were used in some countries in the early 2000s but are no longer followed by any public networks. Globalstar adapted this standard to satellite networks because it simplified the development of dual-standard phones which could operate in a terrestrial network when available and switch to satellite service when not.

The particular type of CDM used in Globalstar is based on Walsh sequences, which are orthogonal sequences. Hence, by multiplying each symbol by a full period of the Walsh sequence (128 chips in the case of Globalstar) the signal has zero correlation with any of the other signals, as long as the two signals are synchronized. There are 128 Walsh sequences of this length, hence up to 128 signals can be transmitted free of interference. To protect against interference from other beams or satellites, the signals of each beam are also multiplied by a unique PN sequence with the same chip rate. These other signals are not completely suppressed,

TABLE 19.1 Globalstar Frequency Bands

Link	Forward Link Band (MHz)	Return Link Band (MHz)	Polarization
Satellite – user	2483.5–2500	1610–1626.5	LHCP
Satellite – gateway	5091.0–5250	6875–7055.0	RHCP/LHCP

but their power is attenuated on average by the ratio of chip rate to data symbol rate, which is called the processing gain. In our case, the interfering signals are attenuated by 21 dB.

The frequency assignments for the user and gateway links are shown in Table 19.1 (Navarra, 2008; Long and Sievenpiper, 2015). The user uplink is in the L-band, the user downlink in the S-band, and the gateway up- and downlinks in the C-band. The bandwidth available on the user links is 16.5 MHz in each direction. On the gateway links, 159 MHz are available on the forward link and 180 MHz on the return link. The full user link bandwidth is available for each beam and all use the same polarization. The separation of the user signals in a beam is based on the spreading code, not on frequency, polarization, or time (see above and in Section 1.3). The gateway link carries a frequency multiplex of eight user link signals on each of the two polarizations, so the signals of all 16 beams can be transmitted. The reason for the larger return link bandwidth is most likely that the return link signals have Doppler frequency shifts.

The 16.5 MHz bandwidth is divided into 13 subchannels of 1.23 MHz. The first nine subchannels are full-duplex (forward and return link) channels while the last four are only forward link channels. The purpose of the latter is not explained. They may be used for call alerts to idle user terminals. Globalstar does not offer paging services like Iridium.

For Globalstar, the chip rate is 1.2288 Mcps and the data rate for the original Globalstar is 9.6 Kbps. The data are protected by a rate-½ constraint-length 9 convolutional code, resulting in a symbol rate of 19.2 Ksps. User rates of less than 9.6 Kbps are also available, which are increased to 19.2 Ksps after convolutional coding through repetition coding, that is a repetition of source symbols. The binary symbol stream is then multiplied by the Walsh sequence. Then the same symbol stream is multiplied by two different PN sequences and modulated onto quadrature carriers to produce a QPSK signal. This is shown in Figure 19.9.

The return link also uses a direct-sequence spread-spectrum signal with the same chip rate and the same forward error-correcting (FEC) coding. However, the maximum data rate is only 4.8 Kbps and, instead of binary signaling, it combines six bits to select one of 64 orthogonal Walsh sequences (Schiff and Chockalingam, 2000).

Globalstar 2 offers higher data rates than the 4.8 or 9.6 Kbps of Globalstar 1. According to Globalstar (2017), the maximum data rate is 256 Kbps. There are

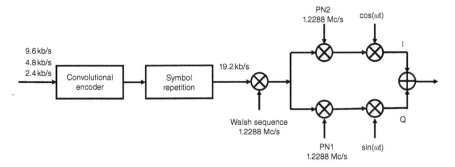

FIGURE 19.9 Globalstar forward link modulator. (Adapted from Schiff and Chockalingam (2000) but much simplified. Reprinted by permission from Springer (© 2000)).

various ways of achieving the higher rate. IS-95 assigns more than one spreading code to the user terminal, so a high-rate user looks like more than one user. Having 25 despreaders/demodulators in the receiver seems excessive, however. It seems more likely that some subbands are reserved for high-rate users and provide a higher data rate, combined with a lower processing gain. Another approach would be higher-order modulation. Any combination of these solutions is possible.

CDM has two advantages over frequency-division multiplexing (FDM) and time-division multiplexing (TDM):

1. In a multipath environment, that is if in addition to the signal from the line-of-sight path there are delayed signal copies from distant reflecting objects such as mountain sides, FDM, and TDM systems suffer from signal distortion. The despreader of a CDM system suppresses the delayed copies of the signal if the delay exceeds one chip duration, so it does not suffer this degradation. If the receiver uses a so-called Rake-receiver, with multiple despreaders, the reflected signals can be constructively added to the direct signal and lead to improved performance. This is a major plus for terrestrial mobile radio systems and may be useful for Globalstar users at the edge of coverage where the elevation angle is close to 10°. The user terminal antenna patterns are basically hemispherical, so echoes can arrive from any side.
2. If a user terminal can receive signals from more than one beam, from the same satellite or a different one, the gateway can transmit its signal through all these beams. The user terminal can combine the signals in the same way as described above. This is used for the handover of a connection from one beam to the next.

The above description applies to the forward link. However, the same method can be applied to the return link. Globalstar uses this approach and can provide a smooth handover between beams and satellites.

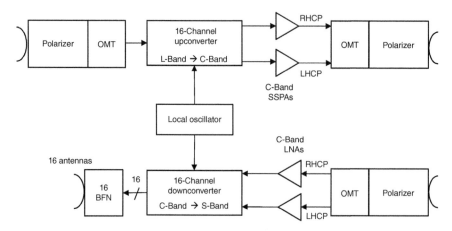

FIGURE 19.10 Simplified block diagram of Globalstar payload. (adapted from Dietrich et al. (1998). (© 1998 IEEE. Reprinted with permission).

19.3.5 Payload Architecture

The payload of the Globalstar satellites has a bent-pipe architecture. A simplified diagram of the repeater is shown in Figure 19.10. The L-band LNAs are not shown because they are integrated into the phased array antenna, making it an active array. The S-band signal, on the other hand, is amplified by a multiport amplifier (Section 11.12.3) which is placed inside the spacecraft body in order to provide more radiation shielding (Globalstar, 2017; Croq et al., 2009). The multiport amplifiers allow for flexible power sharing between the 16 beams, depending on the user distribution.

There are separate antennas for receive and transmit at C-band.

19.3.6 Services

Both system generations provide the same kind of services, that is voice, two-way data, and one-way messaging. Dual-mode phones (satellite and terrestrial network) are no longer offered. The maximum data rate available as of this writing is 72 Kbps with the Sat-Fi2 satellite hotspot (Globalstar Inc., 2020). The messaging service can be used to track persons and goods.

19.4 O3B

19.4.1 Overview

The O3b system was designed and is being operated by O3b Networks Ltd., which is a wholly owned subsidiary of SES S.A. The system provides communication (Internet and cell phone backhaul) services to the underserved "other three billion" people in the equatorial regions of the earth to whom low-priced high-speed fiber, microwave, or satellite connections are unavailable (Barletta, 2012) Its customers are telecom operators and Internet service providers rather than individual consumers. The latitudes served are within 50° of the equator (SES, 2019).

As of 2020, the system consists of 20 satellites in MEO equatorial orbit at an altitude of 8062 km. Their revolution is in the same direction as the earth's. The ground period is 360 minutes or 6 hours (Barletta, 2012) and the actual period is 288 minutes. Four of these satellites are in-orbit backups (Rosenbaum, 2016). Three of the backups are from the first batch of four satellites and have degraded signal characteristics (Spaceflight 101, 2019).

In September 2017, SES announced the next generation of O3b satellites and placed an order for an initial seven from Boeing Satellite Systems. Expected to launch in 2021, the O3b mPower constellation of MEO satellites for broadband internet services will "be able to deliver anywhere from hundreds of megabits to 10 gigabits to any ship at sea through 30,000 spot beams". Software-defined routing will direct traffic between the mPower MEO satellites and SES's geostationary fleet (Henry, 2018). The number of spot beams probably applies to the whole fleet, (SES, 2020) mentions 5000+ spot beams per satellite. Also, it is not clear whether all these beams are active in parallel or whether beam-hopping is used.

In June 2018, SES announced that it had FCC approval to increase the size of its O3b fleet to a total of 42 satellites. Some of the new satellites will be in inclined orbits, so that the constellation will provide service at all latitudes (Henry, 2018).

This section focuses on the operational O3b satellite communication system as of 2020, that is the system based on the 20 satellites manufactured by Thales Alenia Space and nine gateway earth stations.

Each satellite is equipped with 10 steerable spot beam antennas for user beams, with a throughput of 1.25 Gbps (uplink plus downlink) each, for a total throughput of 12.5 Gbps. Two more antennas are for the gateway links.

19.4.2   Spacecraft

The spacecraft has the general shape of the Globalstar-2 LEO spacecraft and claims heritage from it (O3b, 2009). Figure 19.11 shows a simplified representation of the satellite and a photograph is provided in Figure 2.8. Its dry mass is 700 kg. Twelve reflector antennas are mounted on the largest spacecraft surface, the earth deck. The long direction of this rectangular surface is oriented east-west on orbit.

19.4.3   Antennas

The 12 antennas, visible in Figure 19.11, are identical (Amyotte et al., 2010). Two antennas are used for gateway links, while the other 10 support user links. These antennas will move constantly to track the spots on the earth they are intended to serve. They could potentially point at any time to anywhere within 26° of the satellite's nadir direction, which covers the entire visible earth.

FIGURE 19.11 O3b spacecraft (Barletta (2012)). (Used with permission of SES).

The diameter of the antennas is 27 cm (SatSig, 2016). At 18 GHz the diameter of the resulting beam at nadir is 700 km at the −4 dB point. This matches the value given in Spaceflight 101 (2019).

The antennas of earlier launches are not the same as those of later ones. An early O3b antenna (Amyotte et al., 2010) is shown in Figure 19.12. The feed does not move, just the dish, so there is no need for rotary or flex waveguide. The developer, MDA of Canada, designed the antenna to minimize crosspol generation. The feed horn is unusually compact. The waveguide and bends are non-standard, with some of the waveguide serving as struts. The diplexer is behind the reflector. Performance is good, and relatively constant across the whole steering range of ±26° in each direction. In Figure 2.8 the later antenna model is visible, which also seems to have

FIGURE 19.12 O3b steerable antennas, early model (Amyotte et al., 2010). (Used with permission of MDA Corporation. COPYRIGHT © 2020 MacDonald, Dettwiler and Associates Corporation and/or its affiliates, subject to general acknowledgments for the third parties whose images have been used in permissible forms. All rights reserved.).

a fixed feed. It has an offset subreflector which seems to be fixed to the main reflector.

Obviously, there is no need to hand connections over from one beam to the next of the same satellite since each antenna generates a single beam which can be pointed anywhere within the field of view. However, there is a handover from one satellite to the next. All connections are handed over simultaneously. Users and gateways have two antennas, one pointing at the old satellite, the other at the new one. In a first step, the forward link signal is transmitted through both satellites, then the return link signal also. Finally, the signals through the old satellite are switched off (Moakkit, 2014). There is no break in service during the handover. The satellite that just handed over its user connections has already a link to the next gateway at that point, through its second gateway antenna. It is ready to start serving the next set of users once the antennas for user links have been repositioned according to the new user locations.

19.4.4 Services

Before the radio interface can be described we have to describe the services that are offered since the radio interface is not identical for all services.

There are currently two groups or tiers of users. Tier 1 comprises national telecom operators and large Internet service providers. They have large base stations with 3.5 m antennas and they are served by a whole beam with data rates of about 600 Mbps in each direction. Tier 2 users are cellular operators which link their base stations through the satellite to the rest of the terrestrial infrastructure (cellular backhaul) and enterprise VPNs. They use antenna diameters of 1–2 m and have data rates up to 155 Mbps. Typical services use symmetrical data rates of 2–10 Mbps. Also in tier 2 are ship terminals with antenna diameters of 1.2 or 2.2 m and data rates up to 155 Mbps (Read, 2013). Such terminals use regular dish antennas and cost at least USD 100,000 (Farrar, 2016).

There may be a future tier 3 service for consumers and small businesses, with antenna sizes 50–100 cm and data rates of 1–2 Mbps on the forward link and 256–510 Kbps on the return link (Sihavong, 2009).

With dish antennas, these satellites are unable to do beam hopping. Hence, only customers within the 10 circles of 700 km diameter can be served at any time. This limitation will presumably be solved with the mPower satellites.

The design of the current generation of satellites requires that all links be between a gateway station and a user. So, one-hop very small-aperture terminal (VSAT) networks are not an option.

19.4.5 Radio Interfaces

The frequency bands licensed by the FCC to O3b are shown in Table 19.2 (Dortch, 2018). Four different satellite generations are shown: Satellites 1–16 are the ones in operation as of 2020. They are discussed in this book. Satellites 17–20

TABLE 19.2   FCC Frequency Allocation for O3b (Dortch, 2018)

O3b Satellites 1–16	O3b Satellites 17–20	O3b Satellites 21–30	O3b Satellites 31–42
(a) Space-to-earth (downlink bands)			
17.8–18.6 GHz (FSS)	17.8–18.6 GHz (FSS)	17.8–18.6 GHz (FSS)	17.8–18.6 GHz (FSS)
18.8–19.3 GHz (FSS)	18.8–19.3 GHz (FSS)	18.8–19.3 GHz (FSS)	18.8–19.3 GHz (FSS)
		19.3–19.7 GHz (MSS FL)	19.3–19.7 GHz (MSS FL)
	19.7–20.2 GHz (FSS/MSS)	19.7–20.2 GHz (FSS/MSS)	19.7–20.2 GHz (FSS/MSS)
			37.5–42.0 GHz (FSS)
(b) Earth-to-space (uplink bands)			
27.6–28.4 GHz (FSS)	27.6–28.4 GHz (FSS)	27.5–28.4 GHz (FSS)	27.5–28.4 GHz (FSS)
28.6–29.1 GHz (FSS)	28.6–29.1 GHz (FSS)	28.4–29.1 GHz (FSS)	28.4–29.1 GHz (FSS)
		29.1–29.5 GHz (MSS FL)	29.1–29.5 GHz (MSS FL)
	29.5–30.0 GHz (FSS/MSS)	29.5–30.0 GHz (FSS/MSS)	29.5–30.0 GHz (FSS/MSS)
			47.2–50.2 GHz (FSS)

add MSS to the offering in an additional frequency band. There is no indication of how this band would be used: with the current bandwidths on the user links and the current number of antennas it does not provide additional capacity.

Satellites 21–30 will be in two orbit planes with 70° inclination (Rosenbaum, 2017) and they add capacity to the feeder link (FL). This will expand coverage all the way to the poles. Satellites 31–42 will be in equatorial orbit. They add links to GEO satellites owned by SES (Henry, 2017). It should be emphasized that the table only shows the frequency bands licensed for service to the United States. It may not fully reflect the capability of the satellites or the operational frequency bands elsewhere. According to Dortch (2018), satellites 21–42 will be able to use the entire 17.7–20.2 GHz band.

The rest of this section focuses on the satellites 1–16 which use just one frequency band each for uplink and downlink. The same frequency bands are used for user and gateway links. Both circular polarizations are used, but only one for each antenna. In total there are 5 subbands of 216 MHz each in the above bands with a separation of 44 MHz and a 200 MHz gap between the third and fourth subband (O3b, 2009). Each user link uses one of these subbands and the gateway links use all five. The separation between user and gateway links is provided by not pointing two antennas with the same polarization to the same location.

For tier 1 users the signal format in both directions is based on the DVB-S2 standard (Barnett, 2013a) which is described in Section 13.4.1. There is a single carrier on a beam serving such a user. According to Barnett (2013a), the signal format for tier 2 users is also based on DVB-S2. However, the sample link budgets do not provide a clear picture of how signals are multiplexed. While the budgets in the above reference show the same number of carriers in both directions, which seems to indicate that the signals for individual users are on separate carriers in both directions, there are others that show fewer forward link than return-link carriers (Barnett, 2013b). This suggests that the forward link uses a mix of FDM and TDM.

The beams between gateways and satellites are always a frequency-multiplex of five user beams.

DVB-S2 uses pulse shaping with a spectrum of raised cosine. From some sample link budgets (Barnett, 2013a), we calculate that O3b uses the rolloff-factor (or excess bandwidth) α of 0.2, the smallest of the DVB-S2 values. Adaptive coding and modulation (ACM) is used in both directions. The modulation format goes all the way to 32APSK (Barnett, 2013a).

The small roll-off factor means that the signals are closely spaced and the frequency shift due to Doppler may be a concern. However, since all signals in one user beam originate from roughly the same geographic area, they have practically the same Doppler shift. There is only a frequency offset between the different subbands on the return-link downlink.

19.4.6 Payload Architecture

The payload has a bent-pipe architecture, so it provides one-hop transparent communications between user installations and a gateway (Rubin, 2012).

The high-level payload architecture of the spacecraft is shown in Figure 19.13. Details of the hardware are provided in the following subsection. This payload diagram was derived from the channel and beam plan given in Read (2012). Each antenna uses one circular polarization for both transmit and receive. The payload has two identical halves, each for half of the antennas. It is quite possible that two antennas, on orthogonal polarizations, communicate with the same gateway or user. Each payload half uses the same frequency plan, in which the user beams use non-overlapping frequency bands and the gateway beam uses the five multiplexed bands. In principle, it is possible for all 10 user beams to serve the same user. One local oscillator (LO), split, is used for all up- and down-conversions. The payload is rather simple.

A failure that would render half of the represented payload useless would be the failure of one of the two receivers for the gateway signals, since the receivers appear to be provided with no redundancy (see Section 19.4.7). So probably there are switches in the payload that allow user-beam hardware chains to be switched for gateway-beam hardware chains. The user beams provide redundancy automatically since they can all point to any spot within their identical field of view.

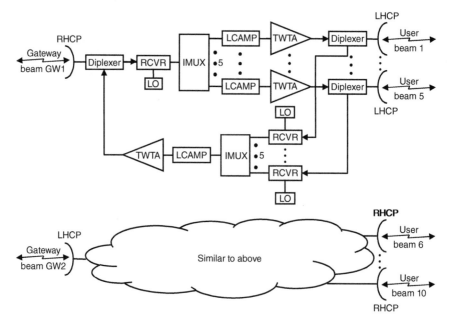

FIGURE 19.13 O3b high-level payload diagram.

19.4.7 Repeater

The repeater, namely the payload minus the antennas, uses at least some Thales hardware: the receivers (TAS, 2012a), the linearizer-channel amplifiers (LCAMPs) (TAS, 2012b), and the traveling-wave tube amplifiers (TWTAs) (Thales, 2012).

The Ka-band receivers (Figure 19.14) are built in assemblies of six as shown in TAS (2012a). From the number of such assemblies delivered for O3b (Barletta

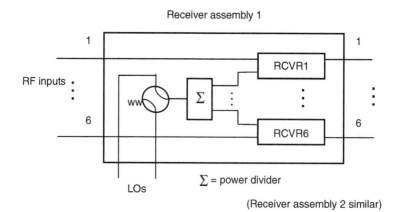

FIGURE 19.14 Details of receiver.

et al., 2013), it appears that no redundant units are provided in a payload. Receiver performance data are as follows: at ambient the noise figure is 2.1 dB; the gain is temperature-compensated; and the bandwidth is 1.5 GHz, wide enough to cover the initial-constellation combined frequency bands.

The same LO (with redundancy) is used for all the frequency conversions. An unusual thing about the LO generation unit is that it also provides the secondary voltages for the LCAMPs (Barletta et al., 2013).

The LCAMPs, all at K-band, probably have a bandwidth of 2 GHz, enough for satellites 1–16, but not to cover the new FSS/MSS band planned for later satellites. It is nowhere stated what LCAMP Thales used for the O3b satellites but 2 GHz was a typical bandwidth of its K-band LCAMPs at the time (TAS, 2012b).

Thales' K-band traveling-wave tubes (TWTs) can have a bandwidth of 2.5 GHz (Thales, 2012), which would be enough to cover O3b's K-band needs for all four satellite generations.

The linearizer plus TWTA forms the linearized traveling-wave tube amplifier (LTWTA). It appears that O3b expects to normally run the forward-link LTWTAs with output power backed off (OBO) about 4 dB from saturated output power, even for one carrier (Barnett, 2013a). This makes sense since the modulation order can be up to 32 (Barnett, 2013a). The nonlinearity of a LTWTA at such an operating point is remarkably small, for example typically at 4 dB OBO the noise-power ratio (NPR) is 21 dB and the ratio of carrier-to-third-order intermod (C/I3) is 34 dB (TAS, 2012b). For the return link, where the five user signals share the high-power amplifier (HPA), the OBO is about 6 dB (Barnett, 2013a).

19.4.8   Ground Segment

19.4.8.1   Gateway Earth Stations   The gateway earth stations provide the interface to terrestrial networks. They are located in places with good access to fiber-optic cables, which is typically the case where submarine cables terminate. An additional consideration for the siting was low average and peak rain losses (O3b, 2009). There are currently nine gateways: Sunset Beach, Hawaii (USA), Vernon, Texas (USA), Lima (Peru), Hortolandia (Brazil), Lisbon (Portugal), Nemea (Greece), Karachi (Pakistan), Perth (Australia), Dubbo (Australia).

The gateways use two 7.3 m antennas to support one satellite. The terrestrial interface is internet protocol (IP)-based. The IP router keeps track of the user IP addresses it is serving for each satellite it is connected to and routes the packets to the appropriate satellite beam.

19.5   ONEWEB

19.5.1   Overview

The OneWeb satellite constellation is owned and operated by the company of the same name. It was originally called WorldVu Satellites Limited. The satellites are licensed by the communications regulatory authorities of the United Kingdom. After going bankrupt in 2020 its major shareholder is now the United Kingdom.

The system concept has evolved over the years, which is reflected in four separate license applications to the FCC between 2016 and 2018. The long-term plan is to have one satellite constellation in LEO and one in MEO: The LEO constellation would consist of 1980 satellites (36 orbital planes with 55 satellites each) in near-polar (inclination 87.9°) circular orbits at an altitude of 1200 km, corresponding to an orbital period of approximately 109 minutes. They will be communicating with the user terminals in the Ku- and V-bands and with the gateways in the Ka/K- and V-bands. As of this writing, 720 satellites are being built and launched, which will be part of this LEO constellation but which do not communicate in the V-band. They will be in 18 orbital planes with 40 satellites each (Barnett, 2016).

The MEO constellation would consist of 2560 satellites (32 orbital planes with 80 satellites each) in orbits at 44–46° of inclination and an altitude of 8400–8600 km, corresponding to an orbital period of approximately 5 hours. They would communicate with the user terminals in the Ku-, Ka/K-, and V-bands and with the gateways in the Ku-, Ka/K-, V-, and E-bands (Barnett, 2018). While in bankruptcy, OneWeb applied to the FCC to increase the constellation size to almost 48,000 satellites (Weimer, 2020). Under the new ownership, the plans are likely to change.

Our discussion will be limited to the first-generation LEO system; that is the satellites without V- and E-band links. The source for all the material in the rest of this section is Barnett (2016), unless noted otherwise.

According to a corporate presentation, the constellation will go live in 2021 with 588 active satellites in 12 orbital planes of 49 satellites each, plus in-orbit spares, for a total of 650 satellites (OneWeb, 2019a). This smaller constellation provides already full global coverage according to the presentation, so it can be assumed that the additional satellites will be put into service as required when the demand justifies it. The schedule will most likely slip, however.

19.5.2 Spacecraft

The satellites are small, approximately mini-refrigerator-size and weighing roughly 150 kg with half of it being payload. They use all-electric propulsion (Henry, 2019a). An artist's rendering is shown in Figure 19.15 (Airbus, 2020). The solar panels generate 50 W of power. Since the satellites are in near-polar orbits they must be able to rotate and tilt. The tilting mechanism is not visible in this picture. The two small dishes are part of the gateway antennas and the structures visible on the earth-facing side are the user link antennas. Both will be discussed below.

The satellites are manufactured by OneWeb Satellites, a joint venture of OneWeb and Airbus. The large number of satellites planned for this constellation called for new production methods: The facilities near Kennedy Space Center in Florida use aircraft-style mass production techniques. The weekly production capacity is given as 15 satellites (OneWeb Satellites, 2020). The production costs are estimated at

FIGURE 19.15   OneWeb satellite. (Used with permission of Airbus).

one million USD (New York Times, 2020). The satellites are launched in batches of 32 aboard Russian Soyuz rockets operated by Europe's Arianespace Consortium (Gebhardt, 2020)

19.5.3   Antennas

The satellite has two gateway antennas so that it can establish a link to the next gateway station before tearing down the link to the last one. The minimum elevation angle for the gateway station is typically 15°. Each antenna has a subreflector which is configured such that its rotation axis is aligned with the feed and such that the image of the feed is at the focus of the main reflector. By aligning the main reflector rotation axis with that of the image of the feed (behind the subreflector), the whole system remains perfectly focused, no matter how the subreflector and main reflector are rotated, thus achieving the wide scan angle (Amyotte, 2020). Section 3.5 has further discussion.

Downlink power control keeps the power flux density on the ground constant when the beam is moved from nadir to the edge of coverage. Based on the antenna footprint (Barnett, 2018) the antenna diameter can be estimated as 40 cm. These antennas are discussed in Section 3.5.

The Ku-band antenna ("Venetian blind antenna") for the user links is discussed in Section 11.10.3. The beam-pattern is shown in Figure 19.16. The ovals are the −3 dB-curves of the 16 linear arrays.

The user-link antennas are fixed to the earth-pointing face of the satellite and therefore their pointing directions are fixed relative to the pointing direction of the

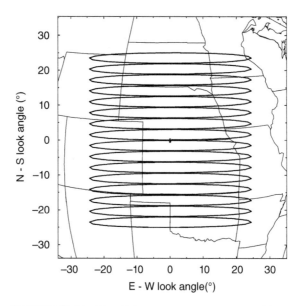

FIGURE 19.16   User link antenna pattern (Barnett, 2016).

satellite. However, the attitude control system of the satellite allows the pointing of the satellite to be adjusted so the beam pattern can be moved in the pitch direction (i.e., north–south), relative to the nominal nadir pointing direction. This feature, called Progressive Pitch™ (OneWeb, 2019b), is used to avoid interference to GEO satellites at latitudes close to the equator (Barnett, 2016). As the satellite approaches the equator, it is pitched more and more so that the area around the equator can be served without creating interference to users of GEO services. At the same time, some of the trailing beams are shut off for the same purpose. When the satellite is very close to the equator its service is shut off completely and it adjusts the pitch in the other direction before service is restarted. Since the constellation provides full-earth coverage, the two neighboring satellites in the same orbit have to provide the missing coverage when a satellite is turned off. When a satellite is exactly over the equator, its neighbors have to pitch by about 7°, so that each one shifts one edge of its coverage area to the equator. Obviously, their neighbors in the orbital plane have to shift their coverage areas, too, to cover areas no longer covered due to the pitching and the turning off of trailing beams of the first satellite.

Since the satellites travel almost exactly north or south, the beam supporting a particular user will change rapidly. Based on the orbital period of 109 minutes, a single beam provides support for no more than 12 seconds. This switching is unlikely to affect the continuity of the service but it will require some control traffic to the user terminal since it has to change channels. Turning the antenna by 90° would increase the connection time dramatically, but it is not an option since as part of the progressive pitch maneuver some beams need to be turned off.

The minimum elevation angle for user terminals is generally greater than 50°. It varies with latitude because of the progressive pitch of the satellites and the lower satellite density near the equator.

19.5.4 Services

As of this writing, the company does not offer any services yet. The Vice President of Regulatory Affairs of OneWeb defined the future service as follows: "highspeed broadband connections to Internet anywhere in the world, from a plane or a ship as well as on land" (Pritchard-Kelly, 2019).

The addressed markets include mobility (maritime, aviation, land mobility, Internet of things), satellite broadband (corporate enterprise, small and medium businesses, consumer residential), government (emergency response, local government, civil defense), and cellular backhaul (OneWeb, 2019a). Mobile operators should be able to offer speeds of up to 50 Mbps with a latency of less than 50 ms to their Internet users (Barnett, 2016).

19.5.5 Radio Interfaces

The frequency bands used by OneWeb are shown in Table 19.3. The link from the gateway to the satellite accommodates eight subbands of 250 MHz in a total of 2 GHz of bandwidth (the TT&C channel is at the low end of the frequency range shown in Table 19.3). Both circular modulations are used, so there are in total 16 subchannels which carry the signals for the 16 user links. Both gateway antennas use the full frequency band but they always point in different directions, so there is no interference problem. There are eight subbands on the user downlink, all using RHCP, so each frequency is used twice. The spatial frequency re-use pattern differs among OneWeb satellites.

The return uplinks have a total bandwidth of 1 GHz, which is divided into eight subbands of 125 MHz, all LHCP, so each subband is used twice. Each group of eight subbands is modulated onto one polarization of the downlink signal to the gateway. In the United States, the frequency band 12.75–13.25 GHz (user to satellite) is not available, so each frequency is re-used four times. The frequency band 19.7–20.2 GHz (satellite to gateway) is not used either, but that still leaves enough bandwidth for the 16 channels.

TABLE 19.3   OneWeb Frequency Bands

Link	Frequency Ranges (GHz)	Bandwidth (Ghz)
Gateway to satellite	27.5–29.1, 29.5–30.0	2.1
Satellite to gateway	17.8–18.6, 18.8–19.3, 19.7–20.2	1.8
User terminal to satellite	12.75–13.25, 14.0–14.5	1.0
Satellite to user terminal	10.7–12.7	2.0

The forward link's user symbol rate on each subchannel is not given. It is probably about 200 Msps to allow for a modulation roll-off of 0.2. The signal consists of a single TDM carrier which carries the data for all active users in a beam. ACM is used. The entire description suggests that the DVB-S2 standard is being followed, probably in its time-slicing mode. This also opens the opportunity to use the fairly inexpensive DVB-S2 components in the mass-produced user terminals. The return channels use MF-TDMA with channel bandwidths in the range of 1.25–20 MHz.

As pointed out above, the satellite antenna's gain variation is about 4 dB over the field of view. In addition, there is about one dB of range variation. The user terminals probably transmit at a constant equivalent isotropically radiated power (EIRP), so the ACM has to take care of the signal power variation. In the forward direction, the gateway station adjusts the EIRP for each burst to compensate these losses.

19.5.6   Payload Architecture

The satellites are pure bent pipes, so the payload is relatively simple. Figure 19.17 shows a likely realization based on the description in the FCC filing (Barnett, 2016). On the forward link, one of the two dish antennas receives the signal from the serving gateway station. The polarizer and orthomode transducer (OMT) separate the two polarizations and the ganged switches forward the two signals to the LNA and downconverter bank. The input multiplexers (IMUXs) split each signal into the signals for eight of the 16 user link antennas. The signals are then amplified by solid-state power amplifiers (SSPAs) and radiated through the Venetian blind antenna.

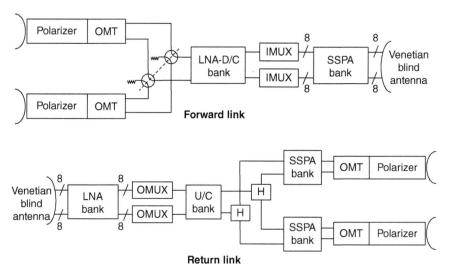

FIGURE 19.17   OneWeb payload diagram.

There is a statement in Barnett (2016) which suggests that the payload may be more complex than assumed here: "in the unlikely event of a case where the FS interference in the 10.7–11.7 GHz band is problematic, for that specific location one option may instead be to use mostly the 11.7–12.7 GHz band for service to the user terminal." This would mean that the 16 user channels can be re-routed to different antennas on the fly.

On the return link, the 16 signals from the Venetian blind antenna are amplified by LNAs and combined into two signals for the downlink by the output multiplexers (OMUXs). Then, they are upconverted to Ka-band and amplified. The hybrids send the composite signals to both polarizers and the antennas, whereby all SSPAs are active simultaneously only for a very brief time during handover.

While the FCC application explicitly says that the return link is briefly active to both gateways, there is no such statement for the forward link.

19.5.7 Ground and User Segments

Since the satellites have no crosslinks, they must have a link to a gateway station wherever they provide service. OneWeb foresees 50 or more gateways across the globe to ensure that the satellites have a visible gateway station from all parts of their orbits. The antenna diameter is typically 2.4 m which results in a beamwidth of less than 0.5°. There may be more than 10 antennas at each site. The minimum elevation is typically 15°. Antenna farms with 10 antennas or more could greatly benefit from phased array technology, with one array producing many beams. OneWeb has a deal with ThinKom for the provision of such gateway antennas (Holmes, 2020).

The Ku-band user terminals will have small antennas, typically in the 30–75 cm range. In the short term, they will use mechanically steered parabolic reflectors. This technology is expensive, requiring two dishes if continuous communication is required. They will eventually be replaced by phased arrays. According to de Selding (2015), OneWeb is developing a user terminal antenna which combines mechanical steering and a phased-array antenna and measures 36 cm by 16 cm. The terminal would provide Internet access at 50 Mbps. The minimum elevation angle for user terminal antennas is about 50°, so the scan angle for a phased array is no more than 40°.

19.6 STARLINK

19.6.1 Overview

Starlink is the name of a constellation of non-geostationary satellites being put into operation by SpaceX. The complete constellation may have close to 12,000 or even 42,000 satellites at various altitudes. It is being deployed in stages.

The first stage, being launched as of this writing, consists of 1584 satellites at an altitude of 550 km. The orbit inclination is 53° (Albulet, 2019). The satellites will be in 72 orbital planes with 22 satellites each (Albulet, 2019). Coverage is not global. Limited commercial service will start once the constellation consists of 800 satellites (Albulet, 2016). Limited beta-testing by users located at latitudes around 50° started in August 2020. The constellation consisted of 720 satellites at that time and provided continuous coverage at this latitude.

The next step would add 2825 satellites at similar altitudes. Of these satellites, 1600 would be in orbits with a 53.8° inclination and the rest would be in orbits with 70°, 74°, and 81° inclinations (Albulet, 2016; Henry, 2020a). This constellation would provide global coverage. The satellites in these first steps would communicate in the Ku and Ka/K bands (very early satellites only in the Ku band) (Albulet, 2018).

Next, 7518 satellites at altitudes between 335 and 345 km would be added. At that stage, V-band frequencies would be added for user and gateway links. Satellites of the first and second generation would be replaced by satellites which also communicate in all of the above bands when they are retired after five years of operation (Albulet, 2017a).

This is as far as the FCC filings go. However, in 2019 the ITU received 20 filings on behalf of SpaceX, each for a set of 1500 satellites, all at altitudes between 328 and 580 km (Henry, 2019b). No information is available about the characteristics of these satellites.

The services to be provided are residential Internet access as well as services for commercial and government organizations. Data rates up to 1 Gbps per user are to be offered (Albulet, 2016). Service should start in late 2020. SpaceX has not published any information on the pricing of the service or equipment.

The entire system, including user and gateway terminals, is being developed and built by SpaceX, which has no track record in communication technology. This may be one reason why the information available on this system is very sketchy, even for the satellites being currently deployed. We have to resort to educated guesses to fill in some of the gaps. We will limit the presentation to the first stage, that is the first 1584 satellites. They are all part of the first FCC filing in 2016, with some modifications.

19.6.2   Spacecraft and Payload

A spacecraft only weighs 260 Kg (SpaceX, 2020) and is quite small, rectangular, and very flat. A Falcon 9 rocket can carry 60 satellites of the current generation into orbit. The authors were not given permission to include an artist's rendering of a satellite, but such figures can be found readily on the Internet, for example (Coldeway, 2019). The figures show four square areas which are phased arrays. The satellite has a single solar array.

Unusually, the FCC documents contain no information on the payload architecture except that it is a regenerative payload.

19.6.3   System

The system concept is based on a data network in the sky, just like Iridium. This means that the data are processed in the satellite to determine their destination and to forward them either to the next gateway station or through an optical crosslink to another satellite until they can be sent to the user terminal. If the communication is between two Starlink users, the data never go through terrestrial networks. Optical signals travel about 50% faster in free space than in optical fibers, so the latency can be smaller than in terrestrial networks, which has advantages for certain applications. The target latency is 25–35 ms (Albulet, 2016). Since optical communication is not under the jurisdiction of the FCC, the inter-satellite links are not discussed in the FCC applications and there is no further information available. According to Press (2019), each satellite will have four links, just like Iridium. Originally, a fifth link to satellites in orbits with different inclinations was planned. There are speculations that Mynaric will be the supplier of the optical link hardware (Henry, 2019c). The company is developing a 10 Gbps infrared laser system for space applications which should become commercially available in 2020 (Mynaric, 2020).

The early satellites have no crosslinks: they may be added at the end of 2020, at the earliest. Instead, they may use relaying through gateway stations (Henry, 2019a). There are even speculations that the user terminals may be used to forward signals from one satellite to the next (Mosher, 2020). There are also claims that the current generation of satellites has no processing at all, that they operate as bent pipes (Farrar, 2020).

Despite the crosslinks, SpaceX plans to install sufficient gateway stations around the globe that the satellites always have a link to a gateway (Nyirady, 2019).

The satellite antennas form "a large number of narrow beams" (Albulet, 2017b) for the up- and downlinks to the user terminals. The FCC filing does not provide any more detailed information. The original concept was to cover the earth with a hexagonal pattern of cells with a diameter of 45 km (Albulet, 2016). It is not clear to what extent this has survived, as will be discussed in Section 19.6.5. The surface of the earth accommodates 390,000 such cells, so if all cells are to be covered with the 4409-satellite constellation, each satellite needs to provide at least 88 beams. This is obviously a very conservative estimate since the satellites are never evenly spread out. On the other hand, it is a very large number of beams for such a small satellite. The author's suspicion is that beam hopping is used to cover the entire area. Such a feature is nowhere mentioned, however.

There is another way of estimating the number of beams: each satellite provides a downlink capacity of 17–23 Gbps, depending on the gain of the user terminal involved (Hughes, 2016). The most complex signal format is 64 quadrature amplitude modulation (QAM) (Hughes, 2019). We do not know what FEC coding rates are being used, but the highest spectral efficiency is probably close to the modulation formats of DVB-S2X, where the 64-point constellation with the highest coding rate results in a spectral efficiency of 4.9 bps/Hz. In Albulet (2016) a channel bandwidth of 250 MHz was given, so each beam could support a throughput of 1.225 Gbps. To achieve the claimed downlink capacity, 19 such beams would be required, which is still quite a large number for such a small satellite. However, SES claims to implement large numbers of beams, too.

When the system becomes operational, in parts of the world, with about 400 satellites in orbit (Henry, 2020b), the user terminal and gateway beams will have a minimum elevation of 25°. This angle will increase to 40° as the constellation is deployed more fully (Albulet, 2018).

TABLE 19.4 Starlink Frequency Plan (Albulet, 2018)

Link	Frequencies (GHz)	Polarization
Satellite to User	10.7–12.7	RHCP
Satellite to gateway	10.7–12.7	RHCP
	17.8–18.6	RHCP, LHCP
	18.8–19.3	RHCP, LHCP
	19.7–20.2	RHCP, LHCP
User to satellite	12.75–13.25	LHCP
	14.0–14.5	LHCP
Gateway to satellite	14.0–14.5	RHCP
	27.5–29.1 GHz	RHCP, LHCP
	29.5–30.0 GHz	RHCP, LHCP

19.6.4 Radio Interfaces

The frequencies used for the various satellite links are shown in Table 19.4. The gateway links use the Ka/K band frequencies, but they also use the Ku-band frequencies. This is possible because all beams are quite narrow. On satellites launched before November 2019, the gateway links use only the Ku-band frequencies (Wikipedia, 2020).

All beams use 250 MHz of bandwidth. This can be computed from the EIRP and EIRP density in Albulet (2018). The modulation format is described as "BPSK up to 64 QAM digital data" (Hughes, 2019) for gateway as well as user links. DVB-S2X uses APSK for 64-point constellations, which results in a lower ratio of peak to average power. This may not be a consideration for the Starlink system because the power levels are so low that linear amplification is not a problem. This will be shown in the next section. The range of data rates suggests that ACM is used.

19.6.5 Antennas

The satellites have four phased array antennas of approximately equal size. There is no description available of their purpose or their design, so the author had to make educated guesses.

It is likely that one pair of antennas serves the Ka/K-band beams, with separate antennas for receive and transmit. The other pair would then be for the Ku-band beams, again with separate antennas for the two directions. According to Gazelle (2018), the cost of the antenna is dramatically reduced if it need not support full-duplex transmission. If the transmit and receive beams use hopping patterns that are offset in time the user antennas need not support full-duplex communication, either.

The FCC documents (Albulet, 2018) allow us to do some analysis. The antenna footprint of the gateway beams at boresight is practically identical for both transmission directions. The 4-dB beamwidth can be calculated to be 1.6°. This requires an antenna aperture of approximately 0.8 m for the downlink and 0.5 m for the uplink. We can also estimate the number of radiating elements. The antenna gain is 41 dBi for both directions. The individual antenna elements must have a more or less constant gain over the whole range of scan angles, which means up to 56.55° in the early stages of system deployment. This limits the element gain to about 5 dBi. This is consistent with the gain of patch antennas, a likely choice of technology for these antennas, which have gains in the range 5–7 dBi (Bevelacqua, 2016). Then the array gain is about 36 dB. This means that the array must consist of about 4000 radiators. If they were regularly spaced over the above antenna apertures, the spacing would be just about equal to the wavelength, which would result in grating lobes, given the wide scanning range. So, the antennas must have an irregular spacing of elements. SpaceX has filed a patent which shows a "space tapered" layout of the elements as shown in Figure 19.18, so this might be used. The maximum EIRP is 39.5 dBW, so the transmit power per element and per beam is −7.5 dBm or 0.2 mW.

A similar analysis can be performed for the user beams, again based on Albulet (2018). Here, the boresight footprints are not the same size for both directions. The calculated 4-dB beamwidths are 3.6° for the downlink and 3.05° for the uplink. The required antenna sizes computed from these numbers are about the same, considering that the exact frequency used for generating the plots in the FCC documents is not known: 0.57 m for transmit and 0.55 m for receive. The antenna gains are 34 dBi for transmit and 35.7 dBi for receive. Using again an element gain of 5 dBi (the scan angle is the same as above), the array gains would be 29 and 30.7 dB, respectively. This means that these antennas have around 1000 radiating elements.

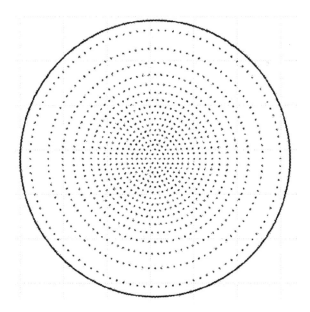

FIGURE 19.18 Space-tapered array (Mazlouman and Schulze, 2018).

The original proposal of SpaceX to the FCC (Albulet, 2016) stated that the user beams would be shaped when steered off-boresight to maintain a circular footprint to cover a single cell in the hexagonal grid described in Section 19.6.3. This would have required the use of additional radiators. In the new FCC documents (Albulet, 2018) this concept has been abandoned. The footprints are now highly elliptical when the scan angle is 57°. The user beams are still called "shapeable", but it is nowhere explained what that means. The maximum EIRP is 32.7 dBW, so the power per element and per beam is −2.3 dBm or 0.6 mW.

In both frequency bands, the transmit antennas can perform sidelobe nulling to reduce interference by about 10 dB in a ±2° zone around the GEO arc (Albulet, 2016).

SpaceX apparently developed the phased-array technology in-house and the company has not published any information about it. Implementing such a large array in the conventional way, that is with individual phase shifters at each element, seems too complex. Add to this the need to produce "a large number of beams", where each beam requires its own phase shifter for each element, seems unrealistic. Satixfy Ltd., has a phased-array technology which allows the implementation of large multi-beam antennas in a more elegant way. The company developed a baseband processing chip and an RF chip as building blocks. In this solution, the data signal timing and the carrier phase are controlled for each antenna element separately. These beamforming operations are performed digitally on complex baseband signals. Up to 32 separate beams can be generated with one processor and one RF chip (Rainish and Freedman, 2016). Since not just the carrier phase, but also the signal timing is adjusted, there is no defocusing of the beam when it is scanned. In 2020, antennas with 1000 elements are being produced, but 4000-element arrays are planned to be available in 2021 and the technology is designed for up to 100,000 elements (Gat, 2020). The transmit power per element that the associated RF chip can supply is 10 mW, so it would be more than adequate for the SpaceX spacecraft antenna. As pointed out above, SpaceX has its own technology but we may guess that it is similar to the Satixfy solution.

19.6.6   Ground and User Segments

SpaceX has recently released pictures of user terminals. They do look like a "UFO on a stick", as described by Elon Musk (Mosher, 2020). They have a phased-array antenna with a diameter of 48 cm. The antenna is motorized to optimize the look-angle to the sky and to tilt the antenna when pointing at a satellite at an elevation angle below 40°. The antenna gain is 33.2 dBi in the receive band at 11.83 GHz and 34.6 dBi in the transmit band at 14.25 GHz and the EIRP is 38.2 dBW (Hughes, 2019).

Assuming again an element gain of 5 dBi, the array gain at the transmit frequency is 29.6 dB, which corresponds to about 900 elements. However, if we space the elements regularly, we need almost 3000 elements to cover the circle with 48 cm diameter if we use a spacing of half a wavelength to avoid grating lobes. So, again, an irregular arrangement of elements must be used. The 3-dB beamwidth of this antenna at boresight is 2.8° in the transmit mode and 3.5° in receive mode. The beamwidth increases to 4.5° and 5.5°, respectively, at the maximum scan angle of 40° (Hughes, 2019). The user terminal corrects for range variations to the satellite and for the gain variation of its own antenna when the beam is scanned by adjusting its transmit power (Hughes, 2019). The user terminals provide the same level of sidelobe nulling as the satellites (Albulet, 2016).

In a speech in 2015, Elon Musk predicted a price of USD 100–300 for the user terminal (Musk, 2015). Satixfy expects the price of such antennas to remain at several thousand dollars in the foreseeable future (Gazelle, 2018). This would preclude the use of terminals by individual users and favor O3b's business model, particularly if unserved populations in less wealthy countries are the target customers.

Some speed and latency data are available from beta testers: downlink speeds were in the range of 10–60 Mbps, uplink speeds were 5–18 Mbps, and latencies (ping round trip time) of 31–94 ms (Brodkin, 2020). This is a far cry from the 1 Gbps per user (Hughes, 2016) and latency below 20 ms (Brodkin, 2020) the system should ultimately support. One can argue that it is early in the system deployment, but if it is a hardware limitation it may take many years before the promised data rate will be achieved.

The gateway terminals will eventually use phased arrays, also, but the first generation is using dish antennas. For Ku-band a commercial product (Cobham) is used with an antenna diameter of 1 m. The EIRP is 53.44 dBW and the polarization is RHCP for both signal directions. The 3-dB beamwidth is 1.5° on transmit and 1.8° on receive. A phased-array antenna would require about 3000 (RX) to 5000 (TX) elements to achieve the same gain. The Ka/K-band antenna is built by SpaceX. The diameter is 1.5 m. The EIRP is 66.5 dBW and both circular polarizations are used in both signal directions. The 3-dB beamwidth is 0.5° on transmit and 0.74° on receive. A phased-array antenna would need about 20,000 (RX) to 45,000 (TX) elements to achieve the same gain. In both cases, the element spacing would be around 1.2 wavelengths with these numbers of elements. There will be several hundred gateway ground stations within the United States alone, all sited near major Internet peering points (Albulet, 2016).

19.7 TELESAT LEO

19.7.1 Overview

Telesat LEO is a satellite constellation the Canadian operator Telesat is planning to launch and operate. It will be implemented in two phases. In a first phase, a total of 78 satellites will be placed in 6 almost polar (98.98°) orbital planes at an altitude of 1015 km and 290 satellites in 20 inclined (50.88°) orbital planes at 1325 km altitude. In a second phase, the almost-polar planes will be increased to 27 and the inclined planes to 40. The number of satellites in almost-polar orbits will increase to 351 and of those in inclined orbits to 1320 for a total of 1671 satellites. The satellites will have optical crosslinks. In the first phase, there will be about 50 gateway stations globally (Beck, 2020).

The project is in an early stage, so very little is known about the technology. A first test and demonstration satellite was put into orbit in January 2018. It has a simple bent-pipe payload with a single beam (Telesat, 2019a). Global service should start in 2023 (Telesat, 2019a).

19.7.2 Services

The constellation is designed to provide broadband services based on OSI layer-2 Carrier Ethernet connectivity. A carrier Ethernet network is operated by a network operator and used for wide-area communications. As opposed to standard Ethernet, the network is shared by many user organizations. It allows each user organization to connect networks in different locations. The Metro Ethernet Forum (MEF) defined attributes for this service, such as scalability, reliability, quality of service, and service management.

Telesat LEO will serve commercial, government, and military users. Examples from the commercial market are backhaul for cell phone network operators and Internet service providers, commercial aircraft and cruise ships, and extending corporate networks to remote facilities (Telesat, 2019b). Individual residential Internet users are not shown as potential customers. So, the business model is the same as for O3b.

Telesat calls the system a fully interconnected global mesh network with full digital modulation, demodulation, and data routing in space (Telesat, 2020). All connection diagrams provided show the traffic going through one of the gateway stations (called landing stations), however, see, for example Beck (2020).

19.7.3 Satellite

The following information is available about the spacecraft (Neri, 2020):

- Four steerable spot beams for the links to the gateways.
- A set of direct-radiating array (DRA) antennas providing up to 24 fully independent, shapeable, and steerable beams for user links.
- One user beam is dedicated to the network entry process. It scans the field of view periodically and signs in newly active terminals.
- Hopping user beams provide capacity where needed (Telesat, 2019a).
- Four optical crosslinks with a data rate of 10 Gbps, two to satellites in the same orbital plane and two to satellites in neighboring planes.
- Full on-board processing of the signals.

19.7.4 Radio Interfaces

The system uses the same Ka/K frequencies for both user links and gateway links. A total of 1.8 and 2.1 GHz of bandwidth is available for downlinks and uplinks, respectively. DVB-S2X will be used on the up- and downlinks for the gateways and on the downlinks to the users. DVB-RCS2 will be used on the user uplink (Telesat, 2019a) (Section 13.4.3).

19.7.5 Antennas

All payload antennas will be phased arrays. Their gain will be about 32 dBi, except for user receive antennas, which will have a gain of 35 dBi. Assuming an element gain of 5 dB, the number of elements is about 500 and 1000, respectively. Obviously, the number of elements will also depend on the kind of beam shaping that will be applied. The antennas should be small, presumably about 25 cm, so it will be possible to fit several user link antennas on the earth-facing panel (Neri, 2020).

REFERENCES

Airbus (2020). Photos for OneWeb. On https://www.airbus.com/search.image. html?q=oneweb. Accessed Nov. 4, 2020.

Albulet M for SpaceX (2016). Technical Attachment. Nov. 15. Part of application to FCC for satellite operation. FCC file number SAT-LOA-20161115-00118.

Albulet M for SpaceX (2017a). Technical Attachment. Mar. 1. Part of application to FCC for satellite operation. FCC file number: SAT-LOA-20170301-00027.

Albulet M for SpaceX (2017b). Technical Information. July 26. Part of application to FCC for satellite operation. FCC file number: SAT-LOA-20170726-00110.

Albulet M for SpaceX (2018). Technical Information. Nov. 8. Part of application to FCC for satellite operation. FCC file number: SAT-MOD-20181108-00083.

Albulet M for SpaceX (2019). Technical Attachment. Aug. 30. Part of application to FCC for satellite operation. FCC file number: SAT-MOD-20190830-00087.

Amyotte E (2020). Personal communication.

Amyotte E, Demers Y, Hildebrand L, Forest M, Riendeau S, Sierra-Garcia S, and Uher J (2010). Recent developments in Ka-Band satellite antennas for broadband communications. *28th AIAA International Communications Satellite Systems Conference (ICSSC-2010)*. Anaheim, CA, USA.

Barletta F, Colucci P, Guerrucci R, Pace G, Perusini G, Ranieri P, and Suriani A (2013). Ka band communication receiver banks with centralized LO and power supply. *31st AIAA International Communications Satellite Systems Conference*. Florence, Italy.

Barnett R (2012). O3b's Non-Geostationary Satellite/Constellation Design. *On* https://www. itu.int/dms_pub/itu-r/md/12/iturka.band/c/R12-ITURKA.BAND-C-0010!!PDF-E.pdf. Accessed March 5, 2021.

Barnett JR for O3B (2013a). Technical information to supplement the existing schedule S for the Virginia Earth Station. July 24. Part of application to FCC for earth station operation. FCC file number SES-LIC-20130528-00455.

Barnett RJ for O3B (2013b). Legal narrative and response to questions 35: Waiver of the Rules. Part of application to FCC for earth station operation. FCC file number SES-LIC-20130528-00455.

Barnett RJ for O3B (2016). Attachment A: technical information to supplement schedule S. Apr. 28. Part of application to FCC for satellite operation. FCC file number SAT-LOI-20160428-00041.

Barnett RJ for OneWeb (2018). OneWeb non-geostationary satellite system, Amendment of the MEO Component. Jan. 4. Part of application to FCC for satellite operation. FCC file number SAT-AMD-20180104-00004.

Beck M (2020). Telesat LEO affordable fiber quality connectivity, everywhere. *CanWISP Conference and Annual General Meeting 2020.* On https://www.canwisp.ca/resources/ Conference%202020/Presentations/11-CanWISP%202020%20Telesat%20Leo%20in%20 partnership%20connecting%20everyone%20everywhere.pdf. Accessed June 17, 2020.

Bevelacqua PJ (2016). Microstrip (patch) antennas. On http://www.antenna-theory.com/ antennas/patches/antenna.php. Accessed May 6, 2020.

Brodkin J (2020). SpaceX now plans for 5 million Starlink customers in US, up from 1 million. Aug. 3. On https://arstechnica.com/information-technology/2020/08/spacex-now-plans-for-5-million-starlink-customers-in-us-up-from-1-million. Accessed Jan. 14, 2021.

Buntschuh F for Iridium (2013). Engineering Statement. Dec. 27. Part of application to FCC for satellite operation. FCC file number SAT-MOD-20131227-00148.

Coldeway D (2019). SpaceX reveals more Starlink info after launch of first 60 satellites. May 24. On https://techcrunch.com/2019/05/24/spacex-reveals-more-starlink-info-after-launch-of-first-60-satellites.

Croq F, Vourch E, Reynaud M, Lejay B, Benoist C, Couarraze A, and Mannocchi J (2009). The GLOBALSTAR 2 antenna sub-system. *2009 3rd IEEE European Conference on Antennas and Propagation.* Berlin, Germany.

Dietrich FJ, Metzen P, and Monte P (1998). The Globalstar cellular satellite system. *IEEE Transactions on Antennas and Propagation;* 46 (6); 935.

Dortch MH (2018). Commission grants O3b modification of U.S. Market Access. June 6. On https://www.fcc.gov/document/commission-grants-o3b-modification-us-market-access. Accessed Dec. 3, 2019.

Farrar T (2016). What about the dish?. From Nov. 22. TMF Associates blog. On http:// tmfassociates.com/blog/category/services/broadband/page/2. Accessed May 27, 2020.

Farrar T (2020). SpaceX and the FCC's $16B problem. From May 08 TMF associates blog. On http://tmfassociates.com/blog/2020/04. Accessed May 8, 2020.

Gat, Y. (2020). private communication.

Gazelle D (2018). LEO constellation for broadband applications, system design considerations. *24th Ka and Broadband Communications Conference.* Niagara Falls, Canada.

Gebhardt C (2020). The 50th Arianespace, Starsem mission completes OneWeb launch. Feb. 20. On https://www.nasaspaceflight.com/2020/02/50th-arianespace-starsem-mission-oneweb-launch/#:~:text=The%2050th%20Arianespace%2C%20Starsem%20 mission%20completes%20OneWeb%20launch,-written%20by%20Chris&text=The%20 European%20launch%20services%20provider,OneWeb%2. Accessed July 18, 2020.

Globalstar (2014). Coverage. On https://www.globalstar.com/en-us. Accessed Sep. 18, 2019.

Globalstar (2017). Globalstar overview. On https://www.globalstar.com/Globalstar/media/ Globalstar/Downloads/Spectrum/GlobalstarOverviewPresentation.pdf. Accessed Oct. 26, 2019.

Globalstar, Inc. (2020). Sat-Fi2 satellite Wi-Fi hotspot. On https://www.globalstar.com/en-us/products/personnel-safety/sat-fi2. Accessed June 7, 2020.

Henry C (2017). SES building a 10-terabit O3b 'mPower' constellation. Sep. 11. On https:// spacenews.com/ses-building-a-10-terabit-o3b-mpower-constellation. Accessed Nov. 16, 2019.

Henry C (2018). SES, with FCC's blessing, says O3b constellation can reach global coverage. June 9. On https://spacenews.com/ses-with-fccs-blessing-says-o3b-constellation-can-reach-global-coverage. Accessed Nov. 16, 2019.

Henry C (2019a). OneWeb's first six satellites in orbit following Soyuz launch. On https:// spacenews.com/first-six-oneweb-satellites-launch-on-soyuz-rocket. Accessed Feb. 20, 2020.

Henry C (2019b). SpaceX submits paperwork for 30,000 more Starlink satellites. Oct. 15. On https://spacenews.com/spacex-submits-paperwork-for-30000-more-starlink-satellites. Accessed Aug. 22, 2020.

Henry C (2019c). Mynaric raises $12.5 million from mystery constellation customer. Mar. 19. On https://spacenews.com/mynaric-raises-12-5-million-from-mystery-constellation-customer. Accessed June 19, 2020.

Henry C (2020a). SpaceX seeks FCC permission for operating all first-gen Starlink in lower orbit. Apr. 20. On https://spacenews.com/spacex-seeks-fcc-permission-for-operating-all-first-gen-starlink-in-lower-orbit. Accessed Apr. 27, 2020.

Henry C (2020b). Starlink passes 400 satellites with seventh dedicated launch. Apr. 22. On https://spacenews.com/starlink-passes-400-satellites-with-seventh-dedicated-launch. Accessed May 11, 2020.

Holmes M (2020). Top 10 hottest companies in satellites. Via satellite. On http://interactive.satellitetoday.com/via/march-2020/the-top-10-hottest-satellite-companies-in-2020. Accessed Aug. 22, 2020.

Hughes T for SpaceX (2016). Legal narrative. Nov. 15. Part of application to FCC for operation Starlink network. FCC file number: SAT-LOA-20161115-00118.

Hughes T for SpaceX (2019). Application for blanket licensed earth stations. Feb. 11. Part of application to FCC for operation of user terminals. FCC file number: SES-LIC-20190211-00151.

IEC Telecom (2021). IRIDIUM 9555. On https://iec-telecom.com/en/product/9555. Accessed Jan. 14, 2021.

Iridium (2003). Iridium Satellite Data Services White Paper. June 2. On http://www.stratosglobal.com/~/media/Documents/irid/Public/irid_whitePaper_satelliteDataServices.pdf. Accessed Sep. 14, 2019.

Iridium (2014). Iridium pilot land station. On https://www.iridium.com/products/iridium-pilot-land-station. Accessed June 7, 2020.

Iridium (2018). Iridium certus. On https://www.iridium.com/services/iridium-certus. Accessed Sep. 14, 2020.

Iridium (2019a). Follow the 8 launch missions. On https://www.iridiumnext.com. Accessed July 28, 2019.

Iridium (2019b). Iridium short burst data (SBD). On https://www.iridium.com/services/iridium-sbd. Accessed Sep. 13, 2019.

Iridium (2020). Iridium Certus® 700 upgrade brings the fastest L-band speeds to the industry. Feb. 27. On https://investor.iridium.com/2020-02-27-Iridium-Certus-R-700-Upgrade-Brings-the-Fastest-L-band-Speeds-to-the-Industry. Accessed June 7, 2020.

Lafond J, Vourch E, Delepaux F, Lepeltier P, Bosshard P, Dubos F, and Bassaler J (2014). Thales Alenia Space multiple beam antennas for telecommunication satellites. *The 8th European Conference on Antennas and Propagation (EuCAP 2014)*. The Hague, The Netherlands.

Long J and Sievenpiper DF (2015). A compact broadband dual-polarized patch antenna for satellite communication/navigation applications. *IEEE Antennas and Wireless Propagation Letters*; 14.

Mazlouman J and Schulze K (2018). Distributed phase shifter array system and method. Patent Application No. US 2018/0241122 A1.

Meza M (2006). Report on the validation of the requirements in the AMS(R)S SARPs for Iridium. On https://www.icao.int/safety/acp/inactive%20working%20groups%20library/acp-wg-m-iridium-6/ird-swg06-wp06-validation%20report%20v0.3_mm102406.doc. Accessed Sep. 22, 2020.

Moakkit H (2014). O3b. an innovative way to use Ka band. *ITU Workshop on the efficient use of the spectrum/orbit resource*; Limassol, Cyprus, Apr. 14–16, 2014.

Mosher D (2020). How Elon Musk's 'UFO on a stick' devices may turn SpaceX internet subscribers into the Starlink satellite network's secret weapon. Jan. 10. On https://www.businessinsider.com/spacex-starlink-satellite-ufo-terminals-how-network-works-2020-1?r=US&IR=T. Accessed May 8, 2020.

Murray P, Randolph T, Van Buren D, Anderson D, and Troxel I (2012). High Performance, High Volume Reconfigurable Processor Architecture. *2012 IEEE Aerospace Conference.* Big Sky, MT, USA.

Musk E (2015). Presentation by Elon Musk. Seattle. On https://www.youtube.com/watch?t=12m37s&v=AHeZHyOnsm4. Accessed Aug. 22, 2020.

Mynaric (2020). Flight terminals (space). On https://mynaric.com/products/space. Accessed June 19, 2020.

N2YO.ORG (2019). Globalstar satellites. On https://www.n2yo.com/satellites/?c=17. Accessed Nov. 19, 2019.

Navarra AJ for Globalstar (2008). Application for mobile satellite service by Globalstar Licensee LLC. Apr. 4. Part of application to FCC for satellite operation. FCC file number SAT-MPL-20200526-00053.

Neri M for Telesat (2020). Exhibit 5 technical narrative. May 26. Part of application to FCC for satellite operation. FCC file number SAT-MPL-20200526-00053.

New York Times (2020). Britain Gambles on a Bankrupt Satellite Operator, OneWeb. July 10. On https://www.nytimes.com/2020/07/10/business/britain-oneweb.html. Accessed July 10, 2020.

Nyirady A (2019). FCC approves lower orbit for SpaceX starlink satellites. Apr. 29. On https://www.satellitetoday.com/launch/2019/04/29/fcc-approves-lower-orbit-for-spacex-starlink-satellites. Accessed May 9, 2020.

O3b (2009). O3b networks. Presentation. On www.itu.int/ITU-D/asp/CMS/Events/2009/PacMinForum/doc/PPT_Theme-2_O3bNetworks.pdf. Accessed Sep. 2, 2013.

OneWeb (2019a). Corporate presentation. Sep. 2. On https://www.oneweb.world/assets/news/media/OneWeb-Corporate-Presentation.pdf. Accessed Aug. 22, 2020.

OneWeb (2019b). OneWeb's Progressive Pitch™ solution for the efficient use of Space and Spectrum. Aug. 19. On https://www.oneweb.world/media-center/onewebs-progressive-pitch-solution. Accessed Aug. 22, 2020.

OneWeb Satellites (2020). Revolutionizing the economics of space. On https://onewebsatellites.com. Accessed Feb. 20, 2020.

Press L (2019). Inter-satellite laser link update. On http://www.circleid.com/posts/20190906_inter_satellite_laser_link_update. Accessed June 19, 2020.

Pritchard-Kelly R (2019). ITU interviews @ WRC-19: Ruth Pritchard-Kelly, VP regulatory affairs, OneWeb. ITU WRC, Sharm El-Sheikh. On https://www.youtube.com/watch?v=qGWhApppFrQ&feature=youtu.be. Accessed Mar. 4, 2020.

Rainish D and Freedman A (2016). Low-cost digital beamforming array structure and architecture. *22th Ka and broadband communications conference.*

Rämö A (2010). Voice quality evaluation of various codecs. *35th International Conference on Acoustics, Speech, and Signal Processing (ICASSP)*; Dallas, Texas, USA.

Read J (2012). Letter from O3b to Patel M of UK's Ofcom spectrum policy group regarding Ofcom call for input, spectrum review 2012. Apr. 12. On https://www.ofcom.org.uk/__data/assets/pdf_file/0029/49907/o3b_limited.pdf. Accessed July 29, 2020.

Read J for O3b (2013). Legal narrative and response to questions 35: waiver of the rules. Part of application to FCC for earth station operation. FCC file number SES-LIC-20130528-00455.

Rosenbaum Z for O3b (2016). Attachment A: technical information to supplement schedule S. June 24. Part of application to FCC for satellite operation. FCC file number SAT-MOD-20160624-00060.

Rosenbaum Zfor O3b (2017). O3b amendment attachment A technical annex. Part of application to FCC for satellite system operation. FCC file number SAT-AMD-20170301-00026.

Rubin T (2012). Ofcom spectrum review - input from O3b limited. Apr. 30. On https://www.ofcom.org.uk/__data/assets/pdf_file/0029/49907/o3b_limited.pdf. Accessed Nov. 22, 2019.

Satcom Direct (2019). Iridium certus gateway. On http://web.satcomdirect.com/cn/akhx0/Apr-NL-Certus. Accessed July 29, 2019.

SatSig (2016). O3b satellite orbit and beams. Apr. 17. On http://www.satsig.net/O3b/O3b-orbit.htm. Accessed Dec. 10, 2019.

Schiff L and Chockalingam A (2000). Design and system operation of Globalstar versus IS-95 CDMA - similarities and differences. *Wireless Networks*; 6 (1); 47.

Seakr Inc. (2016). Cronus_iridium. On https://www.seakr.com/wp-content/uploads/2016/08/Cronus_Iridium.png. Accessed Sep. 11, 2019.

de Selding P (2015). Virgin, Qualcomm invest in OneWeb satellite internet venture. Jan. 15. On https://spacenews.com/virgin-qualcomm-invest-in-global-satellite-internet-plan. Accessed Dec. 12, 2019.

SES (2019). O3B MEO. On https://www.ses.com/networks/networks-and-platforms/o3b-meo. Accessed Nov. 16, 2019.

SES (2020). Investor Presentation 2020. On https://www.ses.com/sites/default/files/2020-05/Roadshow%20Presentation_May_2020_FINAL.pdf. Accessed Aug. 22, 2020.

Sihavong NS (2009). Presentation O3b networks. On www.itu.int/ITU-D/asp/CMS/Events/2009/PacMinForum/doc/PPT_Theme-2_O3bNetworks.pdf. Accessed Sep. 22, 2013.

Spaceflight 101 (2019). O3b satellite overview. On http://spaceflight101.com/spacecraft/o3b. Accessed Dec. 10, 2019.

SpaceX (2020). Starlink mission. On https://www.spacex.com/sites/spacex/files/fifth_starlink_press_kit_0.pdf. Accessed Apr. 20, 2020.

TAS (2012a). Receiver - LNA - DOCON. On https://www.thalesgroup.com/sites/default/files/database/d7/asset/document/Receiver-LNA-Docon102012.pdf. Accessed Dec. 4, 2019.

TAS (2012b). CAMP - SSPA. On https://www.thalesgroup.com/sites/default/files/database/d7/asset/document/CAMP_SSPA-2012.pdf. Accessed Dec. 4, 2019.

Telesat (2019a). Telesat Global LEO Constellation. *MilCIS Conference*. On https://static1.squarespace.com/static/5274112ae4b02d3f058d4348/t/5e0c5916a5b0832cb78c2db4/1577867573707/2019-3-2f-3.pdf. Accessed June 15, 2020.

Telesat (2019b). Telesat LEO, affordable and secure fiber-quality connectivity, everywhere. Sep. 26. On https://www.itu.int/en/ITU-R/space/workshops/2019-SatSymp/Presentations/211%20-%20Satellite%20Innovation%20and%20WRC-23%20TELESAT%20-ENGLISH.pdf. Accessed June 15, 2020.

Telesat (2020). Telesat LEO – transforming global communications. Mar. 9. On https://www.telesat.com/wp-content/uploads/2021/02/Lightspeed-Brochure.pdf. Accessed March 5, 2021.

Thales (2012). TWTA data sheet. Thales microwave & imaging sub-systems. On www.thalesgroup.com/Markets/Security/Documents/Space_K-Ka_band_TWT. Accessed Sep. 17, 2013.

Thales (2020). Iridium Certus, Thales MissionLINK Brochure. On www.thalesgroup.com/LINKsatcom. Accessed June 7, 2020.

Thales Alenia Space (2018). The Iridium Next project. Dec. 11. On https://academieairespace. com/wp-content/uploads/2019/04/Maute.pdf. Accessed Aug. 14, 2019.

Union of Concerned Scientists (2020). UCS satellite database. On https://www.ucsusa.org/ resources/satellite-database. Accessed June 7, 2020.

Weimer B for OneWeb (2020). Application for modification. May 26. Part of application to FCC for satellite operation. FCC file number: SAT-MPL-20200526-00062.

Wikipedia (2019). Iridium communications. On en.wikipedia.org. Accessed July 29, 2019.

Wikipedia (2020). Starlink. On en.wikipedia.org/wiki/Starlink. Accessed May 13, 2020.

Zehl SS and Schneider (2016). Iridium satellite hacking – HOPE XI 2016. 17 Aug. 2019. On https://www.youtube.com/watch?v=cvKaC4pNvck. Accessed Nov. 19, 2019.

CHAPTER 20

MOBILE SATELLITE SYSTEMS IN GEO

20.1 INTRODUCTION

20.1.1 Regulations

Mobile satellite service (MSS) systems have mobile users. The user terminals may be portable or movable. Mobile users may be on land, on ships, or in airplanes. The International Telecommunication Union (ITU) defines the service as follows (ITU, 2016a):

> *A radiocommunication service between mobile earth stations and one or more space stations, or between space stations used by this service; or between mobile earth stations by means of one or more space stations. This service may also include feeder links necessary for its operation.*

Table 20.1 shows the frequency bands from 1 to 31 GHz which the ITU has allocated to MSS on a primary basis (Section 17.3). These frequencies are in the L-, S-, K-, and Ka-bands. The allocations contain many notes specializing the allocations, which are not given here. All the systems described in this chapter use these bands for user communications, except for Inmarsat-5 on the downlinks to the user. Almost all the bands are additionally allocated to other services on a primary basis. Section 14.6 presented a discussion on interference into geostationary-orbit satellites (GEOs). All the systems use bands with primary allocation to the fixed satellite service (FSS) for their feeder links.

Satellite Communications Payload and System, Second Edition. Teresa M. Braun and Walter R. Braun.
© 2021 John Wiley & Sons, Inc. Published 2021 by John Wiley & Sons, Inc.

TABLE 20.1   Frequency Bands Where MSS Is Primary (ITU, 2016a)

Region 1		Region 2		Region 3	
Frequency Band (GHz)	Direction	Frequency Band (GHz)	Direction	Frequency Band (GHz)	Direction
1.518–1.559	Down	1.518–1.559	Down	1.518–1.559	Down
1.610–1.675	Up	1.610–1.675	Up	1.610–1.675	Up
1.980–2.010	Up	1.980–2.025	Up	1.980–2.010	Up
2.170–2.200	Down	2.160–2.200	Down	2.170–2.200	Down
2.483–2.500	Down	2.483–2.500	Down	2.483–2.500	Down
–	–	–	–	2.500–2.520	Down
–	–	–	–	2.670–2.690	Up
5.000–5.150[a]	Both	5.000–5.150[a]	Both	5.000–5.150[a]	Both
20.1–21.2	Down	19.7–21.2	Down	20.1–21.2	Down
29.9–31.0	Up	29.5–31.0	Up	29.9–31.0	Up

[a] For users of internationally standardized aeronautical systems.

20.1.2   Summary of Some MSS Systems

Some MSS systems have GEO satellites but some have non-GEO. In this chapter, we describe five GEO MSS systems:

- Thuraya, two spacecraft, first one launched in 2003
- Inmarsat-4 series, three spacecraft, first one launched in 2005. Alphasat, one spacecraft, an upgrade to Inmarsat-4, launched in 2013
- TerreStar, two spacecraft. First one launched in 2009, later known as EchoStar T1, operation stopped in 2013. Second one launched in 2017, now known as EchoStar XXI. Now owned by EchoStar Mobile.
- SkyTerra, one spacecraft, launched in 2010. Owned by Ligado Networks, formerly known as LightSquared.
- Inmarsat-5 series, four spacecraft, first launch in 2013, also known as Global Xpress or simply GX.

The first four systems are very different from the fifth. For one thing, the first four use L- or S-band frequencies for the user links, while the fifth uses Ka/K-bands. The low frequencies have traditionally been used for user communication because the users do not have to point their antennas and the user equipment is inexpensive to make (Evans et al., 2005). The services of the first four systems are low-data-rate with their air interfaces (Section 10.1.2) fundamentally based on telephony standards, while some services of the fifth are broadband. At least some of the services of every system are IP-based. The first four systems each have one huge reflector, because of the low frequencies, while the fifth has numerous, much smaller dishes.

The major characteristics of the five satellite systems can be summarized. For the first four satellites systems:

- The satellites may better be called "GSOs" since they do not keep their inclinations to the usual ±1°, but they enjoy the ITU primary status accorded to GEOs.
- The services provided are narrowband communications over spot beams. *Personal communications service (PCS)* and *personal mobile communications (PMC)* are both names for such communications.
- The payloads have reflector antennas offset-fed with a phased array.
- The ground stations have gateways with terrestrial telephone systems and the Internet.

In contrast, the characteristics of Inmarsat-5 are as follows:

- The satellites are GEOs, station-kept to within ±0.05° east–west and north–south.
- The services provided are voice over IP and broadband IP.
- The payload has multiple reflectors fed with many horns, a single feed per beam.
- The ground stations have gateways with terrestrial telephone systems and the Internet.

Table 20.2 provides a short summary of further characteristics of the five systems. Thuraya and Inmarsat-4/Alphasat offer user-user links in addition to the usual user-ground station links. The low-frequency payloads all have some kind of digital onboard processor (OBP) except TerreStar. OBPs are described in general in Chapter 10, with particularities given in this chapter. Thuraya and Inmarsat-4/Alphasat have onboard beam-forming (OBBF) for the user beams, while TerreStar and SkyTerra have ground-based beam-forming (GBBF) (Section 11.9.2). Inmarsat-5 has no beam-former. TerreStar and SkyTerra were planned to work together with cell phone networks. An **auxiliary terrestrial component (ATC)** consists of terrestrial base stations and mobile terminals licensed to the operator of the MSS system, that reuses the satellite frequencies for the terrestrial communications integrated with the satellite communications (FCC, 2014).

Figure 20.1 gives a summary description of the satellite air interfaces and their correspondence with terrestrial air interfaces, for the four low-frequency systems. For a terrestrial system, the air interface is the communications standard between two stations. The left-hand third of the figure shows generations of standards within the Global System for Mobile Communications (GSM) family. GSM itself is the second generation (2G) of standard for digital cellular networks. (The generation before GSM, namely 1G, was for analog cellular networks (Wikipedia, 2020).) The middle and right-hand thirds of the figure show the correspondence between the GSM family and two families of satellite standards based on the GSM family.

TABLE 20.2 Brief Summary of Five Mobile Satellite Systems

System	Launch Dates	Has Direct User-User Links besides User-Ground Station	Payload Type	Beam-Forming Type	ATC Planned
Thuraya	2000–2008	Yes	Digital transparent	OBBF by digital processor	No
Inmarsat-4 and Alphasat	2005–2013	Yes	Digital transparent	OBBF by digital processor	No
TerreStar/ EchoStar XXI	2009–2016	No	Bent-pipe	GBBF	Yes
SkyTerra	2010	No	Digital transparent	GBBF	Yes
Inmarsat-5	2013–2017	No	Bent-pipe	Single feed per beam	No

One family of satellite standards, represented in the second column of the figure, is GEO-Mobile Radio-1 (GMR-1). It has had a few releases, which correspond to various GSM generations. Release 2 added packet-switched data (Geo-Mobile Packet Radio Service (GMPRS)) to circuit-switched voice and data (ETSI TS 101 376-3-21, v1.1.1, 2001). Thuraya uses release 2, which is akin to the 2.5th generation of the GSM family. TerreStar and SkyTerra use release 3, which added compatibility with the ITU's terrestrial Universal Mobile Telecommunications System (UMTS) (ETSI TS 101 376-1-3, v3.1.1, 2009). This release is also known as GMR-1 3G and GMR1-3G. It is akin to the 2.9th generation of the GSM family, known as Enhanced Data Rates for GSM Evolution (EDGE). The other satellite standard is GEO-Mobile Radio-2 (GMR-2) (ETSI TS 101 376-3-21, v1.1.1, 2001). Inmarsat-4 and Alphasat use it for telephony (but the Broadband Global Area Network (BGAN) standard for data services—Section 20.3.2) (Vilaça, 2020a).

Table 20.3 shows the manufacturers of the main features of the satellites of all five systems. US manufacturers are in 2016 the only ones that can make the huge unfurlable reflectors used for the L- and S-band user links.

20.2 THURAYA

Thuraya is a company, founded in 1997, in the United Arab Emirates (UAE). It originally focused on land mobile users but in 2013 extended its focus to maritime mobile users (de Selding, 2013). Thuraya was acquired by Yahsat in 2018, also located in the UAE (Yahsat, 2018). Thuraya's two satellites provide coverage in all

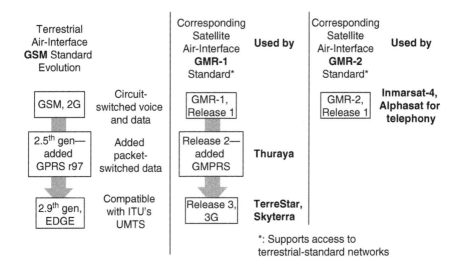

FIGURE 20.1 Air interfaces for the four L- or S-band mobile satellite systems, except for Inmarsat and Alphasat data services.

TABLE 20.3   Key Contractors of Five Mobile Satellite Systems

System	Spacecraft Contractor	Unfurlable Reflector Manufacturer	Digital Processor Manufacturer
Thuraya	Boeing	TRW, now Northrop Grumman	Boeing
Inmarsat-4 and Alphasat	EADS, now Airbus	TRW, now Northrop Grumman	EADS Astrium UK, now Airbus
TerreStar/ EchoStar XXI	SS/L, now Maxar SSL	Harris, now L3Harris	N/A
SkyTerra	Boeing	Harris, now L3Harris	Boeing
Inmarsat-5	Boeing	N/A	N/A

of the eastern hemisphere except for the southernmost tip of Africa, much of Russia, and some of the oceans.

Key characteristics of the Thuraya system are the following:

- Two operating spacecraft
- User-user links besides user-ground station links
- Gateways to public switched telephone networks (PSTNs) and public land mobile networks (PLMNs) (Section 1.1)
- Air interface GMR-1, release 2, which has circuit-switched telephony and packet radio
- Digital transparent payload
- OBBF, the first commercial spacecraft to employ this (Matolak et al., 2002).

20.2.1 Spacecraft

Thuraya has two satellites, Thuraya-2 at 44° E longitude, launched in 2003, and Thuraya-3 at 98.5° E longitude, launched in 2008. The very first satellite in the series, Thuraya-1, was launched in 2000 but failed on orbit.

In 2020 Yahsat announced the contract signing for a new Thuraya-4 to replace Thuraya-2 and an optional second satellite, Thuraya-5, to replace Thuraya-3. Thuraya-4 will have significantly more capacity than the current satellites. Yahsat will also refresh Thuraya's ground network and update its suite of mobile communications products. Work on Thuraya-4 started immediately and launch is expected by late 2023. Work on Thuraya-5 would start at a later date, to be announced by Yahsat (Henry, 2020).

A Thuraya spacecraft is depicted in Figure 20.2. The earth, not shown, is below and a little to the left of the spacecraft. The large L-band reflector attached to a boom is the most obvious spacecraft feature. The phased array that feeds it is the square, shiny object. The spacecraft has an unusual shape for a GEO, not being a rectangular box. The two flat, more-or-less square, light-grey objects attached side-to-side are thermal radiator panels (Applied Aerospace Structures, 2008). These two panels were deployed together from either the north or south spacecraft panel. One of the two radiator panels on the other side is visible as a dark grey object (Pon, 2002).

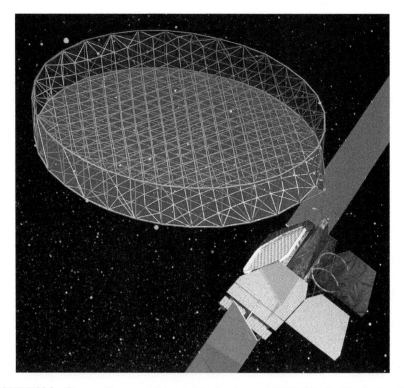

FIGURE 20.2 Thuraya-2 spacecraft image (Boeing Images, 2001). (Credit: The Boeing Co).

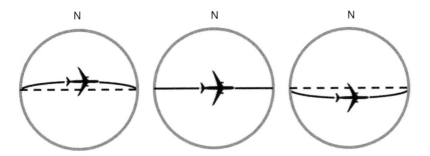

FIGURE 20.3   Thuraya spacecraft pitch orientation over lifetime.

The spacecraft inclination is allowed to vary over life within about ±6°. The spacecraft performs no north–south station-keeping, which saves fuel. Typically satellites perform both north–south and east-west station-keeping burns. The north–south burns take much more fuel (Lim and Salvatore, 2013). Figure 20.3 illustrates the changing orbital inclination over life. (Actually, the author does not know if it starts positive and goes to negative or vice versa.) An airplane represents the spacecraft, whose body axes were illustrated in Figure 2.6. When the orbital inclination is non-zero, the airplane body is meant to be rolled, that is, tilted from the plane of the paper, illustrated by the foreshortened wings.

The system accommodates the slowly varying inclination as follows (Thuraya Satellite Telecommunications, 2003):

- Each spacecraft keeps its pitch axis normal to orbital trajectory
- Each spacecraft keeps its body pointed toward its center of coverage area by means of an uplink beacon and onboard sun sensors
- Primary ground station continuously uploads beam coefficients that compensate for beam deformation.

20.2.2   Coverage Area and Services

An indicative map shows that Thuraya-2 coverage is of Europe, Africa except the very southernmost part, the Atlantic Ocean coastal area west of Europe and northern Africa, and the western half of Asia. Thuraya-3 coverage is of eastern Asia except Russia, Australia, and the Pacific Ocean coastal area east of them. It overlaps that of Thuraya-2 over India (Global Satellite France, 2014). The future satellite Thuraya-4 will provide coverage in the Middle East, Africa, Europe, and Central Asia, while the optional Thuraya-5 would provide coverage in the Asia-Pacific area (Henry, 2020).

The system provides two kinds of services to land and maritime users:

- Circuit-mode GSM-type voice services, which include voice, fax at up to 9.6 Kbps, low-rate data at up to 9.6 Kbps, and short message service

(SMS—text messages) (Thuraya Telecommunications, 2014c). These traffic channels are bidirectional (ETSI TS 101 376-1-3, v3.1.1, 2009).

- Packet-mode services, also called *IP data services* or *Internet connectivity*, similar to GPRS. The packets are in bursts. These traffic channels are unidirectional (ETSI TS 101 376-1-3, v3.1.1, 2009). There are two types of this service (Thuraya Telecommunications, 2014b):
  - Standard or variable-bit-rate IP with up to 444 Kpbs on both forward and return, in a shared mode
  - Streaming or constant-bit-rate IP with up to 384 Kbps, in a dedicated mode.

The system supports two types of communication paths: between a user and a ground station in either direction via the satellite, and between any two users via the satellite. The user-ground station communications can be circuit-mode or packet-mode, while user-to-user is only circuit-mode (Sunderland et al., 2002).

20.2.3 Ground and User Segments

The primary ground station, in the UAE, is responsible for the entire network. It has the satellite control facilities and one of the two uplink beacon stations. The station is collocated with gateways for PSTNs and PLMNs. The station continually computes all coefficients for the OBBF and transmits them to the spacecraft for implementation (Thuraya Satellite Telecommunications, 2003).

Other entities besides Thuraya are free to make their own regional ground stations that are independent of the primary ground station. They would similarly be connected to PSTNs and/or PLMNs (National Satellite Telecommunications Services, 2014). Currently, in 2020, there are no regional ground stations (Vilaça, 2020c).

Thuraya has roaming agreements with mobile network operators in 161 countries around the world (Thuraya Telecommunications, 2020).

The system is designed to be used with dual-mode terminals (satellite/terrestrial) that allow the users to roam between the satellite and terrestrial networks (ETSI TS 101 376-1-3, v3.1.1, 2009). The terminals are equipped with GPS receivers. When a terminal accesses the system, it starts sending its position periodically to the primary ground station so the system can select satellite and beam (ETSI TS 101 376-3-21, v1.1.1, 2001). One of the terminal types is the Satsleeve Hotspot, which can "transform" an iPhone or Android smartphone into a Thuraya terminal (SatPhoneStore, 2020). All terminals have circular polarization (CP).

20.2.4 Frequencies and Beams

The Thuraya satellite frequencies and polarizations are shown in Table 20.4. The user links use L-band frequencies and single CP. Thirty-four MHz is the bandwidth on the uplink (return) and on the downlink (forward). Satellite links with the ground stations are feeder links. The feeder links use C-band frequencies and dual CP. For

TABLE 20.4   Thuraya Satellite Frequencies and Polarizations

	Uplink		Downlink	
	Frequencies (GHz)[a]	Polarization	Frequencies (GHz)[a]	Polarization
User links	1.6265–1.6605	LHCP[b]	1.5250–1.5590	LHCP[b]
Feeder links (with ground stations)	6.4250–6.7250	Dual[c] circular (?) polarization	3.4000–3.6250	Dual[c] circular (?) polarization

[a] Thuraya Satellite Telecommunications (2003).
[b] Thuraya Telecommunications (2014a).
[c] EngineerDir (2014).

both polarizations together, 2×300 MHz is the bandwidth on the uplink (forward) and 2×225 MHz is on the downlink (return). The future satellite Thuraya-4 and the optional satellite Thuraya-5 will also operate in L-band (Henry, 2020).

The onboard digital signal processor (DSP) creates 256 transmit-and-receive beams (Roederer, 2010). The beams are reconfigurable in their number, shape, and size, with the smallest size being $0.7°$ in diameter (Martin et al., 2007). The future Thuraya-4 will create 250 user beams (Henry, 2020).

The primary ground station computes the coefficients for the OBBF and transmits them to the spacecraft for implementation. This goes on continually. The spacecraft's non-zero inclination causes the spacecraft to trace a diurnal figure-8 in the sky, relative to the ground coverage. Additionally, the inclination slowly varies over the spacecraft lifetime. The relative motion causes beam deformation, which the coefficients compensate (Thuraya Satellite Telecommunications, 2003). This method is used for the user beams as well as the feeder beams (Dutta, 2010).

20.2.5   Air Interface

Thuraya follows a subset of the communications standard GMR-1, release 2, for both the user and feeder links. The standard requires L-band for user links and C or Ku for feeder links (Wyatt-Millington et al., 2007). Both circuit-switched connections and packet-mode Internet are possible. Not only user-gateway communication is possible but also user-user in one hop (Matolak et al., 2002).

Thuraya uses a combination frequency-division multiplexing (FDM)/time-division multiplexing (TDM) multiplexing on all links. On the return link, the TDM is actually time-division multiple access (TDMA) (ETSI TS 101 376-1-3, v3.1.1, 2009). The FDM/TDM scheme is illustrated in Figure 20.4 (ETSI TS 101 376-1-3, v3.1.1, 2009, Sunderland et al., 2002). The same scheme is used on both forward and return link (ETSI TS 101 376-1-3, v3.1.1, 2009). Each modulated carrier has a channel bit rate of 46.8 Kbps.

All control channels are synchronized, which makes for fast handovers (Matolak et al., 2002).

Of the various GMR-1 release 2 modulation schemes, Thuraya uses only coherent π/4-quaternary phase-shift-keying (QPSK) (Section 12.8.1) (Thuraya Satellite Telecommunications, 2003). For the pulse filter, Thuraya uses the root raised-cosine (RRC) filter with rolloff $\alpha = 0.35$ (ETSI TS 101 376-1-3, v3.1.1, 2009).

20.2.6 Satellite-System Architecture

The following description of the Thuraya satellite-system architecture comes from Sunderland et al. (2002). Each spacecraft handles the traffic of four ground stations. A spacecraft can handle 160 subbands per ground station. The spacecraft can handle 3140 carriers on the forward link, 3140 on return, and 3140 on user-to-user links. On the forward link the OBP can route any of 640 feeder subbands to any one

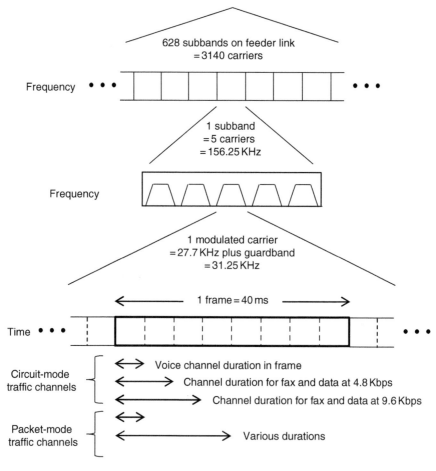

FIGURE 20.4 Thuraya FDM/TDM scheme.

of 219 user-subbands. On return the OBP can route any of 219 user-subbands to any one of 628 feeder subbands. Any subband can be assigned to any beam, and more than one subband can be assigned to a beam. User-to-user, the OBP can route any input carrier, time slot, and beam to any output carrier, time slot, and beam.

20.2.7 Payload

20.2.7.1 Payload Architecture A high-level diagram of the Thuraya payload is given in Figure 20.5. The signal path for user-to-user communications is visible, along with the paths between ground station and user. The abbreviations have been explained in earlier chapters.

20.2.7.2 Antennas and Amplifiers Thuraya has what, at the time, was the largest L-band reflector ever deployed on a commercial satellite (Matolak et al., 2002). It has a round 12.25-m aperture (Thomson, 2002). It is offset-fed by a phased array of 128 elements (Roederer, 2010), all diplexed patch-excited cups (Section 11.8.5) of one-wavelength cup-opening diameter (Roederer, 2005). The high-power amplifiers (HPAs) are 17-W solid-state power amplifiers (SSPAs) (Martin et al., 2007). The equivalent isotropically radiated power (EIRP) at beam center lies between 36 and 48 dBW (Thuraya Telecommunications, 2015).

The future Thuraya-4 will also have a 12-m L-band antenna (Henry, 2020).

The digital beam-forming network (BFN) is part of the digital processor. We describe the BFN here, though, and the rest of the digital processor in the next section. Beam-forming coefficients include amplitudes and phases, not delays (Brown et al., 2014), which is sufficient since the percentage bandwidth of the channels

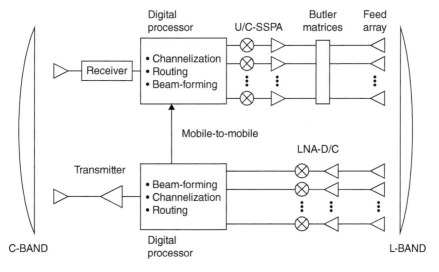

FIGURE 20.5 Thuraya payload top-level diagram (Alexovich et al., 1997).

compared to the carrier frequency is so small. The digital beam-forming allows for optimizing the beams over changing traffic demand (Matolak et al., 2002). The BFN is a low-level one on both transmit and receive (Section 11.12.1).

The phased array's amplification is provided by multimatrix amplifiers (MMAs) (Section 11.12.5); thus, the phased array is semiactive (Section 11.12.1) (Roederer, 2010). The first bank of Butler matrices is included in the BFN, and the second bank comes after the HPAs. The relative powers of the beams are the same as the relative power levels of the beam signals entering the BFN. Twenty percent of total radio-frequency (RF) output power can be allocated to any spot beam (Thuraya Satellite Telecommunications, 2003).

On receive from the users, the same radiating elements are used but with low-noise amplifiers (LNAs), one per radiating element.

The feeder-link C-band antenna has a shaped reflector with diameter of 1.27 m. It handles dual polarization (Martin et al., 2007), probably CP since CP is not affected by Faraday rotation at C-band but linear polarization (LP) is (Recommendation ITU-R, P.531-10, 2009).

The C-band antenna is powered on transmit by four 125-W traveling-wave tube amplifiers (TWTAs) of which two are active. One TWTA is used per polarization (Martin et al., 2007).

20.2.7.3 Onboard Processor

A block diagram of the Thuraya OBP is given in Figure 20.6. The following description comes from Sunderland et al. (2002) except where noted. The OBP has three main parts: the forward processor for ground station-to-user communications, the return processor for user-to-ground station communications, and the mobile-to-mobile switch (MMS) (shown embedded in the forward processor) for user-to-user communications. The OBP is a DSP (Chapter 10). The functionality has been described in the section above on the satellite-system architecture. This article tells of four ground stations, while the figure shows processing for three. Two of the ground stations are redundant. The figure shows 144 radiating elements for return, but in fact this means 144 LNAs in a redundancy ring for the 128 radiating elements (Vilaça, 2020a).

We now describe the blocks in the diagram of the forward processor (the similarly named blocks in the return processor perform the opposite functions). The BCA is the baseband converter assembly, which converts L-band to a near-baseband frequency (Chie, 2014). The rest of this description comes from Sunderland et al. (2002). In the ground station section of the forward processor, subband processing is dividing an input band of frequencies into subbands. This is done with polyphase filters (Section 10.2.3.1) and the fast Fourier transform (FFT). Grouping and ungrouping is the process of dividing subbands into constituent carriers and re-forming subbands. Beam-forming was described above. In the radiating-element section of the forward processor, subband processing is the combining of subbands into bands. The forward and return processors, as well as the MMS, are completely configured from the ground.

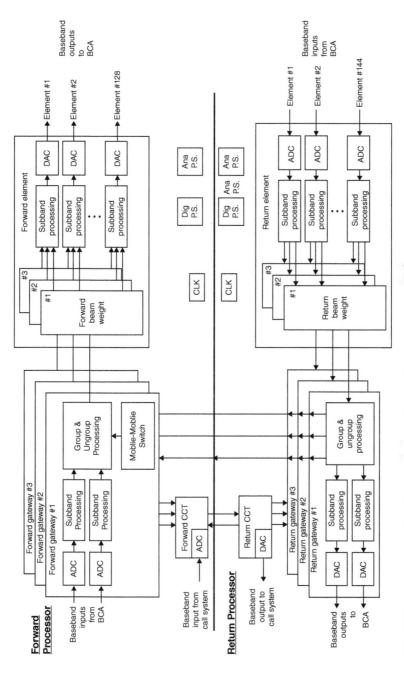

FIGURE 20.6 Thuraya OBP block diagram. (©2002 IEEE. Reprinted, with permission, from Sunderland et al. (2002)).

The MMS is a non-blocking circuit switch. It switches incoming time slots of individual carriers to any outgoing time slot and carrier (Sunderland et al., 2002). The MMS includes surface acoustic-wave (SAW) filters for input and output filtering (Kongsberg Defence Systems, 2008).

The future Thuraya-4 will also have a payload processor (Henry, 2020).

20.3   INMARSAT-4 AND ALPHASAT

Inmarsat is a British satellite telecommunications company offering global mobile service. It owns and operates 12 GEOs in 2020. The company was developed by an intergovernmental organization established in 1979 by a United Nations directive. Its purpose was to provide communications for maritime users. It was privatized in 1999, becoming based in the UK (Wikipedia, 2017a). The company is currently flying its fifth-generation of satellites (Section 20.6) and expects to have the first of its sixth generation launched in 2021.

Inmarsat-4 and Alphasat represent the company's fourth generation of satellites. These satellites form part of the world's first global, IP-based satellite 3G network (Gizinski III and Manuel, 2015). Perhaps the best known of the services provided is the *BGAN* for terrestrial users. The fourth-generation satellites are part of the overall Inmarsat system.

Alphasat represents a partial upgrade to the Inmarsat-4 satellites. It has previously been known as Alphasat I-XL and Inmarsat XL (Gunter's Space Page, 2014a). Alphasat provides additional power and performance margins compared to Inmarsat-4 but a somewhat reduced coverage area (Khan et al., 2013). The key properties of the part of the Inmarsat system served by Inmarsat-4 and Alphasat are as follows:

- Four operating spacecraft
- User-to-user links (Vilaça, 2004), user-ground station links, and ground station–ground station links
- Gateways to the Integrated Services Digital Network (ISDN) and the Internet at the ground stations. ISDN offers circuit-switched voice and data and packet-switched data over the PSTN.
- Proprietary communications standard, offering UMTS-like services
- Digital transparent payload
- OBBF.

In the sections that follow, if Inmarsat-4 and Alphasat have different features then Inmarsat-4 is described first followed by Alphasat.

20.3.1   Spacecraft

There are three Inmarsat-4 spacecraft, which are identical (EADS, 2005), and one Alphasat, which has higher capability and capacity than the Inmarsat-4 spacecraft (Inmarsat, 2017b; Gunter's Space Page, 2014a, 2017):

- Inmarsat-4 F1, also known as I-4 Asia-Pacific, at 143.5°E and launched in 2005.
- Inmarsat-4 F2, also known as I-4 MEAS (Middle East and Asia), at 63.9° E and launched in 2005. Until 2015 known as I-4 EMEA (European, Middle East, and Africa), located at 25°E longitude.
- Inmarsat-4 F3, also known as I-4 Americas, located at 98.4° W and launched in 2008 (Inmarsat, 2013a).
- Alphasat, at 24.9° E longitude, launched in 2013. Completely replaced F2 at its former location by 2015 (Inmarsat, 2015b). Provides better service to the system's busiest service region.

An Inmarsat-4 spacecraft is illustrated in Figure 20.7 and Alphasat in Figure 20.8. For the Alphasat spacecraft, the earth deck is the long side facing front and partially left (Witting et al., 2012). This represents a 90° rotation of the usual Alphabus on-orbit orientation (ESA, 2011). This is also the case for Inmarsat-4 (Vilaça, 2020a). The phased-array feed system is on the east or west panel, and the L-band reflector is deployed off the same panel.

The inclination of all the spacecraft is allowed to vary within ±3° (Guy et al., 2003; Gabellini et al., 2010). This reduces the need for station-keeping fuel, as we have seen for Thuraya. At least for I-4 Americas, the longitude is maintained within ±0.1° of its nominal longitude (McNeil, 2004).

FIGURE 20.7 Inmarsat-4 spacecraft image (Gunter's Space Page, 2017). (Used with permission of Airbus SE).

FIGURE 20.8 Alphasat spacecraft image. (© ESA. From ESA (2017)).

20.3.2 Coverage Area and Services

All the spacecraft have earth-coverage (EC), regional, and spot beams. The previous generation of spacecraft, Inmarsat-3, had only a few wide spot beams (Satbeams, 2007).

Figure 20.9 shows the expected coverage areas of the Inmarsat-4 and Alphasat spacecraft. The Alphasat coverage is not oval like the others. Next east of the Alphasat coverage area is the MEAS area, which largely overlaps the Alphasat coverage and about half overlaps the Asia-Pacific coverage. It covers practically all of Africa, Europe, and Asia and most of Australia.

The Inmarsat system with the Inmarsat-4 spacecraft provides two kinds of UMTS-like services to land, maritime, and aeronautical users:

- Circuit-mode services like those of ISDN, namely voice, video, fax, messaging, and data, at up to 64 Kbps.
- Packet-mode services. There are two types of this service (Inmarsat, 2009b, 2011a, 2012b):
  - Standard IP with up to 492 Kbps on both forward and return
  - Streaming IP with up to about 450 Kbps.

The system with Alphasat provides an improvement of the packet-mode services:

- Circuit-mode same as for Inmarsat-4
- Packet-mode services. Additions of
  - Higher-rate symmetric streaming IP (Inmarsat, 2014b)
  - Asymmetric streaming IP (Irish and Marchand, 2013)
  - Lower-rate IP, interactive and streaming (Khan et al., 2013).

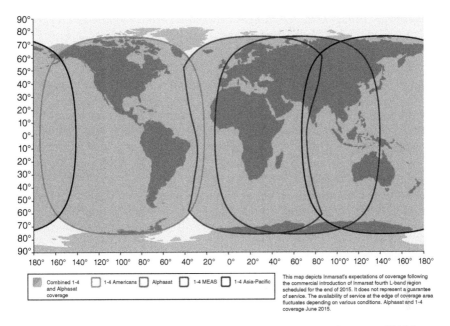

FIGURE 20.9 Inmarsat-4 and Alphasat map of expected coverage (Inmarsat, 2013c). (Used with permission of Inmarsat).

For both Inmarsat-4 and Alphasat, circuit-mode service can be obtained alone, in which case it is called "Global Satellite Phone Service (GSPS)," available on the Isatphone. This service follows the GMR-2 standard. Direct user-to-user service is circuit-mode only (Orbitica, 2013). Or the two kinds of service can be gotten simultaneously on a broadband terminal (Inmarsat, 2009a). Broadband service has three names:

- BGAN for land users
- FleetBroadband for maritime users
- SwiftBroadband for aeronautical users.

20.3.3 Ground and User Segments

The satellites are controlled from the satellite control center (SCC) at company headquarters in London. Four TT&C stations support the SCC (Inmarsat, 2009a). The network operations center (NOC), also in London, monitors the ground stations continuously (Inmarsat, 2013b).

Inmarsat owns three ground stations for Inmarsat-4 and Alphasat, in Hawaii, the Netherlands, and Italy called "satellite access stations (SASs)" (Inmarsat, 2017d). They are all of similar design. The one in Paumulu, Hawaii, serves the spacecraft I-4 Americas and I-4 Asia, while the two in Burum, Holland, and Fucino, Italy, are redundant for Alphasat and I-4 EMEA. The ground stations have gateways into the Internet and the ISDN networks (Inmarsat, 2009a). Sometimes Inmarsat uses the term "land earth station (LES)" for ground station, but usually LES is only for the services provided by earlier-generation spacecraft.

In 2013 a new SAS opened in China, owned by a Chinese company and the Chinese government (Inmarsat, 2014a). The Chinese government required such a SAS before it would allow Inmarsat to market its services in China.

The ground stations upload the beam-forming coefficients plan daily to the spacecraft, since the spacecraft inclination is allowed to vary within ±3° but the beam coverages are meant to stay the same (Guy et al., 2003). The beams are reconfigured throughout the day (Guy, 2009).

Connected to the ground stations are *points of presence (PoPs)*, through which companies or governments can access the Inmarsat network (NordicSpace, 2006; Inmarsat, 2012c). These PoPs may themselves be connected to the Internet and ISDN.

Inmarsat's Data Communications Network (DCN) links the ground stations and the NOC. It supports the signaling among ground network elements as well as the transport of traffic data (Inmarsat, 2016).

Inmarsat sells the vast majority of its bandwidth wholesale to distribution partners, which sell directly or indirectly to consumers (Inmarsat, 2014a).

The system allows dual-mode telephones for roaming between satellite service and terrestrial UMTS networks (TS2 Technologie Satelitarne, 2009). Broadband terminals are satellite-only (Inmarsat, 2012b). When they access the Inmarsat-4 spacecraft they must have a directional antenna. For Alphasat, they can have an omnidirectional one, due to the latter's L-band reflector being bigger (ESA, 2011). They can also have a lower signal-to-noise ratio (SNR) (Khan et al., 2013).

The user terminals are equipped with GPS receivers. When a terminal accesses the system, it starts sending its position periodically to the system (Inmarsat, 2006, 2011b). Details of terminal registration are provided below. Once a terminal has service it must point to the spacecraft, unless it uses Alphasat's low-data-rate service with an omni antenna (Khan et al., 2013). Terminals use right-hand circular polarization (RHCP).

20.3.4   Frequencies and Beams

The information in this section comes from Vilaça (2004), except where noted otherwise.

The Inmarsat-4 frequencies are shown in Table 20.5. The user links have L-band frequencies and RHCP. Thirty-four MHz is the bandwidth on the uplink (return) and on the downlink (forward). The feeder links have C-band frequencies and dual CP. With the two polarizations the total bandwidth on the uplink (forward) is 2×90.8 and 2×106.4 MHz on the downlink (return). The ground station-to-ground station links use C-band and RHCP. These links are for administrative traffic between the ground stations (Martin et al., 2007).

TABLE 20.5   Inmarsat-4 Frequencies and Polarizations (Vilaça, 2004)

	Uplink		Downlink	
	Frequencies (GHz)	Polarization	Frequencies (GHz)	Polarization
User links	1.6265–1.6605	RHCP	1.5250–1.5590	RHCP
Feeder links for user communications	6.4250–6.5158	Dual CP	3.5514–3.6578	Dual CP
Ground station–ground station links	6.5240–6.5290	RHCP	3.6606–3.6656	RHCP

Alphasat takes advantage of the MSS L-band extension granted to ITU Regions 1 and 3 (Section 17.3) at the ITU's World Radiocommunication Conference (WRC) of 2003. The extension is the addition of the uplink frequency band 1.6680–1.6750 GHz and the downlink frequency band 1.5180–1.5250 GHz. The addition is 2×7 MHz on the uplink and the same on the downlink, a little over 20%. Region 2 had its MSS allocation decreased to match the allocation for the other two regions (Vilaça, 2004). Also, the C-band frequencies have doubled (Irish and Marchand, 2013).

On the Inmarsat-4 and Alphasat feeder links, the band is divided into subbands of 12.6 MHz with consecutive center frequencies separated by 15.6 MHz. The sub-band-frequency plan is the same on the two polarizations. On both polarizations the uplink band is divided into six subbands and the downlink band into seven; in each direction, up to five subbands of these are for user communications (Vilaça, 2004). On both polarizations the rest of the subbands, up to three on the uplink and up to four on the downlink, are for aircraft communications addressing and reporting (ACARS), a service which transmits short messages between aircraft and ground station (USA-Satcom, 2012).

Each spacecraft has three sizes of L-band beams all formed with the phased array-fed reflector antenna: global, regional, and spot (Vilaça, 2004). The 19 regional beams cover the visible globe, as shown in Figure 20.10. The global beam and the regional beams are only for user-terminal registration (McNeil, 2004; Inmarsat, 2009a). The registration process starts with the terminal calculating its position from GPS and acquiring the global beam's signal. That signal gives it information which allows the terminal to know which regional beam's signal to acquire. Then the Inmarsat system hands the terminal over to the correct spot beam for a communications session. When the session is over, the system moves the terminal back to its regional beam (McNeil, 2004).

The "universal" set of Inmarsat-4 spot beams is shown in Figure 20.11. Adjacent beams' centers are 1.1° apart (Vilaça, 2004), and beam diameter at the subsatellite point is about 1200 km (Wang and Liu, 2011). From this set of about 300 beams that cover the visible globe, is chosen the set of 200 or so spot beams that is constructed for the particular orbital slot and current traffic pattern (Guy et al., 2003).

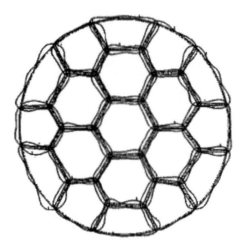

FIGURE 20.10   Inmarsat-4 regional beams. (©2003 IET. Used with permission, from Guy et al. (2003)).

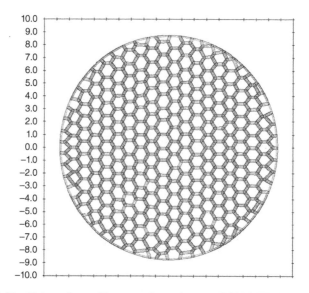

FIGURE 20.11   Universal set of Inmarsat-4 spot beams. (©2006 IEEE. Reprinted, with permission, from Stirland and Brain (2006)).

The maximum number of simultaneous spot beams per satellite is variously stated as being between 156 and 228. The digital beam-former can make up to 256 beams (McNeil, 2004). Of these, one is the global beam and 19 are the regional beams. The number of spot beams is limited to 200 by what the ground-station hardware can deal with (Cobham Satcom, 2012). Actually, 193 spot beams are produced (Vilaça, 2020b).

The Alphasat beams must be the same size as the Inmarsat-4 beams, since again there are about 200 and they provide the same coverage.

The regional beams come in three colors (frequency and polarization combinations) and the spot beams in seven (Guy et al., 2003).

20.3.5 Air Interface

The system built around Inmarsat-4 and Alphasat follows a subset of the communications standard Inmarsat Air Interface 2 (IAI-2) proprietary to Inmarsat (Howell, 2010). IAI2 uses FDM on the forward links and TDMA/FDM on the return links (Plass, 2011). Inmarsat calls a carrier with specific bandwidth, modulation, and coding a "bearer." Each user terminal has a forward bearer and/or a return bearer. The network monitors the return link for each user terminal so the bearer's characteristics can be optimized for the terminal properties, the link quality, and the system's available power and bandwidth (Khan et al., 2013).

For Inmarsat-4 the modulation schemes for the spot beams on the forward link are QPSK and 16QAM (Khan et al., 2013) and π/4-QPSK and 16QAM on the return link (Howell, 2010). There are about ten coding rates from 1/3 to 9/10. E_s/N_0 for both forward and return links is between −6 and 21 dB. The channel bandwidth is 200 KHz. On the forward feeder link, in each subband of 12.6 MHz there are 630 such channels; the same holds true for the return feeder link. On the forward links the channel bit rate is up to 512 Kbps. The data is in frames and in bursts of 80 ms. On the return links the channel bit rate is up to 492.8 Kbps (Khan et al., 2013). The bearer bandwidth can be 25, 50, 100, or 200 KHz. The data is in bursts of 5 or 20 ms (Howell, 2010).

For Alphasat the possibilities have been extended with the addition of the new "High Data-Rate (HDR)" and "Low Data-Rate (LDR)" capabilities. The HDR capability is faster symmetric IP-streaming and the introduction of high-rate asymmetric IP-streaming. The capability is enabled by the addition of the 32QAM and 64QAM constellations, with 32QAM being preferable on the return link. There are about ten coding rates. The bearer bandwidth can be either 100 or 200 KHz. Each bearer is dedicated to a terminal. On 100 KHz the bit rate is up to 858 Kbps (Khan et al., 2013). On the HDR's symmetric variety the forward and return channels and rates are equal; on the asymmetric, envisioned for video streaming back to a studio, the forward rate is only 64 Kbps (Inmarsat, 2014b). The LDR capability is for disadvantaged terminals, such as those at edge of coverage. The coding rates are lower, ranging from 1/9 to 1/3. The channel bandwidth is either 25 or 50 KHz. On the return links the bit rate on a burst over a 50-KHz channel is 56 Kbps, with the burst length being 80 ms. The channel may be shared (Khan et al., 2013).

20.3.6 Payload

20.3.6.1 Payload Architecture A top-level payload diagram of the Inmarsat-4 payload is given in Figure 20.12. It shows the signal flow for all three types of signal paths through the payload, namely between the ground station and users, directly between users, and between ground stations. One thing in the diagram that will only be discussed briefly is the navigational payload. It sends GPS L1 and L5 signals to the SAS, indicating the satellite location. These signals transmit integrity and correction data that can be used to improve the accuracy of users' position determination (Martin et al., 2007).

We briefly describe the payload according to signal flow, using (McNeil, 2004) as the principal source for information. Other sources are noted as applicable.

Let us start with the C-band to L-band signals. The C-band global receive horn passes two signals, one from each polarization, to the C-band receive section. The receive section has, for each polarization, a preselect filter, a LNA, and a multiplexer (MUX) to separate the full band into two parts. The receive section's gain is controlled by an automatic level control (ALC) feedback loop using a measured level from the SSPAs. The gain has a 16-dB range. The C-band downconverter demultiplexes 2×5 subbands of user signals from the ground station and converts all ten to an intermediate frequency (IF). After this conversion, there is a further conversion to near-baseband, which we presume is also represented by the C-band downconverter in the figure. The digital forward processor, called the "mobile

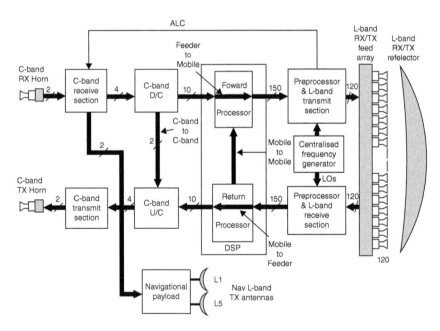

FIGURE 20.12 Inmarsat-4 top-level payload diagram (Vilaça, 2004). (Used with permission of Inmarsat).

processor" by the manufacturer, performs channelization, routing, and digital beam-forming (Airbus Defence and Space, 2014b). It dynamically allocates 100 KHz-wide half-channels to the various spot beams, allowing the satellite to manage a varying traffic pattern. It contains the analog-to-digital converters (ADCs) and the digital-to-analog converters (DACs) (Biglieri and Luise, 1996). It outputs 150 signals to the radiating elements; why 150 instead of 120, which is the number of radiating elements in the phased array, is unknown to the author. The post-processor upconverts the signals to L-band. The L-band transmit section amplifies the signals with SSPAs arranged in multiport amplifiers (MPAs) (Section 11.12.3), which then pass the signals to the L-band phased array. The array forms the feeds for the L-band reflector. The array is semiactive but the other kind from Thuraya's, which uses MMAs.

The units for the L-band to C-band signals correspond to the forward-link units but perform reverse functions.

The units for the L-band to L-band signals are among those already described. The only thing new is that the return processor's IF outputs go to the forward processor. The connection may bypass the return processor's DACs and the forward processor's ADCs.

The units for the C-band to C-band signal are among those already described. What is new is that the C-band downconverter's outputs go to the C-band upconverter. The downconversion to IF and the upconversion from IF may be bypassed.

The payload has 510 SAW filters, 680 frequency converters, and 28 switch matrices (Kongsberg Defence Systems, 2005). The SAW filters are used for bandpass filters in the forward and return processors and in the pre- and post-processors (Airbus Defence and Space, 2014d).

The units, except for the C-band receive section, will be described in some detail in the following sections.

The Alphasat payload is different from that of Inmarsat-4 in four main ways:

- Increased L-band bandwidth, enabling 20% more channels
- Doubled C-band bandwidth
- Greater gain on the user links
- Faster digital processing.

20.3.6.2 *Antennas*

The Inmarsat-4 communications payload has three antennas, one C-band dual-CP global horn for receive, one C-band dual-CP global horn for transmit, and one L-band single-reflector antenna fed by a semiactive phased array. We will start with C-band and finish with L-band.

Inmarsat-4's key performance parameters are given in two tables, C-band in Table 20.6 and L-band in Table 20.7. The given parameters are antenna gain, EIRP for downlinks, and G/T_s for uplinks. Note that there is no return L-band global beam.

The L-band antenna is a single-reflector offset-fed by a defocused phased array (see below). The reflector is 9 m in diameter with a surface of gold-plated molybdenum mesh having 558 facets (Guy et al., 2003). The lack of a subreflector means

TABLE 20.6   Inmarsat-4 C-Band Key Performance Parameters of Each Polarization (McNeil, 2004)

C-Band Forward Link		C-Band Return Link	
Peak Antenna Gain (dBi)	Peak G/T_s (dB/K)	Peak Antenna Gain (dBi)	Peak EIRP (dBW)
22	−6.4	22	35

TABLE 20.7   Inmarsat-4 L-Band Key Performance Parameters (McNeil, 2004)

	L-Band Forward Link		L-Band Return Link	
	Nominal Peak Antenna Gain (dBi)	Maximum Attainable EIRP (dBW)[a]	Peak Antenna Gain (dBi)	Peak G/T_s (dB/K)
Spot beam	42	70	40–42	12.3–14.3
Regional beam	34	58	34	3.0
Global beam	22	43	N/A	N/A

[a] Cannot all occur simultaneously.

that not just the subreflector but also its tower do not need to be accommodated on the spacecraft (Ueno et al., 1996). The antenna f/D is 0.53 (Guy, 2009). Very large reflectors like this one often have such a small f/D because a larger f would mean an even longer reflector boom. However, such a f/D causes high phase errors in the outer beams, leading to increased sidelobe levels, reduced beam gain, and increased beamwidth. The phased array feed can fix this (Gallinaro et al., 2012).

The phased array has the shape of an eight-sided polygon measuring 2.5 m by 2.5 m and 0.6 m deep, as shown in Figure 20.13. It has 120 elements, each consisting of a radiating helix in a cup, a transmit bandpass filter, and a receive bandpass filter (Stirland and Brain, 2006). The beam-forming is performed digitally and at a low level on both transmit and receive. Sixteen to 20 elements are used per spot beam. Of these, only three to five have high amplitude coefficients, with the rest having low amplitude coefficients to shape the beam and improve C/I. The groups of elements used for spot beams overlap. The regional beams use more elements than 20, and the global beam uses pretty much all the elements (Vilaça, 2020b).

Unusually, the reflector is in the near field of its feeds, namely the phased array (Guy et al., 2003).

The phased array is defocused as a feed, that is, it is not in the reflector's focal plane (Guy, 2009). This offers better beam reconfigurability. Inmarsat-4 uses the first method for defocusing, namely without flexing the reflector, described in Section 11.5.3.

To test the Inmarsat-4 L-band antenna at the factory there were two key considerations: first, it was impractical to do it with the reflector deployed and, second,

FIGURE 20.13   Inmarsat-4 array feed. (©2006 IEEE. Reprinted, with permission, from Stirland and Brain (2006)).

the whole payload had to be used. The payload was installed in a spherical near-field test range and the fields were measured on the surface of a sphere that surrounded the feed array. The fields were transformed to the far field taking into account the reflector (Section 3.12). Only a subset of beams were measured (Stirland and Brain, 2006).

Two ground beacons aid in pointing the L-band antenna. For each beacon, the spacecraft has two orthogonal pairs of narrow receive beams, each beam 1° off the beacon direction, so that orthogonal difference signals can be formed. This is illustrated in Figure 20.14. The spacecraft attitude, particularly yaw, is improved (Guy et al., 2003). At least for I-4 Americas, antenna pointing is maintained within ±0.1° of nominal (McNeil, 2004).

The Alphasat L-band antenna differs from the Inmarsat-4 antenna primarily in its larger reflector, 11 m in diameter instead of 9 m. Another difference is that the feed array is 7-sided instead of 8-sided. The radiating elements of the phased array are the same, and the diplexer on each element is very similar. The radiating elements' transmit and receive bandpass filters were reoptimized for the wider bandwidths (Dallaire et al., 2009).

20.3.6.3 *High-Power Amplifiers*   For Inmarsat-4 each polarization of the C-band transmit horn is fed by four combined SSPAs each with 10.4 W output power. The redundancy is 6:4; thus, there are 12 SSPAs of which 8 are active at any one time. Each SSPA compensates its gain and phase shift over temperature, and each

FIGURE 20.14   Illustrative Inmarsat-4 beams that aid in pointing of L-band antenna. (Guy et al. 2003). (©2003 IET. Used with permission).

has an ALC circuit and a linearizer. The SSPA is designed for more than 40 W so is operated at about 7 dB below its 2-dB compression point P2dB. In the ALC range of RF input powers, the NPR is about 21 dB and the power efficiency is about 15% (Kiyohara et al., 2003).

The L-band high-power amplification seems to be performed the same way for both Inmarsat-4 and Alphasat (Vilaça, 2004). It uses 150 SSPAs of which 120 are active (EADS, 2005). The SSPAs are arranged in 15 MPAs, one of which is illustrated in Figure 20.15. In each MPA are two banks of four active SSPAs, each bank a 5:4 redundancy ring. Each active SSPA is followed by a transmit filter. Recall that the presence of MPAs in the payload is transparent to the rest of the payload, except for the fact of amplification, and that the MPA outputs connect one-to-one with radiating elements. For Inmarsat-4, the maximum available power input to the L-band antenna is 700 W. The combined power can be dynamically apportioned among the three size-categories of L-band beams (McNeil, 2004).

On forward-link transmit, each of the eight inputs to a MPA consists of eight contiguous channels 200 KHz wide. The MPA outputs correspond to the inputs one to one, and each output connects to one radiating element (Mallison and Robson, 2001).

The L-band SSPAs are able to meet the stringent requirements on amplitude and phase tracking with a digital control scheme that compensates for variations in driver and output stage characteristics over temperature, SSPA dynamic range, and operating frequency range (GlobalSpec, 2014).

It seems that the only difference in L-band amplification for Alphasat relative to Inmarsat-4 is that the transmit filters are different and the SSPAs have increased

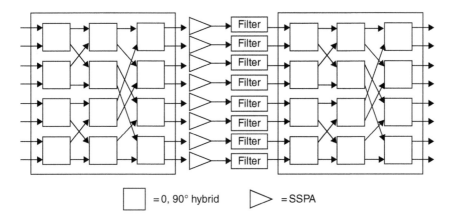

FIGURE 20.15   Inmarsat-4 and Alphasat multiport amplifier in L-band transmit section (SSA redundancy not shown) (Mallison and Robson, 2001). (©2001 Airbus Defence and Space. Used with permission).

output power (Vilaça, 2004). The SSPAs have nominal output power of 15 W and power efficiency of 31% (Lohmeyer et al., 2016).

20.3.6.4 Forward and Return Onboard Processors

An Inmarsat-4 spacecraft can route any uplink feeder-channel to any mobile downlink beam. More power and bandwidth can be allocated to certain beams, in order to handle fluctuations in traffic (EADS, 2005). The forward feeder-link contains 10 sub-bands, each 12.6 MHz wide, of data that will be transmitted to users. The return feeder link also contains such subbands. Thus, altogether 2×126 MHz of bandwidth between the users and the ground station is processed by the payload (Vilaça, 2004). Additionally, the forward and return processors together can route direct user-to-user calls.

Specifically, each digital processor routes half-channels of 100 KHz bandwidth (Mallison and Robson, 2001). There are 1260 such half-channels in each direction. Each half-channel has independent gain control. Since the user links in each direction have 34 MHz of bandwidth altogether and the user beams come in seven frequencies, each beam has about 5 MHz of bandwidth. The user-to-user links comprise 84 duplex 100-KHz half-channels.

Now for implementation details. The following description is the result of combining three papers, which are nearly consistent if the terminology is unified (Leong et al., 1996; Biglieri and Luise, 1996; Mallison and Robson, 2001). Other sources are named where used. There may well be small errors in what follows.

We start with the Inmarsat-4 forward processor, illustrated in Figure 20.16. Somewhere, which we have assumed to be in the C-band downconverter, the 10 subbands from the ground station are converted to near-baseband. The forward

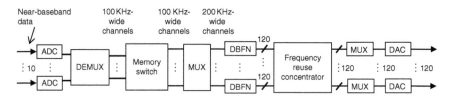

FIGURE 20.16   Inmarsat-4 forward processor block diagram. (After Biglieri and Luise (1996), Mallison and Robson (2001)).

processor puts the ten signals through ADCs. The forward-link ADCs have bandwidth 27 MHz (Vilaça, 2004). The demultiplexer separates the individual 100 KHz-wide half-channels and shifts them all to the same near-baseband frequency band. The digital samples of each 100-KHz half-channel are decimated to a rate consistent with this bandwidth. A time-domain "memory switch" maps them all to the signal-processing paths of their various downlink frequencies and beams. It also performs level control on the half-channels. The following MUX combines channels into 200 KHz-wide full channels. Each full channel is replicated to number 120 copies, which are fed to the channel's own digital beam-forming network (DBFN). The DBFN can form a beam with potentially every radiating element. Now, the signals for each radiating element must be combined. First, the signals on the same downlink frequency are summed in the "frequency reuse concentrator." This is done for every different frequency. Then a frequency MUX combines the signals at different frequencies, increasing the digital sample rate to account for the full bandwidth of the combined signal that will go down the beam. The last thing the processor does is DAC.

The Inmarsat-4 return processor is practically the same, just performing the reverse operations. The return-link ADCs and DACs have a bandwidth of 29 MHz (Vilaça, 2004).

The Inmarsat-4 user-to-user signal processing is illustrated in Figure 20.17. The user transmissions, in half-channels 100-KHz wide, are formed in the DBFNs from the radiating-element signals. The half-channels then go into the user-to-user memory switch, which allocates each to the correct downlink frequency and beam. The memory switch also performs level control. From there they combine with the forward half-channels into 200 KHz-wide full channels, which then go to the DBFNs for the downlink. The user-to-user signals also go down on the feeder link so the ground station can monitor them (Mallison and Robson, 2001).

The Alphasat processors differ much from those on Inmarsat-4. Alphasat carries eight integrated processors, which work in parallel. Together they perform the same functions as the two Inmarsat-4 processors (Airbus Defence and Space, 2013). The processor architecture is different, as illustrated at a high level in Figure 20.18. Switching-and-beam-forming is performed between demultiplexing and multiplexing. The port bandwidth is much wider, 250 MHz compared to 27 MHz (Hili and Malou, 2013). The much wider processor inputs mean that there is some

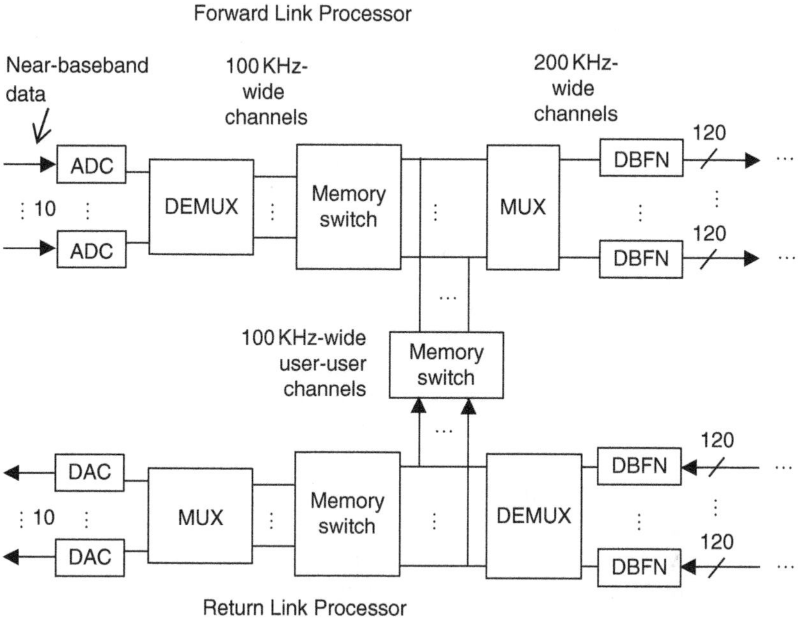

Forward Link Processor

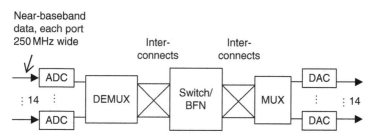

FIGURE 20.17 Inmarsat-4 forward and return processors linked by user-user processing. (After Mallison and Robson (2001), Biglieri and Luise (1996)).

FIGURE 20.18 High-level architecture of each of eight Alphasat processors. (After Hili and Malou (2013)).

duplication of uplink channels before input to the processor. The processors belong to generation 3 of the Airbus-UK developments (Brown et al., 2014). The ADCs and DACs have a wider bandwidth, 48.5 MHz (Vilaça, 2004).

20.3.6.5 *Analog Frequency Converters and Oscillators* The analog frequency converters are the C-band downconverter and the L-band post-processor for the forward link, as well as the L-band pre-processor and the C-band upconverter for the return link.

As stated above, the Inmarsat-4 C-band downconverter demultiplexes the ten subbands of user signals from the ground station and converts them to IF.

The Inmarsat-4 L-band pre-processor performs frequency conversion from IF to L-band, filtering, local oscillator (LO) distribution, and redundancy switching. The post-processor performs similar functions (Kongsberg Defence Systems, 2005). Both the pre- and post-processors have SAW bandpass filters (Airbus Defence and Space, 2014d).

The Inmarsat-4 centralized frequency generator, called the "synchronized clock generator unit" by the manufacturer (Airbus Defence and Space, 2014c), feeds the L-band pre- and post-processors. It contains two dual synthesizer modules and one reference module. The unit provides two synchronized frequency outputs for each of the forward and return processors. The reference oscillator is a quartz oven-controlled crystal oscillator (OCXO) at 10 MHz (Airbus Defence and Space, 2014a).

20.4 TERRESTAR/ECHOSTAR XXI

TerreStar Networks, a US company founded in 2002, planned to have the TerreStar-1 satellite on orbit offering mobile service combined with an ATC to the US. The idea of an ATC is that the satellite company reuses its satellite frequencies for terrestrial coverage that it integrates with the satellite coverage. An ATC is known as a **complementary ground component (CGC)** in Europe.

TerreStar Networks went bankrupt in 2010. EchoStar Mobile bought the backup satellite, TerreStar-2, and had it launched in 2017. The satellite was renamed EchoStar 21, then EchoStar T2, then Echostar XXI. The company plans a similar system as TerreStar did but with coverage for the European Union (EU). EchoStar Mobile is an Irish company and a subsidiary of the US company EchoStar Corporation. EchoStar XXI is identical to TerreStar-1.

Key properties of the system that TerreStar planned and that EchoStar Mobile now operates are the following:

- One operating spacecraft
- User-ground station links
- ATC (not a reality yet in Europe)
- GMR-1 release 3 as air interface standard
- Bent-pipe payload
- GBBF.

20.4.1 TerreStar History

We present here a short history of the TerreStar company and its satellites, just because the story is dramatic and unlike that of Thuraya or Inmarsat.

TerreStar Networks wanted to have a satellite for IP-based voice, data, and video, along with a third-generation cellular network that would reuse the satellite frequencies. This all-new ATC would cost a vast amount. The system would allow the smallest dual-mode user terminals (Satnews, 2010). The phones were obliged by the Federal Communications Commission (FCC) to have both satellite and ATC capabilities (FCC, 2010). In 2009, TerreStar developed a dual-mode cell phone (Epstein, 2010) that had the normal cell phone-size, and AT&T released the phone to the market in 2010 (Phonearena, 2010).

TerreStar contracted for TerreStar-1, which launched in 2009. TerreStar also contracted for the obligatory ground-spare (SpaceNews, 2010). The MSS portion of the integrated system was obliged to be commercially available in the entire US before TerreStar could offer the ATC portion (FCC, 2010).

In 2010 TerreStar entered bankruptcy (TerreStar Networks, 2010), and in 2013 it emerged from bankruptcy (Securities and Exchange Commission, 2013). In 2012, Dish Network bought substantially all the assets of the company including the on-orbit TerreStar-1, the almost-built TerreStar-2, two ground stations, and spectrum licenses (Dish Network, 2013a). (It is not clear that TerreStar had built any cellular base stations (de Selding, 2010b).) Dish paid $1.4 billion for these assets. In the same year, it also paid $1.4 billion for the DBSD company, formerly known as ICO, which owned a GEO satellite and spectrum licenses. Neither acquisition fit in with the rest of the Dish business (Dish Network, 2013a).

In 2013 the FCC released Dish from the three obligations: (1) the ground spare, (2) that the phones provide satellite service along with the cellular, and (3) that satellite services be fully operating before terrestrial (Dish Network, 2013a).

In the second quarter of 2013 Dish decided it had no use for the TerreStar-2 satellite and the ICO satellite and wrote off most of their value. It also stopped operating TerreStar-1 (Dish Network, 2013b).

In December 2014 EchoStar Mobile bought the unlaunched TerreStar-2 for $55 million. In January of that year, EchoStar had bought the Solaris Mobile company and renamed it to EchoStar Mobile (EML). Solaris Mobile had a payload in orbit that by 2015 had not generated much business (Advanced Television, 2015).

20.4.2 Spacecraft

The two TerreStar spacecraft are identical (SpaceNews, 2010). TerreStar-1, launched in 2009, was at the time the heaviest commercial satellite ever launched (Semler et al., 2010). It was licensed in Canada (TerreStar Networks, 2009). It was at $111.1°$ W longitude (Dish Network, 2013a). When Dish Network bought the satellite, they renamed it EchoStar T1 (Gunter's Space Page, 2014b). Later Dish took it out of operation. The second spacecraft, originally named TerreStar-2 and now named EchoStar XXI, launched in 2017. It is located at $10.25°$ E longitude (EchoStar Mobile, 2018).

Figure 20.19 depicts one of the spacecraft. It has the typical rectangular-box shape. The 18-m S-band reflector off the east panel is the most obvious feature. The phased array that forms its feeds is also on the east face, tilting out a little. The reflector on the west panel is part of the Ku-band antenna (Semler et al., 2010).

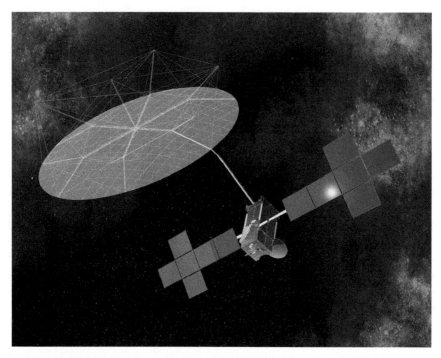

FIGURE 20.19   TerreStar-1 spacecraft image (Gunter's Space Page, 2014b). (©2009 Maxar Technologies. Reprinted with permission).

TerreStar-1's inclination was allowed to vary over life within ±6°. The spacecraft performed no north–south station-keeping, which saved fuel. The spacecraft used 22 uplink beacons, two at the ground stations and 20 at beacon stations, to help maintain spacecraft pointing (Semler et al., 2010).

20.4.3 TerreStar Coverage Area and Services

TerreStar-1 was licensed to provide service in the US and Canada (TerreStar Networks, 2009). Its coverage area was all the US states, Puerto Rico, the US Virgin Islands, and the territorial waters (AT&T, 2010). Coverage also included all of Canada but at a lower EIRP and G/T_s.

The TerreStar system provided the following satellite services to land and maritime users (AT&T, 2010):

- Packet-mode services (IP) similar to GSM, GPRS, and UMTS. Traffic channels were unidirectional (ETSI TS 101 376-1-3, v3.1.1, 2009). The main applications were data, video and voice (Satnews, 2010). The author could find no information on the rates.

20.4.4 TerreStar Ground and User Segments

The TerreStar system provided communication paths between a user and a ground station in either direction (TerreStar Networks, 2010).

The ground infrastructure was as follows. There were two ground stations which seem to have been identical, in North Las Vegas, Nevada, and Allan Park, Ontario. Each ground station was collocated with a gateway to the PSTN. The Canadian ground station provided TT&C for the satellite, including maintenance of the proper orbital location. The NOC was at Richardson, Texas. There were 20 calibration earth stations (CESs), 15 in the US and 5 in Canada, to aid in the GBBF. Each CES collected data from the satellite and transmitted it to the ground stations for processing (Epstein, 2010).

The ground stations performed the beam-forming, GBBF (Epstein, 2010). The BFN coefficients included both amplitude and phase. The GBBF also equalized the transponder channels and compensated for their amplitude and phase variation (Semler et al., 2010). The ground stations continually adapted the beam-forming to account for the diurnal movement of the spacecraft relative to its ground coverage. The diurnal movement itself slowly changed as the spacecraft inclination changed over spacecraft lifetime (Semler et al., 2010).

Both ground stations and all 20 CESs transmitted beacons to the spacecraft to help it maintain pointing (Semler et al., 2010).

The TerreStar system was designed to work with dual-mode terminals (satellite/ terrestrial) that allowed the user to roam between the satellite and cellular networks. They were equipped with GPS receivers (Phonearena, 2010). The user terminals had LP on both receive and transmit (Semler et al., 2010). Hughes Network Systems developed a chipset for both TerreStar and SkyTerra handsets that could handle GMR-1 release 3 as well as cellular network standards including GSM, GPRS, and WCDMA (PR Newswire, 2009).

20.4.5 Frequencies and Beams

The frequencies of TerreStar and EchoStar XXI are shown in Table 20.8. The frequencies with the users are in S-band and the frequencies with the ground stations are in Ku-band. There was a 20 MHz-wide band for user uplinks and another 20 MHz for user downlinks. However, TerreStar-1 was authorized by the FCC to use just 10 MHz in each direction, the lower half of the uplink band and the upper half the downlink band (FCC, 2010). The ground stations transmitted on a 250 MHz-wide band and received on two 250 MHz-wide bands. The satellite employed dual

TABLE 20.8 TerreStar and EchoStar XXI Frequencies and Polarizations (Semler et al., 2010)

	Uplink		Downlink	
	Frequencies (GHz)	Polarization	Frequencies (GHz)	Polarization
User links	2.000–2.020	LHCP	2.180–2.200	LHCP
Feeder links	12.75–13.00	Dual CP	10.70–10.95 and 11.20–11.45	Dual CP

CP on all links except for the downlink to the users, which was left-hand circular polarization (LHCP) only.

The TerreStar-1 coverage of the continental US, Alaska, and Canada was called the "composite beam" and consisted of up to 550 S-band spot beams. These beams could be manipulated in size, shape, location, and power within the coverage area with a few hours' notice. A spot-beam coverage map is provided in National Association of Broadcasters (2010). The small "Hawaii beam" was contoured and covered Hawaii. The small "Puerto Rico beam" was also contoured and covered Puerto Rico and the US Virgin Islands (SatStar.net, 2014; Epstein, 2010).

20.4.6 Air Interface

The TerreStar system used the communications standard GMR-1 release 3 for both the user and feeder links (FCC, 2010). Release 3 evolves the packet-mode services of release 2 to 3rd-generation UMTS-compatible services. GMR-1 release 3 uses a combination FDM/TDM multiplexing on all links. On the return link the TDM is actually TDMA. The same scheme is used on both forward and return link. The standard allows for a selection of modulation, coding, and bit rates. The standard also allows for multiple carrier-bandwidths, power control, and link adaptation. The standard calls for the RRC pulse-shaping filter with roll-off factor $\alpha = 0.35$. All control channels are synchronized, which makes for fast handovers (Matolak et al., 2002). TerreStar planned to use the W-CDMA air interface for the ATC system (FCC, 2010).

EchoStar XXI also uses the standard GMR-1 release 3 (EchoStar Mobile, 2018).

20.4.7 Payload

20.4.7.1 Payload Architecture   The TerreStar/EchoStar XXI payload is simple compared to that of Thuraya, Inmarsat-4, and Alphasat. No OBBF is necessary since the radiating elements' signals are transmitted to and from the ground station, which performs GBBF. All the elements' signals are multiplexed in the frequency domain. The payload would have a LNA for each polarization. The downconverter banks would shift each element's signal to its appropriate downlink band. The return-link processing would be the reverse. No credible information is available that says the payload has a processor.

20.4.7.2 Antennas and High-Power Amplifiers   The S-band antenna is one reflector fed by three phased arrays. The reflector is 18 m in diameter and of gold-plated mesh. At the time of TerreStar-1 launch, this was the largest reflector on a commercial satellite (Harris Corporation, 2009). The size of it is shown in Figure 20.20. Each radiating element is a stacked, cupped, microstrip disk (Simon, 2007). The composite beam used the array of 78 elements, the Hawaii beam used a separate array of eight elements, and the Puerto Rico/US Virgin Islands beam used yet another array of eight elements. All three arrays are visible in Figure 20.21, which is a closeup of Figure 20.19. The composite beam used all 78 elements on receive and 62 of them on transmit (Semler et al., 2010).

FIGURE 20.20 Photo showing size of TerreStar-1's S-band reflector (Harris Corporation, 2009). (Used with permission of L3Harris Technologies).

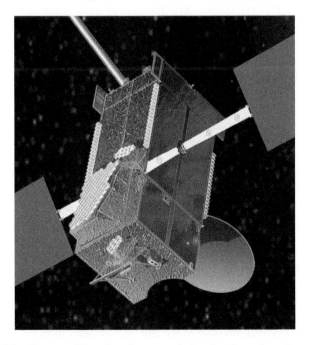

FIGURE 20.21 Closeup of TerreStar-1 spacecraft showing three phased arrays (Gunter's Space Page, 2014b). (© 2009 Maxar Technologies. Reprinted with permission).

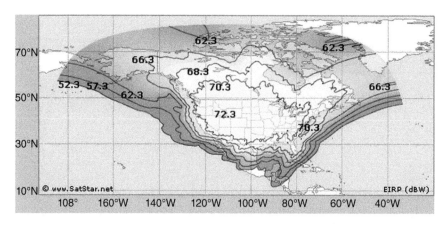

FIGURE 20.22 TerreStar-1 EIRP map of composite beam (SatStar.net, 2014). (Used with permission of SatStar Ltd., www.satstar.net).

The S-band HPAs are 100-W TWTAs (Military & Aerospace Electronics, 2005). There is no available information that says MPAs are used.

G/T_s is above 21 dB/K (Semler et al., 2010).

An EIRP map of TerreStar-1's composite beam is given in Figure 20.22. The EIRP to southern Canada was roughly the same as to the continental US but the EIRP to the far reaches of Canada was about 8 dB less than that. EIRP to Alaska was a few dB less than that to the contiguous US (CONUS). EIRP to Hawaii was 59–61 dBW and to Puerto Rico and the US Virgin Islands was about 64 dBW (SatStar.net, 2014).

The 1.5-m Ku-band antenna for the feeder links appears to have a single reflector. It probably has two feeds, one for each ground station. The feeds are dual CP. The reflector and the feed(s) are located on the west panel (Semler et al., 2010).

Twelve 24-W Ku-band nonlinearized TWTAs were made for TerreStar-1. This seems to imply that there are four TWTA redundancy rings, each 3:1 redundant, which in turn implies that there are two banks of LNAs, TWTAs, and frequency converters for each ground station.

20.4.8 EchoStar Mobile's Use of EchoStar XXI

20.4.8.1 General A representative coverage map of EchoStar XXI is given in Figure 20.23. The coverage area is all of Europe except the very northernmost part of Scandinavia but including Iceland, the westernmost part of the Near East, and the part of Africa bordering the Mediterranean. The beam is contoured. It appears that the two small phased arrays, which on TerreStar provided coverage outside CONUS, are not in use.

EchoStar XXI provides mobile voice and data services to Europe (SatBeams, 2020b). The EML Advantage service is provided by the spot beams of EchoStar XXI. It offers voice that is not over IP and a configurable streaming speed and is

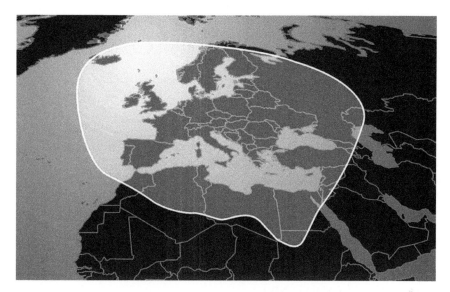

FIGURE 20.23   Representative coverage map for EchoStar XXI (EchoStar
Mobile, 2018). (Used with permission of EchoStar Mobile Limited).

aimed at business. The EML Momentum service is for fixed or mobile connectivity and is aimed at machine-to-machine (M2M) communications (EchoStar
Mobile, 2017). The author could find no statement of coverage or whether the
satellite providing the service is Echostar XXI or Eutelsat 10A, on which EchoStar
Mobile owns the S-band payload.

EchoStar XXI has its primary ground station in Griesheim, Germany. The spacecraft operations center is there, as well as the GBBF. EchoStar Mobile has 16 CESs
(EchoStar Mobile, 2017). The secondary ground station is at Rambouillet, France
(Russell, 2017b).

EchoStar Mobile offers two types of user terminal, both IP-based. The Hughes
4200 portable data terminal is for use with the EML Advantage service. It has an
integrated, flat phased array and can be used in all EU states and Norway and
Switzerland (Echostar Mobile, 2019a). The Hughes 4500 S-band terminal seems to
offer the EML Momentum service. It has an integrated omni antenna. There is no
statement about where it works or what beams it relies on (EchoStar Mobile, 2019b).

Solaris Mobile already owned the S-band payload of Eutelsat 10A before the
company was acquired by EchoStar. During in-orbit test, the antenna was found
to have failed. EchoStar reported that the payload is not fully operational
(EchoStar, 2020). The author was unable to find any map or statement of coverage
or the identification of the ground station for it, but in 2020 EchoStar Mobile is still
claiming to use it.

EchoStar Mobile's business model is not to operate the entire system itself but
to offer services throughout Europe on a wholesale basis to regional and local

operator partners. The partners might offer services such as mobile broadband and support for private networks and mobile virtual network operators (Analysys Mason, 2015).

20.4.8.2 *Complementary Ground Component* EchoStar Mobile wants to develop a system with integrated satellite and terrestrial communications, as TerreStar Network wanted (EchoStar Mobile, 2018). It is not clear who will provide the CGC. EchoStar has been working to include satellite communications in the 5G mobile-network air-interface standard, so that the integrated network will be able to provide 5G (EchoStar Mobile, 2019e).

For the system, EchoStar Mobile needs official approvals. In 2010, the European Commission awarded two companies, Solaris Mobile and Inmarsat, licenses to operate an S-band MSS integrated with an optional terrestrial mobile service throughout the 28 countries of the EU. In that agreement, every EU country has to allow EchoStar Mobile to operate the CGC (DotEcon Ltd, 2017). However, each country has to license it and each country is allowed to lay its own conditions (Analysys Mason, 2015). In 2020, the author can find no specific information about the CGC (EchoStar Mobile, 2019c).

EchoStar Mobile has rights to 15 MHz up and 15 MHz down (EchoStar Mobile, 2019d).

20.5 SKYTERRA

Ligado Networks owns and operates the SkyTerra satellite as part of its wireless broadband offering over North America. The first precursor company was founded in 1988. Since at least 2008 the successor companies planned to develop an ATC, but only in April 2020 did it gain FCC approval.

The SkyTerra system is similar to TerreStar in many ways. Its key features are as follows:

- One operating spacecraft
- User-ground station links
- ATC (its own cellular network)
- GMR-1 release 3 as satellite communications standard
- Digital transparent payload
- GBBF.

20.5.1 History

SkyTerra has a history even more dramatic than TerreStar's.

In 1988 the company American Mobile Satellite Corporation was founded. It later changed its name to Motient Corporation. After a merger, it became Mobile Satellite Ventures (MSV) (Wikipedia, 2017b). It contracted in 2006 to have three satellites built (Gunter's Space Page, 2014c). In 2008, MSV changed its name to

SkyTerra (Wikipedia, 2015). The company wanted to build an L-band satellite communications network fully integrated with its own ATC, a third-generation cellular network throughout North America. The company would sell its services wholesale to other companies (LightSquared et al., 2012). The services would be all-IP and the interfaces with the ATC would be standards-based (Mitani, 2007). The terminals would be comparable to current cell-phones (Segal, 2007). They were required to have both terrestrial and satellite capability (SpaceNews, 2011). In May 2007, it was reported that the order for the third satellite, which would have served South America, was canceled (Analytical Graphics, Inc., 2014).

In 2010 SkyTerra became part of a new company, LightSquared. Later that year SkyTerra 1 launched (Wikipedia, 2015). SkyTerra 2 was also completed. It was identical to SkyTerra 1 and would serve the same market and would serve as in-orbit spare. The ATC portion of the system was required to be ancillary to the satellite portion (Segal, 2007). LightSquared had to pay hundreds of millions of dollars to Inmarsat, for Inmarsat to adjust its use of L-band so that LightSquared could have a contiguous piece of spectrum. This same year LightSquared signed a $7 billion contract with Nokia Siemens Network to build and install the terrestrial network including about 36,000 terrestrial base stations. Use of the ATC was conditioned on maintaining the satellite communications (de Selding, 2010a).

In 2011 the FCC permitted LightSquared's customers to sell phones that offered only the terrestrial service, if the company could prove that this would not interfere with GPS signals (SpaceNews, 2011).

The L1 GPS signal, at 1.575 GHz, is relied on by virtually all GPS applications (Parkinson, 2018). It lies near the two bands 1.526–1.536 and 1.545–1.555 GHz that LightSquared was authorized by the FCC to use for terrestrial in addition to satellite forward link to user (Parkinson, 2018). A panel of US experts concluded that there would be interference into GPS (Ferster, 2012a). The European Commission said that Egnos and the future Galileo would be rendered useless (de Selding, 2011).

LightSquared switched from Nokia Siemens to Sprint Nextel, thereby saving an estimated $13 billion; LightSquared would piggyback onto existing cell-phone base stations (Godinez, 2011).

In 2012 the FCC announced it would revoke LightSquared's conditional authority for the ATC because it would interfere too much into GPS signals. It was thought that LightSquared had already invested $3 billion in the ATC (Ferster, 2012b). Sprint canceled its deal with LightSquared (Lawson, 2012). LightSquared filed for bankruptcy (LightSquared, 2012). The satellite manufacturer repurposed SkyTerra 2 for another customer (Gunter's Space Page, 2014c).

In March 2015 LightSquared emerged from bankruptcy, after its creditors agreed that EchoStar's Charles Ergen could get his $1 billion loan back in cash. The company renamed itself New LightSquared and was allowed $1.25 billion of operating funds (Brown, 2015). In 2016, the company settled its lawsuits against three GPS equipment suppliers (Divis, 2016). It renamed itself Ligado Networks.

Iridium raised its concern in 2016 that the Ligado terrestrial forward link would interfere into its user terminals (Russell, 2017a).

Serious concerns of the GPS community continued to haunt Ligado's proposed ATC, including the aviation industry (Divis, 2016) and the Air Force (Capaccio and Shields, 2016). In return for giving up the upper 10 MHz of its forward-link ATC spectrum (Goovaerts, 2016), Ligado wanted to share a different 5 MHz that the US's National Oceanographic and Atmospheric Agency uses. Spearheaded by the American Meteorological Society, 22 organizations sent a letter to the FCC saying that none of the concerns raised by the ATC have been resolved (Myers et al., 2017). In May 2018, Ligado notified the FCC it would reduce the power in its terrestrial forward link by a factor of more than 1000 (Alleven, 2018). However, the leading American GPS experts showed that it would take a further reduction by a factor of more than 10,000, down to 0.00022 W (Divis, 2018).

In April 2020 the FCC unanimously approved Ligado's application, in which Ligado framed the ATC as a potential 5G network (Pelkey, 2020). The US Departments of Defense and Transportation immediately called on the FCC to reverse the decision (Erwin, 2020). Three months later a committee charged with giving the US government advice on space-based navigation services concluded it was a "high-risk" decision that jeopardizes GPS (Foust, 2020).

20.5.2   Spacecraft

Ligado has one spacecraft, SkyTerra 1, at 101.3° W longitude. It was launched in 2010 (LightSquared, 2015), at which time it was the largest commercial satellite ever launched (Segal, 2007).

The spacecraft is depicted in Figure 20.24, with the earth deck facing down. It is generally rectangular in shape but has the oddity of a large tower on the anti-earth deck, which holds the boom for the huge L-band reflector. The Ku-band reflector for feeder links is visible on the east or west panel. The outboard radiator panels contain heat pipes, as do the ones in the spacecraft body (LightSquared, 2006).

The spacecraft's inclination is allowed to vary over life within ±6° (LightSquared, 2006). The spacecraft performs no north–south station-keeping, which saves fuel (Mitani, 2007). It is kept at its nominal longitude within ±0.05° (LightSquared, 2006).

To compensate for L-band antenna mispointing due to thermal effects of the reflector, each ground station transmits four beacons to the spacecraft. From the received signals an estimate is made of the thermal deformation, which is used to compensate for mispointing (Koduru et al., 2011).

The spacecraft generates and sends down power-control beacons to the ground stations (LightSquared, 2006).

20.5.3   Coverage Area and Services

SkyTerra 1 is licensed to provide service in the US and Canada (Mitani, 2007). Figure 20.25 shows the coverage area of SkyTerra 1 and an example of the user spot beams. The coverage area includes all the US states, Puerto Rico, the US Virgin

FIGURE 20.24   SkyTerra 1 spacecraft image (Mitani, 2007). (Used with permission of Ligado Networks LLC).

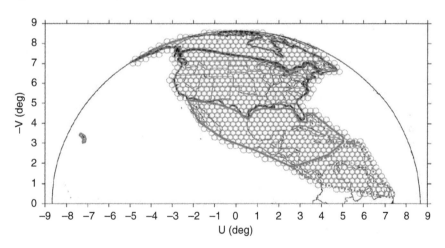

FIGURE 20.25   SkyTerra 1's L-band coverage area and spot beams example (Mitani, 2007). (Used with permission of Ligado Networks LLC).

Islands, and the territorial waters (Mitani, 2007). It also includes Canada at a lower EIRP and G/T_s. Alaska has complete full-time coverage when the spacecraft inclination is $0°$ but otherwise it has complete coverage only about 50% of the day (LightSquared, 2006).

The system provides services to land and maritime users (LightSquared, 2006):

- Packet-mode (all-IP) services similar to GSM, GPRS, and UMTS, up to 150–300 Kbps in the forward link and 9–38 Kbps in the return link (LightSquared, 2006). Traffic channels are unidirectional (ETSI TS 101 376-1-3, v3.1.1, 2009). The main applications are Internet, video, and voice (Mitani, 2007).

20.5.4 Ground and User Segments

The SkyTerra system provides communication paths between a user and a ground station in either direction (Mitani, 2007).

There are four ground stations, located at Napa, California; Dallas, Texas; Saskatoon, Saskatchewan; and Ottawa, Ontario (LightSquared, 2006). Each is collocated with gateways to the PSTN and Internet (Koduru et al., 2011). The ground stations at Napa and Ottawa provide TT&C. The satellite system is coordinated through the network management system located at the NOC, as would be the ATC (Koduru et al., 2011).

The ATC network would also have gateways into the PSTN and the Internet (Koduru et al., 2011). The ATC network would belong to LightSquared (Epstein, 2010).

The four ground stations perform beam-forming, GBBF. They form signals for the L-band antenna's radiating elements on the forward link and receive signals from the elements on the return link. More information on this is provided in the section below on satellite-system architecture. The ground stations can form a great variety of shapes and sizes of user beams (LightSquared, 2006). The GBBF performs amplitude and phase weighting for the satellite feed array's radiating elements. It also aligns the delays of the signals and equalizes the channels (Koduru et al., 2011). The ground stations adapt the set of beam-forming coefficients to account for the slowly varying spacecraft inclination over spacecraft lifetime (LightSquared, 2006).

The ground stations receive beacons generated by the spacecraft and use them to set the power level for the feeder links (LightSquared, 2006).

The ground stations also manage the high power on the spacecraft so the high-power devices do not become damaged. These include the diplexers, MMAs, and cables (Koduru et al., 2011).

Cobham offers the Explorer 122 user terminal for IP-based push-to-talk communications that include voice and data. It has a GPS receiver. The data speeds are up to 1 Mbps download and 10 Kbps upload (Cobham Satcom, 2017). Cobham also offers the Explorer MSAT-G3 user terminal, which incorporates the Explorer 122 and is also push-to-talk. It can connect to up to two cellular networks (Cobham Satcom, 2016).

20.5.5 Frequencies and Beams

The following SkyTerra 1 information comes from (LightSquared, 2006) except where otherwise indicated. The SkyTerra 1 frequencies are shown in Table 20.9. The frequencies with the users are in L-band, and the frequencies with the ground

TABLE 20.9   SkyTerra 1 Frequencies and Polarizations (LightSquared, 2006)

	Uplink		Downlink	
	Frequencies (GHz)	Polarization	Frequencies (GHz)	Polarization
User links	1.6265–1.6605	Dual CP	1.5250–1.5590	RHCP
Feeder links	12.75–13.25	Dual CP	10.70–10.95 and 11.20–11.45	Dual CP

stations are in Ku-band. There is a 34 MHz-wide band for user uplinks and another such for user downlinks. In partial contradiction to this is the source (Koduru et al., 2011), which speaks of only 30 MHz in each direction. The ground stations transmit (uplink) on a 500 MHz-wide band and receive (downlink) on two 250 MHz-wide bands. The satellite employs dual CP on all links except for the downlink to the users, which is only RHCP.

An example of SkyTerra 1's L-band spot beams was shown in Figure 20.25. The beams have a diameter of 0.40° (Mitani, 2007). There are about 500 of them in both directions, each with a nominal service area 100–150 miles in diameter (Segal, 2007). The system can generate not just spot beams but also a great variety of beam sizes and shapes (LightSquared, 2006). A small contoured beam covers Hawaii. Another small contoured beam covers Puerto Rico and the US Virgin Islands (Mitani, 2007).

There are three colors (frequencies) on the forward link and seven on the return link (Koduru et al., 2011).

The motion of the spacecraft relative to the ground would deform the beams but the beam-forming coefficients compensate. The fact that the inclination varies from zero means that over the course of the day the ground stations have to continually update the coefficient set.

20.5.6   Air Interface

SkyTerra 1 follows a subset of the communications standard GMR-1 release 3 (Section 20.4.6) for the user links. SkyTerra does not use circuit-mode services but only packet-mode. SkyTerra was also going to follow the standard Geostationary Mobile Satellite Adaptation (GMSA) (LightSquared, 2006), which was being developed by Qualcomm (Tian et al., 2013). Since the Qualcomm Web site does not mention it in 2015 and 2020, it seems to have died.

The standard allows for a selection of modulation, coding, and bit rates. SkyTerra uses only binary phase-shift -keying (BPSK) and QPSK, channel bit rates up to 160 Kbps, and carrier bandwidth from 31.25 to 156.25 MHz, in both directions (LightSquared, 2006).

20.5.7 Satellite-System Architecture

The SkyTerra 1 system needs 4 GHz of feeder links in each direction (forward and return). Since only 500 KHz is available, both polarizations are used and four ground stations are necessary (LightSquared, 2006). The EIRP per beam is flexible, and the channelization and routing are flexible. The GBBF and the digital channelizer provide rapid reconfiguration of coverage and reallocation of bandwidth (Mitani, 2007).

Thirty MHz is available for the downlinks to the users. Each of the four ground stations forms the radiating element signals for 7.5 MHz of this. The 7.5 MHz is in three segments of 2.5 MHz, one segment of 2.5 and another of 5, or one contiguous 7.5 MHz segment. The ground stations can share the payload L-band capacity in any ratio. Each spot beam supports numerous carriers (LightSquared, 2006).

The ATC cells would use the same frequency bands as the satellite beams. The terrestrial cells would be so much smaller than the satellite beams that the cells could use the frequencies that the satellite beam is not using. The system could change frequency assignments dynamically to address change in demand (Mitani, 2007).

20.5.8 Payload

20.5.8.1 Payload Architecture The payload is simple compared to that of Thuraya, Inmarsat-4, and Alphasat. A top-level payload diagram is given in Figure 20.26.

In the return-link hardware chain, shown in the upper half of the diagram, the receive filters are analog. In the block labeled "Converter L-IF," the ADC takes place. In the block labeled "Converter IF-Ku," the DAC takes place. Similarly for the forward-link hardware chain. Each hardware chain contains a digital channelizer. In the forward-link, the block labeled "HPA" is actually a set of MPAs

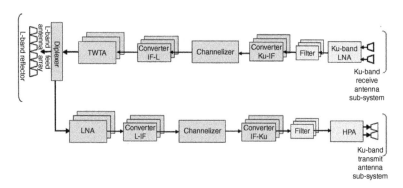

FIGURE 20.26 SkyTerra 1 payload diagram (LightSquared, 2006).

(Koduru et al., 2011), probably ten of them. The payload offers a gain adjustment of ±6 dB, in both directions (LightSquared, 2006).

20.5.8.2 *Antennas and High-Power Amplifiers*

The L-band antenna is one reflector fed by a phased array. The reflector is 22 m in diameter (Mitani, 2007) and is made of mesh (Segal, 2007). At the time of launch it was the largest reflector on a commercial satellite (Aerospace Technology, 2010). The size of it is shown in Figure 20.27. The radiating elements are dual-polarization patch-excited cups (Section 11.8.5). There are 82 radiating elements in the array that feeds the reflector (Mitani, 2007).

The L-band HPAs are SSPAs, arranged in MPAs (Section 11.12.3) (Mitani, 2007). As much as 10% of the satellite's transmit power can be allocated to one spot beam (LightSquared, 2006).

Information on the L-band antenna and amplifiers is given in Table 20.10 and an EIRP plot is given in Figure 20.28.

The 1.6-m Ku-band antenna, shown in Figure 20.29, provides four beams to the four ground stations for feeder links (Mitani, 2007). Characteristics of the Ku-band antenna and TWTAs are provided in Figure 20.29 and Table 20.11, respectively.

20.5.8.3 *Digital Channelizer*

The payload's digital channelizer channelizes the near-baseband signals for processing and switching. The channelization is partly fixed and partly tunable (LightSquared, 2006). It can be programmed to select variable-bandwidth channels up to a contiguous bandwidth of 7.5 MHz

FIGURE 20.27 Photo showing size of SkyTerra 1 L-band reflector (Harris Corporation, 2010). (Used with permission of L3Harris Technologies).

TABLE 20.10   Characteristics of SkyTerra 1 L-Band (LightSquared, 2006)

	Uplink (Return Link)	Downlink (Forward Link)
Polarization	Dual CP	RHCP
Peak antenna gain	30–47 dBi	30–47 dBi
System noise temperature	650–400 K	N/A
Peak G/T_s	2–21 dB/K	N/A
Power into antenna	N/A	4000 W
Total EIRP at peak, max per beam	N/A	80 dBW

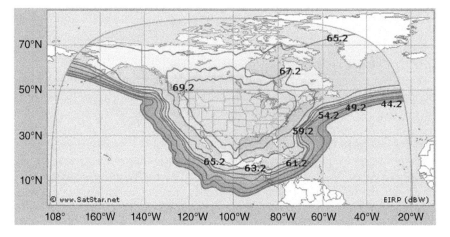

FIGURE 20.28   SkyTerra 1 L-band EIRP plot (SatStar.net, 2017). (Used with permission of SatStar Ltd., www.satstar.net).

TABLE 20.11   Characteristics of SkyTerra 1 Ku-Band (LightSquared, 2006)

	Uplink (Forward Link)	Downlink (Return Link)
Polarization	Dual CP	Dual CP
Peak antenna gain	42 dBi	42 dBi
System noise temperature	780 K	N/A
Peak G/T_s	11 dB/K	N/A
Power into antenna	N/A	50 W
Total EIRP at peak, max per beam	N/A	51.5 dBW saturated

anywhere in the 30 MHz-wide transmit band (LightSquared, 2006). Channelization is based on the FFT (Koduru et al., 2011). The L-band filters in the channelizer are digital. Most filtering of the user signals is done by the channelizer (LightSquared, 2006). The port bandwidth is on the order of the port bandwidth of the Alphasat processor (Brown et al., 2014).

FIGURE 20.29   Closeup of SkyTerra 1 spacecraft image showing Ku-band dish (Mitani, 2007). (Used with permission of Ligado Networks LLC).

20.6 INMARSAT-5 (GLOBAL XPRESS) F1-F4

The first four Inmarsat-5 or Global Xpress satellites operate in the Ka/K-bands. The satellites are closely integrated with the Inmarsat-4 L-band BGAN service as a backup in case of extremely heavy rain (Inmarsat, 2017a). The key properties of the part of the Inmarsat system served by Inmarsat-5 are as follows:

- Four operating spacecraft
- User-ground station links (Inmarsat, 2012a)
- Gateways at ground stations and other network access points, to the Internet and telephone companies.
- Digital Video Broadcasting (DVB)-S2 on forward link and proprietary iDirect protocol on return link, both adaptive to link conditions
- Bent-pipe payload (Spaceflight 101, 2017)
- Single-feed-per-beam (SFPB) reflector antenna.

A fifth Global Xpress satellite was launched in 2019 to deliver focused capacity across Europe and the Middle East. Its capacity is greater than that of the first four satellites combined (Inmarsat, 2020). It is not described below.

The satellites and ground segment offer extra provisions for the military, especially that of the North Atlantic Treaty Organization (NATO) member states (Gizinski III and Manuel, 2015), besides the commercial provisions, but here we only discuss commercial or mixed-use.

20.6.1 Spacecraft

The four Inmarsat-5 spacecraft are identical (Spaceflight 101, 2017):

- F1, also known as Global Xpress 1 and I-5 EMEA (European, Middle East, and Africa), at 63° E and launched in 2013
- F2, also known as Global Xpress 2 and I-5 Americas and Atlantic Ocean Region, at 55° W and launched in 2015
- F3, also known as Global Xpress 3 and I-5 Pacific Ocean Region, located at 179° E and launched in 2015
- F4, also known as Global Xpress 4, located at 56° E in 2020, to provide extra capacity over Asia, launched in 2017 (Inmarsat, 2017a; Gunter's Space Page, 2017; Satbeams, 2020a). It may happen that the satellite is repositioned, as its role could change over its life (Henry, 2017b).

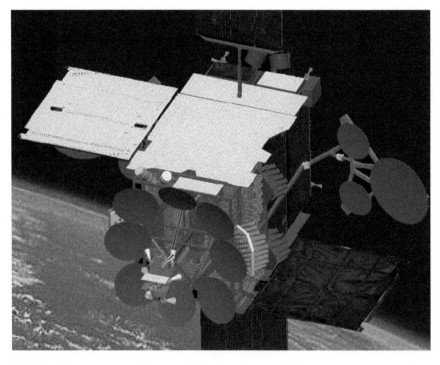

FIGURE 20.30 Inmarsat-5 spacecraft image (Inmarsat, 2017b). (Used with permission of Inmarsat).

An Inmarsat-5 spacecraft is illustrated in Figure 20.30. On the earth deck are eight reflectors. On the visible east or west panel are three reflectors, and the panel opposite has a similar arrangement. Extended off of the north and south panels are deployed radiator panels.

The spacecraft is station-kept to within ±0.05° east-west and north–south, at least F2 is (Inmarsat, 2012a).

20.6.2 Coverage Area and Services

Figure 20.31 shows the expected coverage areas of the spot beams of the Inmarsat-5 spacecraft. (The EC beam is not shown.) F4's coverage overlaps that of F1 and F3.

The Inmarsat system with the Inmarsat-5 spacecraft provides services including broadband Internet access, multimedia, and voice over IP (Inmarsat, 2012a). It is principally used for mobile broadband communications. Some specific non-military applications are as follows (Inmarsat, 2015a):

- Live full-motion video
- Video teleconferencing
- Broadband IP network interconnectivity
- IP multicast
- Disaster recovery
- Emergency response.

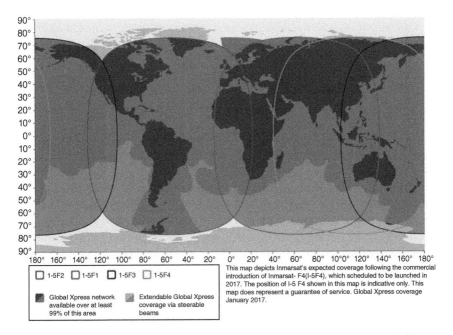

FIGURE 20.31   Inmarsat-5 map of expected coverage (Inmarsat, 2017b). (Used with permission of Inmarsat).

A service called "Synthetic Virtual Network" allows a customer to purchase IP bandwidth, measured in Mbps, not MHz (Vilaça and Bath, 2011). It is akin to a virtual hub network (Section 17.4.2).

Until 2017 the maximum bit rate per channel on the user downlink was 50 Mbps (Henry, 2017a), but that year iDirect introduced a user terminal in its iQ series which achieved 330 Mbps on the forward link with the DVB-S2X standard. Inmarsat and iDirect intended to gradually introduce these improved capabilities to the GX system during 2018 and 2019 for all user sectors (Digital Ship, 2017).

20.6.3 Ground and User Segments

The satellites are controlled from the SCC at company headquarters in London (Inmarsat, 2017c). The TT&C stations are in the gateways (Inmarsat, 2012a). The NOC, also in London, monitors the ground stations (Inmarsat, 2013b). The management systems for ground stations and network are provided by iDirect (Hibberd, 2012).

The ground segment has multiple non-military components. One component is the ground stations or gateways, which are called "SASs" as for Inmarsat-4 and Alphasat broadband. All Inmarsat-5 ground stations have the same capabilities. There are two per satellite, a primary and a hot-standby redundant one to provide site diversity in case of heavy rain (Hadinger, 2015). One pair of ground stations is at Fucino, Italy, and Nemea, Greece; another pair is at Lino Lakes in Minnesota, USA, and Winnipeg in Manitoba, Canada; the third pair is at two sites near Auckland, New Zealand (Inmarsat, 2014d). The hot-standby one is within the same gateway beam as the primary, so less than 1500 km away. The primary and redundant SAS are connected by terrestrial lines (Hibberd, 2012). Government and large commercial customers can have their equipment located right at the SASs. Another component of the ground network is the three "meet-me points (MMPs)" in Europe, Asia-Pacific, and the US (Gizinski III and Manuel, 2015). At these points, customer networks can access the Inmarsat-5 system. The MMPs also have gateways to the Internet and public telephone systems (Hadinger, 2015). Inmarsat-5 connects to most international telecommunications carriers for phone calls and SMS (Inmarsat, 2013d). Another terrestrial component is the Inmarsat DCN, which consists of leased circuits connecting the SASs and the MMPs to the Inmarsat IP backbone (Gizinski III and Manuel, 2015).

The services are not aimed at the consumer market but at the very small-aperture terminal (VSAT) market of the maritime, enterprise, government, and aeronautical sectors (Vilaça and Bath, 2011).

Inmarsat-5 offers two models of commercial service. One is that a value-added reseller sells service packages that Inmarsat has put together. The other is that a virtual network operator sells bandwidth to value-added resellers for the latter to customize and subdivide before reselling (Hibberd, 2012).

A current terminal for airborne use is Orbit's GX46, whose commercial market is business jets. The terminal operates over 19.2–20.2 GHz on the forward link and 29.0–30.0 GHz on the return link (as well as over higher bands in each direction for military use). It can switch between the two CPs. The reflector diameter is 46 cm (Orbit Communications Systems, 2019). A current terminal for maritime use is Intellian's GX60NX, whose market is commercial vessels, fishing fleets, large workboats, and leisure yachts. Inmarsat reported in 2020 that more than 9000 vessels worldwide are using Global Xpress satellites (Wingrove, 2020). The terminal operates over 19.2–20.2 GHz on the forward link and 29.0–30.0 GHz on the return link, with LHCP for receive and RHCP for transmit. The reflector diameter is 65 cm. The antenna is on deck and the rest of the terminal is below deck, connected by cable (Intellian, 2020).

The user terminal initially acquires the system with help from the system. The terminal knows where it is located from its GPS receiver and it applies this information to the preloaded satellite beam map that it has obtained from the signaling channel (below). The system indicates a broadcast channel, over which the terminal receives configuration parameters from the system (Inmarsat, 2014c). The user terminal has a continual conversation with the Inmarsat-5 system to allow optimal operation of the terminal in the system. Handover between channels in the same beam or between overlapping beams of one satellite are seamless, but handover between satellites is not seamless (Hibberd, 2012).

20.6.4 Satellite Beams

An Inmarsat-5 satellite has various kinds of beams, all of which are used on both forward and return (Inmarsat, 2012a):

- 89 global payload (GP) spot beams that cover the visible earth, of which 72 are active at any given time. For user communications.
- Six high-capacity payload (HCP) spot beams that are independently steerable to any visible location on the earth. For user communications.
- Two gateway spot beams that are steerable to any visible location on the earth. For gateway (feeder) communications.
- One EC beam. For the signaling that user terminals need to initially acquire the system.

The 89 GP spot beams are about 2° across, as illustrated in Figure 20.32. Everywhere in the combined coverage area of the beams, the EIRP and G/T_s are within 2 dB of peak (Hadinger, 2015).

The six steerable HCP spot beams are narrower than the GP spot beams. The beams are able to be radially symmetric because their horns are separated from each other. Any number of these beams can overlay any GP spot beam

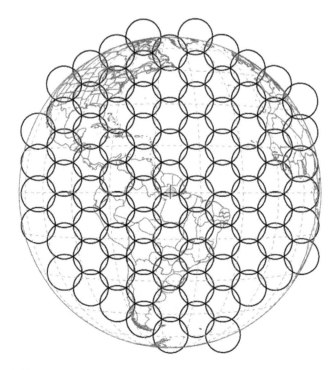

FIGURE 20.32 Inmarsat-5's 89 global payload spot beams, representative coverage (Inmarsat, 2012a).

(Hibberd, 2012), to permit a capacity equal to eight times that of a GP beam (Hadinger, 2015).

The two gateway spot beams are a little larger than the HCP beams. Each satellite communicates with one gateway, either to the primary for that satellite or to its hot-standby one. The redundant one is used in the event of high rainfall at the primary. Both gateway spot beams point to the same location on the earth (Inmarsat, 2012a).

20.6.5 Frequencies

Inmarsat-5 is different from the other mobile satellite systems addressed in this chapter in that it is in the FSS (Section 17.4), not the MSS. Specifically, its HCP frequency bands are allocated to the FSS (ITU, 2016a). However, the ITU's 2015 WRC allowed mobile users to communicate with FSS satellites in the bands 19.7–20.2 and 29.5–30 GHz (ITU, 2016b). The feeder bands are allocated to the FSS, as they are for the other MSS systems in this chapter.

Table 20.12 gives the frequency bands of each satellite's communication links. These links comprise the user and feeder links and are carried on spot beams.

TABLE 20.12 Inmarsat-5 Frequencies and Polarizations of Communication Links (Inmarsat, 2012a)

	Uplink		Downlink	
	Frequencies (GHz)	Polarization	Frequencies (GHz)	Polarization
User links—high-capacity payload (HCP) spot beams	29.0–29.5	Dual CP	19.2–19.7	Dual CP
User links—global payload (GP) spot beams	29.5–30.0	RHCP	19.7–20.2	LHCP
Feeder links for HCP	27.5–28.0	Dual CP	17.7–18.2	Dual CP
Feeder links for GP	28.0–29.5	Dual CP	18.2–19.7	Dual CP

The user links as well as the feeder links are at Ka-band for the uplink and at K-band for the downlink. The Ka-band uplink band of 2.5 GHz is divided into three subbands: 0.5 GHz wide for the HCP forward feeder channels, 1.5 GHz wide for the GP forward feeder channels, and 0.5 GHz wide for the GP return user channels. The HCP return user channels share the highest 0.5 GHz of the GP forward feeder channels. The K-band downlink band of 2.5 GHz is similarly divided into three subbands: 0.5 GHz wide for the HCP return feeder channels, 1.5 GHz wide for the GP return feeder channels, and 0.5 GHz wide for the GP forward user channels. The HCP forward user channels share the highest 0.5 GHz of the GP return feeder channels.

The combined feeder-link bandwidth in each direction is 2 GHz. Taking into account the dual CP used on all feeder links, the combined feeder-link bandwidth in each direction is 4 GHz.

The user links on the HCP spot beams use dual CP for both up and down, with all channels on one polarization (LHCP on forward and RHCP on return) and two channels repeated on the other polarization. The user links on the GP spot beams use only one CP on up and the orthogonal polarization on down. One GHz is the bandwidth of the combined user links on the uplink (return) and also on the downlink (forward). Half of this is shared with GP feeder links.

The GP user beams employ six-fold frequency reuse on both forward and return (Inmarsat, 2012a).

Table 20.13 gives the frequency bands of each satellite's signaling links, both forward and return. The links are very narrowband and are carried on the EC beam. The channels are within the GP feeder and user bands, at their lower edges. Signaling makes known the frequencies and locations of every beam to the user terminals. It allows automatic network configuration and rapid network login (Nicola and Plecity, 2013).

The channel structure of the forward and return links is as follows. On both feeder and user links, on both forward and return, the GP channels are 40 MHz wide and contain a modulated signal in 32 MHz. Similarly, all HCP channels are 125 MHz

TABLE 20.13   Inmarsat-5 Frequencies and Polarizations of Signaling (Inmarsat, 2012a)

Link	Frequencies (GHz)	Polarization
Up	28.0045–28.0095	Dual CP
	29.5045–29.5095	RHCP
Down	18.2045–18.2095	Dual CP
	19.7045–19.7095	LHCP

TABLE 20.14   Inmarsat-5 Possible GP Channel-to-Beam Allocation (Inmarsat, 2012a)

Number of Available Channels	Number of GP Spot Beams	Number of Channels Per GP Spot Beam
48	48	1 (fixed)
24	12	1 or 2
	29	Up to 1

wide and contain 100 MHz of signal (Inmarsat, 2012a). Note that the guard bands are 20% instead of the usual 10%.

The channel assignment to beams is variable, which gives higher functionality to the payload. For the GP beams, there are 72 channels and 72 active beams. Most beams carry one channel but some carry two or none. The GP channel usage is shown in Table 20.14. For the HCP beams, there are eight channels and six beams. Most beams can be assigned one of two possible channels (Inmarsat, 2012a).

20.6.6   Air Interface

The system is adaptive in order to provide the maximum throughput for a user terminal consistent with the propagation conditions at the terminal. The air interface supports adaptive modulation and coding on both forward and return link, as well as multiple carriers with different symbol rates on the return link (Hibberd, 2012).

Inmarsat-5 uses the DVB-S2 standard including adaptive coding and modulation (ACM) for the forward link. It uses a proprietary iDirect protocol for the return link, which is similar to DVB-Return Channel Satellite (RCS) Next Generation (NG) (Hibberd, 2012). The iDirect protocol uses multi-frequency time-division multiple access (MF-TDMA) (Hadinger, 2015). iDirect describes A-TDMA as "a channel access method that allows the return channel configuration to optimally change based on link conditions and spectral degradation" (iDirect, 2017). The system provides acceleration of the Internet protocols Transmission Control Protocol (TCP) and Hypertext Transfer Protocol (HTTP) (Hibberd, 2012).

The forward link has 32 MHz-wide modulated carriers in 40 MHz channels for the GP beams and 50 MHz-wide carriers on the HCP beams. The modulation schemes available are QPSK through 16APSK (Hibberd, 2012).

The return link has multiple carriers within each 32 or 50 MHz. The multiple carriers can have different symbol rates and bandwidths. The modulation schemes available are BPSK, QPSK, and 8PSK. Spread spectrum is an option. The resource access method is TDMA for the user terminal, with user terminals assigned time slots (Hibberd, 2012).

A duplex channel consists of one channel on the forward link and one multiple-frequency channel on the return link (Hibberd, 2012).

In 2017 Inmarsat reported that it was going to upgrade its system in 2018 and 2019 to use the DVB-S2X standard (Digital Ship, 2017).

The system allows a user terminal to have seamless beam handover in less than 250 ms with about one air-interface frame of data interruption. The system may trigger a channel handover due to traffic loading, which was expected in 2012 to take about 5 seconds. Both of these handovers are fast because of the dual receiver in each user terminal, which allows make before break. Satellite handover is not seamless and requires up to about 45 seconds, including repointing the user terminal antenna (Hibberd, 2012).

The system employs many ways to mitigate rain fade on the various links. The mitigation on the feeder links occurs on two levels. Both are based on monitoring the uplink fade level with a pilot signal that is retransmitted by the satellite. Fades of 3–6 dB can be overcome by increasing the uplink transmit power. When the fade is stronger, the other SAS paired with the current one, on hot standby, is switched in to replace the current one. The switch takes tens of ms, and the downlink traffic to the user terminals remains synchronized. The mitigation on the user K-band forward links is based on the user terminal's SNR that the terminal constantly signals back. The system changes the terminal's forward-link DVB-S2 MODCOD (Section 13.4.1), which defines the modulation and coding. This was expected in 2012 to be able to deal with about 15 dB of fading. The mitigation on the user Ka-band return links occurs in four levels. All are based on measurement of the return signal's C/N. The first mitigation method is to increase the user terminal's transmit power a little if possible. The second is to change to a channel that needs lower C/N. The third is to change the modulation and coding of return channels. The fourth method, under consideration in 2012, is to replace the carriers with multiple carriers of lower symbol rate. These four methods together were expected to provide at least 30 dB of rain fade mitigation on the user return links (Hibberd, 2012). As a last backup, a user terminal that integrates both Ka/K- and L-band capabilities can go over to L-band in extreme rain (Inmarsat, 2014c).

20.6.7 Payload

20.6.7.1 Payload Architecture The payload is a bent-pipe payload with sophisticated switching and filtering capabilities (Vilaça and Bath, 2011).

The payload does not modify any channel bandwidths. For the GP channels, it merely frequency-converts blocks of 6 or 12 consecutive channels. For the HCP channels, it merely frequency-converts pairs of channels at the same frequency (Inmarsat, 2012a).

FIGURE 20.33   Inmarsat-5 feed horn arrays on one side panel with stowed reflectors (Boeing Images, 2017). (Credit: The Boeing Co).

The payload has 61 active TWTAs on the forward link. The GP spot beams are produced by a bent-pipe repeater and have TWTAs in a 60:48 redundancy configuration. The HCP spot beams are powered by twelve 130-W TWTAs (Spaceflight 101, 2016). The EC beam has the remaining TWTA.

The payload has six active TWTAs on the return link (Inmarsat, 2012a).

20.6.7.2 Antennas All reflectors are gimbaled. The antennas were made by Harris Corporation, now L3Harris (Harris Corporation, 2011).

The earth-deck antennas can be seen in Figure 20.30. There are six reflectors of the same size and two smaller ones of the same size. The two smaller ones are opposite each other. The six larger ones create the six HCP spot beams, and the two smaller ones create the gateway beams. All eight reflectors have only the main reflector and a feed horn which provides dual CP on both transmit and receive. Such feed horns do exist, which can receive on Ka-band and transmit on K-band (Chan and Rao, 2008; Lee-Yow et al., 2010).

The six reflectors deployed off the east and west panels, shown in Figure 20.30, create the GP spot beams. Off of each of these panels are two smaller reflectors for receive and one large one for transmit. All employ the SFPB scheme without sub-reflectors. The 89 feeds for receive are split among the four smaller reflectors. The two transmit antennas are oversized with a long focal length, and their 89 feeds are split among the two reflectors. All six reflectors are shaped in the same way (Mugnaini and Benhamou, 2019). The reflectors are shown stowed in Figure 20.33.

The global beam, transmitting and receiving, is created by one or two horns on the antenna tower.

The maximum EIRP that the payload will transmit in the 17.7–19.7 GHz frequency band is 57 dBW for a modulated signal of 50 MHz. The maximum EIRP in the 19.7–20.2 band is 56.1 dBW for a modulated signal of 32 MHz (Inmarsat, 2012a).

REFERENCES

Advanced Television (2015). EchoStar Mobile to launch 2016. *Advanced Televison*; May 11. On advanced-television.com/2015/05/11/EchoStar-mobile-to-launch-2016. Accessed 2016 May 13.

Aerospace Technology (2010). SkyTerra 1 telecommunications satellite, United States of America. On www.aerospace-technology.com/projects/skyterra1telecommuni/. Accessed 2014 Oct. 7.

Airbus Defence and Space (2013). A communications satellite with vision. News release. June 18. On www.space-airbusds.com/en/news2/alphasat-designed-and-built-by-airbus-defence-and-space.html. Accessed 2014 Oct. 13.

Airbus Defence and Space (2014a).Quartz crystal oscillators OCXO-F. Product sheet. On www.space-airbusds.com/en/equipment/quartz-crystal-oscillators-ocxo-f.html. Accessed 2014 Oct. 10.

Airbus Defence and Space (2014b). Mobile processor. Product sheet. www.space-airbusds.com/en/equipment/mobile-processor-cmc.html. Accessed 2014 Dec. 28.

Airbus Defence and Space (2014c). Synchronised clock generator unit (SCGU). Product sheet. On www.space-airbusds.com/en/equipment/synchronised-clock-generator-unit-scgu.html. Accessed 2014 Dec. 28.

Airbus Defence and Space (2014d). Surface acoustic wave bandpass filters. Product sheet. On www.space-airbusds.com/fr/equipments/surface-acoustic-wave-bandpass-filters. html. Accessed 2014 Dec. 31.

Alexovich J, Watson L, Noerpel A, and Roos D (1997). The Hughes Geo-Mobile Satellite System. *Proceedings of International Mobile Satellite Conference*; June 16–18.

Alleven M (2018). Ligado proposes license modification to protect GPS aviation devices. News article. May 31. On www.fiercewireless.com/wireless/ligado-proposes-license-modification-to-protect-gps-aviation-devices. Accessed 2019 Apr. 5.

Analysys Mason (2015). Socio-economic benefits of harmonisation of the S-band CGC in Europe. Nov. 6. On EchoStarmobile.com/en/Newsroom.aspx. Accessed 2017 Oct. 20.

Analytical Graphics, Inc. (2014). Mobile Satellite Ventures. *Spacecraft Digest*. Oct. 15. On www.agi.com/resources/downloads/data/spacecraft-digest/display.aspx?i=2960. Accessed 2015 Feb. 17.

Applied Aerospace Structures (2008). Thuraya-3 successfully launched. News article. Feb. 1. On www.aascworld.com/Thuraya/news--1202746423.newsitem.html. Accessed 2014 Oct. 19.

AT&T (2010). TerreStar Genus™ dual-mode cellular/satellite smartphone now available from AT&T. Press release. Sep. 21. On www.att.com/gen/press-room/?pid=18505&cdvn=news&newsarticleid=31218&mapcode=. Accessed 2015 Feb. 2.

Biglieri E and Luise M. editors (1996). *Signal Processing in Telecommunications, Proceedings of 7th International Thyrrhenian Workshop on Digital Communicatons*, Sep. 10–14, 1995. London: Springer. Paper in Chapter 4 by Craig AD and Petz FA, Payload digital processor hardware demonstration for future mobile and personal communication systems.

Boeing Images (2001). Drawing of a Thuraya satellite. Image 01pr-01515e. July.

Boeing Images (2017). Photo of an Inmarsat-5 satellite in factory. Image 16403811395_2973d57b91_k.

Brown N (2015). LightSquared creditors to vote on latest bankruptcy exit plan. News article. Jan. 20. Reuters. On www.reuters.com/article/2015/01/20/us-lightsquare-bankruptcy-idUSKBN0KT28H20150120. Accessed 2015 Feb. 14.

Brown SP, Leong CK, Cornfield PS, Bishop AM, Hughes RJF, and Bloomfield C (2014). How Moore's law is enabling a new generation of telecommunications payloads. *AIAA International Communications Satellite Systems Conference*; Aug. 4–7.

Capaccio A and Shields T (2016). Air Force wary of GPS interference from LightSquared's successor. *Bloomberg*; Mar. 16. On www.bloomberg.com/news/articles/2016-03-16/air-force-wary-of-gps-interference-from-lightsquared-s-successor. Accessed 2016 May 14.

Chan KK and Rao SK (2008). Design of high efficiency circular horn feeds for multibeam reflector applictions. *IEEE Transactions on Antennas and Propagation*; 56 (1) (Jan.); 253–258.

Chie CM, former Boeing employee (2014). Private communications. Oct 26 and 28.

Cobham Satcom, formerly Thrane and Thrane (2012). TT-6900 Inmarsat BGAN RAN (3G radio access network). Product sheet. On www.ttvms.com/sitecore/content/www,-d-,thrane,-d-,com/Systems/Products/TT-6900.aspx. Accessed 2014 Dec. 8.

Cobham Satcom (2016). Explorer MSAT-G3. Product sheet. Nov. On www.groundcontrol. com/MSAT/Explorer_MSAT-G3_Brochure.pdf. Accessed 2020 Oct. 25.

Cobham Satcom (2017). Explorer 122. Product sheet. Mar. On www.cobhamsatcom.com/land-mobile-satcom-systems. Accessed 2020 Oct. 25.

Dallaire J, Senechal G, and Richard S (2009). The Alphasat-XL antenna feed array. *Proceedings of the European Conference on Antennas and Propagation*; Mar. 23–27.

de Selding PB (2010a). LightSquared cash jump-starts Inmarsat spectrum clearing. *SpaceNews*; Aug. 18.

de Selding PB (2010b). TerreStar files for bankruptcy, court okays EchoStar cash infusion. *SpaceNews*; Oct. 20.

de Selding PB (2011). European Commission adds voice to LightSquared opposition. *SpaceNews*; July 25.

de Selding PB (2013). Thuraya ringing up higher sales after four-year slide. *SpaceNews*; Mar. 22.

Digital Ship (2017). Latest iDirect modem pushes GX throughput to 330 Mbps. On www.thedigitalship.com/news/maritime-satellite-communications/. Accessed 2020 Oct. 26.

Dish Network (2013a). Form 10-K, annual report, year ending 12/31/12. Filed with SEC. Feb. 20. On dish.client.shareholder.com/annuals/cfm. Accessed 2015 Jan. 30.

Dish Network (2013b). Form 10-Q, quarterly report, period ending 06/30/13. Filed with SEC. Aug. 6. On dish.client.shareholder.com/results.cfm. Accessed 2015 Feb. 13.

Divis DA (2016). LightSquared seeks FCC approval for GPS "coexistence" plan. *Inside GNSS*; Jan. 15. On www.insidegnss.com/node/4785. Accessed 2016 May 13.

Divis DA (2018). GPS experts vote unanimously to oppose Ligado's newest proposal. On insidegnss.com/gps-experts-vote-unanimously-to-oppose-ligados-newest-proposal/. Accessed 2019 Apr. 5.

DotEcon Ltd (2017). Pricing of satellite complementary ground component, prepared for ComReg [Irish Commission for Communications Regulation]. Mar. On www.comreg.ie/media/dlm_uploads/2017/03/ComReg-1719a.pdf. Accessed 2017 Oct. 20.

Dutta S, inventor; ATC Technologies LLC, assignee (2010). Methods of ground based beam-forming and on-board frequency translation and related systems. U.S. patent 7,706,748 B2. Apr. 27.

EchoStar (2020). 2019 annual report for year ended December 31, 2019. Mar. 18. On ir.echostar.com/index.php/financial-information/annual-reports. Accessed 2020 Oct. 24.

EchoStar Mobile (2017). The future is now. Services brochure. On www.echostarmobile.com/Services/SatelliteServices.aspx. Accessed 2020 Oct. 23.

EchoStar Mobile (2018). About. On www.echostarmobile.com/en/echostarmobile/about.aspx. Accessed 2020 Oct. 23.

EchoStar Mobile (2019a). Hughes 4200 portable data terminal. On www.echostarmobile.com/Products/ProductsOverview/hughes-4200.aspx. Accessed 2020 Oct. 23.

EchoStar Mobile (2019b). Hughes 4500 S-band terminal. On www.echostarmobile.com/Products/ProductsOverview/hughes-4500.aspx. Accessed 2020 Oct. 23.

EchoStar Mobile (2019c). Converged services. On www.echostarmobile.com/Services/Converged.aspx. Accessed 2020 Oct. 23.

EchoStar Mobile (2019d). Regulatory. On www.echostarmobile.com/Innovation/Regulatory.aspx. Accessed 2020 Oct. 23.

EchoStar Mobile (2019e). 5G standardization. On www.echostarmobile.com/Innovation/5GStandardization.aspx. Accessed 2020 Nov. 11.

EngineerDir (2014). Thuraya-2, 3. Product description. On www.engineerdir.com/product/catalog/3797/. Accessed 2014 Oct. 21.

Epstein JW, CEO of TerreStar Networks (2010). Declaration of Jeffrey W. Epstein pursuant to local bankruptcy rule 1007-2 in support of first day pleadings. To US bankruptcy court, southern district of New York. Oct. 19. On www.terrastarinfo.com/pdflib/3_15446.pdf. Accessed 2015 Feb. 9.

Erwin S (2020). DoD issues new rebuke of FCC's decision to allow Ligado 5G network. *Space News*; Apr. 18.

European Aeronautic Defence and Space Co (EADS) (2005). First Inmarsat-4 satellite ready for launch. Press release. Mar. 8. On www.defense-aerospace.com/articles-pres/307/pres_release_ar.html. Accessed 2014 Oct. 8.

European Space Agency (ESA) (2011). Alphasat, framework for the Alphasat mission. Fact sheet. June 10. On telecom.esa.int/telecom/media/document/Alphasat%20factsheet%20 10-5-2011%20JH.pdf. Accessed 2014 Dec. 1.

European Space Agency (ESA) (2017). Alphasat I/Inmarsat-XL (Inmarsat-Extended L-band payload)/InmarSat-4A F4. *Earth Observation Portal.* On directory.eoportal. org/web/eoportal/satellite-missions/content/-/article/alphasat#overview. Accessed 2019 May 12.

ETSI TS (European Telecommunications Standards Institute Technical Specification) 101 376-3-21, v1.1.1 (2001). GEO-mobile radio interface specifications; part 3: network specifications; sub-part 21: position reporting services; stage 2 service description; GMR-1 03.299. Mar.

ETSI TS 101 376-1-3, v3.1.1 (2009). GEO-mobile radio interface specifications (release 3); third generation satellite packet radio service; part 1: general specifications; sub-part 3: general system description; GMR-1 3G 41.202. July.

Evans B, Werner M, Lutz E, Bousquet M, Corazza GE, Maral G, Rumeau R, and Ferro E (2005). Integration of satellite and terrestrial systems in future multimedia communications. *IEEE Wireless Communications*; Oct.

Federal Communications Commission (FCC) (2010). Grant of authority to TerreStar Networks to operate dual-mode mobile terminals that can be used to communicate either via TerreStar's geostationary-orbit MSS satellite, TerreStar-1, or via ATC base stations. Jan. 13. On apps.fcc.gov/edocs_public/attachmatch/DA-10-60A1.txt. Accessed 2015 Feb. 9.

Federal Communications Commission (FCC) (2014). Ancillary terrrestrial component. Sep. 29. On transition.fcc.gov/ib/sd/ssr/atc.html. Accessed 2015 May 11.

Ferster W (2012a). Revised LightSquared plan still interferes into GPS. *Space News*; Jan. 13.

Ferster W (2012b). FCC to pull LightSquared license. *Space News*; Feb. 15

Foust J (2020). GPS committee calls FCC Ligado order a "grave error." *SpaceNews*; July 1.

Gabellini P, D'Agristina L, Dicecca L, Di Lanzo D, Gatti N, and Angevain J-C (2010). The electrical design and verification of the Alphasat TDP# 5 antenna farm. *ESA Antenna Workshop on Antennas for Space Applications*; Oct.

Gallinaro G, Tirrò E, Di Cecca F, Migliorelli M, Gatti N, and Cioni S (2012). Next generation interactive S-band mobile systems, challenges and solutions. *Advanced Satellite Multimedia Systems Conference and Signal Processing for Space Communications Workshop*; Sep. 5–7.

Gizinski III SJ and Manuel R (2015). Inmarsat-5 Global Xpress: secure, global mobile, broadband. On gobcss.com/wp-content/uploads/2015/06/gizinski.stephen.paper_.pdf. pdf. Accessed 2016 May 14.

Global Satellite France (2014). Téléphones satellites Thuraya. On www.telephonesatellite. com/wp-content/uploads/2010/01/Thurayacouverture.jpg. Accessed 2014 Oct. 19.

GlobalSpec (2014). Satellite communications equipment from Airbus Group, solid state power amplifier (SSPA) – L/S band. Product sheet. On www.globalspec.com. Accessed 2017 July 17.

Godinez V (2011). LightSquared picks Sprint to build 4G network. *The Dallas Morning News*; July 28. On www.dallasnews.com/business/technology/headlines/. Accessed 2015 Feb. 14.

Goovaerts D (2016). Ligado unveils mid-band 5G network plan to FCC. *Wireless Week*. May 24. On www.wirelessweek.com/news/2016/05/ligado-unveils-mid-band-5g-network-plan-fcc. Accessed 2017 Oct. 21.

Gunter's Space Page (2014a). Alphasat (Inmarsat-4A F4). Mar. 25. On space.skyrocket.de/doc_sdat/alphasat.htm. Accessed 2014 Oct. 13.

Gunter's Space Page (2014b). TerreStar 1,2 → EchoStar T1, T2. Aug 8. On space.skyrocket.de/doc_sdat/TerreStar-1.htm. Accessed 2014 Oct. 10.

Gunter's Space Page (2014c). SkyTerra 1, 2 (MSV 1, 2, SA). Apr 5. On space.skyrocket.de/doc_sdat/SkyTerra-1.htm. Accessed 2014 Oct. 7.

Gunter's Space Page (2017). Inmarsat-4 F1, 2, 3. Mar. 25. On space.skyrocket.de/doc_sdat/inmarsat-4.htm. Accessed 2014 Oct. 17.

Guy RFE (2009). Potential benefits of dynamic beam synthesis to mobile satellite communication, using the Inmarsat 4 antenna architecture as a test example. *International Journal of Antennas and Propagation*; 2009; 1–5.

Guy RFE, Wyllie CB, and Brain JR (2003). Synthesis of the Inmarsat 4 multibeam mobile antenna. *IET International Conference on Antennas and Propagation*; Mar. 31–Apr. 3.

Hadinger PJ (2015). Inmarsat Global Xpress: the design, implementation, and activation of a global Ka-band network. *AIAA International Communications Satellite Systems Conference*; Sep. 7–10.

Harris Corporation (2009). Harris Corporation antenna reflector for TerreStar communications satellite successfully deployed. Press release. Sep. 28. On harris.com/view_pressrelease.asp?pr_id=2809. Accessed 2015 Feb. 12.

Harris Corporation (2010). Photo of SkyTerra 1 L-band antenna deployed in factory. On orbitrax.com/wp-content/uploads/2010/12/Picture-179-e1291495841623.png. Accessed 2015 Feb. 16.

Harris Corporation (2011). Harris Corporation awarded satellite antenna contract by Boeing for Inmarsat-5 satellites. Press release. Mar. 3. On www.harris.com/press-releases. Accessed 2017 Sep. 11.

Henry C (2017a). Inmarsat CEO hints at more advanced Global Xpress satellites. *SpaceNews*. Mar. 7.

Henry C (2017b). Inmarsat undecided on how it will use the satellite SpaceX is launching next week. *SpaceNews*. May 11.

Henry C (2020). Yahsat begins Thuraya fleet refresh with Airbus satellite order. *SpaceNews*. Aug. 28.

Hibberd C (2012). Inmarsat Global Xpress network—meeting the challenges of providing a seamless global Ka-band service to mobile terminals. *AIAA International Communications Satellite System Conference*; Sep. 24–27.

Hili L and Malou F (2013). ESA-CNES deep sub micron program ST 65nm. Viewgraph presentation. On escies.org/download/webDocumentFile?id=60164. Accessed Oct. 10, 2014.

Howell A of Inmarsat (2010). Broadband global area networks. Viewgraph presentation. *Standards and the New Economy Conference* led by Cambridge Wireless; 2010 Mar. 25. Accessed 2014 Nov. 29.

iDirect (2017). Glossary of satellite terms. On www.idirect.net/Company/Resource-Center/Satellite-Basics/Glossary-of-Terms.aspx. Accessed 2017 Sep. 17.

Inmarsat (2006). Troubleshooting BGAN, v 1.0. Aug. 6. On www.inmarsat.com/wp-content/uploads/2013/10/Inmarsat_Troubleshooting_BGAN.pdf. Accessed 2014 Dec. 8.

Inmarsat (2009a). Inmarsat Group Ltd Form 20-F; annual report to Securities and Exchange Commission. Apr. 29. On www.inmarsat.com/wp-content/uploads/2013/11/Inmarsat_Group_Ltd_Form_20F_2009.pdf. Accessed 2014 Dec. 9.

Inmarsat (2009b). BGAN X-Stream™ FAQs. June. On www.inmarsat.com/wp-content/uploads/2013/10/Inmarsat_BGAN_X-Stream_FAQs.pdf. Accessed 2014 Dec. 4.

Inmarsat (2011a). 2010 annual report and accounts. On www.inmarsat.com/wp-content/uploads/2013/10/Inmarsat_plc_Annual_Report_and_Accounts_2010.pdf. Accessed 2014 Dec. 8.

Inmarsat (2011b). IsatPhone Pro user guide. Oct. On www.inmarsat.com/wp-content/uploads/2014/04/IsatPhone_Pro_UG_Oct_2011_EN.pdf. Accessed 2014 Dec. 8.

Inmarsat (2012a). Schedule S technical report and Attachment A, technical annex. Part of application to FCC for earth station authorizations to use with Inmarsat-5 F2. FCC file no SES-LIC-20120426-00397.

Inmarsat (2012b). Broadband terminals, a quick reference guide. June. On www.inmarsat.com/wp-content/uploads/2013/10/Inmarsat_BGAN_Terminal_Comparison.pdf. Accessed 2014 Nov. 10.

Inmarsat (2012c). Seven questions to ask before building your satellite SCADA network. On www.inmarsat.com/wp-content/uploads/2013/10/Inmarsat_Satellite_SCADA_Network.pdf. Accessed 2014 Dec. 9.

Inmarsat (2013a). Our coverage. Oct. On www.inmarsat.com/about-us/our-satellites/our-coverage/. Accessed 2014 Nov. 10.

Inmarsat (2013b). Our network. On www.inmarsat.com/about-us/our-satellites/our-network/. Accessed 2014 Dec. 8.

Inmarsat (2013c). I-4 satellite coverage map. Oct. On www.inmarsat.com/wp-content/uploads/2013/10/Inmarsat_I-4_Satellite_Coverage.pdf. Accessed 2016 May 19.

Inmarsat (2013d). Carriers. On www.inmarsat.com/support/carrier/. Accessed 2017 Oct. 11.

Inmarsat (2014a). 2013 annual report and accounts. On www.inmarsat.com/inm/downloads/financial-reports/2013/annual-report/sources/index.htm/. Accessed 2014 Dec. 2.

Inmarsat (2014b). BGAN HDR. Product information. On www.inmarsat.com/wp-content/uploads/2014/04/BGAN_HDR_April_2014.pdf. Accessed 2015 Jan. 11.

Inmarsat (2014c). Global Xpress land terminal features, raising the bar on VSAT. Product brochure for VSAT manufacturers. On hmstelcom.com/wp-content/uploads/2017/06/Inmarsat-GX-Land-Terminal-Features-March-2014-EN-LowRes.pdf. Accessed 2020 Oct. 26.

Inmarsat (2014d). Inmarsat completes Global Xpress ground network. News release. Nov. 7. On www.inmarsat.com/news/inmarsat-completes-global-xpress-ground-network/. Accessed 2017 Oct. 8.

Inmarsat (2015a). Inmarsat Global Xpress; global, mobile, trusted. Brochure. Jan. On www.inmarsat.com/wp-content/uploads/2015/01/Inmarsat_USG_GX_Brochure_January_2015_EN_LowRes.pdf. Accessed 2017 Oct. 11.

Inmarsat (2015b). Inmarsat to create fourth full-service L-band region. News release. On www.inmarsat.com/news/inmarsat-create-fourth-full-service-l-band-region/. Accessed 2017 Oct. 18.

Inmarsat (2016). The European aviation network. Apr. On www.inmarsat.com/wp-content/uploads/2016/01/Inmarsat_European_aviation_network_April_2016_EN_LowRes.pdf. Accessed 2017 Oct. 19.

Inmarsat (2017a). Global Xpress. On www.inmarsat.com/service/global-xpress/. Accessed 2017 Sep. 4.

Inmarsat (2017b). Our coverage. On www.inmarsat.com/about-us/our-satellites/our-coverage. Accessed 2017 Sep. 26.

Inmarsat (2017c). Inmarsat-5 F4 mission. On www.inmarsat.com/i5f4. Accessed 2017 Sep. 26.

Inmarsat (2017d). Our network. On www.inmarsat.com/about-us/our-satellites/our-network/. Accessed 2017 Oct. 19.

Inmarsat (2020). Global Xpress. www.inmarsat.com/services/global-xpress. Accessed 2020 Sep. 15.

Intellian (2020). GX60NX. Product datasheet. On www.intelliantech.com/products/inmarsat-gx-maritime-terminal/gx60nx/#. Accessed 2020 Oct. 26.

International Telecommunication Union (ITU) (2016a). *Radio Regulations*, vol. 1, Articles.

International Telecommunication Union (ITU) (2016b). Final acts WRC-15; resolution 156. *World Radiocommunication Conference*. On handle.itu.int/11.1004/020.1000/4.297.43. en.100. Accessed 2017 Sep. 21.

Irish D and Marchand G of Inmarsat (2013). SwiftBroadband technical workshop. Viewgraph presentation. June. On www.inmarsat.cm/wp-content/uploads/2013/10/Inmarsat_APC_ 2013_06_Dale_Irish.pdf. Accessed 2015 Jan. 11.

Khan AH, Febvre P, and Fines P (2013). Low data rate and high data rate technologies for next-generation BGAN. *AIAA International Communications Satellite Systems Conference*; Oct. 15–17.

Kiyohara A, Kazekami Y, Seino K, Tanaka K, Shirasaki K, Fukazawa S, Iwano N, Kittaka Y, and Gill R (2003). Superior tracking performance of C-band solid state power amplifier for Inmarsat-4. *AIAA International Communications Satellite Systems Conference*; Apr. 17–19.

Koduru C, Tomei B, Sichi S, Suh K, Ha T, and Gupta R (2011). Advaced space based network using bround based beam former. *AIAA International Commuications Satellite Systems Conference*; Nov. 28-Dec. 1.

Kongsberg Defence Systems (2005). Inmarsat-4 F1 launched. News release. Apr. 14. On www.kongsberg.com/en/kds/kns/newsarchive. Accessed 2014 Nov. 29.

Kongsberg Defence Systems (2008). Thuraya-3 brings another 335 Norspace SAW filter modules into orbit. News release. On www.kongsberg.com/en/kds/news/newsarchive. Accessed 2014 Oct. 8.

Lawson S (2012). Sprint cancels LightSquared LTE deal. *ComputerWorld*; Mar. 16. On www.computerworld.com/article/2502746. Accessed 2015 Feb. 14.

Lee-Yow C, Scupin J, Venezia P, and Califf T (2010). Compact high-performance reflector antenna feeds and feed networks for space applications. *IEEE Antennas and Propagation Magazine*; 52 (4) (Aug.); 210–217.

Leong CK, Mathur RP, Craig AD (1996). Payload digital processor hardware demonstrator for satellite communications systems. *Proceedings, International Conference on Communication Technology*; vol. 1; May 5–7.

LightSquared (about 2006). Technical appendix, to FCC, seeking authority to communicate with SkyTerra 2. On licensing.fcc.gov/myibfs/download.do?attachment_key=845640. Accessed 2015 Feb. 17.

LightSquared (2015). About LightSquared. On www.lightsquared.com/satellite-services/. Accessed 2015 Feb. 14.

LightSquared and 19 other companies (2012). Debtors' motion for entry of order directing joint administration of related chapter 11 cases. To US bankruptcy court southern district of New York. May 14. On www.kccllc.net/lightsquared/document/list/3244. Accessed 2015 Feb. 14.

Lim W and Salvatore J (2013). Method and system for maintaining communication with inclined orbit geostationary satellites. U.S. patent application 2013/0309961 A1. Nov. 21.

Lohmeyer WQ, Aniceto RJ, and Cahoy KL (2016). Communications satellite power amplifiers: current and future SSPA and TWTA technologies. *International Journal of Satellite Communications and Networking*; vol. 34; Mar/Apr.

Mallison MJ and Robson D (2001). Enabling technologies for the Eurostar geomob‚le satellite. *AIAA International Communications Satellite System Conference*; Apr. 17–20.

Martin DH, Anderson PR, and Bartamian L (2007). *Communications Satellites*, 5th ed., El Segundo, CA: The Aerospace Press; and Reston, VA: American Institute of Aeronautics and Astronautics, Inc.

Matolak DW, Noerpel A, Goodings R, Vander Staay D, and Baldasano J (2002). Recent progress in deployment and standardization of geostationary mobile satellite systems. *Proceedings, IEEE Military Communications Conference*; vol. 1; Oct. 7–10.

McNeil SD (about 2004). Inmarsat 4F2 attachment 1 technical description. FCC filing. On licensing.fcc.gov/myibfs/download.do?attachment_key=-94644. Accessed 2014 Dec. 28.

Military & Aerospace Electronics (2005). Space Systems/Loral buys amplifiers for communications satellites. News article. Dec. 1 On www.militaryaerospace.com/home/article/16707994/space-systemsloral-buys-amplifiers-for-communications-satellites.

Mitani B of MSV Canada (2007). MSV's next generation satellite system. Viewgraph presentation. *ITU-T Workshop, Satellites in NGN?*; July 13. On www.itu.int/dms_pub/itu-t/oth/06/08/T06080000110001PDFE.pdf. Accessed 2015 Feb. 15.

Mugnaini S and Benhamou M (2019). In-orbit test strategy and results for GX multibeam antenna. On usnc-ursi-archive.org/aps-ursi/2019/abstracts/2344.pdf. Accessed 2020 Oct. 26.

Myers JN, McEntee C, Bahrami A, Seitter KL, Baker M, Root SA, Nolen B, Hutchison K, Marotto R, Block JH, and 12 more authors (2017). Ex parte letter to FCC from 22 organizations. June 27. On www.ametsoc.org/ams/index.cfm/about-ams/ams-position-letters/joint-letter-to-fcc-on-sharing-of-satellite-spectrum/. Accessed 2017 Oct. 21.

National Association of Broadcasters (2010). New satellite phones *still* on the horizon. *TV TechCheck*. July 26. On www.nab.org/xert/scitech/2010/TVTechCheck/TV072610.asp. Accessed 2015 Feb. 5.

National Satellite Telecommunications Services (2014). Ground segment. On www.thuraya.com.kw/sagroundsegment.html. Accessed 2014 Oct. 20.

Nicola G and Plecity M (2013). GX aviation services and functionality. On www.inmarsat.com/wp-content/uploads/2013/10/Inmarsat_APC_2013_05_George_Nicola.pdf. Accessed 2017 Sep. 21.

NordicSpace (2006). Broadband for everybody. Article. Jan. 1. On nordicspace.net/2006/01/01/broadband-for-everybody. Accessed 2014 Dec. 9.

Orbit Communications Systems (2019). GX46 airborne satcom terminal. Product information. On orbit-cs.com/wp-content/uploads/sites/3/2019/05/Orbit-GX46-DS-v9.06.pdf. Accessed 2020 Oct. 25.

Orbitica (2013). Tarif des communications Inmarsat Fleet 77/55/33. On www.francesatellite.com/inmarsat/com_fleet.html. Accessed 2014 Dec. 8.

Parkinson B (2018). A grave threat to GPS and GNSS. Jan. 16. On www.gpsworld.com/a-grave/threat-to-gps-and-gnss. Accessed 2019 Apr. 6.

Pelkey T (2020). FCC unanimously approves Ligado's application to facilitate 5G and Internet of things services. Apr. 20. *News from the Federal Communications Commission*. On docs.fcc.gov/public/attachments/DOC-363823A1.pdf. Accessed 2020 May 27.

Phonearena (2010). TerreStar Genus™. Product spec sheet. On www.phonearena.com/phones/TerreStar-Genus_id4914. Accessed 2015 Feb. 2.

Plass S, editor (2011). *Future Aeronautical Communications*. Published online by InTech. Sep. 26.

Pon R, inventor; Space Systems, Loral, assignee (2002). Aft deployable thermal radiators for spacecraft. U.S. patent 6,378,809 B1. Apr. 30.

PR Newswire (2009). Hughes announces agreement with SkyTerra and TerreStar to implement GMR1-3G satellite air interface on chipset for wireless handsets. News article. Apr. 2. On www.prnewswire.com/news-releases/. Accessed 2015 Feb. 1.

Recommendation, ITU-R, P.531-10 (2009). Ionospheric propagation data and prediction methods required for the design of satellite services and systems. Geneva: ITU-R.

Roederer AG (2005). Antennas for space: some recent European developments and trends. *Proceedings, International Conference on Applied Electromagnetics and Communications*; Oct. 12–14.

Roederer AG (2010). Semi-active satellite antenna front ends: a successful European innovation. *Proceedings, Asia-Pacific Microwave Conference*; Dec. 7–10.

Russell K (2017a). Iridium, Ligado dispute over spectrum heats up. *Via Satellite Magazine*. Mar. 31. Accessed 2017 Oct. 21.

Russell K (2017b). EchoStar completes in-orbit testing of EchoStar 21 satellite. *Via Satellite Magazine*. Aug. 29. Accessed 2017 Oct. 20.

Satbeams (2007). Inmarsat-3 F4. Information site. On www.satbeams.com. Accessed 2014 Dec. 2.

Satbeams (2020a). Inmarsat GX4 (Inmarsat 5F4). Information site. On www.satbeams.com. Accessed 2020 Sep. 15.

Satbeams (2020b). EchoStar 21 (EchoStar T2, TerreStar 2). Information site. On www.satbeams.com. Accessed 2020 Sep. 15.

Satnews (2010). Command center—Dennis Matheson, CTO, TerreStar. Interview article. *MilsatMagazine*; Jan. On www.milsatmagazine.com/story.php?number=1505720585. Accessed 2015 Jan. 31.

SatPhoneStore (2020). Thuraya SatSleeve Hotspot. Product information. On www.satphonestore.com/tech-browsing/old/thuraya-nav/thuraya-satsleeve-hotspot.html. Accessed 2020 Oct. 12.

SatStar.net (2014). Composite beam; Hawaii beam; and Puerto Rico beam. EchoStar T1 beams. On www.satstar.net/beams/echot1_comp.html; www.satstar.net/beams/echot1_hawaii.html; www.satstar.net/beams/echot1_puerto.html. Accessed 2015 Jan. 30.

SatStar.net (2017). L1 – Conus beam. Skyterra 1 beams. On www.satstar.net/beams/skyterra1-l1.html. Accessed 2017 Oct. 24.

Securities and Exchange Commission (2013). Form 8-K, Current report, TerreStar Corporation. On www.getfilings.com/sec-filings/130308/TERRESTAR-CORP_8-K/. Accessed 2015 Feb. 12.

Segal RS of MSV (2007). Application for limited waiver, before the FCC, in the matter of Mobile Satellite Ventures Subsidiary LLC. May 23. On licensing.fcc.gov/myibfs/download.do?attachment_key=-130082. Accessed 2015 Feb. 15.

Semler D, Tulintseff A, Sorrell R, and Marshburn J (2010). Design, integration, and deployment of the TerreStar 18-meter reflector. *AIAA International Communications Satellite System Conference*; Aug. 30 to Sep. 2.

Simon PS (2007). LinkedIn page. Jan. 1. Accessed 2012 Sep. 17.

SpaceNews (2010). TerreStar-2 in final payload integration. Apr. 25.

SpaceNews (2011). Editorial: Regulatory relief for LightSquared. Feb. 8.

Spaceflight 101 (2016). Inmarsat 5-F3. May 14. On spaceflight101.com/spacecraft/inmarsat-5-f3. Accessed 2017 Oct. 11.

Spaceflight 101 (2017). Inmarsat 5-F4 satellite overview. May 2. On spaceflight101.com/falcon-9-inmarsat-5-f4/inmarsat-5-f4/. Accessed 2017 Sep. 11.

Stirland SJ and Brain JR (2006). Mobile antenna developments in EADS Astrium. *European Conference on Antennas and Propagation*; Nov. 6–10.

Sunderland DA, Duncan GL, Rasmussen BJ, Nichols HE, Kain DT, Lee LC, Clebowicz BA, Hollis IV RW, Wissel L, and Wilder T (2002). Megagate ASICs for the Thuraya satellite digital signal processor. *Proceedings, IEEE International Symposium on Quality Electronic Design*; Mar. 21.

TerreStar Networks (2009). Exhibit 1, Request for extension of special temporary authorty [*sic*]. On licensing.fcc.gov/myibfs/download.do?attachment_key=775105. Accessed 2015 Jan. 30.

TerreStar Networks (2010). Case administration Website. On www.TerreStarinfo.com/. Accessed 2015 Feb. 12.

Thomson MW (2002). Astromesh™ deployable reflectors for Ku- and Ka-band commercial satellites. *AIAA International Communication Satellite Systems Conference*; May 12–15.

Thuraya Satellite Telecommunications (2003). Technology. On thuraya.com.pk/space.html. Accessed 2014 Oct. 19.

Thuraya Telecommunications (2014a). Cobham flat panel fixed antenna 1426. Product sheet. On www.thuraya.com/cobham-flat-panel-fixed-antenna-1426. Accessed 2014 Oct. 21.

Thuraya Telecommunications (2014b). Knowledge center. On www.thuraya.com/faqs/41. Accessed 2014 Oct. 17.

Thuraya Telecommunications (2014c). Products. On www.thuraya.com/products-list. Accessed 2014 Oct. 18.

Thuraya Telecommunications (2015). Satellite capacity leasing, assured access of space segment. Service description. On www.thuraya.com/capacity-leasing. Accessed 2015 Jan. 22.

Thuraya Telecommunications (2020). Network coverage. On www.thuraya.com/network-coverage. Accessed 2020 Oct. 12.

Tian B, Jalali A, Jayaraman S, and Namgoong J, inventors; Qualcomm, assignee (2013). Reverse link data rate indication for satellite-enabled communications systems. U.S. patent 8,588,086 B2. Nov. 19.

TS2 Technologie Satelitarne (2009). Inmarsat iSatPhone PRO. Product information. On www.ts2.pl/en/isatphone-inmarsat Accessed 2014 Dec. 2.

Ueno K, Ohira T, Tsunoda H, and Ogawa H (1996). Phased array fed single reflector antenna for communication satellites. *International Symposium on Antennnas & Propagation*; Sep. 24–27. On ap-s.ei.tuat.ac.jp/isapx/1996/pdf/2C3-3.pdf. Accessed 2014 Nov. 30.

USA-Satcom (2012). Inmarsat Aero-P ACARS multi-channel decoder. On usa-satcom.com/inmarsat-aero-p-acars-multi-channel-decoder. Accessed 2014 Dec. 17.

Vilaça M (2004). Using the new L band MSS allocations. *Study Group 8 Seminar on Tomorrow's Technological Innovations*. Sep. 9. On www.powershow.com/view1/5ea89-ZDc1Z/USING_THE_NEW_L_BAND_MSS_ALLOCATIONS_Study_Group_8_Seminar_on_Tomorrows_Technological_Innovations_Ge_powerpoint_ppt_presentation. Accessed 2010 Feb. 5.

Vilaça M (2020a). Private communication. Sep. 11.

Vilaça M (2020b). Private communications. Sep. 26, Sep. 27, and Oct. 5.

Vilaça M (2020c). Private communication. Oct. 15.

Vilaça M and Bath D (2011). Inmarsat Global Xpress - global broadband mobility. *Ka-band Conference*; Oct. 3–5.

Wang J-L and Liu C-S (2011). Development and application of Inmarsat satellite communication system. *International Conference on Instrumentation, Measurement, Computer, Communication and Control*; Oct. 21–23.

Wikipedia (2015). SkyTerra. Accessed 2015 Feb. 14.

Wikipedia (2017a). Inmarsat. Oct. 17. Accessed 2017 Oct. 18.

Wikipedia (2017b). Ligado Networks. Sep. 3. Accessed 2017 Oct. 21.

Wikipedia (2020). GSM. Aug. 27. Accessed 2020 Sep. 14.

Wingrove M (2020). New Inmarsat GX antenna technology launched. *Riviera Maritime Media* Web site. Aug. 5. On www.rivieramm.com/news-content-hub/news-content-hub/new-inmarsat-gx-antenna-technology-launched-60500. Accessed 2020 Oct. 25.

Witting M, Hauschildt H, Murrell A, Lejault J-P, Perdigues J, Lautier JM, Salenc C, Kably K, Greus H, Garat F, and 25 more authors (2012). Status of the European data relay satellite system. *Proceedings, International Conference on Space Optical Systems and Applications*; Oct. 9–12.

Wyatt-Millington RA, Sheriff RE, and Hu YF (2007). Performance analysis of satellite payload architectures for mobile services. *IEEE Transactions on Aerospace and Electronic Systems*; 43 (Jan.); 197–213.

Yahsat (2018). Yahsat completes Thuraya acquisition and appoints new CEO. News release. Aug. 5. On www.yahsat.com/en/news/2018/yahsat-completes-thuraya-acquisition-and-appoints-new-ceo. Accessed 2019 Apr. 5.

APPENDICES

A.1 DECIBEL

Electrical engineers usually give ratios in **decibels** (dB). This is a logarithmic scale. Communications-satellite power and noise levels have a wide range, often several orders of magnitude. Using dB flattens the range to something more humanly graspable. A ratio of powers P_1 and P_2 is given in dB as follows:

$$\text{Power ratio} \frac{P_1}{P_2} \text{ in dB} = 10 \log_{10} \left(\frac{P_1}{P_2} \right)$$

The factor of 10 in front of the logarithm causes the range of P_1/P_2 between 1 and 10 to have a dB range of 0 to 10, that is, almost the same, instead of the small range, without the factor of 10, of 0 to 1.

It is wise to memorize a few power ratios in dB, as shown in Table A.1.

Then, one can calculate most other power ratios mentally by the various relationships below:

$$10 \log_{10} (UV) = 10 \log_{10} U + 10 \log_{10} V \qquad 10 \log_{10} \frac{1}{U} = -10 \log_{10} U$$

$$10 \log_{10} \sqrt[n]{V} = \frac{1}{n} 10 \log_{10} V$$

Satellite Communications Payload and System, Second Edition. Teresa M. Braun and Walter R. Braun.
© 2021 John Wiley & Sons, Inc. Published 2021 by John Wiley & Sons, Inc.

TABLE A.1   A Few Power Ratios In dB to Memorize

Numerical power ratio	dB power ratio
1	0
2	3.0
3	4.8
4	6.0
5	7.0
7	8.5
10	10
100	20
1000	30

For example, 30 in dB is $10\log_{10}(10*3)$, namely about $10+4.8 = 14.8\,\text{dB}$, and 15 in dB is about $14.8-3.0 = 11.8\,\text{dB}$.

The term "numerical ratio" is frequently used to mean a ratio that is not in dB. In this book, the term more often used is "not in dB."

A ratio of amplitudes or voltages has a different definition of dB, and this can be confusing. The ratio of amplitudes A_1 and A_2 is given in dB as follows:

$$\text{Amplitude ratio}\,\frac{A_1}{A_2}\,\text{in dB} = 20\log_{10}\left(\frac{A_1}{A_2}\right)$$

In fact, this is the same definition, since amplitude is proportional to the square root of power (assuming the voltage is measured over the same impedance):

$$\text{Power ratio}\,\frac{A_1^2}{A_2^2}\,\text{in dB} = 10\log_{10}\left(\frac{A_1^2}{A_2^2}\right)$$

What can also be confusing is that sometimes not just ratios but terms with units are given in a dB-type format, such as a frequency in dBHz. The question is, which definition of dB to use, the one with a factor of 10 on the log or the one with a factor of 20. The factor of 10 is used if, in a commonly used expression, the term either directly multiplies or divides a power. By "directly" we mean without taking its square or square root. An example is a bandwidth B, since it is in the common expression $\text{SNR} = P/(N_0 B)$. Similarly, the factor of 10 is also used if, in a commonly used equation, the term is set equal to an expression in which a power or inverse power is directly present, as for the symbol energy E_s in the equation $E_s = P/R_s$. In fact, the definition with factor of 20 is not often used.

A.2   FOURIER TRANSFORM

Familiarity with the Fourier transform and the duality of the time and frequency domains is a prerequisite for understanding some parts of the book. Nonetheless, for completeness, we define the Fourier transform and the inverse transform.

Suppose that $w(t)$ is a function of time. Then, we write its Fourier transform as a function of frequency f as $W(f)$. They are related as follows:

$$\mathcal{F}\big[w(\cdot)\big](f)=W(f)=\int_{-\infty}^{\infty}w(t)e^{-j2\pi ft}\,dt \quad \text{where } \mathcal{F} \text{ is the Fourier transform}$$

$$\mathcal{F}^{-1}\big[W(\cdot)\big](t)=w(t)=\int_{-\infty}^{\infty}W(f)e^{j2\pi ft}\,df \quad \begin{array}{l}\text{where } \mathcal{F}^{-1} \text{ is the inverse}\\ \text{Fourier transform}\end{array}$$

The Fourier transform is usefully applied to signals and filter impulse responses. It is also usefully applied to antennas, for which, instead of time and frequency, the two domains are spatial and pattern. Roughly speaking, the narrower a function is (that is, the smaller the interval over which it is non-zero), the wider its Fourier transform.

Refer to any textbook on probability theory and random processes to see how the Fourier transform of a function changes when various operations are applied to the function.

A.3 ELEMENTS OF PROBABILITY THEORY

A.3.1 Introduction

After Fourier analysis, probability theory is the field of mathematics most commonly applied in specification, development, and sell-off of the communications payload (and indeed the end-to-end communications system). We present definitions and properties, then practical and positively useful tips on applying probability theory. Further topics can be found in any textbook on elementary probability theory.

A.3.2 Definition of Random Variable and Probability Density Function

We summarize here the definitions of the probability terms most commonly needed in communications analysis involving the payload (or indeed the end-to-end communications system).

A *probability space* is made up of two things: a *space* or set $\mathcal{S}$ of elements and a *probability function* Pr defined on *events* or subsets of $\mathcal{S}$ that has the following properties:

$(1)\,\mathrm{Pr}\big(A\big)\ge 0$ where A is a subset of S

$(2)\,\mathrm{Pr}\big(S\big)=1$

$(3)\,\mathrm{Pr}\big(A\cup B\big)=\mathrm{Pr}\big(A\big)+\mathrm{Pr}\big(B\big)$ if $A\cap B=\varnothing$, where A and B are subsets of S,

"$\cup$"means "union," "$\cap$" means "intersection," and $\varnothing$ is the null set

The elements of $\mathcal{S}$ are also events, the *elementary events*. $\mathcal{S}$ could, for example, be the interval [0,1] of real numbers between and including 0 and 1.

Roughly speaking, a **random variable** or **random number** X is a real-valued function defined on the elements of a probability space $\mathcal{S}$ (see a probability textbook for a complete definition). Then, on a subset of $\mathcal{S}$, X takes on a set of values. We can invert this and define a subset of $\mathcal{S}$ by the set of values that X takes on it:

$$\Pr(X = x) \triangleq \Pr(\text{event on which } X = x) \quad \text{where } x \text{ is a value of } X$$

and "$\triangleq$" means "is defined as." Similarly,

$$\Pr(X \le x) \triangleq \Pr(\text{event on which } X \le x)$$

The **(cumulative) distribution function** (cdf) P_X of the random variable X is defined by

$$P_X(x) \triangleq \Pr(X \le x) \quad \text{for all values } x \text{ of } X$$

When there is no ambiguity, the subscript X is usually omitted. If the cdf is continuous, then X is of *continuous type*, and if the cdf is a staircase-type, X is of *discrete type*. Now and then the *mixed type* comes up, where the function is continuous in some regions but has at least one jump in value. The definition of cdf applies to both continuous and discrete random variables.

The **(probability) density function** (pdf) p_X for a continuous random variable is defined by

$$p_X(x) \triangleq \frac{d}{dx}\Pr(X \le x) \quad \text{for all values } x \text{ of } X$$

For a discrete random variable, it is defined by

$$p_X(x_j) \triangleq \Pr(X = x_j) \quad \text{for all values } x_j \text{ of } X$$

Sometimes the subscript X is omitted.

A.3.3 Mean, Standard Deviation, and Correlation

The properties of a random variable X most often used in payload analysis are the **mean** m_X and **standard deviation** σ_X:

$$m_X = \text{mean of } X \triangleq \int_S x p_X(x)\,dx$$

$$\sigma_X = \text{standard deviation of } X = \sqrt{\text{variance of } X}$$

$$\text{where variance of } X \triangleq \int_S (x - m_X)^2 \, p_X(x)\,dx$$

The subscript X may be dropped when the meaning is clear.

Often we are interested in a function g of a random variable X, for example, antenna gain as a function of the E-W component of antenna pointing error. If the function g satisfies certain benign criteria (given in any probability textbook), then $g(X)$ is also a random variable. Its *expected value* $E(g(X))$ is defined by

$$E\big(g(X)\big) \triangleq \int_S g(x) p_X(x)\,dx$$

The mean and variance of X can be expressed in this formulation, where $g(x) = x$ for the mean and $g(x) = (x - m_x)^2$ for the variance:

$$m_X = E(X)$$
$$\sigma_{X^2} = E\Big[\big(X - m_X\big)^2\Big]$$

Suppose now we have two random variables, X and Y, each with any pdf. The *joint pdf* of X and Y is written $p_{X,Y}$ and is defined on all pairs of values of X, Y, respectively. X and Y are said to be **independent** if

$$p_{X,Y}(x,y) = p_X(X)\,p_Y(y) \text{ for all values } x \text{ of } X \text{ and } y \text{ of } Y$$

that is, if the probability that X has any particular value is not influenced by the value that Y has and vice versa. The concept of independence extends to more than two random variables, in which case the term *mutual independence* is more often used for clarity.

A concept related to independence is **correlation**. The amount of correlation between any two random variables X and Y is given by their correlation coefficient $\rho_{X,Y}$:

$$\rho_{X,Y} \triangleq \frac{E\Big[\big(X - m_X\big)\big(Y - m_Y\big)\Big]}{\sigma_X \sigma_Y}$$

Note that the correlation coefficient can be negative. The special case where $Y - m_Y = k(X - m_X)$ for some $k \neq 0$ provides us with the bounding values of the correlation coefficient, namely $+1$ and -1:

$$\rho_{X,Y} = \operatorname{sgn}(k)$$

If X and Y are independent, they are uncorrelated. The inverse does not necessarily hold; however, if X and Y are Gaussian, then it does hold.

A.3.4   Sum of Random Variables

We often have to deal with sums of random variables, for example, line items in performance budgets. Suppose X and Y are random variables. Then their sum $X + Y$ is also a random variable. Obviously, the sum of more than two random variables is also a random variable. The mean of the sum is just the sum of the individual means:

$$m_{X+Y} = m_X + m_Y$$

The standard deviation of the sum is more complicated. It depends on how correlated X and Y are, as follows:

$$\sigma_{X+Y} = \sqrt{\sigma_{X+Y}^2}$$
$$\sigma_{X+Y}^2 = \sigma_X^2 + \sigma_Y^2 + 2\rho_{X,Y}\sigma_X\sigma_Y$$

Examples for when $\sigma_X = \sigma_Y$ are the following:

$$\sigma_{X+Y}^2 = \begin{cases} 4\sigma_X^2 & \text{when } \rho_{X,Y} = 1 \\ 2\sigma_X^2 & \text{when } \rho_{X,Y} = 0 \\ 0 & \text{when } \rho_{X,Y} = -1 \end{cases}$$

The sum of Gaussian random variables is Gaussian, but no other pdf family has this property.

A.3.5 Gaussian Probability Density Function

The Gaussian pdf is the most frequently used pdf in payload and communications system analysis. There are several reasons for this. One is that some variables truly do have a Gaussian pdf or very close to it. A second reason is that some variables have a pdf only partially known but known to be Gaussian-like in some respects. The best reason is that usually analysis is based on the sum of several random variables, and sums are usually Gaussian-like (see Section A.3.9 on Central Limit Theorem). Sometimes sums are of many small terms of roughly equal size; they are of roughly equal size because if there were a term much larger than the others, it would be reduced by way of some design improvement. The last reason for the frequency of the Gaussian assumption is its ease of use. For example, the Fourier transform of a Gaussian pdf is a Gaussian function. Computations with Gaussian pdfs can often be done analytically (i.e., by hand) or are at least simpler than with other pdfs, and estimates can sometimes be made quickly. Simulations are almost always faster because Monte Carlo can usually be avoided (Section A.4). We have to caution here that a Gaussian pdf can only be assumed after careful assessment.

The Gaussian pdf corresponding to a mean m and a standard deviation σ is given by

$$p(X) = \frac{e^{-(X-m)^2/(2\sigma^2)}}{\sigma\sqrt{2\pi}}$$

and is plotted in Figure A.1 for mean of zero.

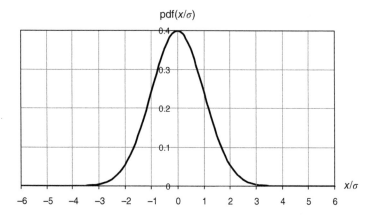

FIGURE A.1   Gaussian pdf with zero mean.

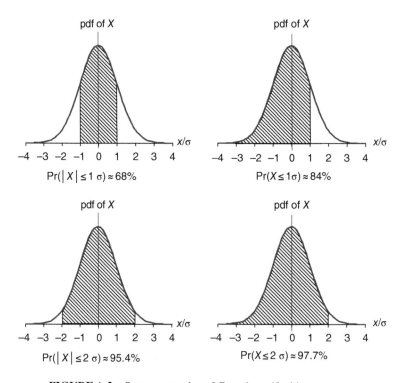

FIGURE A.2   Some properties of Gaussian pdf with zero mean.

The probabilities associated with 1σ and 2σ values are illustrated in Figure A.2. Similar values for 3σ are the following: 99.68% for the probability that a Gaussian random variable is within $\pm3\sigma$ of its mean and 99.83% for the probability that it is less than its mean plus 3σ, respectively.

A.3.6 Uniform Pdf

Besides the Gaussian pdf, the most common pdf we meet is the **uniform pdf**, which has a constant value on some interval and is zero outside. The uniform pdf usually comes in one of two forms: non-zero on an interval centered about 0 or non-zero from 0 to a positive number. The standard deviation for two such cases is given below, and the general pdf is plotted in Figure A.3:

$$\sigma \text{ for uniform density on} \begin{cases} -1\,\text{to}+1\,\text{is}\,\sqrt{\tfrac{1}{3}} \approx 0.577 \\[2mm] 0\,\text{to}+1\,\text{is}\,\sqrt{\tfrac{1}{12}} \approx 0.289 \end{cases}$$

A.3.7 Pdf of Diurnal Illumination Variation of GEO Panels

Another useful pdf is the *panel-illumination pdf* (Section 15.5). Over the day, a GEO's earth deck is in the sun half the day (if it is not in the shadow of an antenna or support structure, anyway) and in darkness the other half. The amount of illumination starts at zero, acts like a sine function until it hits zero again, and then stays at zero for the second half of the day. The same is true for the east and west panels. The diurnal illumination is plotted in Figure A.4a, where the illumination phase is arbitrary. The diurnal illumination can be considered a random variable if the time

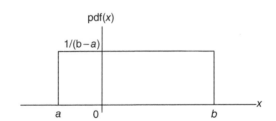

FIGURE A.3 Uniform probability density function.

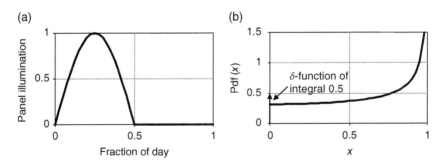

FIGURE A.4 Diurnal variation in illumination of earth deck and east and west panels: (a) illumination over the day (b) pdf.

at which we want to know what the illumination is, is random and uniformly distributed over the day. The pdf for the diurnal illumination variation is drawn in Figure A.4b and stated as follows:

Pdf for diurnal variation in panel illumination at x from 0 to 1 is

$$\frac{1}{2}\delta(x)+\left(\pi\sqrt{1-x^2}\right)^{-1}$$ which has mean $\pi^{-1}\approx0.318$ and std dev $\sigma\approx0.386$

The corresponding random variable is a mixed-type (Section A.3.2) since the cdf is discontinuous at $x=0$, jumping in value from 0 to 0.5.

The use of this pdf is to approximate the diurnal temperature pdf of units on the inside of the earth deck, the east panel, and the west panel, as well as the resultant variation of insertion loss. The pdf's independent axis values would range from diurnal minimum to diurnal maximum temperature.

A.3.8   Pdf of Diurnal Variation in Delta of East and West Panel Illumination

Here we are interested in the delta of the diurnal illuminations of the east and west panels. One panel is in the sun while the other is in darkness. The delta diurnal illumination is plotted in Figure A.5 (a), where the starting phase is arbitrary. The pdf for the delta variation is drawn in Figure A.5 (b) and stated as follows:

Pdf for diurnal variation in delta of east and west panel illumination,

at X from -1 to 1, is $\left(\pi\sqrt{1-X^2}\right)^{-1}$, which has mean 0 and std dev $\sigma\approx0.707$

The use of this pdf is to approximate the diurnal delta temperature of similar units on the inside of a GEO's east and west panels when there are no heat pipes connecting the two panels. Then, the resultant variation of delta insertion loss can be approximated (Section 15.4). The pdf's independent axis values would range from diurnal minimum delta to maximum delta temperature.

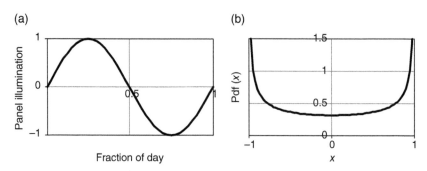

FIGURE A.5   Difference in diurnal variation in east and west panel illumination: (a) difference in illumination over the day (b) pdf.

A.3.9 Central Limit Theorem

The *Central Limit Theorem* is one of the most useful theorems in communications analysis. It says that the sum of a number n of random variables (rvs) that are mutually independent and identically distributed (iid) becomes closer and closer to Gaussian as n becomes larger. Because the rvs are iid, the mean and variance of the sum are, respectively, n times the mean and n times the variance of one such rv.

The convergence to Gaussian is remarkably fast in most cases we come across, luckily for our analysis and simulation. Most of the rvs in our sums have either a Gaussian or a uniform pdf, and usually, there is a mixture (Sections 9.3.2 and 15.5.7). We already know that the sum of Gaussian rvs is Gaussian, so let us look at the sum of rvs with uniform pdfs. We want to find out how many independent rvs have to be in the sum for the sum's pdf to be well approximated by Gaussian within $\pm 2\sigma$ of the mean value. Figure A.6 shows pdf plots for the following cases involving uniformly distributed rvs: (a) sum of two rvs each distributed on ± 0.5, (b) sum of three such, (c) sum of four such, and (d) sum of one distributed on ± 0.5, one on ± 0.35, and one on ± 0.65. Zero mean is assumed for all rvs. The standard deviation of the cases is 0.41, 0.5, 0.58, and 0.51, respectively. In each plot, the Gaussian pdf of the same standard deviation is shown as a dashed line. The convergence to Gaussian with increasing n is visible in the progression of the plots (a), (b), and (c).

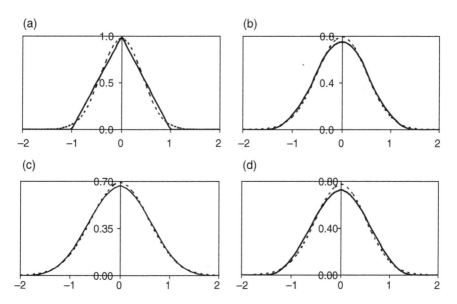

FIGURE A.6 Pdfs of sums of uniformly distributed random variables: (a) sum of 2, (b) sum of 3, (c) sum of 4, (d) sum of 3 of different pdfs. Gaussian pdfs shown with dashed lines.

TABLE A.2   Integrals on ±1σ and ±2σ of Pdfs of Sums of Independent, Uniformly Distributed Random Variables

Random variables in sum	Integral on ±1σ	Integral on ±2σ
Two uniform, both on ±0.5	0.65	0.966
Three uniform, all on ±0.5	0.66	0.957
Four uniform, all on ±0.5	0.67	0.957
One uniform on ±0.5, another uniform on ±0.35, and another uniform on ±0.65	0.66	0.961
Gaussian	0.68	0.954

Table A.2 shows the integrals of the same four sum pdfs over ±1σ and ±2σ as well as the integrals for Gaussian. Even for only three uniformly distributed rvs in the sum, the integrals are very close to the integrals for Gaussian. The case of three uniformly distributed rvs with different standard deviations, corresponding to plot (d) in Figure A.6, is given in the table as well to show that even if the standard deviations of three rvs of the kind likely to be found in a payload analysis are somewhat different, the sum of the rvs is close to Gaussian. However, if one or two of the three are much larger than the others, Gaussian may not be a good approximation. This is something to watch out for.

An extreme but thereby revealing example is the sum of a few randomly phased cosines. The pdf of one cosine, shown in Figure A.7a, could hardly be less Gaussian-like. However, the power of summing a few rvs to yield a Gaussian-like pdf is illustrated in Figures A.7b–d, for three, four, and five, respectively, in a sum. Note the different scales on the various plots.

The conclusion is that in most situations to be encountered in communications analyses involving the payload, the sum of as few as three iid random variables is approximated well on ±2σ by a Gaussian random variable. Furthermore, when we do not know the individual pdfs accurately, which is often the case, there is even more reason to accept the approximation.

A.4   GAUSS–HERMITE INTEGRATION TO APPROXIMATE EXPECTED VALUE OF FUNCTION OF GAUSSIAN RANDOM VARIABLE

Gauss–Hermite integration is a tool that can replace most Monte Carlo simulations over independent Gaussian random variables with a quick calculation. What is being evaluated in either method is a function whose independent variables have independent Gaussian probability density functions. Gauss–Hermite integration is a weighted sum of the function evaluated at a few points, so the computational burden is much less. This is especially a benefit when the function is being evaluated over many random variables including the Gaussian ones or when the function is

(a)

(b)

(c)

(d)

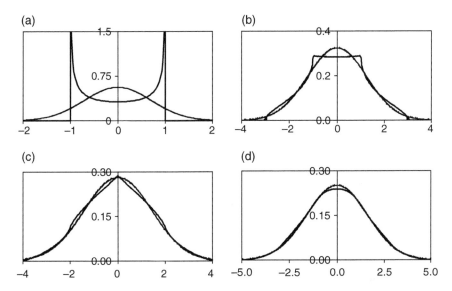

FIGURE A.7 Pdfs of sums of independent, randomly phased cosines: (a) one, (b) sum of 3, (c) sum of 4, (d) sum of 5. Gaussian pdfs of same standard deviations shown with gray lines.

evaluated as just part of a longer calculation. In the book Sections 9.6 and 16.3.9, examples are given of integrals that are candidates for Gauss–Hermite.

The way to perform a Gauss–Hermite integration is to first perform it on say 6 points and then on 10 points and to check if the difference in the answers is within a chosen error bound. If so, the 6 points were enough. If not, the procedure must be repeated with say 10 and 16 points. It has been the author's experience that 6 points is enough for the functions she has used this on, which included the error function and link availability due to rain at Ku-band. If the function to be evaluated is dependent on other variables besides the Gaussian ones, try 6 and 10 points on sets of the other variables which represent the extreme cases of the function; if 6 points is enough here, then it will be enough elsewhere.

The formula for Gauss–Hermite integration is as follows:

$$\frac{1}{\sqrt{2\pi}\sigma} \int_{-\infty}^{\infty} e^{-y^2/2\sigma^2} f(y)\,dy \approx \sum_{i=1}^{n} w_i f(\sigma x_i)$$

where the abscissas x_i and the weights w_i for a few n are given in Table A.3 (derived from Abramowitz and Stegun (1965)). See (Abramowitz and Stegun, 1965) for more sets of abscissas and weights for n up through 20.

The abscissas and weights are not taken from the Gaussian pdf. The abscissas and weights are shown in Figure A.8 for $n = 6$ and in Figure A.9 for $n = 10$.

TABLE A.3 Gauss–Hermite Integration Abscissas and Weights

$\pm x_i$	w_i	$\pm x_i$	w_i	$\pm x_i$	w_i
$n=6$		$n=10$		$n=16$	
0.6167065900	4.088284695e-1	0.4849357074	3.446423349e-1	0.3867606043	2.865685212e-1
1.889175877	8.861574602e-2	1.465989094	1.354837030e-1	1.163829100	1.583383727e-1
3.324257433	2.555784402e-3	2.484325841	1.911158050e-2	1.951980345	4.728475235e-2
$n=8$		3.581823482	7.580709344e-4	2.760245047	7.266937603e-3
0.5390798112	3.730122577e-1	4.859462827	4.310652630e-6	3.600873623	5.259849265e-4
1.636519041	1.172399076e-1			4.492955301	1.530003216e-5
2.802485861	9.635220121e-3			5.472225705	1.309473216e-7
4.144547185	1.126145384e-4			6.630878196	1.497814723e-10

(after Abramowitz and Stegun (1965).

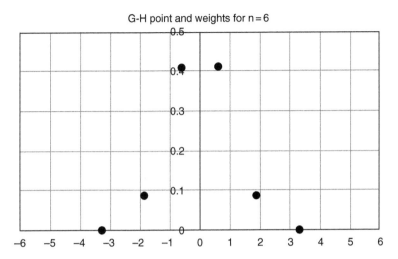

FIGURE A.8   Gauss–Hermite integration's abscissas and weights for $n = 6$.

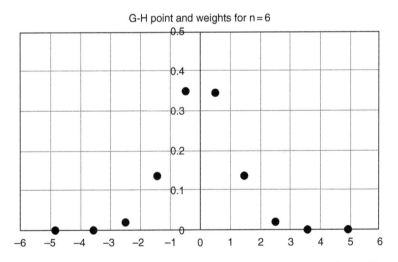

FIGURE A.9   Gauss–Hermite integration's abscissas and weights for $n = 10$.

REFERENCE

Abramowitz M and Stegun IA, editors, (1965). *Handbook of Mathematical Functions, with Formulas, Graphs, and Mathematical Tables*, New York: Dover Publications.

INDEX

Satellite Communications Payload and System, Second Edition. Teresa M. Braun and Walter R. Braun.
© 2021 John Wiley & Sons, Inc. Published 2021 by John Wiley & Sons, Inc.